Pseudo-Paracelsus

Pseudo-Paracelsus

Forgery and Early Modern Alchemy, Medicine and Natural Philosophy

Edited by

Didier Kahn
Hiro Hirai

TUTA SUB AEGIDE PALLAS · 1683 ·

BRILL

LEIDEN | BOSTON

Originally published as special issue of Brill's journal *Early Science and Medicine*, volume 24 (2019), no. 5–6 (published February 2020), now with a revised Introduction. A new Appendix by Julian Paulus, entitled "A Catalogue Raisonné of Pseudo-Paracelsian Writings: Texts Attributed to Paracelsus and Paracelsian Writings of Doubtful Authenticity," has also been added.

Cover illustration: An alchemist in his laboratory. Oil painting on wood by James Nasmyth (1808–1890). Wellcome Library no. 47333i. Wellcome Collection. Public Domain Mark.

Library of Congress Cataloging-in-Publication Data

Names: Kahn, Didier, editor. | Hirai, Hiro, editor.
Title: Pseudo-paracelsus : forgery and early modern alchemy, medicine and natural philosophy / edited by Didier Kahn, Hiro Hirai.
Description: Leiden ; Boston : Brill, 2019, 2022. | Includes index.
Identifiers: LCCN 2021047070 (print) | LCCN 2021047071 (ebook) | ISBN 9789004503373 (hardback) | ISBN 9789004503380 (ebook)
Subjects: LCSH: Paracelsus, 1493–1541. | Medicine—Philosophy—History.
Classification: LCC R147.P2 P74 2019, 2022 (print) | LCC R147.P2 (ebook) | DDC 610.1—dc23/eng/20211109
LC record available at https://lccn.loc.gov/2021047070
LC ebook record available at https://lccn.loc.gov/2021047071

Typeface for the Latin, Greek, and Cyrillic scripts: "Brill". See and download: brill.com/brill-typeface.

ISBN 978-90-04-50337-3 (hardback)
ISBN 978-90-04-50338-0 (e-book)

This book is printed on acid-free paper and produced in a sustainable manner.

Contents

Illustrations

Notes on Contributors

Tobias Bulang
is Professor of medieval and early modern literature at Heidelberg University. He has published on poetry *ca.* 1200 and on late medieval literature, early modern fiction and satire, demonology, herbals and Paracelsianism. He is Editor in Chief of *Daphnis: Journal of German Literature and Culture of the Early Modern Period*. His monographs include *Enzyklopädische Dichtung* (2011) and *Guldîne linge: Fünf Essays zu Gottfrieds Tristan* (forthcoming in 2021).

Dane T. Daniel
is Professor of History at Wright State University, Lake Campus. Most of his publications concern Paracelsus's radical theology, its influence on his cosmology, and its reception. The centerpiece of both his dissertation and current research is Paracelsus's *Astronomia Magna* (1537/38). He studied History and Philosophy of Science at Indiana University, researched in Germany for multiple years with the help of the DAAD (German Academic Exchange Service), served as a Research Ambassador for the DAAD, and has been a research fellow at both the Dibner Institute of the History of Science and Technology and the Chemical Heritage Foundation (now the Science History Institute).

Charles D. Gunnoe, Jr.
is Professor of History at Aquinas College in Grand Rapids, Michigan, where he has served as provost, dean of faculty, and chair. His research centers on the cultural, medical, and environmental history of Southwestern Germany and Switzerland in the sixteenth century. He is the author of *Thomas Erastus and the Palatinate* (2011), co-author of *An Introduction to the Heidelberg Catechism* (L. Bierma, et. al., 2005), and editor of *Paracelsian Moments* (with G. Scholz-Williams, 2003).

Hiro Hirai
PhD in philosophy and history of science, University of Lille 3 (France), is Research Associate at the Center for Science and Society, Columbia University. He has published widely in Renaissance and early modern natural philosophy, medicine and alchemy, including *Le Concept de semence dans les théories de la matière à la Renaissance, de Marsile Ficin à Pierre Gassendi* (2005) and *Medical Humanism and Natural Philosophy* (2011). He also edited *Cornelius Gemma. Cosmology, Medicine and Natural Philosophy in Renaissance Louvain* (2008), *Justus Lipsius and Natural Philosophy* (2011) and *Jacques Gaffarel between Magic and Science* (2014). He is currently preparing his third monograph on Renaissance natural philosophy, medicine and magic.

Didier Kahn
PhD (1998), University of Paris IV (Sorbonne), is Senior Researcher at the CNRS, Cellf 16–18 (Sorbonne). He has published extensively in the history of alchemy, science, medicine and literature, including *Alchimie et paracelsisme en France à la fin de la Renaissance* (2007), *La Messe alchimique attribuée à Melchior de Sibiu* (2015) and *Le Fixe et le volatil. Chimie et alchimie, de Paracelse à Lavoisier* (2016). In 2010 he published a heavily annotated edition of Montfaucon de Villars' witty dialogue *Le Comte de Gabalis* (1670). He is now working on the works of Paracelsus. He is also involved in the edition of Denis Diderot's *Œuvres complètes* (currently published by Hermann).

Julian Paulus
is an independent historian and bibliographer with a special interest in early modern alchemy and Paracelsianism. He teaches mathematics and geography at a school in Baden-Württemberg, Germany. His works include "Alchemie und Paracelsismus um 1600: Siebzig Porträts," in Joachim Telle, ed., *Analecta Paracelsica* (1994) and *Paracelsus-Bibliographie 1961–1996* (1997).

Lawrence M. Principe
is the Drew Professor of the Humanities at Johns Hopkins University in the Department of History of Science and Technology and the Department of Chemistry, and Director of the Singleton Center for the Study of Premodern Europe. His recent books include *The Transmutations of Chymistry: Wilhelm Homberg and the Académie Royale des Sciences* (Chicago, 2020) and *The Secrets of Alchemy* (Chicago, 2013). He is currently at work on critical editions of the alchemical texts by the fourteenth-century alchemist and imprisoned prophet John of Rupescissa.

Martin Žemla
PhD in philosophy, Charles University (Prague), is Research Fellow at the Centre for Renaissance Texts of Palacky University (Olomouc) and teaches at Charles University. He has been working on late medieval and Renaissance philosophy and religion, especially German mysticism, Renaissance Platonism and Paracelsianism. His work includes the Czech cotranslations of Marsilio Ficino's *Three Books on Life* (2020) and Heinrich Khunrath's *Amphitheatre of Eternal Wisdom* (2017). He is cowriting two English monographs, one on Khunrath and another on the Czech physician, astronomer and mathematician Thaddaeus Hagecius. He is also writing a book on the transformations of "German mysticism" from the thirteenth through the seventeenth centuries.

Introduction

Didier Kahn
CNRS, Cellf 16–18 – Sorbonne University
dkahn@msh-paris.fr

Hiro Hirai
Center for Science and Society, Columbia University
hhirai2@gmail.com

Over the last 40 years, a number of major publications have considerably expanded our knowledge of the works of the Swiss physician Paracelsus (1493/94–1541) and his followers. As a result, Paracelsianism appears more clearly as a driving force in widening the horizons of alchemy by offering it a platform for more serious consideration than it had hitherto been afforded, and in exerting considerable influence on the shaping of modern biology and medicine in an environment of fierce scholarly controversies.

Despite much progress, there are still wide gaps in the research literature. The internal chronology of Paracelsus' works is not known in detail. Often, this led historians to deem his system static although closer study indicates that his ideas constantly evolved. The bulk of spurious works attributed to Paracelsus throughout the second half of the sixteenth century has made it even more difficult to establish an internal chronology, either for the authentic or inauthentic works.

The production of forgeries under Paracelsus' name was an integral part of the diffusion of Paracelsianism. Many of those forgeries were widely read and extremely influential, not only in the fields of medicine and "chymistry" (alchemy/chemistry), but also in cosmology, anthropology, theology and magic. For many cases, we know neither when exactly they were first attributed to Paracelsus, nor who their authors were, nor even how much they may derive from authentic works by Paracelsus. A whole strand of German Paracelsianism, although widely influential far beyond the German borders – extending to Central Europe, the Low Countries, France, England and Italy – is thus entirely out of our hands. Charles Webster in 1982 was amongst the first to remark upon the impact of Paracelsian forgeries on the formation of both Paracelsus' popular image and Paracelsianism.[1]

1 Charles Webster, *From Paracelsus to Newton. Magic and the Making of Modern Science* (New York, 1996; 1st ed. 1982), 57. See also Charles D. Gunnoe, "Erastus and Paracelsianism:

 | DOI:10.1163/9789004503380_002

Paracelsian forgeries have yet to be studied in their own right. Discussing their content, sources, topics, potential authorship and composition dates will be beneficial on many levels. It will reveal their distinctive and common features, their mutual connections and their kinship, in style and content, with the genuine works of Paracelsus, thereby compelling scholars to renew their focus on the latter. These reciprocal consequences will result in a much sharper and more detailed knowledge both of Paracelsus himself and of Paracelsianism.

It is, of course, important to understand that most of these forgeries were regarded to be authentic by early modern readers, who considered the works of Paracelsus as a whole without any further consideration into the chronological developments within the works. Although Johann Huser, the first editor of Paracelsus' complete philosophical and medical works (1589–1591), tried to differentiate between authentic and spurious treatises, this, along with the setting of a chronological order, only became a project seriously taken up by Karl Sudhoff between 1894 and 1933. Today's historians need this critical approach, in addition to an understanding of how early modern Paracelsians received the works now deemed spurious.

All these considerations have led us to conceive a wide-ranging, collective research project on both Paracelsus and pseudo-Paracelsus. A special panel on these topics was held in Boston at the annual meeting of the Renaissance Society of America in April 2016. Three months later, we organized another conference at Villa Vigoni (Italy) under the title: "Pseudo-Paracelsus: Alchemy and Forgery in Early Modern Medicine and Natural Philosophy." The papers collected here represent part of the results of this conference.[2] In parallel, a digitization of the Huser edition of Paracelsus, generously funded by the Theophrastus Stiftung, has been made available via the online database THEO, engineered by Urs Leo Gantenbein (University of Zurich/Paracelsus Project).[3]

All papers in this special issue are arranged according to Paracelsian chronology. First, Dane T. Daniel examines the authenticity of two treatises of Paracelsus, the *Brief an die Wittenberger Theologen* (1525) and the *Astronomia*

Theological Motifs in Thomas Erastus' Rejection of Paracelsian Natural Philosophy," in Allen G. Debus and Michael T. Walton, eds., *Reading the Book of Nature: The Other Side of the Scientific Revolution* (Kirksville, MO, 1998), 45–66, on 64, n. 64.

2 Generous support for this project was provided by the Center for the History of Philosophy and Science, Radboud University (Nijmegen); Sorbonne University (Paris), ED III; the Cellf 16–18 (CNRS and Sorbonne University); the OBVIL (Sorbonne University); the Society for the History of Alchemy and Chemistry (London); Amadeo Murase and Seigakuin University (Tokyo); and Clare Hirai.

3 <www.paracelsus-project.org>, accessed 31 October 2019. The database includes Sudhoff's *Paracelsus-Handschriften* and a few vols. of Sudhoff's edition.

Magna (1537/1538). His attempt to establish criteria regarding their authenticity stresses the need to consider theological motifs whenever they are present in Paracelsian texts. Then Didier Kahn discusses one of the earliest spurious treatises ascribed to Paracelsus: the *Philosophia ad Athenienses*. First published in 1564, this treatise is dominated by philosophical, cosmological and theological considerations. In the third article, Tobias Bulang focuses on one of the most interesting and lesser-known alchemical pseudo-Paracelsian treatises: the *Aurora philosophorum* (written before 1569), a main topic of which is the genealogy of knowledge since Adam. This treatise was hitherto known only through its abridged Latin translation by Gerhard Dorn (1577). Bulang presents the critical edition in progress of the still unpublished German text. Hiro Hirai then discusses the whole editorial context behind the first publication of the famous *De natura rerum* (1572), one of the richest and most complex treatises of the pseudo-Paracelsian corpus. This contextualization shows how *De natura rerum* emerged at the crest of a wave of publications of spurious works. The next article, by Charles D. Gunnoe, compares theories of plague in the authentic plague treatises of Paracelsus and in *De pestilitate*, an interesting and elusive spurious text written prior to 1578 and first published in the Huser edition in 1589. In all these plague treatises, plague is associated closely with magic and imagination.

With Martin Žemla, we return to the field of theological works: Žemla discusses two spurious works, in fact written by Adam Haslmayr, published in the collection of Paracelsian and Weigelian treatises *Philosophia mystica* (1618): *Astronomia Olympi novi* and *Theologia Cabalistica*. This important collection helped underline the significance of Paracelsus's theological ideas. By combining his theology with alchemy and the idea of universal reform, it fit well with the ethos of the early Rosicrucian manifestos. Finally, Lawrence M. Principe gives a detailed description of the gradual formation of the alchemical corpus attributed to the fictitious Basilius Valentinus between 1599 and 1717, including a biography for its supposed author. Due to its connection with Paracelsian ideas, the Valentine corpus can be viewed, in Principe's words, as "an ancillary part of the pseudo-Paracelsian literature."

To these papers Julian Paulus adds an extensive catalogue of all known spurious works (or once deemed spurious) which were posthumously ascribed to Paracelsus. The very first of them chronologically speaking is a short tract on plague (§ 3.8). This text was perhaps formerly in possession of Paracelsus but was erroneously believed to be written by him. It seems to have been attributed to him by one or several members of his circle in Salzburg shortly after his death. First printed in 1553 as a *Flugschrift*, it was reprinted in the following year within another plague treatise along with one of the two epitaphs prepared by the friends of Paracelsus, which partly contributed to the formation of a legend about him. This somewhat "accidental" attribution suggests that no treatises composed by Paracelsus himself were available to the Salzburg circle after his death – at least in a reasonably publishable form. As it seems, the only works left behind by the Swiss physician in Salzburg were theological writings, which were certainly considered to be dangerous and impossible to publish.[4]

The plague tract was followed by an impressive number of other publications, which were deliberate forgeries under the name of Paracelsus. Before condemning their authors as counterfeiters according to modern norms, however, we should ask ourselves whether the posthumous pupils of medieval thinkers such as Albertus Magnus and Thomas Aquinas really considered themselves forgers or counterfeiters. Rather, they could have self-identified as the carriers of their master's will by completing the unachieved work on their own terms. Whenever we deal with medieval and early modern forgeries, this kind of mindset should be taken into account to avoid biased retrospective judgments, which can be misguided by our modern worship of the notion of "authorship."

A good example is the case of the *Apocalypsis Hermetis* (§ 4.1), which is an anonymous treatise on quintessence after the manner of Johannes de Rupescissa. Although it was circulated in manuscripts from the 1530s on, it began to be attributed to Paracelsus only around the 1570s. It is true that the notion of quintessence was often employed in the authentic writings of Paracelsus. But this should not mask the real meaning of the attribution of one more treatise on this topic to him. There are two major points to be considered. One is the fact that no genuine monograph of Paracelsus is devoted to this notion. The other, more important fact is that the notion of quintessence was for Paracelsus something entirely different from the alchemical one expounded in the writings of Johannes de Rupescissa or pseudo-Raymond Lull and in the works of their followers, including the *Apocalypsis Hermetis* itself.[5] Thus the purpose of the unknown figure who first ascribed this anonymous

4 Heinz Dopsch, "Testament, Tod und Grabmal des Paracelsus," in Heinz Dopsch and Peter F. Kramml, eds., *Paracelsus und Salzburg* (Salzburg, 1994), 251–277, on 269 (itemized inventory of Paracelsus' possessions): "etliche handschriftliche Hefte zur Theologie, die Theophrastus konzipiert haben soll." Those are the only works left behind explicitly indicated as writings of Paracelsus himself.

5 On the notion of quintessence in Johannes de Rupescissa, see Michela Pereira, "Heavens on Earth. From the *Tabula Smaragdina* to the Alchemical Fifth Essence," *Early Science and Medicine*, 5 (2000), 131–144, on 139–141. For the case of Paracelsus, see Didier Kahn, "Quintessence and the prolongation of life in the works of Paracelsus," *Micrologus*, 26 (2018) [*Longevity and Immortality. Europe – Islam – Asia*], 183–226.

treatise to Paracelsus becomes quite obvious: the integration of a treatise specifically devoted to the Rupescissan or pseudo-Lullian quintessence was certainly a good way to fill what could be perceived as a longstanding gap in the corpus of Paracelsus. Indeed the Swiss physician himself had often referred to one of his works as *De quinta essentia*, although no text was ever identified with it by his followers or modern scholars.[6] Ascribing to him one or several writings on quintessence (Paulus, Appendix, below, § 3.6) could be a way to complete the unfinished work of the master.

From the point of view of the sixteenth-century alchemists, who were among the most assiduous readers of Paracelsus, another obvious gap in their master's corpus was the absence of almost any treatise on alchemy properly speaking.[7] It was quite natural and logical to try to fill this gap by integrating forgeries with a strong flavor of alchemy into the Paracelsian corpus. The process began as early as the beginning of the 1560s. In this domain, financial factors were undoubtedly at work.

More generally, financial gains exerted powerful incentive for the production and circulation of forgeries under the name of Paracelsus. The Basel printer Pietro Perna (1519–1582) repeatedly stated that he acquired manuscripts of Paracelsus (whether containing genuine or spurious works) against payment.[8] In 1574, Adam von Bodenstein also evoked similar circumstances in a preface to his Paracelsian collection containing several apocryphal treatises.[9] So did Johann Huser in 1589 and, later still, an anonymous writer of alchemical verses (ca. 1600).[10] A whole business of "mail-order supply" of alchemical and magical texts, including Paracelsian ones such as *Arcana divina* (§ 4.3), can even be observed in Germany as late as the eighteenth century.

Generally speaking, authors of Paracelsian forgeries have proved hitherto exceedingly difficult to identify. As far as we know (although it may be too early to judge), it does not seem plausible that one single writer composed several treatises, as each text displays distinctive features that make it stand out

6 Paracelsus was most likely referring to the corresponding chapter in the authentic work known to us as *Archidoxis*. See below the comments by Julian Paulus on *De quinta essentia* (§ 3.6).

7 On this topic, see Andrew Sparling, "Providence and Alchemy: Paracelsus on How Knowledge Unfolded, Matter Developed, and Bodies Might Be Perfected," Ph.D. diss. (University of Nevada, Reno, 2018); Sparling, "Paracelsus, a Transmutational Alchemist," *Ambix*, 67 (2020), 62–87.

8 Mentioned by Karl Sudhoff, *Bibliographia Paracelsica* (Berlin, 1894), 241.

9 Wilhelm Kühlmann and Joachim Telle, *Corpus Paracelsisticum*, vol. 1 (Tübingen, 2001), 492.

10 Paracelsus, *Bücher und Schrifften*, ed. Johann Huser (Basel: Konrad Waldkirch, 1589–1591), 6: 421; Wilhelm Kühlmann and Joachim Telle, *Corpus Paracelsisticum*, vol. 1 (Tübingen, 2001), 492.

among the others. The production of forgeries under the name of Paracelsus was probably, then, a collective phenomenon to which each author contributed separately and independently in most cases, gradually altering the original ideas of Paracelsus.

This collection of papers along with the extensive catalogue of the spurious works of Paracelsus will certainly contribute to an advanced understanding of both Paracelsus and Paracelsianism.[11] We hope it inspires other scholars to join our ongoing project.[12]

11 The collection of the first seven articles has been published as the special issue "Pseudo-Paracelsus: Forgery and Early Modern Alchemy, Medicine and Natural Philosophy," *Early Science and Medicine* 24 (2019), 415–572. It was followed by another special issue "Paracelsus, Forgeries and Transmutation," *Ambix* 67 (2020), 1–99. See also the special issue "The Forgotten Reformation," *Daphnis* 48 (2020), 1–327.

12 The Index Nominum of the first seven articles has been processed by Claire Carpentier (CNRS, Cellf 16–21).

The Authenticity of Paracelsus' *Astronomia Magna* and *Brief an die Wittenberger Theologen*: Towards a Diagnostic Rubric Clarifying Authentic and Spurious Elements in Paracelsus' *Oeuvre* on the Basis of Theological Motifs

Dane T. Daniel
Wright State University
dane.daniel@wright.edu

Abstract

The paper is an evaluation of differing opinions regarding the authenticity of two of Paracelsus' works, the *Astronomia Magna* (1537/1538) and *Brief an die Wittenberger Theologen* (1525). Karl Sudhoff – the exceptionally erudite and prolific Paracelsus scholar and editor – considered Paracelsus' letter to Luther and the Wittenberg theologians to be spurious. Others have questioned the extent to which the *Astronomia Magna* (often considered to be Paracelsus' *Meisterstück*) should be placed among the genuine texts of the Paracelsian corpus. Based on historical and conceptual data, and focusing on Paracelsus' theology, the author argues that both works under consideration are authentic. The study thus highlights the need to develop a diagnostic rubric for evaluating the authenticity of works in the Paracelsian *oeuvre* and suggests ideas toward this end on the basis of theological motifs prevalent within Paracelsus' clearly authentic works.

Keywords

Paracelsus – Paracelsians – Martin Luther – Karl Sudhoff – history of science – alchemy – chymistry – history of medicine – Radical Reformation

 | DOI:10.1163/9789004503380_003

Introduction

Scholars have long haggled over the spurious and the genuine in the realm of *Paracelsica*.[1] Although great strides have been made in this regard, some ambiguity remains concerning how scholars can objectively and uniformly separate authentic from inauthentic texts. In fact, the challenge of delineating the spurious from the genuine has been so acute that not even Karl Sudhoff (1853–1938) – the veritable dean of modern Paracelsus studies – has been left unscathed.

Although no scholar has contributed more to our understanding of the Paracelsian corpus than Sudhoff, there has been disagreement among scholars about his judgments.[2] Three examples where these disagreements have arisen are as follows: 1) the dating of some of the works in the massive Paracelsian *oeuvre*; 2) the bifurcation of Paracelsus' collected works into two distinct veins: medico-natural philosophical on the one hand and theological on the other;[3] and 3) the designation given to Paracelsus' *Brief an die Wittenberger Theologen*,[4] which Sudhoff considered to be spurious, and the *Astronomia Magna*, a text that he placed among the authentic writings.[5] The focus in this paper is on the last of these issues, i.e., differences of opinion regarding the authenticity of works within the Paracelsian corpus. My aim is to underscore the need to develop a diagnostic rubric for evaluating the authenticity of works attributed to Paracelsus; I will be suggesting ideas toward this end on the basis of theological motifs from Paracelsus' genuine works. First of all, I will agree with Hartmut

1 The author wishes to thank Didier Kahn and Pia Holenstein for their exceptionally helpful comments and suggestions.

2 See Karl Sudhoff, *Versuch einer Kritik der Echtheit der Paracelsischen Schriften*, vol. 1: *Bibliographia Paracelsica* (Berlin, 1894); vol. 2: *Paracelsus-Handschriften* (Berlin, 1898–1899).

3 Sudhoff divided the modern edition of Paracelsus' collected works into two parts. The first division is Paracelsus, *Sämtliche Werke, 1. Abteilung: Medizinische, naturwissenschaftliche und philosophische Schriften*, ed. Karl Sudhoff, 14 vols. (Munich-Berlin, 1922–1933) (hereafter: Sudhoff, I/1, I/2, etc.). The first volume of the second division is Paracelsus, *Sämtliche Werke, 2. Abteilung: Die theologischen und religionsphilosophischen Schriften*, ed. Wilhelm Matthiessen (Munich, 1923) (hereafter: Matthiessen, II/1). Kurt Goldammer directed the publication of several more volumes of the second division between 1955 and 1986 (Wiesbaden, 1955–1986) (hereafter: Goldammer, II/2, etc.). The sixteenth-century authoritative edition is Paracelsus, *Bücher und Schrifften*, ed. Johann Huser, 11 vols. (Basel, 1589–1591) (hereafter: Huser, 1, 2, etc.).

4 See "Epistola. Philippi Theophrasti Paracelsi ad Theologos Witebergenses," in Paracelsus, *Philosophia De Limbo, Æterno Perpetuoque Homine Novo Secundæ Creationis ex Jesu Christo Dei Filio*, ed. Johann Stariz (Magdeburg, 1618).

5 The *Astronomia Magna* is in volume 12 of the Sudhoff edition, published in 1929. It first appeared in 1571 as Paracelsus, *Astronomia Magna: oder die gantze Philosophia sagax der grossen und kleinen Welt*, ed. Michael Toxites (Frankfurt, 1571).

Rudolph's tentative acceptance of the authenticity of Paracelsus' *Brief an die Wittenberger Theologen* (1525) with the support of historical and conceptual data. Secondly, I will turn my attention to the *Astronomia Magna* (1537/38); owing to doubts I have seen and heard expressed concerning the authenticity of this text, or at least sections of it – most notably in published form by the late Heinrich Schipperges (1918–2003) –, I will focus on this *Meisterstück* of Paracelsus' late career. Thus, this paper highlights the theological contexts and ideas that help determine the genuineness, or absence thereof, of any given text by Paracelsus, recognizing that his writings cannot be properly understood in isolation from his religious ideas.[6]

Paracelsus' theology was so idiosyncratic that one can identify when a work resounds with the *theologica* of the genuine Paracelsus or when a text is clearly dissonant from his religious thinking. Certain theological themes stand out as undoubtedly characteristic of Paracelsus, and these Christological, anthropological, and epistemological motifs can help us to differentiate between the authentic and the spurious, especially when delineating post-1530 works, and, in this case particularly, when dealing with the *Astronomia Magna*.[7] In order to identify a work as authentic within Paracelsus' *oeuvre*, a good starting place is to determine the veracity of specific themes common to his authentic works, whether natural philosophical or theological.[8]

In this paper, I will highlight distinctive theological themes – often either misunderstood or misrepresented by his chief followers and editors – that are common to both his theological and medico-natural philosophical writings,

6 This was a principal point in Andrew Weeks, *Paracelsus: Speculative Theory and the Crisis of the Early Reformation* (Albany, NY, 1997). Also, with pertinence regarding the *Astronomia Magna*, and concerning the unity between religious and natural philosophical, see Hartmut Rudolph, "Prädestination und 'seliges Leben'. Ein Beitrag zur Verhältnisbestimmung von Weltbild und Theologie bei Paracelsus," in Volker Zimmermann, ed., *Paracelsus: Das Werk - die Rezeption*, (Stuttgart, 1995), 85–98, on 86–87: "1) Auch die in der 1. Abteilung der Gesamtausgabe vorhandenen und im wesentlichen bereits in der Huser-Ausgabe (1589–1591) edierten Schriften enthalten eine Reihe bedeutender theologischer Passagen zu Schlüsselthemen der paracelsischen Laientheologie. 2) Besonders auffällig sind Parallelen zwischen den Frühschriften in Band 1 der Sudhoff-Ausgabe und einigen frühen theologischen Traktaten, sodann zwischen längeren theologischen Passagen in der 'Astronomia magna' (1537/38) und in möglicherweise ebenfalls in den 1530er Jahren entstandenen theologischen Schriften, gerade auch den Kommentaren zum Matthäus-Evangelium."

7 Obviously, it is also necessary to develop further philological criteria with regard to the dating of texts; this, in and of itself, could identify works penned after Paracelsus.

8 Of course, one must approach this with the understanding that his ideas were not static, but developed greatly over time, and that there are no shortage of contradictions and confusing synonyms in his thought and vocabulary.

including, respectively, *De Genealogia Christi* and *Opus Paramirum*.[9] These themes include Paracelsus' unique Christology (with its Mariological undertones); the two-creation concept and sacramental theology that Paracelsus ties to it (whereby God the Father created the mortal realm, and God the Son its eternal analogy); his theory of the "limbus" (based on his exegesis of Genesis 2:7, the foundations for his anthropology);[10] and his bipartite epistemology of light of nature vs. light of the Spirit (according to this dichotomy, natural knowledge is revealed to a person's natural sidereal body – which acts as a magnet and which Paracelsus also calls the "spirit" in the context of the body-spirit-soul trichotomy – by the mortal stars, while a special type of revelatory knowledge is given directly to a human's immortal components by the Holy Spirit or the immortal stars);[11] the last of these incorporates within it the dichotomy of natural magic vs. divine (or celestial) magic. All of these occur in the context of his microcosm-macrocosm theory, that is, the analogy between the human and the cosmos. It is indeed impossible to delineate the genuine Paracelsian world-view outside of the context of his theories on the types of bodies, souls, and spirits, which he explicated according to his understanding of Scripture. As I will show, the *Astronomia Magna* definitively presents

9 Carlos Gilly discusses the sidestepping by Toxites and Bodenstein of "specific theological issues in the prefaces to their editions of Paracelsus." See Carlos Gilly, "*Theophrastia Sancta*' – Paracelsianism as a Religion," in Ole Peter Grell, ed., *Paracelsus: The Man and His Reputation, His Ideas and Their Transformation* (Leiden, 1998), 151–186. See *De genealogia Christi*, Goldammer, II/3 and *Opus Paramirum*, Sudhoff, I/9.

10 Paracelsus' understanding of human nature is incomprehensible outside of his exegesis of Genesis 2:7 on the creation of Adam. See Paracelsus' explanation in the *Opus Paramirum* (1530/1531) in Sudhoff, I/9: 193: "Dieweil nun der limbus ist prima materia des menschen, so muß der arzt wissen was der limbus sei. dan was der limbus ist das ist auch der mensch. der den limbus erkent, der weiß was der mensch ist. also sol der arzt geboren werden. nun ist der limbus himel und erden, ober und under sphaer, die vier element und was in ir ist, darumb er bilich den namen hat microcosmus, denn er ist die ganze welt." This passage, and its clear reiteration in the *Astronomia Magna*, is discussed in Dane T. Daniel, *Paracelsus'* Astronomia Magna *(1537/38): Bible-Based Science and the Religious Roots of the Scientific Revolution*, PhD diss. (Indiana University, Sept. 2003), 4.

11 This is a complicated and not yet clearly explicated topic: The immortal soul is given to the human directly by God. Adam received the soul, the divine image, in Genesis 2:7, and other humans receive the soul at conception, per Paracelsus. Another immortal component is the resurrection body received by Christians at baptism; according to the *Astronomia Magna*, they also receive a new conscience, and perhaps even a new celestial spirit (analogous to the mortal spirit, also called the sidereal body). Furthermore, given that Christians receive the Holy Spirit, it is clear that there are various immortal and eternal components within Christians that could participate in divine epistemology, i.e., the reception of celestial knowledge. Such a topic deserves further exploration to determine the exact mechanisms involved in revelatory epistemology.

a picture in accordance with Paracelsus' characteristic Christology, anthropology, and epistemology – all central to his microcosm-macrocosm theory. The text is genuine Paracelsus par excellence, even if there is space for future research to pursue, to determine the extent to which editors may have slightly embellished or altered the prodigious work.[12]

This paper will note differing opinions concerning the *Letter to the Wittenberg Theologians* and the *Astronomia Magna*, and will highlight the aforementioned three motifs – and their characteristic ancillaries – which are common to the authentic Paracelsus' *Weltanschauung* and central to the *Astronomia Magna*.[13] As I will explain, there is insufficient reason – either historically or conceptually – to call into question the veracity of either Paracelsus' *Letter to the Wittenberg Theologians* or the *Astronomia Magna* (1537/38).[14] Also, and again, I am suggesting ways in which the many volumes of the Paracelsian corpus may be approached when evaluating texts or passages of questionable authenticity – a comprehensive study of a topic of that magnitude would deserve a dissertation or monograph-length treatment; in the short space

12 Some of the discussion, especially that related to the types of astronomy, are unique to the *Astronomia Magna* (and its earlier drafts); I am not inclined to find these to be spurious (especially given the seemingly authentic stylistic elements), but – given some of the doubts I have heard – I do think the concepts related to natural magic warrant scrutiny in further research.

13 According to Katharina Biegger, following Sudhoff, Paracelus' *Letter to the Wittenberg Theologians* is among the theological works of dubious authenticity; others are *De diebus ægyptiacis*, *De arte præsaga*, and *Brief an Johannes Œcolampad*. Actually, the inclusion here of *De arte praesaga* is an unfortunate mistake on the part of Biegger, since this text was printed by Johann Huser from Paracelsus' autograph manuscript (which explains why Sudhoff never doubted its authenticity): see Huser, 9: table of contents and 79–111; Sudhoff, I/14: VIII. As for the *Letter to the Wittenberg Theologians*, it is contained in *Philosophia De Limbo*, ed. Stariz. Biegger's dissertation advisor was Kurt Goldammer. She credits Sudhoff for evaluating – in his 1898–1899 *Paracelsus-Handschriften* – not only the medicinal/scientific/magical material, but also 169 theological manuscripts. She adds that Wilhelm Matthießen, prior to editing a volume of the second division of the collected works, published ten of Paracelsus' *Abhandlungen* in the *Archiv für Reformationsgeschichte*, 14 (1917), 1–48, 81–122, and ibid., 15 (1918), 1–29, 125–56.

14 Didier Kahn has shown that the *Astronomia Magna*, as well as Paracelsus' concept of astronomy, developed over several years, and it is difficult to pinpoint precisely when the *Astronomia Magna* was written. See Didier Kahn, "Paracelsus' Ideas on the Heavens, Stars and Comets," in Miguel Ángel Granada, Patrick J. Boner and Dario Tessicini, eds., *Unifying Heaven and Earth: Essays in the History of Early Modern Cosmology* (Barcelona, 2016), 59–116, on 73: "Paracelsus apparently began, or planned, to work on it as early as 1530, if we are to recognize the *Astronomia magna* in an allusion he made in that year to a future work *de potestate astronomica et magica*." Although Kahn questions the reliability of this suggestion, it is certainly viable; see the relevant passage in Huser, 10: *Appendix*, 69.

available here, the discussion must be limited to a few suggestions for moving forward while examining a pair of texts. Before turning to the paper's central theme, i.e., theological motifs in the *Astronomia Magna*, I wish first to briefly address a few problems in Sudhoff's valuations, which indicate the extent to which he and many scholars after him underestimated the significance of Paracelsus' religious writings, and examine the differing opinions regarding the authenticity of the *Brief an die Wittenberger Theologen*.

Firstly, Sudhoff's dating of Paracelsus' works has been addressed by various scholars. Notably, Andrew Weeks vehemently criticizes Sudhoff's dating of Paracelsus' earliest tracts to "around 1520." It is puzzling that Sudhoff made such an estimation in volume one of his edition of Paracelsus' collected medical and philosophical writings, especially having already alluded – in 1922, as he began to publish the collected works – to the fact that tracts such as the *Eleven Treatises* and *Volumen Paramirum* were unlikely to have been written earlier than 1524 in Salzburg (and the textual evidence suggests an even later date).[15] Then, in his Paracelsus biography of 1936, Sudhoff announced that he would prefer to put all of volumes one and two in the year 1526.[16]

Similarly, Udo Benzenhöfer, in his important *Studien zum Frühwerk des Paracelsus im Bereich Medizin und Naturkunde*, agrees that the contents of the first three volumes in the Sudhoff edition cannot be dated with certainty, and he simply places the early writings in the years before and during Paracelsus' Basel period of 1527–28.[17] As Weeks continues, the year 1524 – during Paracelsus' first Salzburg period – marked the beginning of his writing enterprise, when he penned *De Septem Punctis*, a denunciation of the idolatries of the church. Countering the questionable modern division of Paracelsus' works into the medical-philosophical and theological, Weeks maintains that "the naturalism of Paracelsus turns on a religious center from the very beginning."[18]

15 Weeks, *Paracelsus: Speculative Theory*, 37–41. See also Paracelsus, *Essential Theoretical Writings*, trans. Weeks, 78.

16 Weeks, *Paracelsus: Speculative Theory*, 37–41. Incidentally, this is the same year that he placed the *Archidoxis*. Most scholars agree that the *Archidoxis* was among his earliest works; see, for example, Benzenhöfer's discussion (next note), and Didier Kahn's remarks on the *Archidoxis* in his "Quintessence and the prolongation of life in the works of Paracelsus," *Micrologus*, 26 (2018): *Longevity and Immortality. Europe – Islam – Asia*, 183–226, on 186–187 n. 12; 204 n. 81; and 214.

17 Udo Benzenhöfer, *Studien zum Frühwerk des Paracelsus im Bereich Medizin und Naturkunde* (Münster, 2005), 7–10.

18 Weeks, *Paracelsus: Speculative Theory*, 36. Sudhoff's suspect dating and division has had a significant impact on Paracelsus studies to this day, and the medical historian's bias has led to problematic analyses in the historiography of Paracelsus, due to a neglect of the *theologica* and its significance.

1 *Brief an die Wittenberger Theologen*

As a case in point, it is fitting here to turn to another of Paracelsus' earliest works, his "Letter to the Wittenberg Theologians." Several scholars no longer accept Sudhoff's doubts about its authenticity. The letter begins with Paracelsus' warm salutations to Luther, Melanchthon, and Bugenhagen. The purpose of the missive is to urge the Wittenberg theologians to read the accompanying commentary – which Paracelsus had just written – on the first five chapters of Matthew's Gospel, and to correct the content if necessary. Thereafter, he outlines an ambitious list of his forthcoming theological output, including exegeses of the Acts of the Apostles, several of Paul's epistles, and those of St. John, Peter, Jude, and James. With characteristic hubris, Paracelsus portrays himself as an equal to his contemporary theologians, and exclaims that he longs to drink with the Lutherans from the "fountains of truth."[19]

As was common in the period, Paracelsus hoped to engage in dialogue with other reformers (including the Lutherans), though it is most likely that he was disappointed in this case. In fact, we know nothing of the Lutheran response to Paracelsus (which is another reason Sudhoff may have had doubts about the letter). And problematic for his ambitions – as Charles Webster has pointed out –, the commentary that accompanies Paracelsus' letter is overtly radical, and this would have placed Paracelsus in the company of radicals like Müntzer or Grebel in the eyes of the Wittenberg theologians.[20]

Although the letter in question has been lost alongside all of Paracelsus' autographs, we know of it through three sources, two in manuscript form, and one printed. A Görlitz manuscript from 1564 and a Copenhagen manuscript from around 1600 contain the letter alongside Paracelsus' commentary on the first five chapters of Matthew.[21] The only printed edition of the letter is found in a collection of Paracelsus' theological works that features several of his writings on the Lord's Supper.[22] This rare example of an early modern theological work by Paracelsus was produced in Magdeburg in 1618, printed by Johann

19 "Epistola. Philippi Theophrasti Paracelsi ad Theologos Witebergenses," in *Philosophia De Limbo*, ed. Stariz. Paracelsus writes: "[...] hat mich gedürstet zu trincken mit euch den Brunnen der Warheit." The pages of the collection are not numbered; this is from the first of the letter's three pages.

20 Charles Webster, *Paracelsus. Medicine, Magic and Mission at the End of Time* (New Haven, CT, 2008), 76.

21 Hartmut Rudolph, "Einige Gesichtspunkte zum Thema 'Paracelsus und Luther'," *Archiv für Reformationsgeschichte*, 72 (1981), 34–54.

22 *Philosophia De Limbo*, ed. Stariz.

Francke and edited by Johann Stariz, both of whom were enthusiasts of the theologies of Weigel and Theophrastus, i.e., Paracelsus.[23]

A brief evaluation of the letter to determine its authenticity reveals the following: the epistle was purportedly written in 1525 during Paracelsus' first sojourn in Salzburg, where he sought to set up a medical practice (as we have seen, this was potentially the very time when Paracelsus began his writing career). All indications are that he was hoping to settle down, which runs counter to his pronouncement in the *Seven Defenses* of 1537 and 1538 that he was a wanderer by choice. (Also indicative of his effort to find stable employment is the fact that he purchased his citizenship in Strasbourg fairly soon after hurriedly leaving Salzburg; his stay in Strasbourg would be brief only because some contacts and good luck led him to Basel, where he would briefly and provocatively serve as city physician and university lecturer.) Regarding his residence in Salzburg (where he would return at the end of his life), we can surmise that he had hoped to find acceptance and success there via his medical practice and writing enterprise. However, it appears that – in his vituperativeness – he clashed with the authorities and was forced to make a hasty exit.[24] In his Salzburg writings, neither the secular nor clerical authorities had escaped his derision, and he may have been affiliated with revolutionary miners, townsmen, and peasants. It was during the Salzburg revolts that he fled to the Upper Rhine, without even packing his bags.

Assuming Paracelsus' ties to these dissident elements, Sudhoff questioned the authenticity of Paracelsus' letter to the Wittenberg theologians in his monumental *Versuch einer Kritik der Echtheit der Paracelsischen Schriften* of 1894.[25] How could Paracelsus be sympathetic to uprisings by peasants and miners, yet at the same time seek an alliance with Luther – the vocal persecutor of the revolution? Although many scholars have been influenced by Sudhoff's opinion that the letter was wrongly attributed to Paracelsus – and are even seemingly afraid to discuss the work without apologizing for its potential inauthenticity – Hartmut Rudolph has pointed to several reasons to accept the letter as genuine.[26] Foremost among these, as Kurt Goldammer also noted, was that the letter was written in March, whereas the polarization between Luther

23 See Gilly, "*Theophrastia Sancta*."

24 Heinz Dopsch, "Paracelsus, Salzburg und der Bauernkrieg," in Heinz Dopsch and Peter F. Kramml, eds., *Paracelsus und Salzburg* (Salzburg, 1994), 299–308, esp. 304–307; and idem, "Paracelsus, die Reformation und der Bauernkrieg," in Heinz Dopsch, Kurt Goldammer and Peter Kramml, eds., *Paracelsus (1493–1541). "Keines andern Knecht..."* (Salzburg, 1993), 201–215. esp. 210.

25 Sudhoff, *Bibliographia*, 510.

26 Rudolph, "Einige Gesichtspunkte."

and the insurgents became evident only in the following months. This would dispel Sudhoff's main argument against the letter's authenticity. Another clue to the potential authenticity of the short letter was the inclusion of a stylistic element that is prevalent throughout Paracelsus' authentic writings, i.e., the motif of lesser light versus the greater light from above.[27] In addition, as Rudolph notes, there are significant parallels between the respective views of Luther and Paracelsus concerning *sola fide* (faith alone) and the Eucharist that could help explain Paracelsus' early attraction to Luther. In analyzing the letter, Rudolph launches into Paracelsus' ties to Luther, especially concerning *sola fide*, particularly with regard to the Lord's Supper. Paracelsus may even be said to agree with Luther – to a certain extent – on the role of faith in the Lord's Supper, even to the extent of espousing consubstantiation rather than transubstantiation. (Consubstantiation is the actual presence of Christ with – or in addition to – the elements of the mass, and transubstantiation is the actual change of the bread and wine into the actual body and blood of Christ.)

Both Luther and Paracelsus were opposed to transubstantiation – especially in the sense of the Aristotelian hylomorphism of the Thomists – who held that the employment of appropriate matter and form would bring about an efficacious sacramental result. That is, by combining proper matter and form, a sacrament would – as a matter of recipe – bring the intended result. Upon consecration by a duly ordained priest, the elements of the mass (bread and wine) combined with the proper form – i.e., the right words – would substantially change from the bread and wine into the actual body and blood of Christ. Both Luther and Paracelsus – though vastly different in other regards – believed that faith alone would activate the change whereby Christ became present in the bread and wine. (Both would evolve in their thinking on the Lord's Supper – and recall Luther's discussion of God's omnipresence when illustrating consubstantiation – but the point here is generally fitting concerning their early Eucharistic concepts.) Both believed that the bread and wine were still present but, in proper communion, Christ's presence entered invisibly. Thus, consubstantiation occurred only when the believer recognized Christ's presence. Paracelsus believed that Christ does not give us a bread that is a dead flesh; rather, it is a flesh with spirit, a flesh that does not die. He elaborated that Judas Iscariot – although partaking in the flesh and blood – remained in the old creation and the natural light. He did not partake in the Spirit of God and the

27 "Epistola. Philippi Theophrasti Paracelsi ad Theologos Witebergenses," 2nd page: "Darumb so soll sich hierauff gebühren / das weniger Liecht in dem mehrern Lichten zu stercken / wie wol iedoch unser anzünden Irridsch ist / dasselbige kömpt aus dem Kieszling / Ich setze aber nicht also ein Liecht hie/ sondern anzuzünden von dem Liecht des Obern /".

eternal light, and thus the Lord's Supper was of no use to him.[28] Now, it is on the topic of Christ's presence – the very ontology of Christ, his genealogy, the creation by Christ, and the Christian participation in His eternal creation – that Paracelsus takes his point of departure from Luther and – for that matter – everyone else. After all, Paracelsus' Christological views, and the application of his Eucharistic theology to his overall natural philosophy, was exceptionally idiosyncratic.

Before turning to this peculiar theology, I conclude this section with the observation that there seems to be little reason to doubt the authenticity of Paracelsus' letter to Luther and the Wittenberg theologians. Clearly, there are a number of elements of Paracelsus' theology either explicit in the letter or implied in Paracelsus' own theological aspirations – including his perceived consonance with the Lutherans in Wittenberg, especially concerning *sola fide* – that would be characteristic of Paracelsus in 1525, and it is probable that Paracelsus was not yet aware of Luther's admonition and persecution of the revolts (which would address what was perhaps Sudhoff's principal concern). An expressed desire to be tied to Luther, despite Paracelsus' occasional and even scathing criticism of Luther, does not necessarily make spurious any given work attributed to Paracelsus.

Any number of works attributed to Paracelsus, including his letter to the Lutherans, can be deemed to be genuine or spurious on the basis of some very clear theological concepts central to his religious thought. The topic of the authenticity of the "Letter to the Wittenberg Theologians" deserves still much more consideration and can be developed further in future studies. But in this short space, I will turn now to the authenticity of the *Astronomia Magna*. As stated above, Paracelsus' Christological views, which included the application of his Eucharistic theology to his overall natural philosophy, were idiosyncratic. Taken as a whole, these ideas give a distinctive stamp to the *Weltanschauung* of the authentic Paracelsus, and such ideas resound strongly in the *Astronomia Magna*.

2 The *Astronomia Magna*

The most pronounced dismissal of the authenticity of the *Astronomia Magna* was that of Heinrich Schipperges, who, in a 1988 study, stated that – in his

28 See Rudolph, "Einige Gesichtspunkte," esp. 44, and Dane T. Daniel, "Paracelsus – die Sakramentenlehre und das Verhältnis von Religion und Naturwissenschaften in der wissenschaftlichen Revolution," in *Manuskripte, Thesen, Informationen der Deutschen Bombastus-Gesellschaft*, 16 (2000), 17–26.

opinion – the *Astronomia Magna* was wrongly attributed to Paracelsus. To boost his argument, he focused on how the work contradicted Paracelsus' medical teaching.[29] Concentrating on the *Entienlehre*, he highlighted Paracelsus' four pillars of medicine, namely, *philosophia*, *astronomia*, *alchimia*, and *physica*. The four astronomies of the *Astronomia Magna*, on the other hand, are natural astronomy, supernatural astronomy (associated with the spiritual body, i.e., the resurrection body created by Christ), the astronomy of faith, and the astronomy of the infernal powers. Schipperges thus argued that the four types of astronomy are in "clear contradiction" to Paracelsus' teaching on the causes of disease; he could not fathom that astronomy had evolved from being a pillar of medicine – one of the four types of causes of disease – to the "mother of all arts."[30] He did not believe that the authentic Paracelsus would neglect the physical realm in his medical thought and turn the doctor into nothing more than a "magus" skilled in the sidereal powers.[31]

I believe that Schipperges misunderstood the actual scope and goals of the *Astronomia Magna*, for the author of the *Astronomia Magna* clearly writes that he wishes to treat the sidereal impact on the whole of nature, whether in medicine, metallurgy, or human talents. His stated goal in the *Astronomia Magna* is not to treat the elemental body and realm, which is – per Paracelsus – the subject of philosophy; in other words, medicine is also treated in other contexts, e.g., the philosophical. One might see contradiction or evolution in Paracelsus' thought, but Schipperges is overstating his case. In Book I, on the natural sidereal powers, Paracelsus writes:

> So now it will be understood, that firstly there is a body about which the *astronomus* does not speak; it is elemental, belonging to the earthly [topics] in the elemental philosophy. But of the natural spirit in the body,

29 Heinrich Schipperges, *Die Entienlehre des Paracelsus: Aufbau und Umriß seiner Theoretischen Pathologie* (Berlin, 1988), 35: "Vor allem die von uns zur Krankheitslehre des Paracelsus herangezogenen Aussagen stehen in einem deutlichen Gegensatz zu der immer wieder mit Vorliebe zitierten 'Astronomia Magna,' die meines Erachtens zu Unrecht dem Paracelsus zugeschrieben wird, obwohl sie als 'System einer Universalwissenschaft' (Goldammer, 1978) von kaum abzuschätzender Bedeutung geworden ist.

Die 'Astronomia magna' gliedert sich in 'vier Ordnungen', die bereits in deutlichem Widerspruch zur Paracelsischen Krankheitslehre stehen. Als 'naturalis astronomia' zeigt sie den Einfluß des Firmaments auf den siderischen Leib, als 'supera' dient sie der Neugeburt zu einem geistlichen Leben (XII,76); als 'astronomia olympi novi' läßt sie den wahren Glauben erkennen und als 'astronomia inferiorum' offenbart sie die infernalischen Mächte (XII, 76). Der 'astrologus' aber soll und kann den 'summum motorem naturae' erkennen; denn die Sterne und die Menschen sind '*gleich vermögens*' (XII, 90)."

30 Ibid.

31 Ibid.

> the *astronomus* has the force and power to speak; for the body is of the elements, and the spirit given to the elemental body is wedded to it and incorporated into it by the firmament. And thus philosophy is divided into two parts, in the essence of the spirit and secondly in the essence of the body, that is, into body and spirit. So notice therein what the heavenly impacts accomplish, that is, what the ethereal spirits deal with and [what they] do – these powers and virtues (*krefte und tugenden*), what God has given to them, I wish to describe here: as the body of the elements, the body of the beasts, the body of men, the body of stones, the body of metals, the body of herbs, and others like them. These all are bodies and *corpora* that have nothing to do with the sidereal spirit.[32]

Thus, Paracelsus' treatment of "astronomy" includes all topics related to the sidereal bodies of the cosmos, including the sidereal body of humans (also called the human "spirit," which accounts for mental and sensible attributes), and the knowledge that the stars rain down upon the world (also called the "light of nature"). Astronomy, per Paracelsus, is thus fundamental to anthropology and epistemology.

On medicine specifically, Paracelsus adds:

> The great art of medicine, which is so wonderfully created by God, takes its cornerstone, i.e. the recognition of human illness, health, and death

32 *Astronomia magna*, Sudhoff, 1/12:16: "also sol nun verstanden werden, das erstlich ein leib ist, von dem der astronomus nichts redet; er ist elementisch, gehört in die irdisch der elementen philosophei. aber von dem natürlichen geist im leib hat der astronomus zu reden gewalt und macht; dan der leib ist der elementen, der geist, der dem elementischen leib geben ist, der ist im vom firmament vermälet und eingeleibt. und also teilt sich die philosophia in zwen teil, in das wesen des geistes und zum andern in das wesen des leibs, das ist in den corpus und spiritum. darinnen so merkt, was die himlischen würkungen volbringen, das ist, was die etherischen geist handlen und tun, wil ich hie dieselbigen krefte und tugenden beschreiben, was got in inen geben hat, als den leib der elementen, den leib des vihes, den leib der menschen, den leib der steinen, den leib der metallen, den leib der kreuter und anderer dergleichen. das sind alles leib und corpora, die da nichts betreffen den siderischen geist"; see Daniel, "Paracelsus' *Astronomia Magna*," 152 n. 6: "Hence, Paracelsus has introduced an important qualifier, that this book does not treat the whole of cosmology, for astronomy does not deal with elemental theory. And yet, the elemental body *is* treated insofar as the firmament incarnates it spiritually. The discussion here is indeed about the effects of the heavens, that is, what pertains to the 'ethereal spirits' and their actions. The rulers – the predestined parts – originate in the heavens, hence the governance by the stars, the subject of natural astronomy. Thus, Paracelsus will be able, for any object or being, to speak in terms of the interactions between elemental body and sidereal body, the former referring to the tangible corpus, the latter to the invisible-subtle-directing spirits."

> – and up to half of its requisites – from the art of astronomy. When this is lacking, the sick person is misled by his doctor. Indeed, the doctor who is unskilled in astronomy cannot be called a complete doctor, for more than half of all diseases are ruled by the firmament.[33]

It is obvious, contrary to Schipperges' assessment, that astronomy – while playing a significant role – is not the only requisite for medicine in the *Astronomia Magna*.

Furthermore, the author of the *Astronomia Magna* writes in Book 1, Chapter Eight, *Probatio in scientiam medicinae adeptae*: "To know what adept medicine is, note the following: It treats only of diseases that we have from the heavens and from the heavenly impact. For there are two sources from which all diseases grow: one is of the stars, the other of the elements."[34] Thus, the writer of the *Astronomia Magna* is patently not restricting medicine to its astronomical aspects. Instead, he merely restricts his discussion to diseases caused by direct sidereal action (whether natural or "adept").[35]

The manner in which the author of the *Astronomia Magna* treats the light of nature is fundamental to and congruent with Paracelsus' authentic writings. A clear understanding of Paracelsus' epistemology and its anthropological dimensions might have helped dispel Schipperges' reservations about the authenticity of the *Astronomia Magna*. Indeed, the major themes of the *Astronomia Magna* bear the characteristic stamp of the genuine Paracelsus'

33 *Astronomia magna*, Sudhoff, 1/12: 3: "die große kunst der arznei, so von got wunderbarlichen geschaffen ist, nimpt iren eckstein zu erkennen die krankheit des menschen, ir gesuntheit und den tot bis auf den halben teil irer noturft aus diser kunst der astronomei. wo solchs gebricht, da ist der krank verfüret mit seinem arzt. dan der arzt der die astronomei nicht kan, der mag nicht ein volkomener arzt genant werden. dan mer dan der halbe teil der krankheiten wird vom firmament regiret."

34 Ibid., 187: "Zu wissen was medicina adepta sei, so merket das selbig also. sie tractirt alein von den krankheiten so wir vom himel an uns haben und die selbige himlische wirkung. dan zwen ursprung seind, aus denen alle krankheiten wachsen: der ein aus dem gestirn der ander aus den elementen."

35 I concede, however, that Paracelsus also argues – and somewhat confusingly – that all natural arts and knowledge are taught by the natural stars, that is, by the light of nature. See *Astronomia magna*, Sudhoff, 1/12: 3: "First I must praise the *sagax* on the ground that God gave – and still gives – all arts and all that is natural to the human through the firmament. Thus, the firmament is the natural light, and the human learns about natural things from the firmament. Next, I must praise her on account of her great utility, for all essential arts spring forth from her." The German reads, "erstlich muß ich loben die sagacem, von wegen das got dem menschen durch das firmament alle kunst und was natürlich ist geben hat und gibt, also das firmament das natürlich liecht ist und der mensch vom firmament das natürlich hat. weiter so muß ichs loben von wegen irer großen nutzbarkeit, so alle notürftige künst aus ir entspringent."

unique anthropology and epistemology. His distinctive microcosm-macrocosm theory can be encapsulated in his two-creation theory that incorporates his unique Christological, anthropological, and epistemological concepts in one *Weltbild*. It is thus pertinent to summarize this teaching, as follows.

In what is perhaps his theological *magnum opus*, *De Genealogia Christi*, Paracelsus elaborated an idea developed in his Eucharistic tracts, namely the two-creation concept.[36] In this cosmology – based on his unique reading of Scripture – Paracelsus differentiated between the creation by God the Father and the creation by God the Son. In this Father vs. Son dichotomy, Paracelsus argued that only the spiritual body of humans – created by the Son – would resurrect, whereas the corporeality from the elements and stars were destined to dissolve irrevocably. The first creation by God the Father could not exist in eternity, and Christ's body – as well as the bodies of everyone else in his genealogy, from Abraham to Mary – never participated in the Adamic corporeality created by the Father. Thus, while denouncing the earthly church clothed in stone (the *Mauerkirche*), Paracelsus forwarded his belief that the Christian, through baptism, receives the resurrection body while still on earth, and that this invisible body – participating in Christ's new creation and nourished by the Eucharist – lives forever as the house of the soul. To reiterate, the elemental body and the sidereal body are both ordained for dissolution, the latter accounting for certain mental and sensible attributes. The soul – which is a piece of God and the very image of God – lives within the invisible resurrection body created by Christ. And in this new body, Christians possess the same characteristics and miraculous abilities as Christ Himself. They can engage in what Paracelsus calls celestial astronomy (or divine magic).[37]

Paracelsus argues that all men participate in the Adamic body, which was originally created from the *limus terrae* of Genesis 2:7. He calls this "dust of the earth" the "limbus," teaching that it contained the seeds of everything that exists in the natural heavens and earth, from oak trees to comets, hence man as a "microcosm." All men, without the necessity of faith, partake in the natural light and recognize the Father's creation. Paracelsus differentiates the natural "limbus" from the "eternal limbus," which is the creation by God the Son. This eternal limbus is that "soil" from which born-again men – the resurrected – are

36 Goldammer dated *De Genealogia Christi* to ca. 1530, but Urs Leo Gantenbein recently argued convincingly that it should rather be dated to the years 1524/1525; see Urs Leo Gantenbein, "The Virgin Mary and the Universal Reformation of Paracelsus," *Daphnis*, 48 (2020), 4–37, on 21–24.

37 See Kurt Goldammer, "Paracelsische Eschatologie: Zum Verständnis der Antrhopologie und Kosmologie Hohenheims," in idem, ed. *Paracelsus in neuen Horizonten* (Vienna, 1986), 87–152, esp. 90–98; see also idem, "Friedensidee und Toleranzgedanke bei Paracelsus," in idem, *Paracelsus in neuen Horizonten*, 153–176.

made. Only through faith may one participate in this eternal realm, and truly partake of the Lord's Supper, nourished by its eternal limbus.

Given the distinctive nature of these ideas, it seems very probable that the presence of Paracelsus' peculiar Christology – including the genealogy of Christ – and the differentiation between the mortal limbus fashioned by God the Father and the eternal limbus created by God the Son, present a distinguishing feature of the genuine works of Paracelsus, a marker that can help reveal a work's authenticity. The *Astronomia Magna* highlights many themes consistent with concepts developed in his Eucharistic writings and discussed similarly in the genuine work *De Genealogia Christi*.

In fact, the *Astronomia Magna* resembles not only the subject matter, language, and style of Paracelsus' *theologica*, but also the subject headings, especially those in *De genealogia Christi*.[38] One witnesses, for example, Paracelsus' discussion of the unity of bodily and divine components in both works. Note the section of *De genealogia Christi* that is entitled "How the Trinity unites with the human" (*Wie sich die trinitet gegen dem menschen vereinigt*), which begins:

> Now, so that you know why I am writing about the Trinity and the use that comes from it, I am not just treating the genealogy of Christ, but want to show something useful regarding the whole production of the world. Namely, all that we have on earth we have from God. Note too that we have gifts from God: one from the Father, one from the Son, and one from the Holy Spirit. The whole world, all men, require the Trinity. That required by the Father also has to have the Son and the Holy Spirit. So know now that humans are differentiated in that some believe in God the Father not as a Father, but as creator. [...] To believe in [God as not a father, but merely] the creator is correct, but it is not a complete belief.[39]

38 Goldammer, II/3: 69–71.

39 Ibid.: "Nun damit und ir wissent, warumb ich von der trinitet dermaßen schreibe und was nutz daraus entspringt, ist nit allein die ursach von wegen der genealogia Christi, in der wir auch seindt, sonder daß ir zu demselbigen noch mehr verstandent, was nutzbarkeit hier aus uns allen in der ganzen fabrication der welt entstehet und entsprengt. nämblich alles das, so wir haben auf erden, haben wir von gott. iedoch aber so merkent, daß wir von gott die gab haben, nemblich eine vom vatter, eine vom son, eine von heilgen geist. damit so verstehent, daß also alle welt, das ist alle menschen, bedürfen der trinitet. denn der des vatters bedarf, muß auch den son und den heiligen geist haben. der auch des sons bedarf, desgleichen des vatters und des heiligen geists. also auch mit dem heiligen geist. So wissent nun auf das, daß wir menschen undterschiedlich sein, in dem etlich glauben in gott den vatter nit als ein vatter, sonder schöpfer. wiewol sie nun nit weiter glauben, iedoch aber so müessen sie den son und den heiligen geist auch haben. etlich glauben in son; die glauben auch in vatter und in son und in heiligen geist. [...] Darumb so wissent, daß glauben in schöpfer ist recht, aber es ist nit ein volkombner glauben."

Here the theological work focuses on anthropology and cosmogony. In the later *Astronomia Magna*'s Book II, Chapter 3, Paracelsus reiterates the earlier discussion with a section entitled "How God unites Himself to the human, and how He stands in relation to him, [and] how He receives him and operates with and through him, as namely in the world to come."[40]

Turning to another subject, one may question whether Paracelsus' discussion of "celestial astronomy," an analogy to his types of natural magic (or "natural astronomy"), was unique to the *Astronomia Magna* and thus perhaps inauthentic. However, here too we see precedents in other works.[41] The insights of Kurt Goldammer, Hartmut Rudolph, Ute Gause, and Arlene Miller Guinsburg are extremely helpful here; this is, once again, a subject that fully incorporates – and is intelligible only in the context of – the genuine Paracelsus' Christology, anthropology, and epistemology.

First of all, Goldammer observes that Paracelsus had broached such subjects as the "*supercoelestis physica*" and "*corpus praeter naturam*" (topics regarding "celestial astronomy") as early as *De vita longa* (1526?).[42] In this text, he observes, Paracelsus had discussed both celestial magic and the idea of a new body within humans (as he was formulating a basis for his theories on prolonging life). Other scholars have similarly studied Paracelsus' celestial magic, which – according to Paracelsus – is the eternal analogy of natural magic (or astronomy) and its types, and evaluated the eternal body, wherein the celestial magic operates. Most notably, Rudolph writes:

> In his new, heavenly body man can exercise the capacity to do divine works [...] In accordance with the 'two schools on earth' [...] a change of 'schoolmaster' takes place, more precisely, there is a gradual transition, a progression to a higher school. Instruction in the wisdom of heaven [...] is no longer given in the school of nature but is given in the newborn body 'down from above' [...] As heavenly magus, born through baptism, his growth achieved by means of the Eucharist, man now rules over the

40 Sudhoff, I/12: 320: (Wie sich got gegen dem menschen vereinigt, und wie er sich gegen in halt, annimbt und mit im und durch in hantlet, als nemlich in der zukünfitgen welt).

41 This section utilizes my more thorough discussion of celestial magic in Daniel, "Paracelsus' *Astronomia Magna*," 230–236.

42 Kurt Goldammer, *Der Göttliche Magier und die Magierin Natur. Religion, Naturmagie und die Anfänge der Naturwissenschaft vom Spätmittelalter bis zur Renaissance, mit Beiträgen zum Magie-Verständnis des Paracelsus* (Stuttgart, 1991), 48; Paracelsus, *De vita longa*, Sudhoff, I/3. Goldammer notes that "Aus dem übernatürlichen Einfluß kommen auch Bild- und Kristallzauber ("imagines, gamahaeae"), den die "philosophastri" (Ficino gemeint?) auf die Sterne des Firmaments zurückgeführt haben."

> heavenly '*magnalia*' [...] Such a celestial magus can perform miracles: he can survive in the fiery furnace like Daniel, walk on water like Christ, pass through walls, heal the sick like the Apostles, indeed bring the very dead back to life.[43]

For her part, Ute Gause comments that "[...] the 'new birth' through baptism makes him capable of using supernatural powers. Prior to this, of course, the Gospel must be heard and assimilated. [...] The *magia coelestis* encourages the domination of the world in even greater measure than the *magia naturalis*."[44] Arlene Miller Guinsburg adjoins: "The Christian, that is, one who has put on Christ's body and blood, is able to accomplish wonders even greater than those performed by the natural magic."[45] As such, Guinsburg argues that "Paracelsus invested the concept of belief with power."[46] She also notes, however, that it is not certain whether Paracelsus believed all Christians were capable of the same types of miracles as those performed by Christ and his twelve disciples – in one instance, Paracelsus implies that prayer and fasting could help a person expel demons if he or she had some capacity in celestial magic.[47] However, Paracelsus was enthusiastic about such powers, which he discussed in "*Probatio in cœlestem philosophiam adeptam*," a section in one of his Matthew Commentaries. According to Paracelsus, Christ's statement that "this is my body, this is my blood," means that "[t]he person who eats my flesh and drinks my blood, etc., is now competent in celestial philosophy" (but only if one also contemplates that he or she is actually consuming the eternal bread and blood).[48]

---

43 Hartmut Rudolph, "Hohenheim's anthropology in the light of his writings on the Eucharist," in Grell, *Paracelsus: The Man and His Reputation*, 187–206, on 204; see also ibid. 197: "In Part Two of his main work, the *Astronomia Magna*, [...] there is mention of 'two schools on earth' [...]; the school 'from the earth' has its schoolmaster 'from nature, in nature and is nature' [...]; that is to say, it is no longer conceived as the mediator of errors; the other 'school down from above [...] teaches heavenly wisdom in the newborn man through him who is from heaven'."

44 Ute Gause, "On Paracelsus' Epistemology in His Early Theological Writings and in His *Astronomia Magna*," in Grell, *Paracelsus: The Man and His Reputation*, 207–224, on 219.

45 Arlene Miller Guinsburg, "Paracelsian Magic and Theology: A Case Study of the Matthew Commentaries," *Medizinhistorisches Journal*, 16 (1981), 125–139, on 134.

46 Ibid., 135.

47 Ibid. Guinsburg cites Paracelsus' Commentary on Matthew 17:20 in Paracelsus, *Matthäus Kommentar*, in Leiden University Library, Cod. Voss. Chym. F. 25, fol. 220r.

48 Paracelsus writes (Sudhoff, 1/12: 397): "Christus spricht: das ist mein leib, das ist mein blut. Mer: wer do isset mein fleisch und trinket mein blut etc. nun ist das auf coelestem philosophiam geret."

Note how Paracelsus himself sums up the topic, which – and again – corresponds to his discussions in both his earlier Eucharistic tracts and *De Genealogia Christi*:

> For in the same way that the astronomer may use magic in the firmament, and bring the heavenly operation here on earth, by virtue of the fact that he is from the earth, thus, you should also discern that you can know from heaven the same operations just as well.[49] It is only a matter of which heaven you are from, whether that of the earth or from the firmament. For one's power always corresponds to that out of which he is. Thus, they restored health to humans, healed the lame and blind, and raised the dead; and brought health to the sick; execrating and restoring health is all one and the same power: for cursing the opulent is just as necessary as healing the sick, and it is contemptible that such great evil and contempt takes place in Christ and in his new birth by means of the earthly damned. But there is no one there who does what the saints did, but rather it is the case that no one is dutiful in the new birth, and it is lamentable (as the Psalmist speaks)[50] that he looks down from above and sees not one who is good, that is, who has the power to heal, or to preach, or to teach in a manner that is not false.[51]

49 See Goldammer, *Der Göttliche Magier*, 46–47: "Die Erkenntnisfunktion und die Beziehung der Magie auf den gestirnten Himmel mit seinen regulären und außerordentlichen Erscheinungen und damit auf die Astrologie ist auch gegeben, wenn in der 'Astronomia Magna' ('Philosophia Sagax'), etwa zur gleichen Zeit entstanden wie der eben zitierte 'Labyrinthus', die 'Magic' zu definieren versucht wird als '*Auslegung* der natürlichen *zeichen*, sie zu erkennen', von Gott 'ubernaturlich in himel' gestellt. Sie erscheinen aber 'wie natürlich', wie der Stern von Bethlehem […]."

50 Psalms 53:2–3. [Vulgate 52:3–4.]

51 Sudhoff, I/12: 329: "Dan zu gleicher weis, wie der astronomus mag im firmament magicam brauchen, und des himels […] operation hie auf erden, aus der kraft, das er vom firmament ist. also sollent ir auch wissen, das ir vom himel gleich so wol die selbigen operationes mögent erkennen. alein an dem ligt es, so ir vom himel seiet, wie der von der erden oder vom firmament. dan die selbige kraft ist alwegen im selbigen, aus dem er ist. also haben sie menschen gesunt gemacht, die lamen, die blinden, die toten lebendig; ist alles die éin und die gleiche kraft, verfluchen und gesunt machen. dan verfluchen das uppig ist wol so not als gesunt machen den kranken, und ist zu erbarmen, das solch groß ubel und verspottung in Christum und in seiner neuen geburt von dem irdischen verdampten geschicht. aber niemants ist da, der das tut das die heiligen getan haben, sonder alein das ist, das niemant der neuen geburt anhengig ist. darumb so ist zu besorgen (wie der psalmist spricht), das er sicht von oben herab und sicht nit einen der gut ist, das ist, der gewalt hat […] gesunt zu machen oder on falsch predigen oder leren."

In the *Astronomia Magna* – which was both fully natural philosophical and theological – Paracelsus sought to fulfill his project to explicate the entirety of his Christological, anthropological, and epistemological teaching. It was thus necessary to address fully the sidereal and immortal bodies of man, which were topics he had been developing since his earliest writings, including his early Mariological and sacramental tracts.[52] It is surprising that any scholar would judge to be spurious a work elsewhere rightly proclaimed as the *magnum opus* of Paracelsus.

Conclusion

In this paper, which has called into question the veracity of some of Sudhoff's valuations concerning dating and authenticity, and in which I have countered Schipperges' doubts concerning the genuineness of the *Astronomia Magna* while highlighting some of the distinguishing elements of the authentic Paracelsus, I hope to have put forward useful ways in which one could construct a diagnostic rubric with regard to the valuation of the authenticity of works attributed to Paracelsus. Understandably, the particular rubric suggested here covers primarily religious aspects of Paracelsus' expression (and even these should be expanded to include such topics as eschatology, cosmogony, and creation *ex nihilo* versus the concept of uncreated matter, with the genuine Paracelsus favoring the former); however, one may analyze other authentic works for additional constructs to be added to the rubric, or one could propose entirely new rubrics based on, for example, medical and/or alchemical motifs (depending on the work or passage in question).

52 For examples of earlier writings from which the *Astronomia Magna* evolved, see his discussion of the elemental spheres and fruits in *Philosophia de generationibus et fructibus quatuor elementorum* (*Philosophy concerning the generations and fruits of the four elements*), wherein he anticipated his discussion of sidereal impressions with his discussion of the impressions raining down from the "fire sphere"); see also his Eucharistic tracts, wherein he developed his concept of the immortal body created by Christ in the new creation and received by Christians sacramentally. Regarding the Eucharistic tracts, see *Coena Dominj nostrj Jhesu Christj Declaratio*, which is copied among other works about the eternal *limbus* in Leiden University Library, Codex Voss. Chym. F. 24, fol. 12r-29v. *Coena Dominj* is transcribed in Dane T. Daniel, "Paracelsus on the Lord's Supper: *Coena Dominj nostrj Jhesus Christj Declaratio*," *Nova Acta Paracelsica*, N.F. 16 (2002), 107–139; it is also in *Philosophia de Limbo*, ed. Stariz, 88–126. The tie between Paracelsus' early Mariology and his subsequent two-creation theory (also called the "immortal philosophy") is discussed in Michael Bunners, *Die Abendmahlsschriften und das medizinisch-naturphilosophische Werk des Paracelsus*, PhD diss. (Berlin/GDR, 1961).

To conclude, there are of course many factors and motifs to consider when making the final determination of the authenticity of a given manuscript or printed book. I believe the historical factors and theological points presented above make a fairly strong case for the authenticity of the *Brief an die Wittenberger Theologen*. Certainly, Sudhoff understood the genuine Paracelsus as well as anyone and apparently misjudged the letter to Luther and his paladins mostly because of incomplete historical data. I concur with Sudhoff (and most everyone else) about the authenticity of the *Astronomia Magna* and believe the divergent conclusion of Schipperges may have been different if he had considered the text in the context of the genuine Paracelsus' religious world-view, including the Christological, anthropological, and epistemological motifs so present in many of Paracelsus' earlier works.

The *Philosophia ad Athenienses* in the Light of Genuine Paracelsian Cosmology

Didier Kahn
CNRS, Cellf 16–18 – Sorbonne Université
dkahn@msh-paris.fr

Abstract

The pseudo-Paracelsian *Philosophia ad Athenienses* (1564) draws upon many of the ideas of Paracelsus but combines them with many other elements not found in the genuine works of the Swiss physician. After discussing the details of its edition, we summarize its content, then examine its possible sources, including the authentic texts of Paracelsus by which the unknown author will have taken inspiration. We discuss specifically the question of uncreated prime matter and provide a number of criteria allowing us to consider the treatise as spurious. Finally, we argue that the unknown author wrote this text and ascribed it to Paracelsus in order to produce a more Platonist Paracelsian cosmology than that of the genuine treatises of Paracelsus, which apparently disappointed the unknown author in this regard.

Keywords

Paracelsus – Birckmann printing press – pseudepigraphy – cosmology – Creation – uncreated prime matter – separation – Neoplatonism – elemental beings – prophecy

Introduction

Inauthentic in both style and content, the *Philosophia ad Athenienses* is based on several of Paracelsus' genuine ideas, but conflates them with other ideas alien to the authentic works of Paracelsus, thus producing a new treatise of Paracelsian cosmology.[1] The result is a cosmology that is even stranger than that of Paracelsus himself: the *Philosophia ad Athenienses* deals with the origins of the created world and is partly responsible for the accusations of heresy directed against Paracelsus in the years 1560–1570, mainly due to its

1 The author warmly thanks Andrew Weeks and Dane T. Daniel for discussing and criticizing a number of topics of this paper and improving his level of expression. All mistakes are his.

 | DOI:10.1163/9789004503380_004

assertion of an uncreated prime matter.[2] Published in 1564, this work is one of the first (actually the second) pseudo-Paracelsian treatises ever printed.[3] In 1590 Johann Huser reprinted the text from the 1564 edition at the head of the eighth volume of the *Bücher und Schrifften*, a volume devoted to meteorology and cosmology. Huser did not make any comment on the authenticity of the work, which he seemed to have accepted at face value. No manuscript seems to have been preserved, and the author is unknown. In 1894 (and again in 1931) Karl Sudhoff wondered whether the work might be a reworking of one of Paracelsus' original writings, and he concluded that "no one can say whether we even may or should suppose any genuine nucleus for this work."[4] Sudhoff was probably the first to express serious doubts about the authenticity of the work. Others, like Walter Pagel, were far less cautious: in Pagel's view (partly based upon his own reading of Kurt Goldammer, though *not* on Goldammer's actual opinion), the work "expresses genuine Paracelsian thought." This opinion gained further currency, owing to Pagel's authority.[5]

I will first describe the 1564 edition, then summarize the content of the text, and point out some of its obvious sources and distinctive features.

2 Charles D. Gunnoe, *Thomas Erastus and the Palatinate. A Renaissance Physician in the Second Reformation* (Leiden, 2011), 309 and 311–314. On this topic, see also Didier Kahn, "La création *ex nihilo* et la notion d'*increatum* chez Paracelse," in Edouard Mehl and Isabelle Pantin, eds., *"De mundi recentioribus phænomenis." Cosmologie et science dans l'Europe des Temps modernes, XVe-XVIIe siècles. Essais en l'honneur de Miguel Ángel Granada* (Turnhout, forthcoming). Other accusations of heresy were voiced from 1561 onwards by Konrad Gessner and Crato von Krafftheim, on the grounds of Paracelsus' alleged Arianism as deduced from Alexander von Suchten's *Decem et octo propositiones* (before 1561); see Ralf Bröer, "Friedenspolitik durch Verketzerung. Johannes Crato (1519–1585) und die Denunziation der Paracelsisten als Arianer," *Medizinhistorisches Journal*, 37 (2002), 139–182.

3 The first one, much less significant, has been discussed by Jürgen Strein and Joachim Telle, "Deutsche Pseudoparacelsica über die Pest," in Dominik Groß and Monika Reininger, eds., *Medizin in Geschichte, Philologie und Ethnologie. Festschrift für Gundolf Keil* (Würzburg, 2003), 349–370, on 350–352. See below, Appendix, § 3.9.

4 Paracelsus, *Sämtliche Werke*, 1. Abteilung: *Medizinische, naturwissenschaftliche und philosophische Schriften*, ed. Karl Sudhoff, 14 vols. (Munich, then Munich-Berlin, 1922–1933) (hereafter: Sudhoff, 1/1, 1/2, etc.), here 1/13 (1931), XII: "Keine einzige handschriftliche Überlieferung dieser Schrift [...], auch nicht das kleinste Fragment. [...] so weiß kein Mensch zu sagen, [...] ob man überhaupt einen echten Kern annehmen kann oder soll"; see the earlier comment by Karl Sudhoff, *Bibliographia Paracelsica* (Berlin, 1894), 102–103, almost identical.

5 Walter Pagel, "Paracelsus and the Neoplatonic and Gnostic Tradition," *Ambix*, 8 (1960), 125–166, on 143; see also idem, *Paracelsus. An Introduction to Philosophical Medicine in the Era of the Renaissance* (Basel, 1982), 88–91, and Giancarlo Zanier, *L'Espressione e l'Immagine. Introduzione a Paracelso* (Trieste, 1988), 57 n. 17. A most interesting discussion on the historiography of the work is Gunnoe, *Thomas Erastus*, 311–312 n. 141.

1 The Anatomy of the First Edition

The first text of a collection of three treatises, the *Philosophia ad Athenienses* was printed by "the heirs of Arnold Birckmann" [i.e., not necessarily by Theodor Birckmann (1534–1586) himself] in 1564 in Köln.[6] The book did not include any preface, *liminar* poem, or dedicatory epistle. However, in the first edition of Paracelsus ever published by the Birckmanns, which appeared just a few months before the *Philosophia ad Athenienses*, the publisher inserted an admonition to the reader, complaining about the difficult handwriting of Paracelsus and explaining their use of the asterisk ("sternlein") to point out barely readable passages or doubtful readings. This policy was also applied to the *Philosophia ad Athenienses*.[7] Indeed, the publisher's difficulty in understanding the manuscript produced many obvious misreadings.[8]

In the 1564 edition (see fig. 2.1), the text of the *Philosophia ad Athenienses* was printed along with two authentic treatises of Paracelsus: *De caducis* (a treatise on epilepsy), and *De contracturis* (on contractures).[9] The text of *De caducis* is based on the same manuscript later used by Huser, which belonged to the Silesian physician Johannes Scultetus Montanus (1531–1604).[10] According to the 1564 editors, *De contracturis* was based on Paracelsus' autograph manuscript,

6 See Joachim Telle, in Wilhelm Kühlmann and Joachim Telle, *Corpus Paracelsisticum: Der Frühparacelsismus*, 3 vols. (Tübingen, 2001–2004 and Berlin, 2013) (hereafter: CP 1, CP 2, CP 3), here CP 1: 659–660, reacting against Sudhoff's assertions by ruling out (perhaps too categorically) the involvement of Theodor Birckmann in any editorship within the Birckmann publishing house prior to 1576.

7 Sudhoff, *Bibliographia*, 99 (*Vom Podagra und seinen speciebus* and *Elf Traktat*); CP 1: 665; see [pseudo]-Paracelsus, *Philosophia ad Athenienses* (Köln, 1564), e.g., book I, §20: *Sidus* in the margin instead of the puzzling *Stylus*; *newen* instead of *nauen* (both alternate readings were indeed adopted by Johann Huser in his edition, mentioned below, n. 9). See the enlightening linguistic remarks of Goldammer on the problems pertaining to the edition of the Kärntner Schriften by Birckmann in 1564 in Paracelsus, *Die Kärntner Schriften*, ed. Kurt Goldammer et al. (Klagenfurt, 1955), 305–306.

8 Huser also tacitly corrected some other previously unnoticed mistakes, e.g., *conparlisch* into *corporalisch* (book II, end of §8). Yet, the following mistakes were not emended by Huser: in book I, §9, *Truphat* might very well be a misreading for *Frucht*; in §11, *Archaltis* might be *Atlantis*; in §13, *Durdales* is clearly *Dryades*, and *Neufareni* might be *Nympharum*; in §20, *Natalischen* might be *Fatalischen*; in book II, §7, *Chayr* is clearly *Chaos*; see below, notes 44–45, 51–52, 55. Other invented words are more difficult to decipher, but are most probably misreadings as well, such as *Neßder* (I, §14), or *Dalni* (I, §20).

9 Paracelsus, *De caducis, das ist, Von den hinfallenden Siechtagen*, in Paracelsus, *Bücher und Schrifften*, ed. Johann Huser, 11 vols. (Basel, 1589–1591) (hereafter: Huser, 1, etc.), 4: 317–366 (Sudhoff, I/8: 261–308); *De contracturis*, in Huser, 4: 93–116 (Sudhoff, I/2: 457–486).

10 Sudhoff, *Bibliographia*, 103.

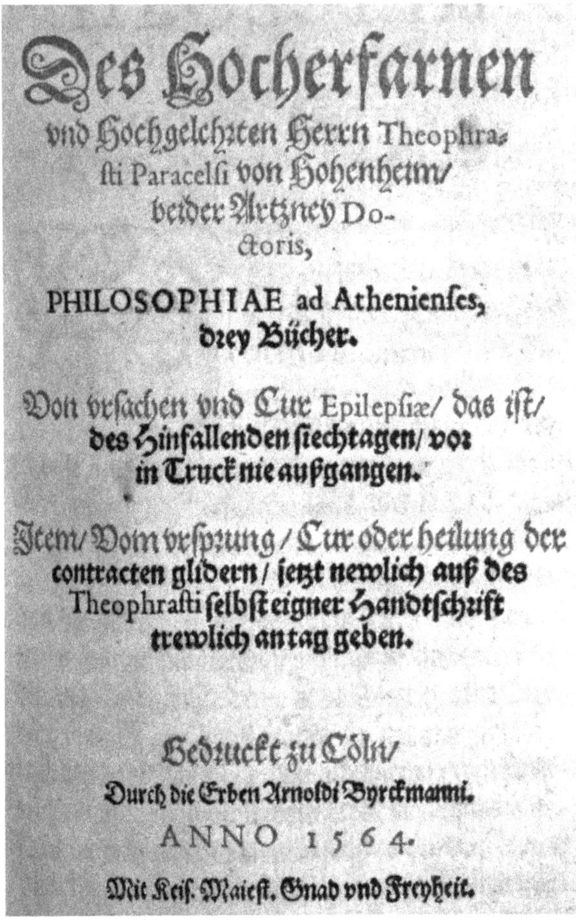

Des Hocherfarnen
vnd Hochgelehrten Herrn Theophra-
sti Paracelsi von Hohenheim/
beider Artzney Do-
ctoris,
PHILOSOPHIAE ad Athenienses,
drey Bücher.
Von vrsachen vnd Cur Epilepsiæ/ das ist/
des Hinfallenden siechtagen/ vor
in Truck nie außgangen.
Item/ Vom vrsprung/ Cur oder heilung der
contracten glidern/ jetzt newlich auß des
Theophrasti selbst eigner Handtschrift
trewlich an tag geben.
Gedruckt zu Cöln/
Durch die Erben Arnoldi Byrckmanni.
ANNO 1564.
Mit Keis. Maiest. Gnad vnd Freyheit.

FIGURE 2.1 Pseudo-Paracelsus, *Philosophia ad Athenienses* (1564), title-page. The initial misprinting "*Phisophia*" has been corrected by a pasted piece of paper. urn:nbn:de:bvb:12-bsb10164687-3

later used by Huser in 1590.[11] Since Huser had no other textual basis for the *Philosophia ad Athenienses* than the 1564 edition, we may deduce that the unknown manuscript used in 1564 did not come from Montanus, whose collections Huser accessed freely.[12]

11 See the title-page in 1564: "jetzt newlich auß des *Theophrasti* selbst eigner Handtschrift trewlich an tag geben"; and see Sudhoff, *Bibliographia*, 103.

12 As explained by Sudhoff, ibid., 102 and idem, I/13: XI. On Montanus, see CP 2: 239–241. The other treatises of Paracelsus published by the heirs of Birckmann in 1564 were mainly drawn from first-hand manuscripts. In the first of these editions, the *Elf Traktat* were obviously printed from the autograph. In the next Birckmann publication (the "writings of Carinthia"), the text was based upon the original manuscripts from Klagenfurt; see Sudhoff, *Bibliographia*, 99 and 100.

The *Philosophia ad Athenienses* contains three books: one is divided into 24 "Texts" (i.e., chapters), the second into 23, and the last only into 6: it is a fragmentary book. Each book has a preface ("Vorred") and a conclusion ("Beschluß") summarizing its contents (thus following the pattern of the *Grosse Wundartzney*, first published in 1536); the only exception is the unfinished third book, which has no conclusion. These "Vorreden" and "Beschlüße" mention Paracelsus in the third person ("Hierinnen beschreibt der Fürst Theophrastus…"), which leads us to assume that there are two writers: the author himself, and an editor.[13] Since the latter proves to be very familiar with the contents of the *Philosophia ad Athenienses* (as we shall see below), it seems that he took part in the editing process of the work *before* it was given to the printer for publication. Therefore, he should not be confused with the editor of the whole printed book (who was perhaps none other than the publisher himself). It seems, then, that three different people were involved in the production of this treatise: a) the unknown author, whom we may call Pseudo-Paracelsus; b) a first editor, who wrote the prefaces and conclusions; and c) a second editor, who gave the whole work to the publisher – unless it is the publisher himself who managed to gather several works of Paracelsus from different origins, including this one. We will return to this question below.

An intriguing detail is the unusual mention of Paracelsus as "Fürst Theophrastus" ("Prince Theophrast") in each preface and conclusion. Paracelsus was once given the title "Fürst" in the 1563 Basel edition of the tract *Von ersten dreyen Principiis*, but it was very seldom used in later editions.[14] Thus, it may be that the editor of the *Philosophia ad Athenienses* was influenced by the most recent publications of Paracelsus' treatises available on the book market, which would date the editorial work – i.e., the writing of the prefaces and conclusions – to the years 1563–1564. This is a mere hypothesis, since the editor of the *Philosophia* did not necessarily need a model for the decision to call Paracelsus "Fürst." Nonetheless, the coincidence is worth mentioning.

The enticing title *Philosophia ad Athenienses* is an obvious allusion to the Acts of the Apostles, chapter 17, where Paul unsuccessfully tried to reveal the truth of Christ to the pagan Athenians who rejected his assertion of the resurrection of the dead.[15] The implication is of course that Paracelsus is going to

13 As noticed by Sudhoff, ibid., 102–103.

14 Its title is: *Deß erfarnesten Fürsten aller Artzeten Aureoli Theophrasti Paracelsi von ersten dreyen principiis* (see Sudhoff, *Bibliographia*, 84–85). The editor was the prominent Paracelsian Adam von Bodenstein, but the printer is unknown.

15 As argued in detail by Richard Bostocke, *The Difference Betwene the Auncient Phisicke* […] *and the Later Phisicke* (London, 1585), chap. 20. On Bostocke, see David Harley, "Rychard Bostok of Tandridge, Surrey (c. 1530–1605), M.P., Paracelsian propagandist and friend of John Dee," *Ambix*, 47 (2000), 29–36.

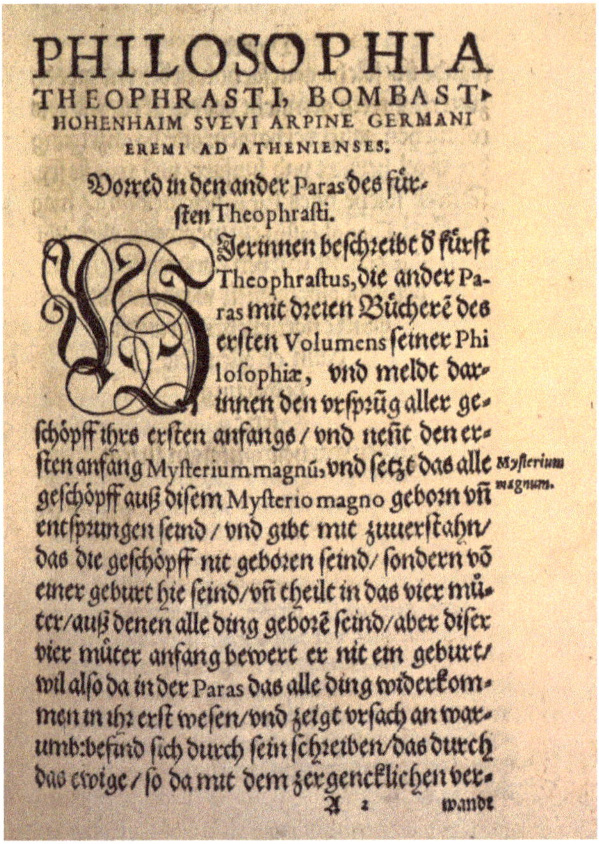
PHILOSOPHIA
THEOPHRASTI, BOMBAST
HOHENHAIM SVEVI ARPINE GERMANI
EREMI AD ATHENIENSES.

Vorred in den ander Paras des fürsten Theophrasti.

Herinnen beschreibt d fürst Theophrastus, die ander Paras mit dreien Bücherē des ersten Volumens seiner Philosophiæ, vnd meldt darinnen den vrsprūg aller geschöpff ihrs ersten anfangs / vnd nēnt den ersten anfang Mysterium magnū, vnd setzt das alle geschöpff auß disem Mysterio magno geborn vñ entsprungen seind / vnd gibt mit zuuerstahn / das die geschöpff nit geboren seind / sondern võ einer geburt hie seind / vñ theilt in das vier müter / auß denen alle ding geborē seind / aber diser vier müter anfang bewert er nit ein geburt / wil also da in der Paras das alle ding widerkommen in ihr erst wesen / vnd zeigt vrsach an warumb: befind sich durch sein schreiben / das durch das ewige / so da mit dem zergencklichen ver-

Mysterium magnum.

A 2 wandt

FIGURE 2.2 Pseudo-Paracelsus, *Philosophia ad Athenienses* (1564), first page. urn:nbn:de:bvb:12-bsb10164687-3 MUNICH, BAYERISCHE STAATSBIBLIOTHEK: 4 Med.g. 164#Beibd.1

reveal the true (i.e., his own) philosophy to those dedicated to standard, classical, scholastic philosophy (regardless of whether or not they were willing to listen to him).

The first page of the work contains several points worth discussing (fig. 2.2). Firstly, this book is almost the earliest edition of Paracelsus in which the name "Bombast" appears.[16] Paracelsus, to the best of our knowledge, adopted "Bombast" as a part of his full name only twice: in the *Intimatio*, his teaching program (now lost) written and printed in Basel in 1527, and in a letter to Bonifacius Amerbach from March 4, 1528 (not printed before 1889). But Konrad Gessner, who had read the *Intimatio*, used this name in the article dedicated to Paracelsus in his *Bibliotheca universalis* (1545).[17] In addition, Adam von Bodenstein, in his second edition of Paracelsus' *De vita longa* (1562),

16 As already noted by Eduard Schubert and Karl Sudhoff, *Paracelsus-Forschungen*, 2 (1889), 92.

17 Konrad Gessner, *Bibliotheca Universalis, sive Catalogus omnium scriptorum locupletissimus* (Zurich, 1545; reprint Osnabruck, 1966, with the *Appendix* of 1555), fol. 614v:

published a puzzling text addressed "to the reader" by a certain, still unidentified Valenti(n)us de Retiis, which contained a number of valuable pieces of information on Paracelsus, beginning with his name: "*Theophrastus Bombast de Hohenheim, ex nobili prosapia Suevigena, apud Eremitas natus.*"[18] On the first page of the *Philosophia ad Athenienses*, Paracelsus is also named *Suevi Arpine* [,] *Germani Eremi*, meaning: "Swabian [in German: *Schwabe*] from Arpinum, German from Einsiedeln." Einsiedeln is Paracelsus' birthplace; "Einsiedler" translates as "Hermit" in German – hence the word (*h*)*eremita* often added to the name of Paracelsus. He himself appended it to "Theophrastus Bombast ex Hohenheim" in the 1527 *Intimatio*.[19]

But why *Arpina*? This epithet recalls an ironic excerpt of a treatise on syphilis by Paracelsus (*Von Blatern / Lähme / Beulen* [...] *der Frantzosen*), where he records that his enemies claim he is not a Latin, but an Arpinate, apparently meaning that while not being able to write in Latin, he is as productive a writer as was Cicero the Arpinate.[20]

Given that this treatise on syphilis first appeared in the *Opus Chyrurgicum* collection, published by Adam von Bodenstein in Strasbourg in 1564 – almost simultaneously with the publication of *Philosophia ad Athenienses* in Köln – it is unclear whether the author (or editor, or publisher) of the *Philosophia* drew this reference to *Arpina* from the *Opus Chyrurgicum*, or from a manuscript of *Von Blatern / Lähme / Beulen* [...] *der Frantzosen* (although the only known manuscript of this work is the one used by Bodenstein, which was already lost, destroyed, or at least not available for Huser's edition in 1589–1591).[21]

"Theophrastus Bombast." Paracelsus, letter to Amerbach: Sudhoff, 1/6: 16 and 36 (with facsimile). On the *Intimatio*, see Huser, 7: fol. α7)iiij recto (Sudhoff, 1/4: 3); CP 3: 508–509.

18 On this text published by Huser himself (Huser, 6: 99), see CP 1: 585–599. Sudhoff believed it to have been written around the Basel period by an enthusiastic contemporary of Paracelsus in order to promote the latter's ideas, especially – but not exclusively – the *Archidoxis* (Sudhoff, 1/3: LII-LIII). However, should we believe it to have been written so early when it presents the works of Paracelsus as huge and covering such diverse fields as philosophy (230 books), medicine (46), alchemy (2), politics (12) and law (*de republica et justitia*), astronomy (*mathematica*: 7), nigromancy (66) and theology (*de summo bono in æternitate*: 1)?

19 See also the letters by Paracelsus to Christoph Clauser from 1527, and by Erasmus to Paracelsus from the same year (Huser, 3: 340 and Huser, 7: fol. α7)v verso; Sudhoff, 1/4: 71; Schubert and Sudhoff, *Paracelsus-Forschungen*, 2, 1889, 105).

20 Paracelsus, *Von Blatern / Lähme / Beulen* [...] *der Frantzosen*, book II, end of chap. XI: "Dem ist er [Theophrastus] kein *Graecus, sed tamen Germanus*. Einem andren / *non est Latinus, tamen Arpinas*" (in Paracelsus, *Chirurgische Bücher und Schrifften*, ed. Johann Huser (Strasbourg, 1605), 265; Sudhoff, 1/6: 349); see the explanations in Schubert and Sudhoff, *Paracelsus-Forschungen*, 2 (1889), 90–91.

21 Bodenstein's dedicatory epistle of the *Opus Chyrurgicum* is dated September 1, 1564. The *Philosophia ad Athenienses* appears in the bookseller Georg Willer's Autumn 1564

However, another unusual detail brings us back to the 1530s, i.e., the description of the three books of the *Philosophia ad Athenienses* as *Paras*: "Vorred in den andern *Paras* des Fürsten Theophrasti [...] Beschluß des andern *Paras* [...] Vorred in die dritt *Paras* [...]," etc. This is obviously derived from the extensive use of this prefix by Paracelsus – in his nickname and in the titles of several of his works: *Paragranum*, *Paramirum*, "Paramirische Schriften," etc. However, Paracelsus used the single word "Para" alone only once, in the title of another work on syphilis published in Nuremberg in 1530: *Von der Frantzösischen kranckheit Drey Bücher. Para.* In the later editions of this work (from 1552 onwards), the puzzling word "Para" was expanded as "Paracelsi" (indicating "Drey Bücher Paracelsi"), although it is far from certain that this was its actual meaning.[22]

Another possibility (although less plausible) is that the author of the *Philosophia ad Athenienses* drew the idea of using this prefix not from the 1530 book on syphilis, but from the aforementioned text "to the reader" by Valenti(n)us de Retiis, which mentioned a work by Paracelsus (unknown to us) under the title of *Parasarchum*. In any case, the use of this prefix in the *Philosophia ad Athenienses* is quite puzzling: the *Philosophia* is introduced in the preface to the first book as the "second *Paras* [*sic*]" of Paracelsus; or, more precisely, as "the second *Paras*, containing three books, of the first volume of the philosophy of Paracelsus," as if a first "*Paras*" should have preceded it.[23] So, how

catalogue of the Frankfurt Book Fair (fol. C recto, under the title *Teutsche Philosophey*); see Sudhoff, *Bibliographia*, 93 and 104.

22 Sudhoff himself, ibid., 10–11, believed "Para." to be an abbreviation for "Paracelsus." Charles Webster, *Paracelsus. Medicine, Magic and Mission at the End of Time* (New Haven, CT, 2008), 41, is of the same opinion. Yet, this is mere hypothesis, not supported by evidence. See the title-page of the work (fig. 2.3): why would Paracelsus have named himself twice in the title ("Durch den Hochgelerten Herrn Theophrastum von Hochenheim / beyder artzney Doctorem / von der frantzösischen kranckheit Drey Bücher. Para[celsi]")? Webster suggests that the name "Paracelsi" was "suggested for inclusion at the draft stage." The elaborate care that has gone into the design and production of this proportionate title-page, however, is suggestive less of such a careless mishap, but rather of a deliberate intention (the meaning of which escapes us). Moreover, the same presentation appears in another edition of the same work published in Köln in the same year (1530): here the word "Para." is not expanded, as it would be from 1552 onwards, but is faithfully reproduced; see Sudhoff, I/7: 10–18 (with facsimiles). For a new interpretation of the name "Paracelsus," see Martin Steinmann, "Paracelsus. Eine neue Deutung seines Namens," *Nova Acta Paracelsica*, N.F. 28 (2018), 217–221: according to Steinmann, the prefix "para" could refer to a medieval Latin etymology of *Paradisus* using the Latin verb *parare*, to give, to offer ("Paracelsus" thus meaning "offering that which is prominent" or "elevated"). How this interpretation may explain the word "Para" alone is yet to be determined.

23 See the first page of the text ("Vorrede"): "die ander *Paras* mit dreien Bücheren des ersten

Durch den Hochge
lerten Herrn Theophrastum
von Hohenheim/beyder artzney Doctorem/
von der Frantzösischen kranckheit
Drey Bücher. Para.
Das Erst von der impostur
der Artzney/ deren zwentzig sindt do durch die kran-
cken verderbt sind worden.
Das Ander vom corrigiren
der selbigen/ynn was weyß sie on verderbung
zu brauchen sindt.
Das Drit von den verderbtē
kranckheiten/ wie den selbigen widerumb zu helffen sey.
Auch wie andere new vner
hört kranckheiten/ aus jrriger vnnd falscher
Artzney entspringen.

FIGURE 2.3
Paracelsus, *Von der Frantzösischen kranckheit Drey Bücher. Para.* (Nuremberg: Friedrich Peypus, 1530), title-page (see note 22). urn:nbn:de:bvb:12-bsb10166394-3
MUNICH, BAYERISCHE STAATSBIBLIOTHEK: 4 Path. 258

might this first "*Paras*" have been constituted? In 1590 Huser published a list of several of Paracelsus' texts from an autograph sheet, entitled "*Volumen primum suae Philosophiae, de Divinis Operibus & factis, & de Secretis Naturae*" ("First volume of his philosophy concerning the divine works and achievements and the secrets of nature").[24] This "first volume of his philosophy" was made up of twenty-three tracts of magic and natural philosophy. It was first published

Volumens seiner *Philosophiae*," and the conclusion ("Beschluß") of the first book: "in diesem andern *Paras* des ersten *Volumen* der *generation rerum*."

24 Huser, 9: after the Table of contents. The name preceding the title is: "P. Theophrasti Bombast ab Hohenheim." Should the initial letter "P." stand for "P[hilippus]" (which seems most likely), then it would be a real problem: this first name appeared for the first time in the epitaph of Paracelsus (CP 3: 227) and was never used by Paracelsus himself (but was soon adopted by most publishers from the second half of the sixteenth century onwards, notably Huser on all title-pages of his edition). Its presence here is quite disturbing, since the sheet was in Paracelsus' handwriting (or was it a "D.," for "Domini," misread by Huser as "P."?). However, the title, per se, cannot be a forgery by Huser. We cannot know for certain which name exactly was written on this autograph sheet; whichever it was, it seems to have been edited by Huser under the same form as his usual title-pages.

in 1567 as Paracelsus' *Philosophia magna* by the Silesian Paracelsian Balthasar Flöter.[25] Paracelsus occasionally referred to it, notably in the treatise *Von den Kranckheiten die der Vernunfft berauben* (*De morbis amentium*).[26] Could it be that the *Philosophia ad Athenienses* was introduced as a sequel to this group (it being "the second *Paras* [...] of the first volume" of the philosophy of Paracelsus)? In that case, we must surmise that the author (or editor) of the *Philosophia ad Athenienses* knew this set of treatises directly from Paracelsus' manuscripts, which he seems to have accessed freely. Following this logic, the works of Paracelsus would have included, in the view of the author (or editor), a first *Paras* – namely "the first volume of his philosophy concerning the divine works and achievements and the secrets of nature" (i.e., the collection later entitled *Philosophia magna*) –, followed by "the second *Paras* in three books of the first volume of his philosophy" (the conclusion to book 1 offers an alternate title: "the second *Paras* of the first volume of the generation of things") – i.e., the *Philosophia ad Athenienses* itself. This is again a mere hypothesis, but it would provide a plausible explanation for this reference to a "second" (rather than a first) "*Paras*."

However – and regardless of this hypothesis – the words "ander Paras" are not used consistently throughout the *Philosophia ad Athenienses*: apart from the excerpts quoted above (n. 23) from the preface and conclusion to the first book, the word "*Paras*" is used in all other conclusions and prefaces of the work as if the "ander *Paras*" referred only to the first *book* of the *Philosophia ad Athenienses*, instead of to the whole work itself. Thus, seeing the second book introduced as "dritt *Paras*," and the third book as "vierdten *Paras*," the reader might conclude that the very first book of the *Philosophia* is missing, and that the available work opens only with the second book of the sequence (the "ander *Paras*").[27] But the publisher had understood it differently, for he chose "Das erst Buch," "Das ander Buch," and "Das dritt Buch" as running titles, which is consistent with the preface to the first book. Furthermore, the beginning of the *Philosophia ad Athenienses* does not read at all as if a real beginning was

25 On this *Philosophia magna*, see Paracelsus, *Theologische Werke*, ed. Urs Leo Gantenbein, vol. 1 (Berlin-New York, 2008), 70–72.

26 This is the "seventh book in the medicine" (Huser, 4: 39–92, here 65): "a[l]s [wir] *de Generatione humana* setzen / *in primo Volumine* unserer Philosophey." This should be a reference to *De generatione stultorum*, number 4 in the autograph list of the *Volumen primum suae Philosophiae, de Divinis Operibus* [...] (Huser, 9: 27–44). The treatise *Von den Kranckheiten die der Vernunfft berauben* was first published in 1567 (Sudhoff, *Bibliographia*, 144–145). Huser published it in 1590 from the autograph manuscript.

27 Thus, in 1583, Gerhard Dorn believed the missing first book to have been suppressed by the enemies of Paracelsus (CP 2: 899).

lacking. It seems, thus, that the publisher did some editorial work, but it is extremely difficult to delineate this work more precisely.[28] Further investigation is problematic because we do not understand who exactly did what in the whole process of editing the work.

The inconsistencies do not end here, however. In the preface to book II, we are told that this book contains statements about the birth of Christ and the Virgin Mary.[29] Unfortunately, there are no such statements in book II, nor in the six surviving chapters of book III. The conclusion of book II does not mention these topics either. Might this suggest that the author of the prefaces and conclusions and the author of the work itself were one and the same person? Might he have intended to discuss the birth of Christ and the Virgin Mary, but then changed his mind and omitted to correct the preface? This kind of inconsistency appears in numerous works of Paracelsus and is, more generally, typical of unfinished works.[30] Furthermore, the preface to book III describes the whole content of the book, although only its first six chapters are extant. After these six chapters we find merely a closing sentence: "The remainder, if there were anything lacking, was missing."[31] In 1590, Johann Huser, looking to the preface of book III, corrected this sentence thus: "The remainder (for the author has undoubtedly gone further) is unavailable."[32] The closing sentence of the 1564 edition was probably written by the publisher, which excludes, by the way, the possibility that he was acquainted with the author of the work (he was apparently perfectly convinced of the authorship of Paracelsus). Or should we believe that the author of the prefaces and conclusions summarized books II and III from another, now lost, version of the work, of which we have not the slightest remnant – except this tiny hint? All of this is certainly conceivable. But the

28 The title of book I is indeed *Textus Philosophiae Paracel. Th. Bombart. Hohaim.* The mistakes on these last two names of Paracelsus, seemingly due to misreadings, are not found on the title-page of the whole work, which leads one to question whether the faulty book I title was composed by a person different from the one who composed the title-page of the whole work. As for books II and III, they are merely entitled *Das ander Buch* and *Das dritt Buch*.

29 *Philosophia ad Athenienses*, fol. E2r: "Der gleichen meldt er [i.e. der Fürst Theophrastus] die geburt Christi / und der Jungfrawen Maria."

30 The most famous cases among the works of Paracelsus include the *Archidoxis* and the *Opus Paramirum*, but we might mention the earlier *Volumen Paramirum* as well: see the entries to these three texts in Udo Benzenhöfer, *Studien zum Frühwerk des Paracelsus im Bereich Medizin und Naturkunde* (Münster, 2005). Other examples include the *Grosse Wundartzney* (see Schubert and Sudhoff, *Paracelsus-Forschungen*, 1, 1887, 44–45), the *Astronomia Magna*, or the fragmentary *Bücher Meteororum* published by Huser after *De Meteoris* (Huser, 8: 277–333).

31 "*Reliqua, si quae deerant, desiderabantur.*"

32 "*Reliqua (sine dubio enim ulterius progressus est Author) non habentur*" (Huser, 8: 47).

least complicated explanation, that the author of the prefaces and conclusions was indeed the author of the work (it being left unfinished for some unknown reason), is perhaps the likeliest. This hypothesis, however, does not explain away the contradictory two different meanings of "ander Paras" throughout the work, and is still not entirely satisfying. The mystery around the different steps of the production of this work seems destined to remain unresolved.

2 Summary of the Work

Now we shall attempt to explain the content of the work (referring to the "Texts," i.e., the chapters of each book, as §1, §2, etc.) in order to discuss its most obvious sources. The first book deals with the beginning of all created things. This beginning was an undifferentiated prime matter called the *Mysterium magnum,* the "mother" of the elements, "grandmother" of the stars, plants, animals, and of everything perishable.[33] The *Mysterium magnum* is uncreated (a point that the author repeatedly mentions) and has been "prepared" by "the supreme artist" (or "supreme craftsman").[34] The created things will not return to the *Mysterium magnum*, but to that which existed before it. What exactly is it that existed before? The author says only that it was something "eternal."[35]

33 Book I, §1, fol. A3r: "Dise *materia* aller ding / ist *Mysterium magnum* […]." English translation: Oswald Croll and Paracelsus, *Philosophy Reformed and Improved in Four Profound Tractates. The I. Discovering the Great and Deep Mysteries of Nature: by* […] *Osw. Crollius. The Other III. Discovering the Wonderfull Mysteries of the Creation, by Paracelsus: Being his Philosophy to the Athenians*, trans. Henry Pinnel (London, 1657), Part 2: 1–62. This translation was not made from the German, but from the Latin translation: Paracelsus, *Operum Medico-Chimicorum sive Paradoxorum, Tomus Genuinus Octavus*, trans. Zacharias Palthen (Frankfurt, 1605), 1–32, which omitted the prefaces and conclusions. A second English translation, far less reliable, appeared in 1894: Paracelsus, *The Hermetic and Alchemical Writings*, trans. Arthur Edward Waite, 2 vols. (London, 1894), 2: 249–281. The translation in modern German by Aschner is not very helpful in the most difficult passages: *Paracelsus. Sämtliche Werke, nach der 10-bändigen Huserschen Gesamtausgabe*, trans. Bernhard Aschner (Jena, 1926–1932, 4 vols.), here 3 (1930): 762–799.

34 Book I, §1, fol. A3v: "Also ist *Mysterium magnum* ungeschaffen von dem höchsten Kunstler zu bereitet" (see also book III, §5: "den obristen Werckmeister"; "den der aller ding ein mächer ist"). For other mentions of the "uncreated," see, e.g., book I, §3, fol. [A4]r: "Also seind zweierley *Mysteria* / das groß als ein *Mysterium increatum* / die andern als kinds kinder / *Mysteria specialia* geheischen"; Book I, §4, fol. [A4]r: "Dieweil und nu auß dem *Mysterio increato* alle andere tödliche seind gewachsen […]." Other occurrences are found in the same §4 and elsewhere, e.g., in book I, §21 (fol. D2r).

35 Book I, end of §2, fol. A3v: "Wiewol das ist / das alle ding widerumb in sein erst wesen kommen / aber nit in das *Mysterium* […] aber in das wol / das vor dem *Mysterium* gewesen ist"; and book I, §21, fol. Dv: "und bleibt allein das / das vor dem *Mysterio magno* gewesen ist / und ewig ist".

In 1571, in a witty dialogue by Thomas Erastus in which two opposing characters, Furnius the alchemist and Erastus himself, discuss the *Philosophia ad Athenienses*, Furnius interprets this sentence as if that which existed before the *Mysterium magnum* were "nothing," thus establishing that the text complied with the Biblical scheme of *creatio ex nihilo*.[36] However, Furnius is playing devil's advocate: he is refuted by Erastus at length and his opinion does not prevail. But if we jump to the end of book 1, we read that although the *Mysterium magnum* is properly "nothing" ("wievol es nichts ist"), we must understand that it is just like a house built out of a word, being the work of the Most High, which tends both to confirm the idea of a creation *ex nihilo* and identify the *Mysterium magnum* with the creative Word, the *Fiat* of Genesis.[37] This is indeed an odd approximation, though less unsettling than the abrupt beginning of the treatise.[38]

Every created thing was already created *within* the uncreated *Mysterium magnum* by the Creator's highest secret and goodness – not in their form, essence and quality, but in the same manner as in a picture which already exists in a piece of wood before it is carved.[39] However, the *Mysterium magnum* was not elemental, even if it contained the four elements (§7). This, of course, is a reworking of the standard idea of the prime matter containing all forms *in potentia*, except that every notion of *potentia* and *actus* has entirely disappeared from the stage.

The Creation occurred through simultaneous separations (*separationes*) and subsequent offspring (*procreationes*) (see Table 1, below). The *Mysterium magnum* was separated into a number of *mysteria specialia*, one for each kind of thing. Thus, there were two kinds of *mysteria*: the uncreated *Mysterium magnum*, and the *mysteria specialia*.[40] Separation (meaning a process similar to alchemical separation) is praised by the author as the most admirable wonder of philosophy (and by 'philosophy,' it is perhaps 'alchemy' that is the intended meaning here, just as it is in alchemical texts), be it the result of a

---

36 Thomas Erastus, *Disputationum de Medicina nova Philippi Paracelsi Pars prima* [...] *Pars quarta et ultima* (Basel, 1571–1573, 4 vols.), 1: 5: "Cum in id corrumpi omnia ait, quod ante mysterium erat, ipsum nihilum intelligit."

37 Book 1, §24: "[...] und nachfolgendts ist zu mercken / das es ein *Mysterium magnum* ist / wiewol es nichts ist / sondern gleich als wenn auß ein wort ein hauß würde / so ist das also zuverstehen solchs dem höchsten allein /".

38 On orthodoxy issues, however, see below, shortly before our conclusion.

39 Book 1, §4: "dann das höchst *Arcanum* / und groß gut des Creators / hat alle ding in das *increatum* geschaffen / nit formlich / nit wesenlich / nit qualitetisch / sondern es ist in dem *increato* gewesen / wie ein Bildt in eim holtz ist [...]"; see also §5.

40 See above, n. 34, quotation from book 1, §3; see also §7 and book 2, §14–15.

TABLE 1 The process of Creation according to the *Philosophia ad Athenienses*

UNCREATED *MYSTERIUM MAGNUM*
↓
[simultaneous separations]
↓
MYSTERIA SPECIALIA

Phase 1 FIRST SEPARATION	Phase 2 SECOND SEPARATION
↙ *VEGETABILISCHEN* ↓ THE FOUR ELEMENTS ↘ EVERYTHING INVISIBLE	
↓ a) FIRE → The heavens *A container for the firmament*	From THE HEAVENS (= FIRE) → The stars, planets, and the whole firmament
b) AIR → A void *A container for the invisible fatalischen* (= elementary beings)	From AIR → *Fata, impressiones, incantationes, superstitiones, maleficia, somnia, divinationes, sortilegia, visiones, apparitiones*, etc. → Elementary spirits: *Melosinie* (human blood), *Dieemae* (hard stones), *Durdales* (trees), *Neufareni* (pores of the earth), *spiritus* (air)
c) WATER → A liquor → near the centre, in its holes, between the other elements and the ethers *A container for the nymphs and wonders of the sea*	From WATER → Fishes and other animals → Salt and sea plants: *Corallen, Trina, Citrones* → Wonders of the sea: nymphs, mermaids, *Drames, Lorindt, Neßder*
d) EARTH → Coagulated into the earthy globe *A container for the things which grow and are fed by the earth*	From EARTH → Wood, metals, minerals, fruits and plants → Animals and humans → Gnomes, *wild leuth* (i.e., *Sylvestres*), *nacht frawen* (i.e., *Lemures*), and giants

Phase 3
(Second separation?)
↓
OFFSPRING (*PROCREATIONES*)
= NEW *MYSTERIA*
↓
OTHER OFFSPRING
("GEBERUNG")
(not exactly produced by separation, but as a result of mistakes, miscarriages, superabundance, or harvests)

Examples:

THE HEAVENS (= FIRE)	AIR	WATER	EARTH
↓	↓	↓	↓
Firmament	Air's spirit (*die Geist des Luffts*)	The *Fatalischen* of water	Men, beasts
↓	↓	↓	↓
Thunder (a harvest of the stars)	Magical storms	*Lorindt*	Feces
	↓	↓	↓
	Poisoning, mortality, poor harvests	Prophecies	Beetles & other creatures

TABLE 1 The process of Creation according to the *Philosophia ad Athenienses* (*cont.*)

Phase 4
(THE END OF TIMES)
FOURTH [sic] SEPARATION

(All things return to what they were in the beginning.
That which has been separated shall not be reunited.)

PERISHABLE THINGS	→	Nothing
ETERNAL THINGS (*Arcana*, human souls, *Evestrum* of the firmament in every element)	→	The eternal
MYSTERIUM MAGNUM	→	∅
→ New *Mysterium magnum*:		*MYSTERIUM MAGNUM ULTIMUM*

How is the world organized with regard to the four elements?

THE FOUR ELEMENTS
↙ ↘
VISIBLE INVISIBLE

(One quality per element.
One element per each single
created thing.)

Four elements mean four different worlds:

THE ETERNAL *ARCANA*:			
ARCANUM	*ARCANUM*	*ARCANUM*	*ARCANUM*
↓	↓	↓	↓
FIRE (Heavens)	AIR	EARTH	WATER
↓	↓	↓	↓
Light, brightness, influence, growth	Nourishment	Fixity, steadiness	Abode
↘	↓	↓	↙
THE FOUR ELEMENTS HOLD TOGETHER AND COHERE			

divine or natural magic – but the author chooses not to draw any conclusions regarding the type of magic.[41]

One part of the *Mysterium magnum* was separated to become the four elements, another part to become everything invisible,[42] and a last part to become everything that grows (§9: "der dritt zu den vegetabilischen").[43] All created things – thus separated in their full majesty, power, and free will – will remain there until the end of the world. At this time there will be a final harvest which is no less admirable than this first separation, and all things will be harvested and put in the attic (§10).

The very first separation – at the beginning of the separation process of the *Mysterium magnum* – was that of the four elements. This separation did not imply any mixing of the elements: fire became the heavens and was a container ("ein kasten") for the firmament; air became a void ("ein lär") where neither substance nor corporeal matter was placed, and was a container for the "invisible *fatalischen*" (which are actually elemental beings, as discussed below); water became a liquor and took its place near the center, in its holes, between the other elements and the ethers – it was a container for the nymphs and wonders of the sea; the element of earth was coagulated into the earthy globe, which is not supported by the other elements, but by the columns of "Archaltis," which are wonders of God;[44] earth was a container for the things that grow and are fed by the earth (§11). This emphasis on the Paracelsian elemental beings at the very beginning of the process of creation is perhaps reminiscent of Plato's

41 Book I, §9: "Dann *separatio* der Philosophien das groste wunder ist [...] Ist es Götlich geschehen durch sein Gotheit / wöllen wir nit Philosophieren / dann die Gottheit uns solches nit zu erkennen gibt: Ists aber natürlich gewesen das *Magi* / so ist es ein wunderbarlich *Magi* gewesen / geschwinder durchgang / behender scheidung / und grosser tugend / deren gleich unmüglich ist der natur mehr solchs zugeben oder machen [...]".

42 Meaning mainly the elemental beings, described as invisible, e.g., in §11 and 13.

43 It seems more fitting not to restrict the meaning of *vegetabilischen* to the vegetable realm, given the words at the end of §11: there earth is said to have been made "a container for the things that grow and are fed by the earth" (*ein kasten der gewaschsen* [*sic*] *so sich vom erdtreich nhären*). However, the treatise is not very consistent on this point: at the end of book I, it states that the *Mysterium magnum* contained all *Mysteria prima*, and in three distinct ways: with respect to the plants ("auff die Vegetabilischen"), the elements, and the sensible beings ("auff die *sensibilia*"). Thus, here the *Mysterium magnum* is said to contain *vegetabilia, elementa, sensibilia*, whereas in §9 it was divided into *elementa, invisibilia* ("unsichtbarn"), and *vegetabilia*. Or is *sensibilia* a misreading for *invisibilia*?

44 "[...] von den seulen Archaltis / und seind *magnalia Dei*." Might this be a misreading for the columns of Hercules (*Herculis*) – or Atlas (*Atlantis*)? References to the "columns of Atlas" (*Atlantis columnae*) are widespread in works from the early sixteenth century, such as the *Lectiones antiquae* by Cœlius Rhodiginus (Venice, 1516), book XVI, chap. 16, or the *Institutiones astronomicae* by Joachim Sterck van Ringelbergh (Köln, 1528), book I, chap. 12.

Timaeus (39e-40a), where the demiurge put as many animals in the world as there are parts of the world, which are four, i.e., the four elements. Thus, the demiurge created four species of animals, one per element.

With this first separation, that of the elements, a second occurred within each element: from fire, which had been changed ("verwandelt") into the heavens, one part became a container and the other part grew ("gewachsen") into stars, planets, and the whole firmament; not as plants grow from the earth, however: the stars were born from the heavens, but via separation (§12: "allein durch scheiden"). From air were separated the *Fata, impressiones, incantationes, superstitiones, maleficia, somnia, divinationes, sortilegia, visiones, apparitiones*, and other beings which happen to be adaptations of Paracelsus' elemental beings, except that those belong to only one element, i.e., air (for air, as explained in §19, is invisible and does not yield anything visible and tangible): they are *Melosinie*, dwelling in human blood (as odd as this may sound); *Dieemae*, living in hard stones; *Durdales*, living in trees; *Neufareni*, living in the air of the earth, i.e., in the pores of earth ("in den *poris terrae*"); and *spiritus*, living in the air which lies in the "chaos" (meaning the atmosphere as we know it).[45] All were put in their proper place and predestined ("gepredestiniert") such that they were invisible, and yet perceptible by us – for the Creator, via supreme skill ("vom höchsten *Arcanem* [*sic*]"), has made no element more subtle than air.[46] From water, fish and other animals were separated, as well as salt, such sea plants as "Corallen, *Trina*, and *Citrones*" (yet two more puzzling words), and wonders of the sea alienated from the ordinary course of nature, such as nymphs, mermaids, and other beings with odd names: *Drames, Lorindt*

45 Book I, §13: "*Neufareni* in dem lufft der erden wonen / als in den *poris terrae*." On *Melosinie*, Goldammer suggested the influence of the fifteenth-century *Volksbuch* by Thüring von Ringoltingen: *Die schöne Melusine*: Kurt Goldammer, *Paracelsus in der deutschen Romantik* (Vienna, 1980), 93. The word *Durdales* is an obvious misreading of *Dryades*. Could *Dieemae* be a misreading of *Gnomae*, *Neufareni* a misreading of *Nymphae* (or rather *Nympharum*), and *spiritus* a misreading of *sylphes*? On the meanings of "chaos," see Didier Kahn, "Paracelsus' Ideas on the Heavens, Stars and Comets," in Miguel Ángel Granada, Patrick J. Boner and Dario Tessicini, eds., *Unifying Heaven and Earth: Essays in the History of Early Modern Cosmology* (Barcelona, 2016), 59–116, 81–84, 86–87, 92, 110. On the elemental beings, see Didier Kahn, "La question des êtres élémentaires chez Paracelse," in Roberto Poma, Maria Sorokina, Nicolas Weill-Parot, eds., *Les Confins incertains de la nature (XIIe–XVIe siècle)* (Paris, 2021), 213–237; Dane T. Daniel, "Invisible Beings in the Natural World: Paracelsus on Ghosts, Angels, and Elemental Creatures," *Nova Acta Paracelsica*, N.F. 29 (2021), 89–129.

46 Book I, §13: "dann kein subtiler Element ist vom höchsten *Arcanem* [*sic*] beschaffen dan denn lufft." Here the phrase "höchstes *Arcanum*" seems to refer to a major quality or skill of the Creator, as in §4: "Dann das höchst *Arcanum* / und groß Gut des Creators / hat alle ding in das *increatum* geschaffen [...]." The word *Arcanum* without the adjective "höchst" is used with an entirely different meaning in book II: see below, n. 57.

(a commotion of water similar to an earthquake), and *Neßder*, along with other rational (yet unnamed) creatures, whether endowed with an immortal soul or not (§14). From earth, wood was separated along with metals, minerals, fruits and plants, animals and humans, as well as gnomes living in the mountains, and "wild leuth" ("wild people") living in the forests – which Paracelsus named *Sylvestres*, but also "Waldtleuten" in his *Liber de nymphis* (and no doubt there was a confusion of "wild" and "waldt" here). Also separated from earth were banshee-like creatures, living in the night ("nacht frawen"), and giants (§15). We can see here a profusion of elemental beings not quite consistent with descriptions given in any of the authentic treatises of Paracelsus.[47] Here the author mixed the elemental beings from the authentic works together with other beings. The same can be said about §24, where the author states that many hundreds of men were created first, including Cyclops, giants, and elemental beings.[48] Here, as is often the case throughout the treatise, the author appears to be especially fond of these elemental beings. It would be foolish, he argues (book II, §11), to limit our philosophy to the element of earth and its noblest creature (i.e., man, obviously): there are more worlds than our own, and we are not even the only people from our element. There are creatures of the night ("die nocturnalischen")[49] and gnomes as well, very different from us, but all were nevertheless created in the *Mysterium magnum*, including other kinds of people of which we are unaware.

As each element was being separated, so too were separated the complexion and nature of each element: warmth and dryness in the heavens (i.e., fire), warmth and moisture in the air, coldness and moisture in water, coldness and dryness in the earth. There were, however, conflicting qualities in every element as well: e.g., coldness and moisture in fire, along with a variety of colors (I, §16). Besides, the fire that burns things to ashes is not the element of fire, but the fire of the earth (*ignis terrae*), which is no element at all, but rather the matter of earth burning – for water can burn as well as anything else, and it can thus be called an aqueous fire (*ignis aqueus*: I, §18). Therefore, each element is more complicated than it may seem from its basic properties: water can have precious stones grow in the earth from its own seeds planted there;

47 See Kahn, "La question des êtres élémentaires."

48 Book I, §24: "[…] dann jeglichs geschlecht hat im *Mysterio magno* sein sonderlichs gehabt / also allein vier auff die Elementen / dann dieselbigen handt nur vier anfang / aber vil hundert auff die menschen: dann *Loripedes* ein besonders / *Ciclopes* ein sonders / *Gigantes* ein sonders / *Mechili* ein sonders / deßgleichen die do in der Erden wonen / die in Lufften wonen / die in dem wasser wonen / und die im feür / und sonderlich alle gewächs / ein jeglich geschlecht sein besonder *Mysterium* im *Mysterio magno* gehabt / und allerley geschöpff do gewesen ist […]."

49 These are most probably the *nacht frawen* already mentioned.

water can also produce flesh, as in the case of nymphs; and earth can be turned into water while still being the element of earth (I, §17–19; II, §5–6).

It was still through the second separation (as far as we can guess) that "offspring" (*procreationes*) appeared, spawning from the diverse things already separated; these offspring became the *mysteria* of other offspring ("andere geberung") yet to appear, not exactly by separation, however, but as mistakes or miscarriages, or as superabundance, or a harvest (see Table 1).[50] Thus, thunder comes from the offspring of the firmament ("auß den *procreationibus firmamenti*"); more precisely, it is "a harvest of the stars," made at the time scheduled by its own nature for it to produce its effect.[51] Other examples of such offspring include magical storms [not coming from air, but from air's *spiritus* (see below, III, §2)], feces (not coming from the earth, but from men or beasts), and "Lorindt" (not coming from water, but from the *Fatalischen* of this element).[52] These offspring generate other things in turn: from the magical storms ("*Fatalischen* gewittern") come the poisoning of a country, huge mortality, and poor harvests; from the feces come beetles and other creatures; from "Lorindt" come prophecies.

Finally, the fourth separation is evoked (although no third separation has been mentioned): in the end, all things will return to that which they were in the beginning. Perishable things will return to nothing, and eternal things to the eternal, such as *arcana* (still a divine quality or divine driving force?) or the human soul. However, that which has been separated cannot be reconstituted: milk, once coagulated, cannot become milk again. Hence the *Mysterium magnum* will not return to that which it came from (I, §21–22). But there will be a new *Mysterium magnum* in the end, after the destruction of this world: the *Mysterium magnum ultimum* (II, §12, 14 and 16).

In the first text of the second book, the author returns to the strong distinction (already made in book I, §23–24) between that which is mortal and that which is immortal in the Creation. Considering this difference, the author states that not everything in Creation has been created by the same god: there

50 Book I, §20: "Also seind die *procreationes* erstanden eins von dem andern durch die grosse separation / so folgen auß den *procreationibus* andere geberung / die dann ir *Mysterium* haben in diesen *procreationibus* / nit uber [Huser corrected this as *nicht aber*] als ein scheidung in gleicher forme der obgemelten / sondern als ein irrung oder mißgeradung / oder als ein uberfluß / oder als ein Ern."

51 Book I, §20: "ein Ern des *Sidus*" (*Sidus* is given in the margin in 1564 as an alternative reading for the obviously erroneous word *Stylus*).

52 The text has *Natalischen*; however, *Lorindt* has been mentioned among elemental beings of water. Since the *Fata* are elemental spirits in the *Philosophia de generationibus et fructibus quatuor elementorum*, and "*Fatalischen* gewittern" are mentioned here five lines below in the same §20, it seems highly plausible that *Natalischen* is but one more misreading.

is a highest and supreme Creator, who is to be the judge of all things and beings in the end. All the *mysteria*, however, have been created by perishable gods working within the *Mysterium magnum*,[53] who had the power to create such imperfect beings as us, human beings. They will be submitted to the final judgment just like us.[54] This is obviously an echo of Plato's *Timaeus* (41–42), where the Creator of the world explains to the inferior gods – the planets and gods of Greek mythology – that he cannot make other creatures himself, lest they be like gods: these creatures have to be mortal, thus they must be made by the inferior gods.

The next chapters are devoted to the origins and nature of the four elements. The elements are the four mothers of all creatures. Every created thing is made up of only one element, not four. That which is visible in a creature is not the element, but its body. The real element is a spirit ("geist"), something like a soul in the body ("wie ein seel im leib"), invisible and intangible ("ungreifflich"): it is the prime matter of the element, residing, however, in all things, for the prime matter of the elements is nothing else than the life that is in every created thing (II, §2 and 6). The element of fire has separated the firmament from itself, not by corporeal means, but by its very essence, and the sky is nothing more than a chaos,[55] which is the vapor ("dunst") of the firmament. The sun has its body from the element of fire, but its warmth and brightness come from the invisible essence of the element. As for the moon and the other stars, they also originate from this element, but only from its ruby color, in which there is no warmth nor any burning power, but a mere dead brightness (II, §7: "ein tödter schein").

Just as every created thing is made up of only one element, every element has only one quality: fire is only hot, not dry or moist, and air is only dry, not hot or cold, etc. As it is the nature of spirits to be onefold, so too are the elements (II, §3). If any disease is found to be both hot and dry, then it is affected by two elements, with each of the two elements in one of the body parts, e.g., one element in the liver and another in the spleen – but there are never two elements in only one body part: such is the peculiar conception of *complexio* in the *Philosophia ad Athenienses* (II, §3 and 4).

53 Book II, §1: "von zergencklichen Göttern / die in *Mysterio magno* haben gewalt gehabt."

54 The author seems, however, to bring some correction to this statement when he later writes (book II, end of §16): "dann da ist nie kein Gott gewesen mehr denn der / der allezeit ein Richter ist gewesen. Und ist ein falsche torheit zu erkennen den tödlichen / den zergencklichen / den faulenden / für ein gewalt haber der dingen der geschöpffen / und deßgleichen zu regieren das ewige: dann was da stirbt hat kein gewalt do zu regieren / darumb allein ein einiger weg ist und ein glaub / und nit mehr zu bedencken."

55 *Chayr* in 1564, loosely corrected as *Chay* by Huser: most probably a misreading for *Chaos*.

There might have been more than four elements, but the world was created as it is. After their destruction it is perhaps the case that other elements will appear, entirely different from them, and a new *Mysterium magnum* will be made (II, §12). The four elements are four worlds, each having their own being and their own diverse species, for our world was not created from the four elements, but rather every world was created from one of the elements and all four worlds hold together as one.[56] This independency is stressed in many ways: every element has its own medicine, not suitable to the beings of the other elements (II, §14). Although the firmament nourishes the world through its elemental fiery force ("krafft") reaching the earth, this nourishment is not essential, for our world (i.e., the earth) is able to sustain itself on its own, just as another world (i.e., another element) is perfectly able to sustain itself without the earth (II, §13). However, the light of the heavens is an extract of the four elements – it is their most noble part in its full and perfect virtue, for it does not stem from the element of fire, but from the *Arcanum*.[57] It is the *Arcana* that make the four worlds hold together and cohere ("stimmen zusamen") in this form, not the nature of the elements, since the four worlds are themselves the elements (II, §13). Our life, our ability to see, hear, etc., does not come from the elements, but from the *Arcana*. Everything eternal comes from, and is identical to, the *Arcanum*: a dog dies, but its *Arcanum* stays; a man dies, but his *Arcanum* stays, as does his soul, which makes him so much nobler than a dog. The *Arcanum*, however, is not an immortal essence, but rather a perfect one.[58] The *Arcana* could very well stay without any element, and the elements without any *Arcanum*. But, as it happens, the element of fire has an *Arcanum* in itself that brings the light, brightness, influence, and growth to the other three; the element of air has an *Arcanum* in itself, from which nourishment comes to the other three; the element of earth has the *Arcanum* of fixity and steadiness in itself, which gives the others the strength to stay fixed and permanent, and water gives all elements the *Arcanum* of abiding ("der auffenthalt"), which preserves them from destruction. Thus, there is a marked difference between the elements and the *Arcana*: while the former are perishable, the latter will remain even in the *Mysterium magnum ultimum*, where everything will be renewed (II, §14).

56 Book II, §13: "nit zuverstehen / das die Welt von vier Elementen müssen enthalten werden / sondern ein jeglichs von eim Element".

57 The phrase *höchsten Arcanum*, or just *Arcanum*, already occurred in book I, §13, 14 and 22 (see above, n. 46), apparently meaning a major quality or skill of the Creator. The meaning seems entirely different here.

58 Book II, §14: "Wir sprechen aber nit das *Arcanum* sey ein wesen wie das ungestorbene / sonder dasselbige volkommen."

There is nothing perishable in the whole world that does not leave something eternal in its wake.[59] Thus the eternal is not the beginning of a created thing: rather, it is indicative of the destruction of its nature.[60] The author notes that although the *Fatales*, Melosines, and Nymphs also leave behind something eternal, he will not go into any details regarding their kinds of corruption, which differ in each of the four worlds (i.e., elements) (II, §17).

The distinction between mortality and immortality is further exemplified by two new notions specific to the *Philosophia ad Athenienses*: the *Evestrum* (most likely inspired by the myth of Plato's cave) and *Trarames* (II, §18).[61] Discussion of these requires the highest wisdom ("der höchsten *Sophiae* zu stehet"). The *Evestrum* is a prophetic shadow; like the shadow of a created thing, it comes into being with it and disappears with it. The *Trarames* is also a shadow, but its essence is invisible and it endows a sharpness of mind ("*Trarames* gibt die scherpffung"); it comes into being with the reason and judgment ("sinnligkeit") of the animals, whether they are endowed with reason or not.[62] Thus, prophecies about what will become of men, animals, wood, etc., come from the light of the *Evestrum* ("auß dem schein *Evestro*"); what concerns reason comes from the *Trarames*.

When all things are destroyed, so too will the *Evestrum* and *Trarames*, but not without leaving something eternal behind ("doch nit ohn ein letz des ewigen," II, §19). The *Evestrum* is the eternal part of the firmament in each of the four worlds, for every element has its own firmament. The firmament that is in the *Evestrum*, however, is separated (*dispersum*): it is not the stars as we see them, but the firmament of the Nymphs, which are not stars, for the Nymphs do not make use of the stars: they have a specific firmament, just like the *Fata*, each of which has its own heaven and earth, its own being and home, its own firmament, stars, and planets – and these do not look anything like the others and are as different from one another as water is from fire. There is a specific *Evestrum* for the *Fatalisch*, another one for the fiery people, and another for the watery, and another for the earthy ones – and their shadow remains after their

59 Book II, §17, fittingly summarized in the margin: "*Omnia creata* verlassen ein ewigs."

60 Ibid.: "und ist also das ewig ein anzeigen der zerbrechung der natur / und nicht ein anfang der geschöpff [...]," which was translated into Latin as follows: "Et sic aeternum est indicium dissolutionis naturae et non principium creatorum [...]."

61 The alternate spelling *Evester* is used in §21 and 23 as well. On *Evestrum* and the myth of Plato's cave, see Ute Frietsch, "*Solve Coagulata!* Verflüssige die Schöpfung! Die pseudoparacelsische Schrift *Philosophia ad Athenienses* als Rätsel und Offenbarung," in Jutta Eming, Volkhard Wels, eds., *Darstellung und Geheimnis in Mittelalter und Früher Neuzeit* (Wiesbaden, 2021), 279–294, on 289.

62 Book II, §18: "*Trarames* aber als ein schatten verstanden wirt unsichtbares wesen / dann auß der ursach das er geboren wirt mit der vernunft / und mit der sinnligkeit der thieren / vernünftigs und unvernünfftigs."

destruction. This *Evestrum* deludes the world, and takes on a cloth of delusion from one world to the other ("schattet sich mit betrügnuß auß einer Welt in die ander"), producing visions, appearances, signs, and shapes; from it come the *Evestrum* of the comets, the *Evestrum* of impressions, and the *Evestrum* of miracles. These three kinds of *Evestrum* are prophetic and "shadowed" ("umbraten") *Evestren*.[63]

It is of utmost importance that we know the *Evestrum prophetum* (II, §20), for the *Turban magna* [*sic*] is an essence which foretells everything that happens in the four worlds. Every event against nature, or out of the ordinary course of nature, is known through the *Evestrum prophetum*, which in turn takes its knowledge from the *Turban magna*. This is the first time this *Turban magna* is mentioned in the work, a puzzling phrase widely accepted among later Paracelsians as meaning "the multitude of all the stars."[64] However, the phrase was almost certainly borrowed by the author of the *Philosophia ad Athenienses* from Revelation 7:9, referring to the Great Multitude in white robes, i.e., God's elect "who came out of the great tribulation" and who "are before the throne of God and serve him day and night in his temple".[65] At the very least this is probably the origin of the phrase; but in the *Philosophia ad Athenienses* it seems rather to refer to some higher essence from which the *Evestrum* draws its prophetic knowledge, as we shall see below. Since the Book of Revelation itself can be regarded as the very essence of prophecy, we might consider that such a borrowing makes some sense. In any event, the *Turban magna* only appears in book II, §20. There we learn that a prophet definitely needs to know the *Turban magna*, for it is made up from sharp reason ("sie ist auß grosser vernunfft") and unites with man's reason, thus enabling a mortal to know the *Turban magna* until his own destruction. All the prophets have always spoken from it, for it contains all the signs of the world and produces all the *Evestren*. From it every comet is sketched ("auß geschettiget"); every

63 *Evestren* is the spelling used by the author. It should be *Evestra* (as in II, §21: see below), but this is not the only occurrence of an inconsistent use of Latin forms in this treatise.

64 See Michael Toxites and Johann Fischart, *Onomastica II* (Strasbourg, 1574), 486: "*Turba magna, omnium astrorum cohors sive multitudo.*" This interpretation is found again in Gerhard Dorn's *Dictionarium Theophrasti Paracelsi* (Strasbourg, 1583) and Martin Ruland's influential *Lexicon Alchemiae* (Frankfurt, 1612).

65 Revelation 7: 13–17. The text in the Vulgate (Rev. 7: 9–10) reads as follows: "*Post haec vidi turbam magnam, quam dinumerare nemo poterat, ex omnibus gentibus, et tribubus, et populis, et linguis: stantes ante thronum, et in conspectu Agni, amicti stolis albis, et palmae in manibus eorum: et clamabant voce magna, dicentes: Salus Deo nostro, qui sedet super thronum, et Agno*" (in the King James version: "After this I beheld, and, lo, a great multitude, which no man could number, of all nations, and kindreds, and people, and tongues, stood before the throne, and before the Lamb, clothed with white robes, and palms in their hands; and cried with a loud voice, saying, Salvation to our God which sitteth upon the throne, and unto the Lamb.").

significant star moving against the course of the heavens and every impression originates from it, not from the firmament or the stars.[66] Those aware of the properties of the *Evestrum* are prophets and foretellers of all things that are to happen, for the Highest of all things ("der Höchst aller dingen") does not speak with mortals, nor does he send his angel from his seat to foretell. Rather, such knowledge, according to the *Philosophia ad Athenienses,* is to be understood from the *Turban magna*, which many heathens and Jews believed to be God himself, blinded as they were in their understanding of the Divinity.

Since the *Evestrum* comes into being with every created thing, it is possible to foretell the fate and life of anything. If someone is about to die, for example, his death will not occur until the *Evestrum* has announced it, either by knocking, rattling, dropping objects, or other such means. After one's death, his *Evestrum* – being united with the eternal – remains on earth.[67] Ordinary people speak of ghosts, of souls wandering, but this is actually the *Evestren* of dead people which are producing signs. The saints produce signs and omens through their *Evestren* alone. Therefore, we must always heed the *Evestra* and their prophecies ("*Evestra praesagientia*"): they rule our sleep and dreams, and show us events that are happening as well as the nature of these events (II, §21).

The *Evestren* have their home in the elements: some in water, others in mirrors, crystals, polished nails, songs, etc. Even God has his *Evestrum mysteriale*, through which all good and all enlightened things are known; conversely, the Devil ("der verdampt"), too, possesses an *Evestrum* in the world, in which every evil and everything breaking the laws of nature is known.[68] We, however, can only know ourselves through our own *Evestrum* (II, §22).

The *Evestrum* is a spirit which teaches astronomy. It is not that nativities and prognostications can be known from the stars: astronomy has its being – its very essence – in the *Evestren*, like an image in a mirror (II, §22). There are innumerable different kinds of *Evestren*. They are the mothers of all things in prophecy, astronomy, and medicine. There is no knowledge of nature without

66 This should have prevented Toxites from interpreting the *Turban magna* as the stars. But Toxites certainly tried to reconcile this passage from the *Philosophia ad Athenienses* with the *Astronomia magna*. In Toxites' fully justified opinion, the *Astronomia magna* was a text of crucial importance. See Dane T. Daniel, *Paracelsus'* Astronomia Magna *(1537/38): Bible-Based Science and the Religious Roots of the Scientific Revolution*, PhD diss., Indiana University, Sept. 2003; CP 2, 208–231.

67 Book II, §21: "*Evestrum* ist auch mit dem Ewigen vereinigt, dann nach dem todt des menschen ist das *Evestrum* bleiblich auff erden: und gibt anzeigung ob der mensch in freuden oder pein ist / mit seiner gebierung." To understand this statement, we should remember that the *Evestrum* draws its knowledge from the *Turban magna*, which is the source of all prophecies. Since prophecies come ultimately from God, the *Evestrum* must be somehow connected to the eternal.

68 We understand "der verdampt" as "the damned one," i.e., the Devil, consistently with the context of this sentence.

an understanding of the *Evestren*. From them the Sibyls and prophets spoke, but it was as if they were sleeping or dreaming. After the destruction and rebirth of the *Evestren*, more marvels will be discovered (II, §23).

In the fragmentary third book (the so-called "vierdt *Paras*") there is no further mention of the *Evestra*. The author returns to the difference between the visible and invisible: in the beginning, every body was a vaporous spirit ("*spiritus fumosus*") which generated a visible body throughout the separation (we do not know which phase of separation the author had in mind), just as vapors of arsenic produce arsenic (III, §1) and a cloud produces hail or lightning while the matter they come from remains invisible – for it was a decree of the *summum bonum* ("das höchst gut"), in the beginning of all things, that every single thing would come from the invisible (III, §2). But this vaporous spirit remained alongside the invisible *spiritus* of the visible body generated by it. Thus, wood and stones – and men too – have both a visible body and an invisible *spiritus*, which is what remains of their original *spiritus fumosus*: from this residual – from this very filth – arises man. He is the most subtle in spirit, and he is a spirit, hinting at a twofold eternity: one is *Caleruth*, which points to "the first eternal," and the other is the "meritorious" ("verdienstlich"), which longs for its own kind, that is, God – and this is the reason why every created thing longs not for its creator, but rather for that from which it comes.[69] Thus, the human body longs for the matter from which it was separated, while the better part ("das verdienstlich") remains eternal, so that each part can eventually return to its original being. This better part, as we learn in III, §2, is actually the soul.

The original vapor, or fume ("rauch"), coagulates as soon as it leaves its *matrix* and comes into contact with extraneous air; everything that comes from the earth is at first a fume from the liquor of quicksilver ("auß dem *liquor mercurii*") – and it becomes plants, trees, etc. The more the fume heats and boils, the more these things grow until the coagulation is completed. Wood is a fume from *Derses*, and plants from *Leffa*. *Stannar* is the mother of metals: from its fume comes their prime matter. *Enur* is the fume of stones. Given that they originate from fumes, thus will all things be destroyed, reduced, and disappear into fumes (III, §3). As for man, he is a coagulated fume of the spermatic matter, coming from the boiling vapors of the body and the seminal body parts ("Samischen glidern": III, §4).

69 Book III, §1: "der mensch dergleichen ist nit anders dann das uberig vom separirten *fumo*: [...] vom selbigen unflat erhept sich der mensch / und das subtilest ist im *spiritu* und ist ein *spiritus* / das ist ein anzeigung zweierley ewiges / Eines Caleruthen / das ander verdienstlich / das Caleruth ist ein anzeigens in dem ersten ewigen: das ander begert zu demselbigen / das ist zu Gott [...]."

According to its preface, this unfinished third book was supposed to go into more depth with regard to theological matters. The author intended to answer the following questions: "Why there are so many religious beliefs; how there is a path to the eternal that does not stem from the religious belief, but from justice and piety alone; then how there is one single path to the celestial, which comes from faith and encompasses all sects of religious belief and all statutes, rules, and canons," so that "this *Paras* is the most effective of all in showing the nature of bliss by a wonderful philosophy."[70]

3 Sources and Oddities

This summary of the *Philosophia ad Athenienses* shows us that the author was much inspired, at least in the first two books, by an authentic treatise of Paracelsus: the *Philosophia de generationibus et fructibus quatuor elementorum*. This treatise (written around 1527) begins with the four elements being produced out of a *materia prima* called *Yliaster*, also called "das grosse *Yliaster*" (compare with *Mysterium magnum*). One cannot help but notice that the odd word *Evester* / *Evestrum* seems to have been modelled on the word *Yliaster* (the spelling *Yliastrum* is also found in the *Philosophia de generationibus et fructibus*). This *Yliaster* was "Nichts," the four elements having been created *ex nihilo*. But giving this *nothing* the name *Yliaster*, of course, lends it a somewhat ambiguous, positive quality. If this *Yliaster* can be thought of as a *something*, rather than a *nothing*, then the Creation cannot strictly be said to have occurred *ex nihilo*. However, other excerpts from the *Philosophia de generationibus et fructibus* clearly insist on creation *ex nihilo*.[71]

70 Book III, "Vorred in den vierdten *Paras*:" "[…] Darbey erscheint die erklerung warumb so mancherley geschlecht seind der leuthen / warumb so vil sprachen / warumb sovil Nation / deß gleichen warumb so mancherley glaub / und wie ein weg ist zu dem ewigen / der kompt nit vom glauben / kompt allein von der gerechtigkeit und frombkeit / darnach wie ein weg sey der allein gehet zu den Himlischen / derselbige auß glauben kompt / darinn alle Secten der glauben vnd Statuten / regelen / und Canonen begriffen werden […]: und ist dieser *Paras* der meist under den andern in anzeigung der seligkeit / mit einer wunderbarlichen Philosophey […]."

71 See Kahn, "Paracelsus' Ideas on the Heavens," 75–77. An interesting discussion on whether *nothing* can simultaneously be regarded as *something* is also found in *De modo pharmacandi* (first published in 1562), tract. IV (Huser, 5: Appendix, 206), although in an entirely different, medical, context. The *Philosophia de generationibus et fructibus* is mentioned in one of the versions of *De Gradibus* dating back to 1527, hence its dating (Huser, 7: 347; Sudhoff, I/4: 92); quoted by Andrew Weeks in Paracelsus: *Essential Theoretical Writings*, trans. Andrew Weeks (Leiden, 2008), 712, note a.

The *Yliaster* was first divided into four particular *Yliastri*, each of which was the prime matter of each of the four elements (compare with the division of the *Mysterium magnum* into *mysteria specialia*). The four elements were produced out of the four *Yliastri* by the vegetative action of the three principles (which were issued from the original *Yliaster*). It must be said that the three principles never appear in the *Philosophia ad Athenienses*. However, the same can be said for a clearly authentic treatise of Paracelsus, *De vita longa*, written in the same period as the *Philosophia de generationibus et fructibus*: thus, it is difficult to infer anything significant from this absence in the *Philosophia ad Athenienses*.

The *Yliaster* was skillfully carved and shaped by the highest *Fabricator* (compare with the *Mysterium magnum* being "prepared" by the "höchsten Kunstler" in I, §1).[72] After its division into four parts, the *Yliaster* was destroyed (in the *Philosophia ad Athenienses* the *Mysterium magnum* will never return to that from which it came); the four elements never stayed together and were never mixed, but each went at once to its own place[73] (in the *Philosophia ad Athenienses* each element also goes to its predestined place, and they do not mix). Air was the first to be taken out of the *Yliaster*; then there were fire, water, and earth: in this treatise it was the element of air, rather than fire, which had a prominent status.[74] The *Philosophia ad Athenienses* is closer to the standard scholastic/Aristotelian cosmology, even if air is said to have been predestined to pervade every vacant place even in the other elements.[75]

The structure of the *Philosophia de generationibus et fructibus quatuor elementorum* is very simple: it consists of four books, one book per element: air, fire, earth, and water. The nature of the elements is described first, and then their "fruits" (compare with the notion of "harvest" in the *Philosophia ad Athenienses*). A major difference between the two treatises is that the

72 *Philosophia de generationibus et fructibus*, 1, I, 6 (Huser, 8: 58–59): "Also ist dises erst *Corpus Yliastri* ein Klotz gewesen / in dem alle Kreütter / alle Wasser / all *Gemmae*, all *Mineral* / all Stein / all *Chaos* gelegen seind / die allein der höchst *Fabricator* hinweg geschieden hatt / und gebildet also subtil / daß das / so hinweg genommen ist worden / ein anderst gewesen ist: Als am ersten hatt er hinweg genommen den Lufft / unnd auß dem ubrigen seind noch drey geworden / das Fewr / Wasser und Erden: Auß diesen hatt er hingenommen das Fewr: noch seind zwey blieben: also biß zum end."

73 *Philosophia de generationibus et fructibus*, 3, I, 2 (Huser, 8: 98).

74 *Philosophia de generationibus et fructibus*, 1, VII-VIII (Huser, 8: 59–60). On the status of air in this treatise, see Kahn, "Paracelsus' Ideas on the Heavens," 81–83.

75 Book I, §13: "denn aller luft ist gepredestiniert in alle Elementen: nit das er in den andern Elementen sey / wie ein vermischung / das ist nit / aber er nimpt ein allerley in allen Elementen / und besitzt das: und besitz nit daß das vor besessen ist."

traditional scheme of concentric spheres for the four elements is preserved and carefully described only in the *Philosophia de generationibus et fructibus*.[76]

In this treatise, there are spirits called the *Fata* within the element of fire. They make omens, such as comets, at the behest of God. These spirits are found again in the *Philosophia ad Athenienses* under the name of "invisible *fatalischen*." Located in air, not fire, they are a species of elemental beings, and this was indeed the implicit status of the *Fata* in the *Philosophia de generationibus et fructibus*. In other treatises, such as the *Liber de nymphis, sylphis, pygmaeis et salamandris*, Paracelsus explained that man is not the only inhabitant of the material world: there are elemental beings as well, inhabitants of the elements (compare with II, §11 above, section 2, para. 6).[77] Among these elemental beings, the *Fata* appear only in the *Philosophia de generationibus et fructibus*.

Other precise details show more striking similarities between the two treatises, such as the idea that burning fire must not be identified with the element of fire, nor is its complexion limited to dryness and warmth.[78] All the same, according to the *Philosophia ad Athenienses* there might have been more than four elements, although the world was created as it is (II, §12); in the *Philosophia de generationibus et fructibus* Paracelsus explains that although God made the world in a fourfold form, he could as well have made it eightfold.[79] As for the remarked absence of the *tria prima* in the *Philosophia ad Athenienses*, it must have been perfectly deliberate: the topic of the *tria prima* was obviously not a major concern for the author of the *Philosophia ad Athenienses*.

This kinship of the two treatises is clearly more like a link between a source and a text based upon it than between two texts written by one and the same author. The strange impression one gets from reading the *Philosophia ad*

76 Note that, notwithstanding this traditional scheme, Paracelsus entirely ignored, or rather removed, the Aristotelian/Ptolemaic separation between the sublunary and aetherial realms in the *Philosophia de generationibus et fructibus*.

77 The mention of "wild leuth", most probably derived from the "Waldtleuten" mentioned in the *Liber de nymphis* (see above, book I, §15), strengthens our impression that the author of the *Philosophia ad Athenienses* knew this treatise quite well.

78 Compare *Philosophia ad Athenienses*, I, 18 with *Philosophia de generationibus et fructibus*, 2, I, 1 (Huser, 8: 65): "Und heist nicht dorumb das Element Fewr / dorumb das nichts wiß als Brennen / als viel gesagt haben: Dann das ist nicht das Element Fewr / das do brennt: Sondern daß / das do brennt / und daß / das ihm wiederwertig ist / das ist / das Gefrieren. Das Element Fewr steht auch nicht in der Complexion Heiß vnd Trucken / sondern auch im Kaltten unnd Nassen / das auch vom Element Fewr ist."

79 *Philosophia de generationibus et fructibus*, 1, VIII (Huser, 8: 60): "Dieser Iliastren seind 4. und nicht mehr / seind auch genug: Also hatt Gott die Welt in ein Gevierdts gesetzt / und lassen genug sein damit / der wol hett mögen 8. machen."

Athenienses is mainly due to four factors: a) the repeated returns to the topic of uncreated matter; b) the abundance of strange words never found in the other works of Paracelsus; c) the insistent, repeated references to elemental beings and related species; and d) significant discrepancies between the *Philosophia* and the authentic works of Paracelsus regarding the questions of the sidereal body, ghosts, and eternity. We will next treat each of these issues in turn.

a) It has been argued, mostly by Walter Pagel, that the idea of an uncreated matter was a genuine idea of Paracelsus.[80] The question is debatable, as the excerpts from authentic works of Paracelsus emphasized by Pagel are rather ambiguous (at least those of *De Matrice* and *De Mineralibus*):[81] there Paracelsus might have interpreted the primordial waters not yet separated by the heavens in Genesis 1:2–7 as uncreated prime matter.[82] On the other hand, creation *ex nihilo* is expressly mentioned in a number of authentic works of Paracelsus.[83] Furthermore, Kurt Goldammer argued that the theological works of Paracelsus fully display an orthodox view of the Creation.[84] The bold, straightforward, insistent assertion of an uncreated *Mysterium magnum* is opposed to both the orthodox view of creation *ex nihilo* and to the scarce, ambiguous, potentially discordant statements on the subject in the authentic works of Paracelsus.

b) A number of strange words appear only in the *Philosophia ad Athenienses*. There are good reasons to suspect several of them to be the results of misreadings, e.g., *Truphat* ("Frucht"?), *Dieemae* ("Gnomae"?), *Durdales* ("*Dryades*"), or *Neufareni* ("*Nympharum*"?). But besides these examples, many others are real *hapax legomena*. Arguably, a penchant for the introduction of such unheard-of notions was a distinctive feature of Paracelsus. Good examples of *hapax* in his authentic works are the *Primum Ens*, only mentioned in the treatise *De renovatione et restauratione*;[85] the puzzling *Labor Sophiae*, only mentioned in *Vom langen Leben*; the *Aquaster* and the *deltica impressio*, only mentioned in *De vita longa*;[86] the *Tristo*, *Agapas*, *Salpalla*, *Zedoch*, *Deneas*,

80 Pagel, "Paracelsus and the Neoplatonic [...] Tradition," 142–143; idem, *Das Medizinische Weltbild des Paracelsus* (Wiesbaden, 1962), 77–85, 80–81.

81 *De matrice* (Huser, 1: 202 and 203), trans. Weeks in Paracelsus, *Essential Theoretical Writings*, 643–645; *De Mineralibus*, near the beginning (Huser, 8: 337).

82 See Kahn, "Paracelsus' Ideas on the Heavens," 112–113.

83 E.g., *Philosophia de generationibus et fructibus quatuor elementorum*, 3, III, 2 (Huser, 8: 115); *De modo pharmacandi*, tract IV (Huser, 5: *Appendix*, 206); *Astronomia magna*, e.g., I, 2 (Huser, 10: 28–29).

84 Kurt Goldammer, *Paracelsus in neuen Horizonten* (Vienna, 1986), 310.

85 And in two spurious works: *De natura rerum* and *Thesaurus thesaurorum*; see Didier Kahn, "Quintessence and the prolongation of life in the works of Paracelsus," *Micrologus*, 26 (2018): *Longevity and Immortality. Europe – Islam – Asia*, 183–226, 195.

86 And in the spurious *Liber Azoth*; see ibid., 199 and 218.

only mentioned in the *Philosophia de generationibus et fructibus*.[87] However, the *Philosophia ad Athenienses* contains more than just a handful of peculiar, unheard-of phrases: nowhere else in the authentic texts of Paracelsus do we find any reference to *Neßder, Dalni, Evestrum, Trarames, Turban magna*, or *Caleruth*, not to mention *Derses, Leffa, Stannar*, or *Enur*. Paracelsus introduced a lot of new terminology in his treatises from the Basel period, but this terminology occurs in a number of authentic treatises.[88] The same cannot be said of the peculiar terminology of the *Philosophia ad Athenienses*. Not only are the words isolated examples, but so too are the notions behind them, especially the *Evestrum* and *Trarames*, which are – again – only found in four spurious and possibly spurious works: the end of book 9 of *De natura rerum*, the *De occulta philosophia, Liber Azoth*, and *De pestilitate*.[89]

c) The extensive reference to elemental beings in the *Philosophia ad Athenienses* is equally striking. In his authentic writings, Paracelsus mentions elemental beings far less profusely – except, of course, in the specific treatise devoted to them, the *Liber de nymphis*.[90] To some degree, the relatively disproportionate focus on elemental beings in the *Philosophia ad Athenienses* leaves the impression that each element, instead of being described by all its "fruits," were rather best described by its inhabitants – especially by its elemental beings.

d) The *Evestrum* described by the author possesses distinctive features absent from the authentic writings of Paracelsus. It is said that after a man's death, his *Evestrum* remains on earth, for it is connected to (more precisely "united with") the eternal. There is only one notion corresponding to part of this description in Paracelsus' anthropology: the sidereal spirit, invisible and distinct from the soul. However, the sidereal spirit is very clearly described as perishable in the works of Paracelsus; it dies with the body, and returns to the stars where it first originated, while the body returns to the earth. No such

87 *Philosophia de generationibus et fructibus quatuor elementorum*, 2, 1, 1; 2, 11 (*De Sole, Luce, Tenebris, & Nocte*); 2, 11 (*De Ventis* and *De Stellis temperatis de Zedoch*) (Huser, 8: 65, 69, 70, 73–74).

88 See Benzenhöfer, *Studien*, passim.

89 Huser, 3: 73, 74, 78; Huser, 6: 361, 362; Huser, 9: 338, 341; Huser, 10, Appendix: 9–10. The authenticity of each of the nine books of *De natura rerum* is carefully discussed by Urs Leo Gantenbein, "Real or Fake? New Light on the Paracelsian *De natura rerum*," *Ambix*, 67 (2020) (forthcoming). Gantenbein accepts the authenticity of book 9 with excellent arguments. I still have doubts, however, fueled (among others) by the mention of both *Evestrum* and *Trarames* in this same book. On *De pestilitate*, see Charles D. Gunnoe, "Paracelsus, the Plague, and *De Pestilitate*," in this volume, and Didier Kahn, "*De Pestilitate* and the Paracelsian Cosmology," *Daphnis*, 48 (2020), 65–86.

90 Kahn, "La question des êtres élémentaires."

thing in the *compositio humana* can be connected with "the eternal," except the soul itself.[91] Equally alien to the works of Paracelsus is the notion of a "twofold eternity" ("zweierley ewiges").[92] Thus the whole theory of prophecy underlying the notions of *Evestrum*, Trarames, and *Turban magna*, and part of the anthropology of the *Philosophia ad Athenienses*, seem to be specific to this treatise.

However, other sources can be identified among the authentic works of Paracelsus. We have already listed those that have inspired several peculiarities on the title-pages. Moreover, a few passages of the *Philosophia ad Athenienses* are perhaps derived from *De Meteoris*, especially three ideas: 1) the stars and planets "grew" ("gewachsen") from the element of fire (I, §12). This idea is developed in chapter 1 of *De Meteoris*. However, the *Philosophia ad Athenienses* insists that no proper vegetable growth, but separation was at work in this process. 2) The real element is not the body of the element, but its spirit ("Geist"), and there, in the spirit, is life.[93] 3) All *Mysteria* have been created by perishable gods ("zergencklichen Göttern") working within the *Mysterium magnum* (II, §1): this might derive from a blend of ideas from Plato's *Timaeus* (42a–e) and Paracelsus' *De Meteoris* (chap. 4) describing spiritual workers such as *Pennates superi*, *Vulcani*, and *Archei* which operate in the heavens to produce winds, rain, lightning, comets, and other meteorological effects by distilling, cooking, and subliming the *tria prima* in the crucibles of the stars.[94]

The notion of *Mysterium* already appears in the *Archidoxis* (not published before 1569), but not with the same meaning nor prominent status as the *Mysterium magnum*.[95] There was, however, an occurrence of *mysterium* as a cosmological essence from which everything has its life in another work of Paracelsus (chap. 6 of *De ente astrorum*, a section of the *Volumen Medicinae Paramirum*, not published before 1575):[96] there it is stated that air gives us life,

---

91 On the sidereal spirit, see Dane T. Daniel, "Invisible Wombs: Rethinking Paracelsus' Concept of Body and Matter," *Ambix*, 53 (2006), 129–142. On the notion of spirit in Paracelsus, see also the fragmentary *Liber de animabus hominum post mortem apparentibus* (Huser, 9: 293–298) and Ernst Wilhelm Kämmerer, "Le problème du corps, de l'âme et de l'esprit chez Paracelse," in Antoine Faivre and Frederick Tristan, eds., *Paracelse* (*Cahiers de l'Hermétisme*) (Paris, 1980), 89–231.

92 I warmly thank Dane Daniel for his remarks on these topics.

93 Compare *Philosophia ad Athenienses*, II, 2 with *De Meteoris*, chap. 2 (Huser, 8: 185 and 188).

94 *De Meteoris*, chap. 4: "*Quid in stellis de viventibus speciebus*" and chap. 9 (Huser, 8: 199–207 and 241–242); Kahn, "Paracelsus' Ideas on the Heavens," 99–102.

95 Huser, 6: 1–2. On the difficult dating of the *Archidoxis*, see Kahn, "Quintessence," 186–187 n. 12.

96 *Volumen Medicinae Paramirum*, "*De ente astrorum*," chap. 6 (Huser, 1: 17–18). See Weeks in Paracelsus, *Essential Theoretical Writings*, 228, note b. On the *Volumen Medicinae Paramirum*, see Benzenhöfer, *Studien*, 33.

but there is something more that contains air and out of which air comes, and which gives life to the firmament: it is called "das *M.*" (which Sudhoff interpreted as "das *Mysterium*").[97] There is nothing created that is above this "*M.*" It encloses all created things in the heavens and earth, and all of the elements live from it and within it, as Paracelsus claims to explain in *De primo creato* (a work which has not come down to us). In the following chapter, Paracelsus even names it "*M. M.*," which Huser interpreted as "*M. magnum.*"[98] This may well be the source of the *Philosophia ad Athenienses*, along with the *Philosophia de generationibus et fructibus*, where on one occasion the *Yliaster* is called "das grosse *Yliaster.*"

It seems that the author must also have known the *Liber de nymphis, sylphis, pygmaeis et salamandris*, which is the only comprehensive discussion of the elemental spirits in the works of Paracelsus. In no other treatise of Paracelsus could the author have found so many names of elemental beings.

In 1564, none of these treatises had been published.[99] The author must have read them in manuscripts, yet unfortunately, this does not help to identify him. The manuscripts of Paracelsus seem to have been accessible to a fair number of enthusiasts, as exemplified by the unknown editor of the two authentic treatises published in 1564 along with the *Philosophia ad Athenienses*, who used a manuscript of Montanus for one of them, and for the other, the autograph or a faithful copy of it.[100]

Besides the treatises of Paracelsus himself, Platonism was an important source of inspiration for the *Philosophia ad Athenienses*. The whole process of the various separations is similar, in some way, to emanationism (even if the similarity is negated by the material, concrete process of the separations and their simultaneity). The *Evestrum*, as mentioned above, may be reminiscent of the myth of Plato's cave. Furthermore, we can point to the obvious aforementioned Timaean source for the "zergencklichen Göttern," the perishable gods in book II, §1, as well as the emphasis on the Paracelsian elemental beings at the very beginning of the process of creation (book I, §11) – perhaps also

97 Sudhoff, I/1: 182 ff.

98 *Volumen Medicinae Paramirum*, ed. Michael Toxites (Strasbourg, 1575), 39; Huser, 1: 19.

99 The *De Meteoris* was first published in 1566: once on Birckmann's printing press in Köln, a few months later in Neisse (Silesia) by Marcus Ambrosius, who also edited, eight days earlier, a fragmentary version of the *Liber de Nymphis* (entirely published only in 1590 by Johann Huser). The *Philosophia de generationibus et fructibus* was never published until the Huser edition, in 1590. A related fragment, however, entitled *De elemento aeris*, was published in Köln (heirs of Arnold Birckmann) in 1567, in the collection *Astronomica et Astrologica*, edited by the Silesian Balthasar Flöter. The *Volumen Medicinae Paramirum* was published in 1575 for the first and last time before the Huser edition.

100 See above, section 1, para. 2.

reminiscent of the *Timaeus*. The idea of an uncreated prime matter falls into the same category: evident sources for it are either *Timaeus* itself or such widespread works as Apuleius, *De Platone et ejus dogmate* (I, 5). As Walter Pagel argued, another possible source might have been Ficino's (or rather Lefèvre d'Étaples') commentary on the Hermetic *Asclepius* (1505).[101] (This commentary was actually written by Jacques Lefèvre d'Étaples, not Ficino, but this was only discovered by P.O. Kristeller in 1937.) Lefèvre d'Étaples was striving to give a Christian reading of the *Asclepius*. As the Hermetic text presents the world as uncreated, Lefèvre stated that the world is both created and not created (*natum* [...] *et non natum*): created with regard to the making of the created things, but uncreated with regard to the truth of the creation, which is nothing other than the word of God (i.e., *Fiat*).[102] Can this statement have been instrumental in the writing of the excerpts on the uncreated *Mysterium magnum*? At any rate, the defensive line used by Gerhard Dorn against anti-Paracelsians in his *De naturae luce physica* (1583) was quite similar to this argument: Dorn boldly stated that the first *Paras* of the *Philosophia ad Athenienses* (supposed missing, as we saw) had perhaps been suppressed by the foes of Paracelsus and should certainly have described the *Mysterium magnum increatum* of the uncreated Unity, whereas the second *Paras* actually described the *Mysterium creatum* of the alterity created from the *increatum*.[103]

However, it must be said that, before Bartholomaeus Reusner (1570) and Thomas Erastus (1571) exposed Paracelsus as heretical due to his negation of *creatio ex nihilo* in the *Philosophia ad Athenienses*, the notion of an uncreated prime matter had been fairly uncontroversial among philosophers, as exemplified (in the words of Andrew Weeks) by "the whole Patristic tradition that included St. Augustine, Johannes Scotus Erigena, and several medieval schoolmen, who likewise equivocated between *creatio ex nihilo* and a creation from the '*rationes aeternae*' present in God before the creation." Thus, "directly or indirectly," philosophers and theologians "borrowed from the Stoics' *logoi spermatikoi* and from the *Logos* of Philo of Alexandria and the Gospel of John"

101 Mentioned by Carlos Gilly in Sebastiano Gentile and C. Gilly, *Marsilio Ficino e il ritorno di Ermete Trismegisto / Marsilio Ficino and the return of Hermes Trismegistus* (Florence, 1999), 304, referring to Pagel, *Das medizinische Weltbild*, 90–91.

102 Isabelle Pantin, "Les '*commentaires*' de Lefèvre d'Étaples au *Corpus Hermeticum*," in Antoine Faivre and Frédérick Tristan, eds., *Cahiers de l'Hermétisme: Présence d'Hermès Trismégiste* (Paris, 1988), 167–183, 172.

103 See CP 2: 899.

to "resolve the issue of whether God created out of something or nothing."[104] Moreover, why would the author of the *Philosophia ad Athenienses* put forward the idea of uncreated matter so boldly if it were highly controversial at the time when he wrote the treatise? Certainly, his goal was not to bring down harsh attacks from theologians on Paracelsus. We must conclude therefore that whenever this treatise was written – before 1564 of course, yet probably earlier still – the idea of uncreated prime matter was not a problematic issue. Accordingly, the author might have been trying merely to use it as a means to bring Paracelsus closer to the Platonic tradition.

Conclusion

The *Philosophia ad Athenienses* was extremely influential. Oddly enough, its fascination with elemental beings – as well as the topic of *Evestrum* – found their way into several prominent pseudo-Paracelsian treatises, especially *De natura rerum*, *De Pestilitate*, and *Liber Azoth* – an influence that could be worth further investigation. If we are to try to explain why the author wrote this forgery, the only easy answer (or hypothesis) is that he came across the cosmological tracts of Paracelsus and was disappointed with what was still traditional in their cosmology, namely, the traditional scheme of concentric spheres for the four elements – even if these tracts were highly unconventional in many other respects.[105] This disappointment might have given him the impulse to compose another more Platonic cosmology, one in which elemental beings would be more prominent, and thus closer to his own understanding of Paracelsus' world-view.

104 Email from Andrew Weeks to Didier Kahn, 12 November 2018; see Andrew Weeks, *German Mysticism from Hildegard of Bingen to Ludwig Wittgenstein. A Literary and Intellectual History* (Albany, NY, 1993), 128, and the discussion in Kahn, "La création *ex nihilo*."

105 See above, n. 76; Kahn, "Paracelsus' Ideas on Heavens," 79–80, 113–114 and passim.

Genealogy of Knowledge and Delegitimization of Universities: The Pseudo-Paracelsian *Aurora Philosophorum*

Tobias Bulang
Germanistisches Seminar, Universität Heidelberg
tobias.bulang@gs.uni-heidelberg.de

Abstract

This article introduces the pseudo-Paracelsian treatise *Aurora Philosophorum* (last third of the sixteenth century) and focusses on the contexts of a specific genealogy of knowledge presented in this text. It reaches from a divine origin of human knowledge to the universities of the author's times. Here, the transfer of science throughout history is described as a process of continuous decline. The concept of knowledge negotiated by this text is conceptualized not as theory alone: knowledge has a genuine instrumental and practical aspect. The article also presents the principles of the forthcoming edition and commentary of the handwritten and printed copies of the *Aurora Philosophorum*.

Keywords

Paracelsus – *Aurora Philosophorum* – pseudepigraphy – Gerhard Dorn – genealogy of knowledge – Adamitic science – edition – commentary

Introduction

The *Aurora Philosophorum* is a treatise about the *transmutatio metallorum*.[1] The earliest surviving manuscripts of it date back to the last third of the sixteenth century.[2] Two of the manuscripts and two printed translations of this

---

1 For the transcription of manuscripts and critical proofs, the author would like to thank Sophie Knapp. For the critical revision of the English text, his thanks go to John Stewart.

2 Prior to Huser's edition from 1605 in Paracelsus, *Chirurgische Bücher und Schrifften*, ed. Johann Huser (Strasbourg, 1605), *Appendix*, 78–92, several handwritten manuscripts and one printed exemplar are known: Universitätsbibliothek Heidelberg, Cod. Pal. Germ. 600 (dated: 1569); Universitätsbibliothek Heidelberg, Cod. Pal. Germ. 303 (dated: 1574);

 | DOI:10.1163/9789004503380_005

text attribute it to Paracelsus himself, and the manuscripts clearly show that it was originally written in German.[3] However, Paracelsus' authorship was rejected in Huser's 1605 edition of the *Aurora*.[4] The text consists of 20 chapters. The first four chapters trace a genealogy of knowledge starting with Adam, who possessed a universal science (later lost), followed by various stations of further transfer of this knowledge to different nations and scholars. This *translatio* is presented as one of continuing decline and corruption up to the present times. The fifth chapter explains the character and the value of Adam's secret knowledge. Chapters six through ten provide a report on failed past attempts to create the philosophers' stone from plants, animal substances or minerals. The report lists all the mistakes made by different groups of alchemists and criticizes the "philosophers" who developed incorrect theories on the subject because they misunderstood the secret language regarding the proper procedures. Philosophers such as Arnaldus de Villanova, Gilbertus Anglicus (the author calls him "the villain"), Johannes de Rupescissa ("the seducer"), Geber ("sophistic"), Aristotle *alchimista*, Rasis, and Polydorus are subjected to harsh criticism. Even Hermes Trismegistus, according to the text, did not grasp the full worth of the Adamic art.[5] Opponents of such philosophers are the so-called "true" philosophers, or the "wise men," who gained a deeper understanding. The only name mentioned here is Morienus ("the wise").[6] The narrator of this history of failure is – obviously – one of these wise people. He presents himself as an older man who admits the mistakes of his youth and explains his multiple misunderstandings and those of his predecessors as having been caused by misreadings of the hieroglyphic language that the ancient texts used to present the deep secrets of God and nature. The history of alchemy appears here as a succession of different sects following false assumptions. The last ten chapters deliver recipes containing the "true" procedures to create the different

Kassel, Landesbibliothek, 4° Ms. chem. 8, 2r-27v (dated: 1575); Sächsische Landes- und Universitätsbibliothek Dresden, Hs. e 7, 247–313 (dated: 1576); [pseudo]-Paracelsus, *Aurora Thesaurusque Philosophorum*, ed. Gerhard Dorn (Basel, 1577). See the description of all texts in Philipp Redl, "*Aurora Philosophorum*. Zur Überlieferung eines pseudo-paracelsischen Textes aus dem 16. Jahrhundert," *Daphnis*, 37 (2008), 689–712. I will quote from the text of the not yet edited Heidelberg manuscript *Cod. Pal. Germ. 600*. Unfortunately, the existence of the New Haven-manuscript (Yale University, Beinecke Library, MS. 309, 51r-71v) came to my knowledge only after this paper was finished. However, the manuscript will be considered in our work on the edition and the commentary.

3 Redl, "*Aurora Philosophorum*," 692–700, 708 n. 7, 710 n. 15.

4 *Aurora*, ed. Huser, 78–92, see Redl, "*Aurora Philosophorum*," 692–693.

5 *Bube, Verfurer, sophistisch*: Cod. Pal. Germ. 600, 25v, 33v.

6 See Cod. Pal. Germ. 600, 29r.

substances necessary for the transmutation of minerals and metals: white tincture of arsenic, red tincture of vitriol, extraction of oil from antimony, and the technique of projection.[7] The last five chapters add more specific information about the fabrication of the philosophers' stone.

Thus, the *Aurora Philosophorum* exposes and legitimizes an expertise which is founded in traditions of knowledge and education beyond the Greek and Latin world of ancient, medieval and humanist scholarship and institutions.

This essay focuses on three aspects. First, the genealogy of knowledge presented in the *Aurora Philosophorum* will be examined with regard to its historical construction, and institutional context in the sixteenth century. Second, I will explore the conceptual content and implications of the Adamitic knowledge constructed in this text. Third, I will present the Heidelberg project of an annotated edition of all early versions of the *Aurora*, handwritten and printed.

1 Genealogy of Knowledge

At the beginning of the first chapter, the *Aurora Philosophorum* states the following:

> Of all sciences and natural secrets in paradise, Adam was the first real, divine and natural inventor. He was powerful and a master of nature and of all creatures thanks to his excellent knowledge and skill, with which God had endowed him, since he knew the entire firmament and all the elements by heart, from the smallest dots to the biggest visible and invisible secrets. He named and called each creature by the right name according to its own qualities and characteristics. And although it is said, that Adam lost the knowledge of natural secrets after the fall, and was entirely deprived of it by God – the truth of this cannot be proven sufficiently.[8]

7 Harald Haferland, "Eine kurze Theoriegeschichte der alchemistischen *multiplicatio* und der *Vorbothe der am philosophischen Himmel hervorbrechenden Morgenröthe* des Johannes de Monte Raphaim (Amsterdam 1703)," in Peter-André Alt, Jutta Emimg, Tilo Renz and Volkhard Wels, eds., *Magia daemoniaca, magia naturalis, zouber. Schreibweisen von Magie und Alchemie in Mittelalter und früher Neuzeit* (Wiesbaden, 2015), 393–417.

8 Cod. Pal. Germ. 600, 3r: "Aller Künsten und natturlicher Geheimniß Wissenschaft ist erstlichen gewesen Adam der erste im Paradeiß, ein rechter, gottlicher und natturlicher Erfinder, welcher der ganzen Natur und Creatur durch sein furtrefflich Erkantus und Geschieckligkeit, darrein in Gott gesetzett, ist mechtig und ein Her gewesen, den er hat deß ganzen Firmaments und aller Elementen von den minsten und kleinesten Puncten biß zu dem großten aller sichbaren und unsichbaren Geheimnuß Erkandnuß und volkommenen Verstandt gehabt, wie ehr

Unlike the common depiction of a man stigmatized by the original sin, Adam appears here as the superior and confident possessor of the full divine and natural knowledge. Adam as the first *astrologus* is a figure which also appears in the *Antiquitates Judaicae* of Josephus Flavius, and his superior knowledge is pointed out in the writings of Jewish and Christian authors, as well as in several legends about the Adamites.[9] In particular, Adam's original language is regarded as the most perfect of all languages.[10] However, compared to such ideas, the *Aurora Philosophorum* highlights a very different aspect. On the one hand, there is the explicit and differentiated description of Adam's knowledge as universal, as divine and natural, and as a knowledge of the secrets of God and the universe. Further, this is not only declarative knowledge (in the sense of mental representations), it is also practical, procedural knowledge. The *Aurora* mentions skill (*geschieckligkeit*), thus indicating instrumental forms of arts and sciences. This genealogy of knowledge finds its *telos* in alchemical operations, which activate natural and divine secrets in the *opus magnum*. Therefore, the original knowledge of Adam is in itself already divine, natural and instrumental. The latter aspect becomes clearer when we consider that some manuscripts of the *Aurora* state that Adam's knowledge was not lost with the Fall. In order to work as a farmer, Adam needed his knowledge to fabricate tools out of metals. That Adam did not use the philosophers' stone to create gold is explained by the simple fact that there was no commerce back then.[11] It

auch einer jeden Creatur, seiner Natur nach seinem rechtem eygengeschaffenen Nahemen gegeben, genennet und geruffen, unnd woll gesprochen wirdt, das Adam die Erkentniß der natürlichenn Geheimniß nach dem Vall im Paradeiß verloheren vnd derselbenn von Gott genzlichen beraubtt worden, so mag es doch mit Wahrheitt (das es also sey) nicht genugsam erwisenn werden." The rules for transcribing the handwriting are explained in part 3 of this study.

9 Flavius Josephus, *Antiquitatum Judaicarum libri I–IV*, ed. Benedictus Niese (Berlin, 1888), 1: 69–71; see Kurt Flasch, *Adam und Eva. Wandlungen eines Mythos* (Munich, 2017), 32–34. Christian notions of Adam's perfection before the Fall are summarized in Leo Scheffczyk, "Adam," *Lexikon des Mittelalters*, 1 (Munich, 2002), 111–114.

10 Umberto Eco, *The Search for the Perfect Language* (Oxford, 1995); Arno Borst, *Der Turmbau von Babel. Geschichte der Meinungen über Ursprung und Vielfalt der Sprachen und Völker*, 4 vols. (Munich, 1995), 955–1262; Tobias Bulang, "Ursprachen und Sprachverwandtschaft in Johann Fischarts *Geschichtklitterung*," *Germanisch-Romanische Monatsschrift*, 56 (2006), 127–148, on 127–136.

11 Cod. Pal. Germ. 600, 5r: "Sintemal do niemandt, den er allein mit seiner lieben Eva, auff einem verfluchten Acker gewesen vnnd gewohenet, bis so lange sie das Ackerwerck, Vihe Zucht, vnnd ander handgrieff in Schwank bracht, auch Kinder gezeucht vnnd fruchtbar beide des Feldes vnnd Leibes geworden."

is clearly implied that the potential to turn metals into gold was part of Adam's universal skills.

The text explains further that Adam predicted disasters of fires and floods. To maintain his knowledge in both cases, Adam's children created tablets or columns of stone and wrote down Adam's knowledge in secret letters and hieroglyphs.[12] And indeed, after the flood, one of those tablets was found at the foot of Mount Ararat, where Noah's ark came to rest. The tablet contained astronomical knowledge of the characteristics and the effects of the planets.

From this point on, a fragmentation of the previously holistic knowledge took place. With the expansion and migration of Noah's offspring all over the world, the original knowledge of Adam became further diversified:

> And since the root and source of all wisdom and philosophy fell deeply, only a few people could maintain parts of this wisdom, which was divided into *particularia* so that one person was an astrologer, the second an arithmetician, the third a geometer, the next a magus, the sixth a cabbalist, the seventh an alchemist etc.[13]

This particularized knowledge was the object of further transfer. It was Abraham who brought his part of that knowledge to Egypt, from where all the nations of the world received wisdom and science. Later, Jacob showed a particular knowledge of *magia naturalis* by placing wooden sticks in the watering place for goats in order to breed spotted offspring (Genesis 30:37–43). For the Hebrews, the Chaldeans, the Persians and the Egyptians, it was crucial that *magia naturalis* was taught together with the theological studies. Priests were also *magi* and medics. The Bible shows that Daniel the prophet, Moses and Elias possessed such science, as well as being masters of Cabbala. In Persia, the kings also used to be *sophi*. They were masters of the "whole nature" of God's gifts. The Egyptians received this knowledge from the Persians and the

---

12 The text differs from the report about Adam's two columns in Josephus Flavius' *Jewish Antiquities*. There, Adam creates one column out of brick and one out of stone to protect the knowledge from water and fire; see Jan Assmann, "Das gerettete Wissen. Flutkatastrophen und geheime Archive," in Martin Mulsow and Jan Assmann, eds., *Sintflut und Gedächtnis. Erinnern und Vergessen des Ursprungs* (Paderborn, 2006), 291–301.

13 Cod. Pal. Germ. 600, 6r: "Und wiwol der Grund und Quelle aller Weisheit und Philosophia ein grossen Fall gethan, also das nicht mehr den bey ezlichen ettwaß vonn solchenn freyen Kunsten gefunden werden, und ist zerteillet worden gleichsam in particularia, das einer ist gewesen ein Astrologus, der ander ein Arithmeticus, der dritt ein Geometra, der ein Magus, der sechst ein Cabalist, der sibendte ein Alchimist, Schmelzer oder der gleichen [...]."

Chaldeans. The further decline of arts and sciences came with the Greeks, who are called idolatrous in the text. Since they did not acknowledge the Mosaic law, they could not grasp the religious foundations of the sciences they received. So, they took home what they got, and made "their own *philosophia*" out of it, trying to interpret everything by reason only. The Latins took over this philosophy and founded the universities, where this degraded knowledge lives on to the present day.

Thus, the Greek and Latin traditions of western universities are delegitimized by the author of the *Aurora*. Whereas the magicians of Persia – the three holy kings – searched for the savior Jesus Christ, western philosophy broke wisdom into pieces. Due to the *superbia* of the Greeks and to devilish seduction, knowledge became secularized and therefore lacked its theological grounds. According to the author of the *Aurora Philosophorum*, Renaissance magi like the Abbot Trithemius or Cornelius Agrippa did not have the grounds of true philosophy at their disposal either.

Although the author explicitly thanks God for the existence of universities, the attempt to delegitimize them is no less drastic. This indicates, however, that hermetic and alchemical sciences cannot be regarded as the dominant source of knowledge in the sixteenth century, but rather as the project of an alternative to the sciences taught in universities. The author of the *Aurora Philosophorum* belongs to a separatist group, which turned against academic traditions of Galenistic medicine and Aristotelian philosophy. Contemporary academic opponents of the hermetics and alchemists addressed them as a heretical sect.[14] On the other hand, the text directs a sharp critique toward famous alchemists who, in other hermetic writings, are treated as prominent ancestors of contemporary adepts in their genealogies of the sciences. By doing so, and with its foundations in Adamic knowledge, the text also provides a Christian alternative to the tradition extending from Zoroaster, or Hermes Trismegistus to the present, implicitly distancing itself from neopagan alternatives. The genealogy of knowledge sketched by the *Aurora* establishes an auctorial position of critical evaluation. Since wisdom and knowledge became

14 See Wilhelm Kühlmann and Joachim Telle, *Corpus Paracelsisticum: Der Frühparacelsismus*, 3 vols. (Tübingen, 2001–2004 and Berlin, 2013) (hereafter: CP 1, CP 2, CP 3), here CP 2: 5–13, 24–27. See also (in an edition of his works donated to him by his pupils) Wilhelm Kühlmann, "Paracelsismus und Hermetismus. Doxographische und soziale Positionen alternativer Wissenschaft im postreformatorischen Deutschland," in Jost Eickmeyer, Ladislaus Ludescher in cooperation with Björn Spiekermann, eds., *Gelehrtenkultur und Spiritualismus. Studien zu Texten, Autoren und Diskursen der Frühen Neuzeit in Deutschland*, 3 vols. (Heidelberg, 2016), 3: 167–190; Wilhelm Kühlmann, "Das häretische Potential des Paracelsismus – gesehen im Licht seiner Gegner," in *Gelehrtenkultur und Spiritualismus*, 3: 235–266.

corrupted, all alchemical attempts to succeed in the *opus magnum* are based on incorrect epistemic assumptions.

2 Adam's Knowledge

The explanation of the original Adamic knowledge clearly follows Neoplatonic ideas of cosmic hierarchy and connection:

> Real *magia* is a science, which is mighty to reveal properties, secrets and effects of all upper and lower elemental matters and fruits. Like the true Cabbala is brought by *anagogia* to God, acts with God and presages from God, and is full of God's secrets, so is *magia* mighty of all natural secrets and able to predict present and future events from natural causes.[15]

As the *Aurora* explains further, *magia* originates in the divine creation; it comes from the "Ternario Divino." *Magia* is "the tie, connecting heavenly powers to terrestrial ones."[16] The connection between the divine and the terrestrial, the effects of divine influences and impressions in the terrestrial *corpora* are called by the magicians the "marriage" of divine and terrestrial entities. Therefore, *magia* reveals the secrets of stars and planets, metals, animals, and plants. The creatures reflect their creator and remain bound to the divine trinity. It is the devil who obstructed this secret knowledge; he robbed the Greeks of this treasure and substituted lies for it. Here again, we find the medieval idea of the connection between the macrocosm and the microcosm. However, the fact that this knowledge empowers the philosopher above all creatures of the earth is an essential aspect in the *Aurora Philosophorum*. The medieval meditation upon the creator and his creation, and the art of reading nature allegorically as a place where the eternal wisdom left its traces, was a passive approach, aimed at the exaltation of God's wisdom and the recognizing of proper behavior in

15 Cod. Pal. Germ. 600, 13r-13v: "Es ist aber die rechte Magia ein Kunst welche gewaltsam ist, aller elementischen untern und obern Corpora und Fruchten Eygenschafftt unnd Geheimniß, Wirckunge zu offenbaren, und gleich wie die rechte Cabala, durch die Anagogia zu Gott gebracht wirt, und mit gott handlet, und auß Gott weissagett, und voller ist der gottlichen Geheimniß, also ist die Magia mechtig aller natturlichen Geheimniß, und Weissagen durch die Nature Gegenwärtiges und Zukunfftiges."

16 Cod. Pal. Germ. 600, 13v: "… sie ist das Band, durch welches zusammen verbunden werden die himmlischen Kraefft zu den irdischen."

the world.[17] In the sixteenth century, this very idea became connected to the pragmatic and instrumental dimensions of human control over nature. Thus, allegorical thought is here embedded in a materialistic approach. In Chapter 5, the theological concept of macro- and microcosmic connections is reformulated to indicate the possibility of the *transmutatio metallorum*:

> The magi, who searched the inner workings of nature, found that all creatures would reach one uniform substance when they were cleared and purged and thus signify holy work and divine secrets, since they recognized that everything became earth and that after the highest, magical and divine separation, a perfect *materia* remained, which during the subtle preparations would change from vegetable substance to mineral and from there into metal substance and from the most perfect metallic matter rises above itself into infinite divine secrets or *essentia* of divine and terrestrial creatures which is called by the magi *Arcanum supercœleste*.[18]

This is a new instrumental approach: the recognition of the secrets of God, the ascent of the intellect from the lowest to the highest is not reached by a process of study and meditation alone; it also requires the process of the "subtile bereytung" (subtle preparation) of various substances. What can be viewed as a technological secularization of epistemic and allegorical speech can also be understood as a re-evaluation of technological work by using originally theological language to circumscribe it. From the perspective of a modern language of science, it is clear that problems of semantic reference are unavoidable when such things happen. The *Aurora* is at once conservative and revolutionary. The text identifies human dominion over nature as the perfect condition in the Biblical accounts of the beginning of mankind. This perfect condition was lost due to a corruption of knowledge. The restoration of human dominion over nature is expected in the near future. That notion matches the ideas of a renewal of science, art and medicine in the works of Paracelsus, although

17 See Herbert Kolb, "Der Hirsch, der Schlangen frißt. Bemerkungen zum Verhältnis von Naturkunde und Theologie in der mittelalterlichen Literatur," in Ursula Hennig and Herbert Kolb, eds., *Mediaevalia litteraria* (Munich, 1971), 583–610.

18 Cod Pal. Germ. 600, 15v-16r: "Die Magi ..., welche der Naturen Innersten nachforscheten, funden, das alle Creaturen in ein eynige Substanz kommen wurden, welche so sie geleuttert und purgiert anzeigtte heylige Werck und gottliche Geheimniß. Dan sie sahen, daß alles Erde wurde vnd nach der hochsten und gottlichen und magyschen Separation ein perfecte Materia dahinden bliebe, welche mit der lenge subtielen Bereittung auß der vegetabilischen Substanz in die mineralisch und auß der mineralischen in die metallische und auß der perfectesten der metallischen Wesenheitt uber siech steiget In ein unendliches gottliches geheimtes Wesen oder Essentia himmlischer und irdischer Creaturn gleichformigkeit zu weisen von denn Magis Arcanum *super coelestis* genennet."

his name is not mentioned at all within any of the *Aurora* texts. The key focus identified in the *Aurora Philosophorum* for the restoration of human dominion over nature, is on the *transmutatio metallorum*, the fabrication of gold.

3 Forthcoming Edition

The annotated edition of the *Aurora Philosophorum*, which is currently a work in progress at Heidelberg University, will contain all extant versions of the text up to and including the version printed in Huser's Paracelsus edition of 1605. This collection includes Latin and German texts, handwritten and printed material, along with transcripts of the texts. Only modest modifications have been made: we silently replace the long 's' with a round 's', we dissolve abbreviations consistently, and provide modern German capitalization and normalized syntax. We keep double consonants and 'ck' according to early modern German orthography even after long vowels (e.g., "Apothecke"), but we reduce the abundance of consonants at the end of words and elsewhere. Other than these changes, all other text-interventions are noted in the critical apparatus.

The production of a critically edited single text is not our aim, nor can it be, as is already obvious given the fact that we have versions in different languages. Furthermore, the German manuscripts not only vary in length, but also present the text very differently in terms of formulation, semantics and syntax. Since they also form very different textual interrelations, we favor the presentation of different versions, which will enable the reader to study different problems of this text's transmission.

The first printed edition of the *Aurora* is the Latin edition from 1577 by Gerhard Dorn. Dorn claimed to have translated a German text originating from Paracelsus himself. Huser's edition of a German *Aurora* in 1605, which is the second printed version, names Dorn as the author, who attributed the text to Paracelsus. Several scholars assume that Huser's German text is a translation of Dorn's Latin text. However, it is not known who translated this text, whether Huser himself or someone else. Two handwritten German manuscripts show great resemblance to Dorn's Latin version. The handwritten manuscript Kassel, Landesbibliothek 4° Ms. chem. 8, dated 1574, presents a German text and also attributes it to Paracelsus. It has been argued, especially by Didier Kahn, that this text is close to the German text Dorn translated into Latin (although probably not the direct source). According to our transcription, the manuscript of the Sächsische Landes- und Universitätsbibliothek Dresden Ms e 7, dated 1576, is also very close to the Kassel manuscript. As not much research has been devoted to the Dresden manuscript, further work is needed.

As far as we can see at present, it seems to make most sense to provide a critical edition of Kassel and Dresden, presenting the Kassel text with the deviations from the Dresden manuscript in the apparatus. This text is then to be presented synoptically together with Dorn's Latin text and Huser's German version. Thus, we plan to make the texts available not only for historical studies about alchemy and Paracelsian thought, but also for translation-studies.

Two handwritten Heidelberg manuscripts, which are the oldest extant versions of the text, have hitherto never been edited. They differ significantly from the manuscripts treated so far. Cod. Pal. Germ. 600 is dated 1569 and is not only the oldest known version, but also the longest. It contains additional text, which is not included in the printed versions nor most other handwritten manuscripts. Philipp Redl has suggested that the second Heidelberg manuscript, Cod. Pal. Germ. 303 (dated 1574), might be a direct copy of Cod. Pal. Germ. 600.[19] According to our transcription and analysis, we can strongly affirm that hypothesis. The obvious errors in Cod. Pal. Germ. 303 are for the most part clearly the result of misreadings of the other manuscript. The comparison of the texts reveals inadvertent mistakes by the copyist, as the following examples illustrate:

Cod. Pal. Germ. 600	*Cod. Pal. Germ. 303*
1) "Firmaments und aller Elementen," 3r	"Ferments und aller elementen," 2v
2) "die Mineralia vnd Metalla, darauß," 3v	"die Mineralia daraus," 3r
3) "solte hernachmalen," 5r	"hernach solte malen," 3v

The first example shows that Cod. Pal. Germ. 600 has "Firmaments," whereas Cod. Pal. Germ. 303 has "Ferments." The text explains that Adam possessed knowledge of the whole world; the "Ferments" version therefore makes no sense whatsoever and must be explained by an error on the part of the copyist. The second example shows that Cod. Pal. Germ. 600 has one word more than Cod. Pal. Germ. 303 ("Mineralia und Metalla, darauß" versus "mineralia, darauß"). Context shows that the metals cannot be dismissed here. The passage explains that Adam used his superior knowledge in order to produce tools for farming after the Fall. Since farming tools are made from metals and not from minerals, the version with minerals only must be regarded as incomplete. Here also, we can explain the mistake as an error by the copyist: after writing "mineralia," he skipped the second word, since it also starts with an 'm' and ends with an 'a,' wrongly supposing himself to have already finished with it. Such slips of the eye (*Augensprung*) are typical mistakes of copyists. The third

19 Redl, "*Aurora Philosophorum*," 698, 704.

example can also be regarded as an error; the uncommon word order in Cod. Pal. Germ. 303 is again explicable as due to the eye of the copyist perceiving the words in reverse order.

At one point the copyist improves the text: Cod. Pal. Germ. 600 has a listing of magical professions: "... that one used to be an astrologer, the second an arithmetician, the third a geometer, the next a magus, the sixth a cabalist, the seventh an alchemist ..."[20]

Obviously, the writer miscounted here; the number five is missing, so that the listing jumps from number four (magus) to number six (cabalist). In contrast, Cod. Pal. Germ. 303 counts correctly: "the second an arithmetician, the third a geometer, the fourth a magus, the fifth a cabalist, the sixth an alchemist."[21] It is quite unlikely that such a mistake in counting could happen if – hypothetically – the second text were copied by the writer of the first one. Skipping the clear numbering and not recognizing the different outcome (seven instead of six) is not even explicable by severe inattention. The other way around is far more plausible: the second writer recognized the wrong count of the original and improves his copy. These observations show the close connection of the manuscripts and indicate that Cod. Pal Germ. 303 might be a direct copy of Cod. Pal. Germ. 600. Thus, a critical edition of both texts based on Cod. Pal. Germ. 600 with the deviations of Cod. Pal. Germ. 303 in an apparatus is the method of choice here. This text will be published separate from the Kassel/Dresden/Dorn/Huser synopsis.

A commentary on the texts is also planned. The starting point for the lemmatization is Cod. Pal. Germ. 600; since it is the oldest copy and has the longest text,[22] reference to the other versions will be provided.

In the commentary, we pursue a number of explicit aims. First, we want to position the text within the field of the knowledge of its time. For the authors and texts mentioned in the *Aurora*, as well as for every segment with intertextual reference, we seek to provide sufficient biographical, institutional and bibliographical information. Technical aspects, religious ideas, and other concepts relevant for the understanding of the text will be thoroughly explicable with regard to recent academic studies.

---

20 Cod. Pal, Germ. 600, 6r: "[...] das einer ist gewesen ein Astrologus, der ander ein Arithmeticus, der dritt ein Geometra, der ein Magus, der sechst ein Cabalist, der sibendte ein Alchimist [...]"

21 Cod. Pal. Germ. 303, 4r: "[...] der ander ein Arithmeticus, der dritt ein Geometra, der vierde ein Magos, der funffte ein Gabalist, der sechste ein Alchamist [...]"

22 Redl, "*Aurora Philosophorum*," 698, states that Cod. Pal. Germ. 303 is a slightly shortened copy of Cod. Pal. Germ. 600.

Second, we will explain the early modern German, wherever it is not clearly understandable to the modern reader. Therefore, the commentary will also provide a contribution to the lexicography of early modern German, providing German terminology for technical and chemical terms and procedures in particular.

Third, we want to focus on the reception of the *Aurora*. Here we hope to show, on the one hand, how the *Aurora Philosophorum* is already, to some extent, the result of a reception and synthesis of various texts. On the other hand, we will mention authors of later times who quoted the *Aurora* or continued working with the contents of this text.

In what ways could such an edition and commentary of the *Aurora Philosophorum* be useful? This question is crucial in any philological project, and I would like to briefly address it here at the end of my paper. First of all, we provide sufficient text and commentary for studies in the field of the history of science. Here, the text is of great importance for the history of Paracelsianism, of alchemy and therefore also of chemistry. Further, we provide a critical edition for literary studies concerning the influence of the text on German poetry. The famous recipients include Jakob Böhme and Johann Wolfgang von Goethe. Third, we provide a German-language document for linguistic studies, especially studies on the development of technical terminology and foreign words in German. Fourth, the *Aurora*, like many alchemical texts of the time, is also part of the historical record for the field of early modern religion and spirituality, and is therefore an important source for research in this field.

Into the Forger's Library: The Genesis of *De natura rerum* in Publication History

Hiro Hirai
Center for Science and Society, Columbia University
hhirai2@gmail.com

Abstract

One of the most popular writings ascribed to Paracelsus, *De natura rerum* appeared in 1572. That was when the movement of forgery production reached its climax, in parallel with the multiple editions of his genuine work *Archidoxis*. This article aims to place the genesis of *De natura rerum* in the context of publication history. It will first reconstruct a "library" by surveying the works ascribed to Paracelsus which could serve as instruments for the "author/reworker/editor" of *De natura rerum*. Then it will examine the evolution of this forgery production by focusing on the divergent editions of *Archidoxis* from 1569 to 1572.

Keywords

Paracelsus – forgery – alchemy – publication – edition – library – *De natura rerum* – *Archidoxis*

Introduction

A strange treatise, entitled *De natura rerum*, or *On the Nature of Things*, was one of the most popular writings published in the name of Paracelsus.[1] First released in seven books (Basel, 1572) and in nine books (Strasbourg, 1584),

1 For the text, see the following editions: Paracelsus, *Bücher und Schrifften*, ed. Johannes Huser, 11 vols. (Basel, 1589–1591), 6: 255–362; Paracelsus, *Sämtliche Werke*, 1. Abt., ed. Karl Sudhoff, 14 vols. (Berlin-Munich, 1922–1933) (hereafter: Sudhoff, 1/1, 1/2, etc.), 1/11: 307–403. My warmest thanks go to Clare Hirai, Didier Kahn, Amadeo Murase and Pamela H. Smith for their kind help and comments in the preparation of the present article.

 | DOI:10.1163/9789004503380_006

it was translated into Latin in seven books (Basel, 1573) and in nine books (Frankfurt, 1605), thus securing a broad readership.[2]

This treatise is remarkable for its highly operative approach to the natural world, even preceding the works of Francis Bacon (1561–1626). The plethora of ideas expounded in the text occupy an elusive realm between magic and science. Some of them are inspired by the tradition of late-medieval alchemy stemming from the corpus attributed to Raymond Lull (c. 1232–c. 1315).[3] Others have more affinity with the genuine teachings of Paracelsus such as the three principles of nature (Salt, Sulphur and Mercury), and nature's internal craftsman called "archeus." Some of the ideas contained in the text attracted the keen attention of many intellectuals of the time; others were severely criticized and subject to religious censorship.[4]

As for the critical reactions of learned physicians, a typical example can be found in Thomas Feyens or Fienus (1567–1631) of Antwerp. Fienus wrote a monograph devoted to the power of imagination in the framework of embryology.[5] In this treatise, entitled *On the Forces of the Imagination* (Louvain, 1608), he criticized Paracelsus and other "Platonic" authors such as Marsilio Ficino (1433–1499) and Pietro Pomponazzi (1462–1525) along with the idea repeated and developed in the first and ninth books of *De natura rerum*. According to this idea, Fienus related, the imagination of pregnant women has a strong and extraordinary power, acting at a distance and imprinting particular images on the fetus in their womb.

2 On the Latin edition of seven books published by Pietro Perna, see Karl Sudhoff, *Bibliographia Paracelsica: Besprechung der unter Hohenheims Namen 1527–1893 erschienenen Druckschriften* (Berlin, 1894), Nr. 145: 243–244. On its translator Georg Forberger (ca. 1543–1604?), see Rudolph Zaunick, *Der sächsische Paracelsist Georg Forberger* (Wiesbaden, 1977); Wilhelm Kühlmann and Joachim Telle, *Corpus Paracelsisticum: Dokumente frühneuzeitlicher Naturphilosophie in Deutschland*, 3 vols. (Tübingen, 2001–2004 and Berlin, 2013) (hereafter: CP 1, CP 2, CP 3), CP 2: 237–239. The Latin edition of nine books was published in Paracelsus, *Operum medico-chimicorum* (Frankfurt, 1605), 6: 198–271. Cf. Sudhoff, *Bibliographia*, Nr. 269: 466–467.

3 On Lullian alchemy, see Michela Pereira, *The Alchemical Corpus Attributed to Raymond Lull* (London, 1989).

4 See Ernst Darmstaedter, "Paracelsus, *De natura rerum*: Eine kritische Studie," *Janus*, 37 (1933), 1–18, 48–62, 109–115; Sepp Domandl, "Hohenheims *De natura rerum*: Eine Einführung in die geistige Welt des Paracelsus," in *Kunst und Wissenschaft um Paracelsus* (Vienna, 1984), 61–88; Hiro Hirai, *Le concept de semence dans les théories de la matière à la Renaissance: de Marsile Ficin à Pierre Gassendi* (Turnhout, 2005), 210–213.

5 See Thomas Fienus, *De viribus imaginationis tractatus* (Louvain, 1608), 25, 41. For the corresponding parts of *De natura rerum*, see Huser, 6: 260, 333; Sudhoff, I/11: 314, 377; see also Hiro Hirai, "Imagination, Maternal Desire and Embryology in Thomas Fienus," in Gideon Manning and Cynthia Klestinec, eds., *Professors, Physicians and Practices in the History of Medicine* (Dordrecht, 2017), 211–225.

In its final form, *De natura rerum* is composed of nine books. Each of the first seven books addresses the following themes: 1) generation; 2) growth; 3) preservation; 4) life or vivification; 5) death or mortification; 6) resuscitation; and 7) transmutation. These topics are heavily influenced by the tradition of medieval alchemy and bear a coherent look as a whole. Added in the second edition of 1584, are the last two books, which discuss: 8) separation; and 9) signatures.

In the body of the work, the author of *De natura rerum* refers to Paracelsus' famous writings such as *Archidoxis* and *Chirurgia magna*. Some of the ideas here developed hint at his familiarity with other works by Paracelsus, such as *Astronomia magna* and *De mineralibus*, among genuine writings and the elusive *De imaginibus*. For instance, the first book of *De natura rerum* on the generation of natural things addresses, among other ideas, the production of artificial life or "homunculus." This is a topic developed in *De imaginibus*, albeit in a quite different dimension.[6] The author's use of the notion of *archeus* also suggests a good knowledge of treatises such as *De mineralibus*.

The style of *De natura rerum* is noticeably "systematic" and "didactic" compared to the genuine writings of Paracelsus. It is therefore reasonable to think that, even if the original core fragments were composed by Paracelsus himself, it was reworked by its editor or another person before its publication. Such was exactly the case with another treatise, entitled *De genealogia mineralium ex Paracelso* (Frankfurt, 1581).[7] Editing and publishing a series of works by Paracelsus in Latin translation, Flemish physician Gerard Dorn (fl. 1566–1584) reworked the genuine treatise *De mineralibus* so as to attain a more coherent and systematic style under the title of *On the Genealogy of Minerals*.[8] However, Dorn notated the work as "from Paracelsus" (*ex Paracelso*) instead of "by Paracelsus" (*Paracelso*) in the title.

6 On the idea of homunculus, see William R. Newman, "The Homunculus and His Forebears: Wonders of Art and Nature," in Anthony Grafton and Nancy Siraisi, eds., *Natural Particulars: Nature and the Disciplines in Renaissance Europe* (Cambridge, MA, 1999), 321–345; idem, *Promethean Ambitions: Alchemy and the Quest to Perfect Nature* (Chicago, IL, 2004), 195–208; Lawrence M. Principe, *The Secrets of Alchemy* (Chicago, IL, 2013), 131–132; Amadeo Murase, "The Homunculus and the Paracelsian *Liber de imaginibus*," *Ambix*, 67 (2020), 47–61.

7 This reworked treatise was included in Gerard Dorn, *Congeries Paracelsicae chemiae de transmutationibus metallorum* (Frankfurt: Andreas Wechel, 1581), 220–277. Cf. Sudhoff, *Bibliographia*, Nr. 186: 324–326. On the publishing activity of Wechel and his collaborators, see Robert J.W. Evans, *The Wechel Presses: Humanism and Calvinism in Central Europe 1572–1627* (Oxford, 1975).

8 Dorn renamed this text as *De mineralium œconomia* and published it with his commentary in his *In Theophrasti Paracelsi Auroram philosophorum…* (Frankfurt: [Christoph Rab], 1583).

In this article my aim is to place the genesis of *De natura rerum* in the context of publication history. It will show the extraordinary flowering of "forgeries" ascribed to the name of Paracelsus. These texts (some were totally invented, others were reworked from genuine fragments) are often endowed with a strong flavor of alchemy and occult philosophy, sometimes straying from the authentic teachings of Paracelsus. Indeed, *De natura rerum* was first presented to readers when this movement of forgery production reached its climax, especially with the multiple editions of Paracelsus' genuine treatise *Archidoxis*.

In what follows, first in section 1, I will try to reconstruct a "library" by surveying and enumerating the works ascribed to Paracelsus which could serve as crucial instruments for the "author/reworker/editor" of *De natura rerum*. Then in section 2, I will analyze the evolution of this movement of forgery production, by focusing on the divergent editions of *Archidoxis* from 1569 to 1572. In doing so, I will shed new light on the genesis of *De natura rerum* in publication history.

1 Paracelsus' Works Published Before *De natura rerum*

Which writings of Paracelsus were available for the "author/reworker/editor" of *De natura rerum*, if he was not Paracelsus himself and if he did not have easy access to the manuscripts of the works attributed to this name? Let us enumerate and examine in chronological order of publication the major works, which address the related themes discussed in *De natura rerum* or which show some connections and resemblance with them.

The treatise entitled *Labyrinthus medicorum errantium* (Nuremberg: Valentin Neuber, 1553) was one of the most important genuine works of Paracelsus, dating from his mature period around 1537/38.[9] It is significant that it was published very early, even predating the beginning of the "Paracelsian revival" movement of the 1560s, animated by the early and fervent German followers of the Swiss physician such as Adam von Bodenstein (1528–1577) and Michael Toxites (1514–1581).[10]

Sudhoff, *Bibliographia*, Nr. 197: 342–344; CP 2: 906–915. On Dorn, see Didier Kahn, "Les débuts de Gérard Dorn d'après le manuscrit autographe de sa *Clavis totius Philosophiae Chymisticae* (1565)," in Joachim Telle, ed., *Analecta Paracelsica: Studien zum Nachleben Theophrast von Hohenheims im deutschen Kulturgebiet der frühen Neuzeit* (Stuttgart, 1994), 59–126.

9 Sudhoff, *Bibliographia*, Nr. 30: 45–47. Cf. Hirai, *Le concept de semence*, 204–205.

10 On the Paracelsian revival movement, see Lynn Thorndike, *A History of Magic and Experimental Science* (New York, 1941), 5: 617–651; Didier Kahn, *Alchimie et paracelsisme en France à la fin de la Renaissance (1567–1625)* (Geneva, 2007), 100–121. On Bodenstein,

The next treatise, widely known as *De vita longa* (Basel: s. n., 1560), was one of the most popular works published in the early phase of the Paracelsian revival movement.[11] It was edited in Latin by Bodenstein, first in the form of four books and then, in 1562, in five books, from the manuscript of Paracelsus' former secretary, Johannes Oporinus (1507–1568). This treatise was dated 1527, which places its composition during Paracelsus' short and notorious career at the University of Basel (spring 1527–1528). The second edition was released by the Basel publisher Pietro Perna (1519–1582).[12]

The treatise entitled *Opus paramirum* (Mülhausen: Peter Schmidt, 1562) was edited in the form of two books by Bodenstein and is generally regarded as the principal work of Paracelsus' middle career, composed around 1531. Bodenstein also edited another popular work, *Von ersten dreyen Principiis* (Basel: s. n., 1563), on the famous theory of the "three principles" (*tria prima*) of nature: Salt, Sulphur and Mercury.[13]

The treatise entitled *Philosophia ad Athenienses* (Köln: Arnold Birckmann's heirs, 1564) was the earliest major printed forgery. Although it contains some striking ideas such as uncreated prime matter, contradictory to Paracelsus' genuine teachings, it exerted a considerable influence on the cosmogony and cosmology of the Paracelsians in the late sixteenth century and beyond.[14]

The genuine manifesto of Paracelsus' new medicine, *Paragranum* (Frankfurt: Christian Egenolff's heirs, 1565), was edited by Bodenstein. Paracelsus developed his new medicine by writing this treatise around 1530 after his bitter experience, failure and public humiliation at the University of Basel. In this work he established the four pillars of medicine: philosophy, astronomy, alchemy and ethics. Thus, his new chymical medicine was a terrestrial astronomy endowed with philosophy and ethics.[15]

---

see CP 1: 104–110. On Toxites, see CP 2: 41–66; Wilhelm Kühlmann, "Humanistische Verskunst im Dienste des Paracelsismus: Zu einem programmatischen Lehrgedicht des Michael Toxites (1514–1581)," *Etudes germaniques*, 50 (1995), 509–526.

11 Sudhoff, *Bibliographia*, Nr. 39: 60–61 and Nr. 46: 70–72; CP 1: 105–146, 203–265.

12 On the date 1527, see Sudhoff, *Bibliographia*, 392. On Oporinus, see Charles D. Gunnoe, Jr., *Thomas Erastus and the Palatinate: A Renaissance Physician in the Second Reformation* (Leiden, 2011), 270–273. On Perna, see Antonio Rotondò, *Studi e ricerche di storia ereticale italiana del Cinquecento* (Turin, 1974), 273–391; CP 2: 729–733, 745–747.

13 Sudhoff, *Bibliographia*, Nr. 47: 72–74; CP 1: 266–285; Hirai, *Le concept de semence*, 199–201. On *Von ersten dreyen Principiis*, see Sudhoff, *Bibliographia*, Nr. 56: 84–85; CP 1: 324–338.

14 Sudhoff, *Bibliographia*, Nr. 65: 101–104; Walter Pagel, "The Prime Matter of Paracelsus," *Ambix*, 9 (1961), 117–135; Hirai, *Le concept de semence*, 213–214; Gunnoe, *Thomas Erastus*, 309–314.

15 Sudhoff, *Bibliographia*, Nr. 66: 104–105; CP 1: 372–386; Hirai, *Le concept de semence*, 197–199.

The next two treatises, *De causa et origine morborum* and *De morbis invisibilibus* (Köln: Arnold Birckmann's heirs, 1565), were to be intimately connected to the major work of Paracelsus, *Opus paramirum*. They are all considered to have been composed around the same period. Birckmann's press became more and more important for the Paracelsian revival movement with the publication of these genuine writings.[16] In the following year, they went further, publishing *Das Buch Meteororum* and *De matrice* (1566). Closely related to *Opus paramirum* in terms of origin and contents, both of these treatises are genuine works of Paracelsus. They are crucial to understand his cosmogony based on a particular interpretation of the Biblical creation story of *Genesis*.[17]

The second major forgery in this list, *Liber vexationum* (Basel: Samuel Apiarius, 1567), also called *De vexatione*, was edited by Bodenstein. The word "vexation" meant a chymical process of "fixation," which suggests the affinity of this work with the medieval tradition of transmutational alchemy. Thus, it was to be suggestively renamed as *Coelum philosophorum* in the monumental edition of Paracelsus' works edited by Johannes Huser (ca. 1545–1600/1) in 1590 and became more popular than ever. Although Huser himself regarded it as genuine, the modern editor Karl Sudhoff (1853–1938) rejected its authenticity by placing it among "spuria" or doubtful writings. Under the fascinating title of *Pyrophilia vexationumque liber*, the Latin translation of this treatise was executed by Gerard Dorn for Pietro Perna's press in Basel in the following year, 1568.[18]

16 Sudhoff, *Bibliographia*, Nr. 70: 111–113. On *Opus paramirum* and related treatises along with a criticism on Sudhoff's editorial choice, see Udo Benzenhöfer, *Paracelsus* (Hamburg, 2002), 88–90. On the Birckmann press, see Leo Norpoth, "Kölner Paracelsismus in der 2. Hälfte des 16. Jahrhunderts," *Jahrbuch des Kölnischen Geschichtsvereins*, 27 (1953), 133–144; CP 1: 656–660; Diethelm Eikermann, "Köln im Jahr 1596 und Leonhard Thurneysser zum Thurn (1531–1596)," *Jahrbuch des Kölnischen Geschichtsvereins*, 81 (2011–2012), 85–126, on 103–107.

17 Sudhoff, *Bibliographia*, Nr. 73: 115; Hirai, *Le concept de semence*, 201–204, 208–210. On the chymical interpretation of Genesis, see Michael T. Walton, "Genesis and Chemistry in the Sixteenth Century," in Allen G. Debus and Michael T. Walton, eds., *Reading the Book of Nature: The Other Side of the Scientific Revolution* (Kirksville, MO, 1998), 1–14; Didier Kahn, "L'interprétation alchimique de la Genèse chez Joseph Du Chesne dans le contexte de ses doctrines alchimiques et cosmologiques," in Barbara Mahlmann-Bauer, ed., *Scientiae et artes: Die Vermittlung alten und neuen Wissens in Literatur, Kunst und Musik* (Wiesbaden, 2004), 641–692; Hirai, *Le concept de semence*, passim.

18 Sudhoff, *Bibliographia*, Nr. 90: 143–144. Cf. CP 3: 329, 708. For the text of *Liber vexationum*, see Huser, 6: 375–395; Sudhoff, I/14: 405–420. On *Coelum philosophorum*, see also Sudhoff, *Bibliographia*, Nr. 221: 389–395, on 393. On Huser, see Joachim Telle, "Johann Huser in seinen Briefen: Zum schlesischen Paracelsismus im 16. Jahrhundert," in idem, ed., *Parerga Paracelsica: Paracelsus in Vergangenheit und Gegenwart* (Stuttgart, 1991), 159–248. On

One of the most famous treatises of a chymical nature by Paracelsus, *Archidoxae libri X* (Cracow: Matthias Wirzbieta, 1569) was first edited in Latin by the Silesian physician Adam Schröter (ca. 1525–ca. 1572). Its German edition immediately followed from Perna's press in Basel under the title of *10 Bücher Archidoxorum* (1570).[19] I will return to these texts more fully in the next section.

In 1570 Toxites in his turn edited a collection of Paracelsus' writings, entitled *Several Treatises* (*Ettliche Tractatus*), including two works *Von den natürlichen Dingen* and *De mineralibus* (Strasbourg: Christian Müller's heirs, 1570). Reprinted in 1582, 1587 and 1597, this collection provided topics related to natural bodies such as plants, minerals and metals.[20] This could serve as a fundamental instrument to the "author/reworker/editor" of *De natura rerum*. Each of the seventeen chapters of *Von den natürlichen Dingen* is devoted to one particular medicinal material such as turpentine, coral and magnet. By contrast, placing the formation of mineral substances in the context of the world creation, *De mineralibus* is highly important to an understanding of Paracelsus' cosmology and cosmogony along with the famous theory of the three principles (Salt, Sulphur and Mercury). It also features the idea of *archeus*, which was a favorite idea of the "author/reworker/editor" of *De natura rerum*.[21]

The treatise entitled *Astronomia magna* (Frankfurt: Sigismund Feyrabend, 1571) was the major product of Paracelsus' mature period.[22] This massive treatise is a crucial piece for the ninth book of *De natura rerum*, which was added in its second edition of 1584. Advanced in this work was the highly unusual understanding of "chiromancy," not as a divinatory reading of the human palm but as an overall science of lines, wrinkles and figures.[23]

Dorn's Latin translation, see Sudhoff, *Bibliographia*, Nr. 101: 160–162. On fixation in chymistry, see Principe, *Secrets of Alchemy*, 152.

19 For Schröter's Latin edition, see Sudhoff, *Bibliographia*, Nr. 108: 170–174; CP 3: 95–174. For Perna's German edition, see Sudhoff, *Bibliographia*, Nr. 116: 183–186. On the contents of *Archidoxis*, see the next section below.

20 Sudhoff, *Bibliographia*, Nr. 120: 196–198; Nr. 192: 334–335; Nr. 214: 365; Nr. 241: 423. On *De mineralibus*, see Hirai, *Le concept de semence*, 183–195.

21 For the notion of *archeus*, see Walter Pagel, *Paracelsus: An Introduction to Philosophical Medicine in the Era of the Renaissance* (Basel, 1958), 104–112; Hirai, *Le concept de semence*, 190, 200, 204–205, 212.

22 Sudhoff, *Bibliographia*, Nr. 131: 219–221; CP 2: 208–231.

23 On Paracelsus's chiromancy, see Karl Möseneder, *Paracelsus und die Bilder: Über Glauben, Magie und Astrologie im Reformationszeitalter* (Berlin, 2009), 122–130, 176–179. On the medieval tradition of chiromancy, see Lynn Thorndike, "Chiromancy in Medieval Manuscripts," *Speculum*, 40 (1965), 674–706; Charles Burnett, "The Principal Latin Texts on Chiromancy Extant in the Middle Ages," in idem, *Magic and Divination in the Middle Ages* (Aldershot, 1991), 1–29; Jean-Patrice Boudet, *Entre science et nigromance: astrologie, divination et magie dans l'Occident medieval* (*XII*[e]*–XV*[e] *siècle*) (Paris, 2006), 336–340.

After the publication of all these writings, Bodenstein produced a new collection of Paracelsus' works. Humanistically and symbolically entitled *Metamorphosis* (Basel: Samuel Apiarius, 1572), it contains eleven books as its first part, followed by two more books as its second part (fig. 4.1).

The first seven books of *Metamorphosis* are grouped under the title of *Von natürlichen Dingen*. Confusingly, this is not the same treatise as the one Toxites published under a similar title two years before in 1570. That was most probably why its title was changed to *De natura rerum* for its second edition in 1584, following the Latin translation of 1573. Strangely enough, the collection has no eighth book but contains *De cementis* as the ninth book, *De gradationibus* as the tenth, and *De projectionibus* as the eleventh (although the latter was said to be unwritten) so as to finish the first part of the collection.[24] Both Huser and Sudhoff regarded *De cementis* and *De gradationibus* to be genuine.

As for the second part, it includes *Manual vom Stein der Weisen* and *Alchimia vom einfachen Fewr*. Later, in the 1605 posthumous publication, Huser retained the first piece as genuine but doubted the authenticity of the second, placing it among a series of spurious works in "Appendix" to his collection of Paracelsus' surgical works.[25] Seen in its original setting, along with its suggestive title of *Metamorphosis*, Bodenstein's editorial and marketing strategy can be grasped without difficulty. There is no doubt that, with the similarity of the title, he was invoking the famous *Metamorphoses* by the Roman poet Ovid, to connect the notion of transmutational alchemy in a humanistic mind just as Giovanni Aurelio Augurello (1441–1524) tried to establish in his famous *Chrysopoeia* (Venice, 1515).[26] As explained earlier, the theme for the seventh book of *De natura rerum* was "transmutation"; what we find are well-known and ordinary chymical operations such as calcination, sublimation, solution, putrefaction, distillation and coagulation. However, most readers would expect the further description of higher and advanced operations for the real transmutation of

24 Sudhoff, *Bibliographia*, Nr. 137: 229–231; CP 1: 481–484; CP 3: 266–278. For the text of *De cementis* and *De gradationibus*, see respectively Huser, 6: 402–410, 411–417; Sudhoff, 1/3: 65–79, 81–88. They were first published with *Archidoxis* in Latin (1570) and in German (1571), both from Perna's press. See my discussion in the next section.

25 For the text of *Manual*, see Huser, 6: 421–436; Sudhoff, 1/14: 421–432. Cf. CP 3: 708. For the text of *Alchimia*, see Huser's posthumous edition of Paracelsus, *Chirurgische Bücher und Schriften* (Strasbourg: Lazarus Zetzner, 1605), Appendix: 71–77. Cf. Sudhoff, *Bibliographia*, Nr. 267: 462–464.

26 Zweder R.W.M. Van Martels, "Augurello's *Chrysopoeia* (1515): A Turning Point in the Literary Tradition of Alchemical Texts," *Early Science and Medicine*, 5 (2000), 178–195; Hirai, *Le concept de semence*, 62–66; Principe, *Secrets of Alchemy*, 180.

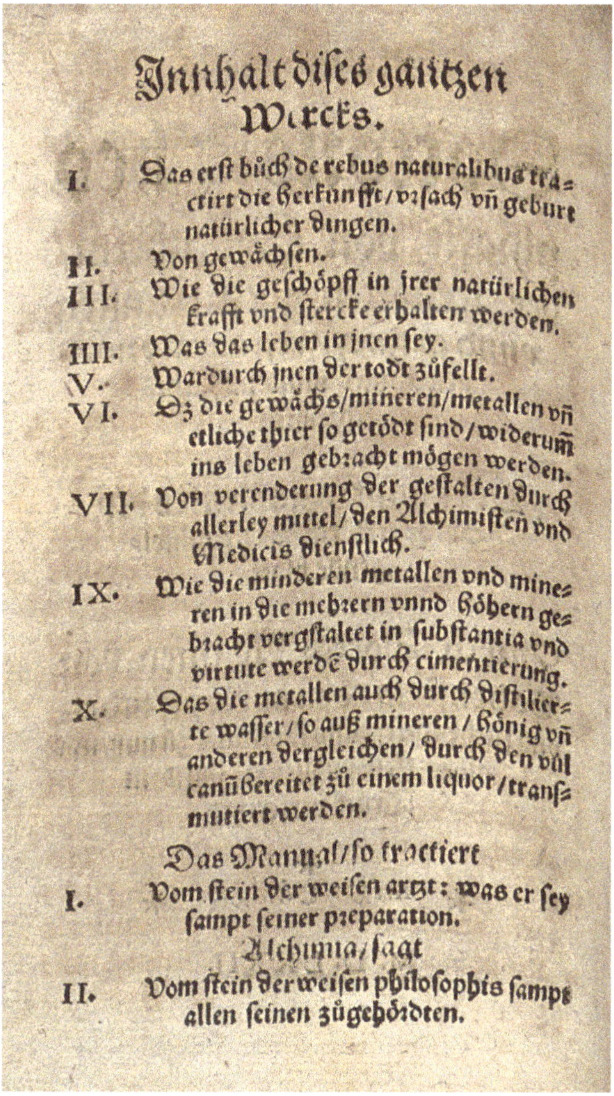
Innhalt dises gantzen Wercks.

I. Das erst büch de rebus naturalibus tractirt die herkunfft/vrsach vñ geburt natürlicher dingen.
II. Von gewächsen.
III. Wie die geschöpff in irer natürlichen krafft vnd stercke erhalten werden.
IIII. Was das leben in inen sey.
V. Wardurch inen der todt züfellt.
VI. Oz die gewächs/mineren/metallen vñ etliche thier so getödt sind/widerumb ins leben gebracht mögen werden.
VII. Von verenderung der gestalten durch allerley mittel/den Alchimisten vnd Medicis dienstlich.
IX. Wie die minderen metallen vnd mineren in die mehrern vnnd höhern gebracht vergstaltet in substantia vnd virtute werdē durch cimentierung.
X. Das die metallen auch durch distillierte wasser/so auß mineren/könig vñ anderen dergleichen/durch den vül canū bereitet zü einem liquor/transmutiert werden.

Das Manual/so tractiert
I. Vom stein der weisen artzt: was er sey sampt seiner preparation.
Alchymia/sagt
II. Vom stein der weisen philosophis sampt allen seinen zügehördten.

FIGURE 4.1 Paracelsus, *Metamorphosis*, ed. Adam von Bodenstein (1572), table of contents. urn:nbn:de:bvb:12-bsb00015387-9 MUNICH, BAYERISCHE STAATSBIBLIOTHEK: Alch. 206 e#Beibd.2

metals. That was what Bodenstein tried to offer by gathering the pieces on "cementation," "gradation" and "projection," along with the "manual" for the philosopher's stone.

To make the situation more complex, this collection contains a dedication letter by Paracelsus to a certain Johann Winckelsteiner in Freiburg. The letter mentions a treatise entitled *De natura rerum* divided into "nine parts" (*neun theil*). It was signed in Villach, Southern Austria, in 1537. However, Paracelsus is not known to have been there in that year. More interestingly, the archival records of Freiburg im Breisgau do not support the existence of this mysterious Winckelsteiner, suggesting that the letter is probably a forgery.[27]

Finally, the second edition of *De natura rerum* (Strasbourg: Bernhard Jobin, 1584) appeared with the addition of the eighth book on "separation" and the ninth book on "natural signs." This edition has no other additional material. Although its editor, Lucas Bathodius (fl. 1584–1597?), is relatively less known, his editorial intention can be easily discerned.[28] He avoided incorporating doubtful materials. This was in diametric opposition to the approach taken by Bodenstein who tried to present this work as the "theoretical preparation" of transmutational alchemy.

The conception of "natural signs" expounded in the ninth book of *De natura rerum* largely differs from the traditional line developed, for example, in the famous *Phytognomonica* (Naples, 1588) by Giambattista della Porta (ca. 1535–1615). Della Porta built his treatise on the basis of ideas advanced by ancient classical authors such as Dioscorides and Pliny.[29] By contrast, the main line of argument on natural signs in the ninth book of *De natura rerum* could be elaborated thanks to the preceding publication of *Astronomia magna*.

2 *Archidoxis* and the Flowering of Forgeries

Now let us return to *Archidoxis*, as seen in its publication history. Indeed, it is necessary to take a closer look at its evolution so as to contextualize the genesis of *De natura rerum* and its first edition in Bodenstein's collection *Metamorphosis* within the flowering of forgeries ascribed to the name of Paracelsus.

Before going into the detail of its publication history, let us take a quick look at the title and structure of *Archidoxis* according to the major collected works of Paracelsus. First, in the monumental edition of ten quarto volumes, its editor Johannes Huser included *Archidoxis* at the head of the sixth volume

27 *Metamorphosis* (Basel, 1572), sig. B[i] r-biii r. See CP 3 : 269, 276, 278.

28 Sudhoff, *Bibliographia*, Nr. 199: 345–346; CP 3: 757–772. On Bathodius, see also Telle, "Huser in seinen Briefen," 200–201.

29 Giambattista della Porta, *Phytognomonica* (Naples, 1588). See Hiro Hirai, "Images, Talismans and Medicine in Gaffarel's *Curiositez inouyes*," in idem, ed., *Jacques Gaffarel between Magic and Science* (Rome, 2014), 73–84; Hiro Hirai and Yohei Kikuchihara, "*Signatura Rerum* Theory," in Marco Sgarbi, ed., *Encyclopedia of Renaissance Philosophy* (Dordrecht, 2015).

(1590) for writings dealing with the "spagyric preparation of natural things."[30] He entitled the treatise *Ten Books of Archidoxis on the Mysteries of Nature* (*Decem libri Archidoxis de mysteriis naturae*). However, immediately after his introductory warning to readers, Huser provided the table of contents with the following title: *The First Part of Archidoxis, Nine Books on the Mysteries of Nature* (*Archidoxis pars prima, novem libri de mysteriis naturae*). This confusion between ten and nine books reflected the complex publication history of the treatise, which will be examined in the present section. In any event, the table tells us that the body of *Archidoxis* consists of nine books:

Book 1, *De mysteriis microcosmi*
Book 2, *De mysteriis elementorum*
Book 3, *De mysteriis quintae essentiae*
Book 4, *De mysteriis arcani*
Book 5, *De mysteriis extractionum*
Book 6, *De mysteriis specificorum*
Book 7, *De mysteriis elixir*
Book 8, *De mysteriis externis*
Book 9, *De praxi, sed non est impressus*

According to the table, the ninth book is said to be "unprinted" (*non est impressus*). Thus, the treatise actually has only eight books. To make the situation more complex, Huser printed these components and titles in the following order, leaving the second book as missing:

Book 1, *De prologo et microcosmo*
[Book 2, left blank with some explanation]
Book 3, *De separationibus elementorum*
Book 4, *De quinta essentia*
Book 5, *De arcanis*
Book 6, *De magisteriis*
Book 7, *De specificis*
Book 8, *De elixiriis*
Book 9, *De extrinsecis*[31]

---

30 For the text, see Huser, 6: 1–98. On the chymical contents of *Archidoxis*, see Ernst Darmstaedter, *Arznei und Alchemie: Paracelsusstudien* (Leipzig, 1931); Reijer Hooykaas, "Die Elementenlehre des Paracelsus," *Janus*, 39 (1935), 175–187; T.P. Sherlock, "The Chemical Work of Paracelsus," *Ambix*, 3 (1948), 33–63, on 43–62.

31 Huser, 6: sig. [a 6 iiij] v.

In his edition of Paracelsus' works, Sudhoff included *Archidoxis* in the third volume for the writings composed around 1525/26 according to his reckoning of dates of composition.[32] Following Huser, he called the treatise *Decem libri Archidoxis de mysteriis naturae*, also adopting his forerunner's choice for the ordering and titles of its components. Thus, Sudhoff's edition, which became standard for modern studies on Paracelsus, consolidated the view in which the second book of *Archidoxis* is considered to be "missing," while the tenth book is left as "unprinted" or missing. This aligned perfectly with the manuscript evidence observed by Huser.

Having established this traditional structure of the treatise, let us now penetrate into the complex story of its publication history. As has been observed in the previous section, the first Latin edition of *Archidoxis* was published in Cracow (1569). Its editor Adam Schröter claimed to have translated it from German. The title *Archidoxae libri X* suggests that it should contain ten books. But Schröter argued that only eight books were extant, and he placed them in the following order: 1) microcosm; 2) elements; 3) quintessence; 4) arcana; 5) magisters; 6) specifics; 7) elixirs; and 8) extrinsics. To remedy the lacuna, Schröter followed his base manuscript, which had been directly copied from Paracelsus' autograph, and added two short treatises: *De renovatione et restauratione* and *De vita longa*.

This is the first appearance of both treatises.[33] The *De vita longa* presented here is short and different from the treatise by the same title in the five books edited by Bodenstein in the early phase of the Paracelsian revival movement. To avoid potential confusion, Sudhoff called the shorter one *Vom langen Leben* in his edition of Paracelsus' works. In any event, these two treatises, *De renovatione* and *De vita longa*, were the early productions of Paracelsus and very closely related to each other. Since *Archidoxis* was also a composition of the Swiss physician's early career, their style and contents bear a strong resemblance. Thus, Schröter's editorial choice could be accepted by readers without major contestation. Note that the main scope of *Archidoxis* is more chymical

32 For the text, see Sudhoff, 1/3: 86–200.

33 For the text of *De renovatione*, see Huser, 6: 100–114; Sudhoff, 1/3: 201–220. For the text of *De vita longa* in one book, see Huser, 6: 115–136; Sudhoff, 1/3: 221–245. On both treatises, see also Didier Kahn, "Quintessence and the Prolongation of Life in the Works of Paracelsus," *Micrologus*, 26 (2018), 183–225. Jan Baptista Van Helmont (1579–1644) established his theory of "alkahest" upon the idea of "sal circulatum" mentioned in *De renovatione*. Cf. Huser, 6: 113; Sudhoff, 1/3: 218; see also Bernard Joly, "L'alkahest, dissolvant universel ou quand la théorie rend pensable une pratique impossible," *Revue d'histoire des sciences* 49 (1996), 305–344, on 318; Paulo A. Porto, "*Summus atque felicissimus salium*: The Medical Relevance of the *Liquor alkahest*," *Bulletin of the History of Medicine* 76 (2002), 1–29, on 11.

and pharmacological than *De renovatione* and *De vita longa* which concern rather regimen and longevity.

In the next year, 1570, Perna's press in Basel published the first German edition, entitled *Archidoxorum* [...] *X Bücher*.[34] Although its title announces ten books, it contains only nine. Perna followed Schröter's choice to include *De renovatione* and *De vita longa*. But he inserted the former in the place of the first text on microcosm, which is in turn placed at the head of the whole treatise as a preface: [0]) microcosm; 1) *De renovatione*; 2) elements; 3) quintessences; 4) arcana; 5) magisters; 6) specifics; 7) elixirs; 8) extrinsics; and 9) *De vita longa*. Reflecting this ordering, Perna spelled out "mysterio microcosmi" in Latin on the title-page and warned readers to take it as the first of ten books: *Archidoxorum* [...] *X Bücher Mysterio Microcosmi für das erste Buch gerechnet*. His editorial choice is striking.

With the first Latin edition and Perna's German edition, perhaps also boosted by the interesting topic of longevity offered by *De vita longa*, the popularity of *Archidoxis* in the book market reached an unimaginable height. In the same year of 1570, two editions were published in Basel, two others in Munich, one in Strasbourg and the other in Köln. As one of these six German editions was executed in the same format by Perna himself, let us set those two editions aside and examine the evolution of the other four.[35]

One German edition was edited by Toxites and printed by Theodosius Rihel (fl. 1555–1608) of Strasbourg.[36] Apparently, this edition did not follow Perna's choice of placing the text on microcosm as a preface, but reverted back to Schröter's ordering as follows: 1) microcosm; 2) elements; 3) quintessences; 4) arcana; 5) magisters or extractions; 6) specifics; 7) elixirs and 8) externals; 9) *De renovatione*; and 10) *De vita longa*. More interestingly in terms of marketing, Toxites presented *Archidoxa 10 Bücher* as the first part of his collection. He thus added two more items, *De tinctura physicorum* as its second part and *De occulta philosophia* as its third. This is the first appearance of both of these treatises. Although Huser took them to be genuine, Sudhoff placed them among forgeries.[37] Thus, this edition by Toxites provided an interesting twist to the evolution of the *Archidoxis* publication.

34 Sudhoff, *Bibliographia*, Nr. 116: 183–186.

35 For the second German edition of Perna, see Sudhoff, *Bibliographia*, Nr. 117: 186–187.

36 Sudhoff, *Bibliographia*, Nr. 118: 187–190; see also CP 2: 162, 415, 490. On Rihel, see *Allgemeine Deutsche Biographie*, 28 (Leipzig, 1889), 429–430.

37 On the text of *De tinctura physicorum* and *De occulta philosophia*, see respectively Huser, 6: 363–374; idem, 9: 329–368; and Sudhoff, I/14: 391–399, 513–542. On the reception of *De tincture physicorum*, see CP 2: 319–320, 1009–1010; CP 3: 276, 714. At the conference on Pseudo-Paracelsus at Villa Vigoni, Italy (25–28 July 2016), Amadeo Murase and Kathrin

Another German edition, prepared by Johannes Albertus Wimpinaeus or Johann Albrecht (ca. 1540–after 1570) of Wimpfen, was published from Adam Berg's press in Munich.[38] Albrecht entitled it *Archidoxorum libri IX*, naming the ninth book as "praxis" but "unwritten" (*non est scriptus*). Its structure is as follows: 1) microcosm; 2) elements; 3) quintessences; 4) arcana; 5) magisters or extractions; 6) specifics; 7) elixirs; 8) externals; and 9) praxis [missing]. Thus, this edition has only eight books. But most probably adopting the editorial choice of his forerunner Toxites, Albrecht presented this set as the first part of his collection, followed by the second part with four other treatises: *Praeparationes*; *De tinctura physicorum*; *De renovatione*; and *De vita longa*. The first item, *Praeparationes*, is a collection of recipes and had already been edited by Bodenstein and published in Strasbourg (1569).[39] Both Huser and Sudhoff classified it as an early work related to *De renovatione* and *De vita longa*. By contrast, *De tinctura physicorum* is a clear forgery of alchemical nature.

Moreover, the press of Adam Berg issued another German edition under the striking title of *Archidoxa 12 Bücher* in the same year. Its structure is as follows: 1) microcosm; 2) [missing]; 3) elements; 4) quintessences; 5) arcana; 6) magisters or extractions; 7) specifics; 8) elixirs; and 9) externals. It also contains four other supplementary materials, which are the same as those of his former edition.[40] Thus the unusual number of "twelve" in the title resulted from the combination of eight extant books and four supplements.

Lastly, *Archidoxorum libri IX* in German was issued from the active and important press of Arnold Birckmann's heirs in Köln. Significantly enough, it bears the suggestive overarching title in Latin, *De mysteriis naturae*, that is, *On the Mysteries of Nature*.[41] As the first part of the collection, it includes eight books: 1) microcosm; 2) [missing]; 3) elements; 4) quintessences; 5) arcana; 6) magisters; 7) specifics; 8) elixirs; and 9) extrinsics. It is followed by two familiar treatises, *De renovatione* and *De vita longa*, as the second part of the collection. Further supplementary materials of a genuine nature are added to this set of texts: *De mineralibus* and short texts on salts, vitriol, arsenic and sulphur

---

Pfister delivered their analysis of this text. As for *De occulta philosophia*, see CP 2: 490–491, 1008; CP 3: 52–56, 61–63, 93–94.

38 Sudhoff, *Bibliographia*, Nr. 119: 190–196. On Albrecht, see Allen G. Debus, *The Chemical Philosophy* (New York, 1977), 135–139; CP 2: 964–1017.

39 Sudhoff, *Bibliographia*, Nr. 104: 164–165; CP 1: 445–451. The text of *Praeparationes* immediately follows *Archidoxis*, *De renovatione* and *De vita longa* in Huser, 6: 212–252; Sudhoff, 1/3: 309–359. For a pharmacological study of its contents, see Wolfgang Schneider, "Über den *Liber praeparationum* des Paracelsus," in idem, *Mein Umgang mit Paracelsus und Paracelsisten* (Frankfurt, 1982), 99–106.

40 Sudhoff, *Bibliographia*, Nr. 129: 215–217.

41 Sudhoff, *Bibliographia*, Nr. 121: 198–203.

extracted from *Von den natürlichen Dingen*. These two components are taken from *Several Treatises* (*Ettliche Tractatus*) edited by Toxites in Strasbourg in the same year of 1570, as is mentioned in the previous section.[42] Thus, this edition by the Birckmann press emphasized a mineralogical dimension along with regimen and longevity, relying exclusively on the genuine texts of Paracelsus.

All these editions prove the unusually high expectation of German readers vis-a-vis *Archidoxis*. The Basel publisher Pietro Perna did not miss this commercial opportunity in the international book market. His collaborator Gerard Dorn prepared for him the second Latin edition as the *Archidoxorum libri X* or *On the Secret Mysteries of Nature* (*De secretis naturae mysteriis*) in 1570 (fig. 4.2).[43]

On this occasion the duo of Perna and Dorn corrected their initial error by placing the text on microcosm as the first book so as to make ten books, as the title announced: 1) microcosm; 2) *De renovatione*; 3) elements; 4) quintessences; 5) arcana; 6) magisters; 7) specifics; 8) elixirs; 9) extrinsics; and 10) *De vita longa*. In this Latin edition, *Archidoxis* is followed by the second part which comprises: *De tinctura physicorum*, *Praeparationes*, *De vexatione*, *De cementis metallorum* and *De gradationibus metallorum*. Although the treatments of regimen and longevity are kept with the inclusion of *De renovatione* and *De vita longa*, greater emphasis is given to the dimension of transmutational alchemy by the supplementary materials, some of which are clear forgeries such as *De tinctura physicorum* and *De vexatione*.

The appeal generated by the inclusion of the fake texts, attributed to Paracelsus and of a strong alchemical flavor, encouraged the Perna-Dorn duo to produce an exclusive collection of such texts in Latin for the international market. Suggestively entitled *Three Books on the Supreme Mysteries of Nature* (*De summis naturae mysteriis libri tres*), this collection was published in the same year of 1570.[44] It includes *De spiritibus planetarum*, which Huser considered to be a forgery, followed by *De occulta philosophia* and *De signis zodiaci*. Remarkably enough, the last item addresses the production of talismans and was to form the first half of the notorious *Archidoxis magica*.[45] Both Huser and

42 Sudhoff, *Bibliographia*, Nr. 120: 196–198. For these texts on salts, vitriol, arsenic and sulphur, see respectively Huser, 7: 141–152, 183–200, 201–209, 163–182; Sudhoff, 1/2: 98–110, 146–165, 166–175, 124–145.

43 Sudhoff, *Bibliographia*, Nr. 123: 205–207.

44 Ps.-Paracelsus, *De summis naturae mysteriis libri tres* (Basel: Pietro Perna, 1570). Cf. Sudhoff, *Bibliographia*, Nr. 125: 209–211.

45 For the text of *De signis zodiaci*, which is the first three books of *Archidoxis magica*, see Huser, 10: Appendix, 67–106; Sudhoff, 1/14: 437–478. On talismans in the Renaissance in general, see Daniel P. Walker, *Spiritual and Demonic Magic from Ficino to Campanella* (London, 1958); Wolf-Dieter Müller-Jahncke, *Astrologisch-magische Theorie und Praxis in der Heilkunde der frühen Neuzeit* (Stuttgart, 1985); Nicolas Weill-Parot, *Les images*

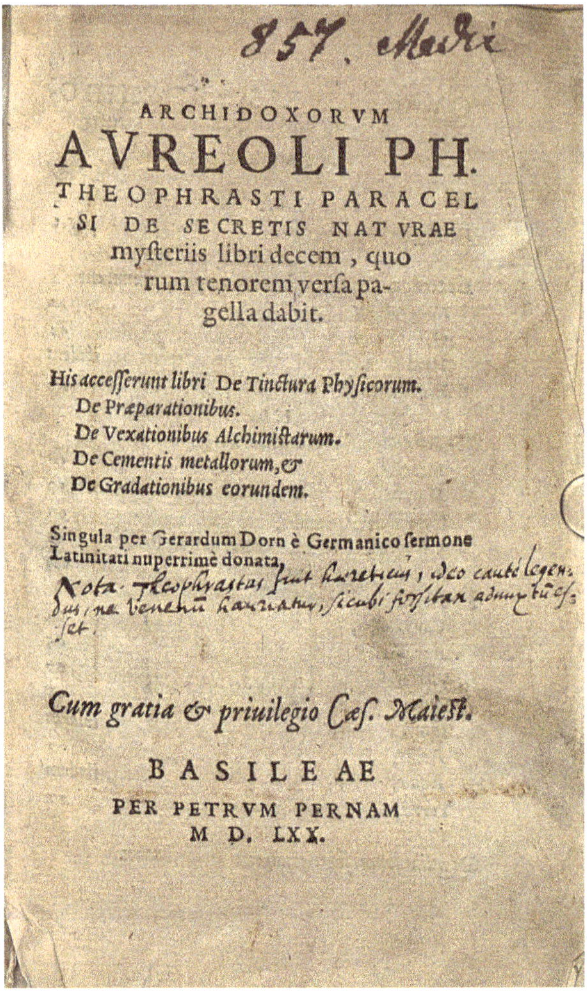

FIGURE 4.2
Paracelsus, *Archidoxorum* [...] *libri decem*, trans. Gerhard Dorn (1570), title-page. The early modern handwritten inscription reads: "Nota. Theophrastus fuit hæreticus, ideo caute legendus, ne venenum hauriatur, sicubi forsitan admixtum esset" ("Notice: Theophrastus was a heretic. Therefore he must be cautiously read, so that the poison did not spread if it happened to be mixed with anything.") urn:nbn:de:bvb:12-bsb00034166-8
MUNICH, BAYERISCHE STAATSBIBLIOTHEK: Alch. 206 d

Sudhoff rejected it as inauthentic. As has been seen above, *De occulta philosophia* was first published as the appendix of *Archidoxis* by Rihel of Strasbourg in the same year, while this is the first appearance of *De signis zodiaci*. As for the short treatise *De spiritibus planetarum*, it was also called *De spiritibus metallorum* in several collections of Paracelsus' works. It turns out to be a translation of *Alchimia*, the German version of which was also included later in Bodenstein's collection *Metamorphosis*.

astrologiques au Moyen Âge et à la Renaissance: spéculations intellectuelles et pratiques magiques, XIIe-XVe siècle (Paris, 2002); Hirai, "Images, Talismans and Medicine."

This exclusive collection in Latin of an alchemical and occultist aspect likely won commercial success in the international book market. Perna thus decided to produce a larger collection of the same sort in German in the following year, 1571. It consists of the texts which had already been published. Endowed with a strong flavor of transmutational alchemy and occult philosophy, it includes *De spiritibus planetarum*, *De tinctura physicorum*, *De gradationibus*, *De cementis* and *De signis zodiaci*.[46]

Upon the success of these collections, it is easy to imagine the motivation of the Perna-Dorn duo to provide a fully enhanced edition of *Archidoxis* the following year (fig. 4.3).[47]

On its title-page, they explicitly announced that "new materials" had been added. These new texts are *De tempore*, *De imaginibus* and *Archidoxis magica*. Among them, *De tempore* is only the first item of the five short tracts which comprise *Five Treatises of Philosophy* (*Philosophiae tractatus quinque*). Here it is actually followed by the other four. Along with this set of five tracts, Huser and Sudhoff accepted *De imaginibus* as a genuine work of an uncertain date.[48] The first appearance of this elusive treatise is of great interest since it develops a remarkable discussion on the production of homunculi. That is the very subject developed in the first book of *De natura rerum* in a different, but no less impressive manner. It is remarkable that these two treatises, *De imaginibus* and *De natura rerum*, were published for the first time in the same town of Basel in the same year of 1572.

As for *Archidoxis magica*, this collection by the Perna-Dorn duo includes its later part: *De compositione metallorum* as Book 6, *De speculi constellatione* as Book 5 and *De sigillis planetarum* as Book 7. Dealing with the production of talismans and with astrological images engraved on stones and minerals, it is highly doubtful that these texts are genuine works by Paracelsus.[49]

---

46 Sudhoff, *Bibliographia*, Nr. 134: 224–226. For the German version of *De spiritibus planetarum*, see Huser's edition of Paracelsus, *Chirurgische Bücher*, Appendix: 71–77. Sudhoff's edition does not contain this text.

47 Sudhoff, *Bibliographia*, Nr. 142: 239–241.

48 For the text of *Philosophiae tractatus quinque*, see Huser, 9: 398–413; Sudhoff, 1/13: 335–358. For the text of *De imaginibus*, see Huser, 9: 369–393; Sudhoff, 1/13: 359–386. I have benefited from the learned Japanese translation of *De imaginibus* by Amadeo Murase, published in Shunichi Ikegami, ed., *Renaissance Natural Philosophy* (Tokyo, 2017), 1: 555–578; see also Möseneder, *Paracelsus und die Bilder*, 71–162.

49 For the text of these three books of *Archidoxis magica*, see Huser, 10: Appendix, 115–138; Sudhoff, 1/14: 479–498; see also Wolfgang Schneider, "Paracelsus, Autor der *Archidoxis Magica*?" in idem, *Mein Umgang*, 107–120. To place this text in its context, it is necessary to execute a rigid philological comparison of its contents, for example, with those of *De occulta philosophia libri tres* (Köln, 1533) by Agrippa von Nettesheim (1486–1535). Cf.

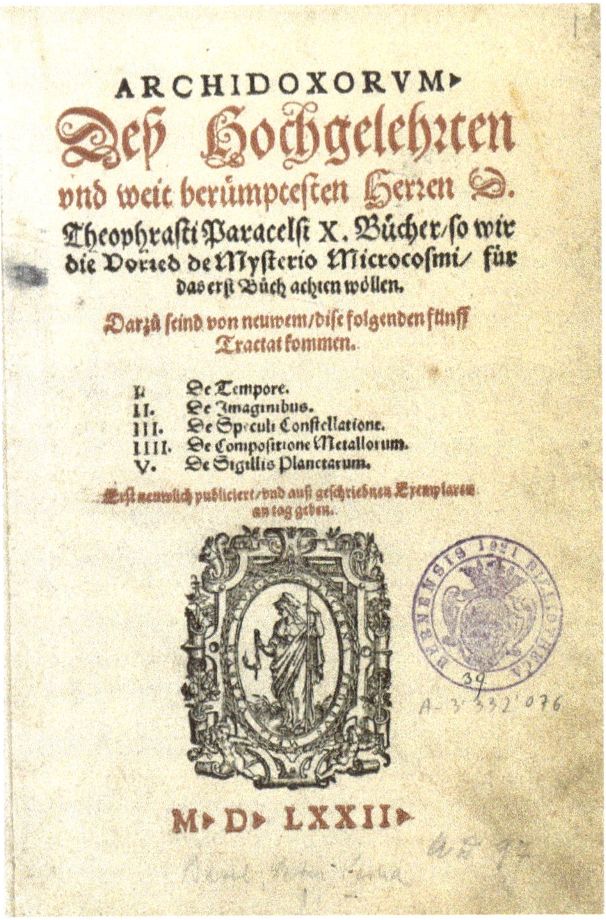
ARCHIDOXORVM.
Deß Hochgelehrten
vnd weit berümptesten Herren D.
Theophrasti Paracelsi X. Bücher/so wir
die Vorred de Mysterio Microcosmi/ für
das erst Büch achten wöllen.
Darzů seind von neuwem/dise folgenden fünff
Tractat kommen.
I. De Tempore.
II. De Imaginibus.
III. De Speculi Constellatione.
IIII. De Compositione Metallorum.
V. De Sigillis Planetarum.
Erst neuwlich publiciert/vnd auß geschriebnen Exemplaren
an tag geben.
M. D. LXXII.

FIGURE 4.3 Paracelsus, *Archidoxorum* [...] *X. Bücher*, ed. Pietro Perna (1572), title-page. BERN, UNIVERSITÄTS-BIBLIOTHEK: MUE AD 97:1

Conclusion

Now it is possible to locate the genesis of Bodenstein's collection *Metamorphosis* in the context of its own publication history. This collection of 1572 included the first edition of *De natura rerum* at its head, followed by the texts which shared a strong inclination toward transmutational alchemy such as *De cementis*, *De gradationibus*, *De projectionibus* and *Alchimia*.

Karl Anton Nowotny, "The Construction of Certain Seals and Characters in the Work of Agrippa of Nettesheim," *Journal of the Warburg and Courtauld Institutes*, 12 (1949), 46–57; Hirai, "Images, Talismans and Medicine."

Although *Archidoxis*, one of the most successful works by Paracelsus, announced ten books in its title, two books were apparently missing in its manuscript. During the evolution of its multiple publications, the editors of *Archidoxis* tried to compensate for its missing parts by complementing diverse texts of an affinitive nature. This genuine work of Paracelsus first appeared in 1569 with its emphasis on regimen and longevity by way of its supplementary materials, namely *De renovatione et restauratione* and *De vita longa*. Then, it gradually absorbed different texts endowed with a dimension of transmutational alchemy as its supplements. More and more works of doubtful nature and origin such as *De tinctura physicorum*, *De occulta philosophia*, *De spiritibus planetarum*, *De vexatione* and *De signis zodiaci* were put together. This tendency, initiated by Toxites and the Strasbourg publisher Rihel, was, however, most enthusiastically pursued by the duo of Pietro Perna and Gerard Dorn, based in Basel. The result was their production of a collection of dubious works and forgeries of an alchemical character in Latin, under the title of *Three Books on the Supreme Mysteries of Nature* (1570) and a larger collection of the same spirit in German (1571). Then the Perna-Dorn duo released the fully enhanced edition of *Archidoxis* in 1572, completely in line with these productions with a strong flavor of transmutational alchemy and occult philosophy.

At the climax of this movement, in the same town of Basel in the same year of 1572, appeared Bodenstein's collection, endowed with a humanistically and alchemically suggestive title of *Metamorphosis*. In this collection *De natura rerum* played the crucial role of providing the "theoretical basis" for those alchemical forgeries. It is worth remembering that Bodenstein was already involved in the edition of an extremely popular fake text *De vexatione* (1567), also renamed *Coelum philosophorum*. Leaving *Philosophia ad Athenienses* (1564) aside, this treatise of a manifestly alchemical intent was the first major forgery attributed to the name Paracelsus.

It is probable that very few readers of the time, except those like Johannes Huser who had access to the manuscripts of Paracelsus and who carefully studied them, were able to distinguish between the genuine and the inauthentic works, amid the flood of publications. This mixture – of the genuine and the fake – contributed to the emergence of Paracelsus' legendary image as the patron of alchemy and occult philosophy.

Paracelsus, the Plague, and *De Pestilitate*

Charles D. Gunnoe, Jr.
Aquinas College
gunnocha@aquinas.edu

Abstract

While *De Pestilitate* is generally regarded as pseudepigraphic by modern scholarship, the treatise occupied a prominent place in Johann Huser's definitive edition of Paracelsus' *Bücher und Schrifften* (1589–1591). The text offers a compelling and generally reliable guide to Paracelsian plague theory with clear resemblances to the authentic *Zwey Bücher von der Pestilentz und ihren zufällen* and *De Peste Libri tres*. The text emphasizes the astrological transmission of the disease, describes a large role for divine retribution and demonological agency, and promotes the utility of folkloric cures. While the plague theory reflects Paracelsus' own thought and is well synthesized with his *tria prima* chemical philosophy, there are sufficient dissonances between the text and the corpus of his accepted writings to maintain its pseudepigraphic status, even if it displays some indications of a reworking of genuine Paracelsica.

Keywords

Paracelsus – plague – spurious works – cosmology – magical imagination – folkloric cures – witchcraft – astrology

Introduction

In his classic article "The Seeds of Disease," and other works in which he undermined the alleged importance of Girolamo Fracastoro and his *seminaria* theory, Vivian Nutton simultaneously held up Paracelsus as "the only innovative medical theorist of the sixteenth century."[1] Nutton based this assertion on the work of Walter Pagel, who argued that Paracelsus' ontological conception

1 Vivian Nutton, "The Seeds of Disease: An Explanation of Contagion and Infection from the Greeks to the Renaissance," *Medical History*, 27 (1983), 1–34. All translations are the author's except where otherwise noted.

 | DOI:10.1163/9789004503380_007

of disease was chief among his scientific contributions.[2] When one thinks of contagious diseases in the late medieval and early modern period, the bubonic plague naturally emerges as the disease that left the largest mark on the era. Putting these two ideas together, one might assume that Paracelsus' writings on the plague would serve as a showcase for his innovative medical theory. Closer acquaintance with his plague writings, however reveals the opposite to be the case; one arrives at a chaotic jumble, so much so that Paracelsus' plague writings have received relatively limited scholarly attention.[3] Rather than serving as a prototype for his purported pathbreaking etiology, the plague treatises are in fact more a model for his spiritualist-magical worldview of the interaction of the microcosm and macrocosm. Moreover, in moving to *De Pestilitate* – a text that is generally regarded as Pseudo-Paracelsian – the level of difficulty only increases. This study will briefly review Paracelsus' historical engagement with the plague, before surveying the plague texts that scholars currently accept as authentic works of Paracelsus. It will then engage in a more thorough analysis of the background and plague theory of *De Pestilitate* itself and conclude with some reflections on where the treatise fits within Paracelsus' larger body of plague writings.

1 Paracelsus and the Plague

While an appreciable number of sixteenth century physicians, such as Paracelsus' critics Conrad Gessner and Thomas Erastus, became infected with plague during their lifetimes, there does not seem to be any clear evidence that Paracelsus himself ever contracted plague.[4] Plague was certainly active in the Upper German/Alpine region during his era, and he is likely to have

---

2 Walter Pagel, *Paracelsus: An Introduction to Philosophical Medicine in the Era of the Renaissance* (Basel, 1958), 137–140.

3 The most important work to date is Rosemarie Dilg-Frank, "Zu Begriff und Bedeutung von '*pestis/pestilentia*' und ihrer Verwendung bei Paracelsus," in Sepp Domandl, ed., *Paracelsus in der Tradition* (Vienna, 1980), 48–66. See also Erik Heinrichs, *Plague, Print, and the Reformation: The German Reform of Healing, 1473–1573* (London, 2018); Niel McDowell, "The Sixteenth-Century German Plague Text and Its Role in Early Modern Medical Discourse." PhD diss. (University of Pennsylvania, 1998); Pagel, *Paracelsus*; Jürgen Strein and Joachim Telle, "Deutsche Pseudoparacelsica über die Pest: Ein 'Begriff' zur Pestdiagnose (1553) und die 'Tabula de pestilitate' von Bartholomäus Scultetus (1578)," in Dominik Gross and Monika Reininger, eds., *Medizin in Geschichte, Philologie und Ethnologie: Festschrift für Gundolf Keil* (Würzburg, 2003), 349–370.

4 For a competent biographical overview, see Udo Benzenhöfer, *Paracelsus* (Reinbek bei Hamburg, 1997; 2nd ed. 2002); see also Charles Webster, "Paracelsus: Medicine as Popular

experienced at least four large waves of plague in his lifetime. There was a major epidemic around the turn of the century (ca. 1500) and another as the Protestant revolt moved to a crescendo ca. 1518–1520. This Reformation plague was followed by another German epidemic in the late 1520s and finally a particularly severe outbreak in the early 1540s near the time of Paracelsus' death. In addition to these larger Central European epidemics, we have evidence that he was present in a few more localized epidemics or aftershocks which were common features of early modern plague distribution.[5] This rough timeline is consistent with the content of his plague writings, and there are many indications that he knew the disease reasonably well and could describe its symptoms with apparent first-hand knowledge.

2 Paracelsus' Plague Treatises

Paracelsus' experience with the plague is reflected in numerous works within the corpus, and he occasionally mentions plague in his theological writings as well.[6] Some of his earliest thinking on plague is reflected in the fragmentary *Volumen medicinae Paramirum*, which Karl Sudhoff dated to ca. 1520, though current scholarship places it closer to 1525 and thus prior to his Basel period.[7] While his views would undergo substantial evolution, central themes from his mature plague writings were already present at this point. As in other texts, Paracelsus used the example of plague to expose the ignorance of academic physicians and the inability of humoral medicine to fully comprehend disease. The reasoning of the treatise is hardly linear – Andrew Weeks has described it as a "curious admixture of *non sequitur* and tautology"[8] – and Paracelsus displays his habitual tendency of incorporating multiple organizing schemes with overlapping competencies and latent contradictions. His driving program in this work is to link the different manifestations of plague to the five *entia* or "substances" of disease. The five *entia* – causes of disease in general and plague

Protest," in Ole Peter Grell and Andrew Cunningham, eds., *Medicine and the Reformation*, (London, 1993), 57–77.

5 Cf. Edward A. Eckert, *The Structure of Plagues and Pestilences in Early Modern Europe Central Europe, 1560–1640* (Basel, 1996).

6 Dilg-Frank, "Zu Begriff und Bedeutung von '*pestis/pestilentia*'," 54–55.

7 Paracelsus, *Sämtliche Werke*, 1. Abt., ed. Karl Sudhoff, 14 vols. (Munich, 1922–1933) (hereafter: Sudhoff, I/1, I/2, etc.), I/1: 163–239. On the difficulty of dating these works, see Andrew Weeks, *Paracelsus (Theophrastus Bombastus von Hohenheim, 1493–1541): Essential Theoretical Writings* (Leiden, 2008), 4–12, 112, and 120; see also Andrew Weeks, *Paracelsus: Speculative Theory and the Crisis of the Early Reformation* (Albany, NY, 1997), 38–39.

8 Weeks, *Paracelsus*, 62.

in specific – are the astral, venal, natural, spiritual and divine substances.[9] While use of *ens/entia* as an explicit framework falls away in his more mature plague treatises, his inclination to view plague as alternatively driven by astral, natural, poisonous, spiritual-imaginative factors, or even as the manifestation of divine retribution does not.[10] The topic of the plague naturally arises in many other of Paracelsus' texts, especially in *Die Bücher von den unsichtbaren Krankheiten* (*Books on Invisible Diseases*)[11] and *Nachschrift aus der Vorlesung über tartarische Krankheiten* (*Postscript from the Lecture on Tartar Diseases*)[12] – portions of which are excerpted wholesale in *De Pestilitate.*[13]

While these references reflect Paracelsus' efforts to integrate plague into his evolving medical theory, he first communicated what may be taken as his mature understanding of plague in a standalone work entitled *Zwey Bücher von der Pestilentz und jhren zufällen* (*Two Books concerning the Plague and its Effects*), written ca. 1529–1530.[14] This twenty-four page work, sometime referred to as the Nördlingen plague treatise, offers a concise overview of Paracelsian plague theory and was the most commonly printed plague treatise of Paracelsus in the sixteenth century.

The understanding of the plague offered in the Nördlingen treatise was further articulated and expanded in the more voluminous yet fragmentary *De Peste libri tres* (*Three Books on the Plague,* hereafter referred to as *De Peste*), also known as the *Commentarius de Peste*, which is generally thought to have been composed ca. 1531-35, although there are few contextual clues in the text of its origins.[15] True to its title, the work consists of three books which respectively run to twenty-one, thirty, and then a brief three pages for the final book.

9 Ibid., 61-69: "*ens astrale, ens venale, ens naturale, ens spiritale,* and *ens deale.*"

10 See Weeks, *Paracelsus,* 49–75, esp. 60–73. [Ps.-]Paracelsus briefly returns to the *entia* theme in the very end of *De Pestilitate,* in a section entitled "De Quinque Entibus aller Kranckheiten"; see Sudhoff, I/14: 654–655; Paracelsus, *Bücher und Schrifften,* ed. Johannes Huser, 11 vols. (Basel, 1589–1590) (hereafter: Huser, 1, 2, etc.), 3: 99.

11 Also known as *Fünf Bücher De causis morborum invisibilium,* Sudhoff, I/9: 249–350.

12 Also known as *De Morbis ex tartaro oriundis,* Sudhoff, I/5: 1-122. This treatise contains a brief chapter "de Peste," at pp. 81-90.

13 The excerpted passages are drawn from *Nachschrift aus der Vorlesung über tartarische Krankheiten,* Sudhoff, I/5: 77–83. There are actually two excerpted passages from the *Vorlesung über tartarische Krankheiten*: broadly, a passage from pp. 77–78 which appears in *De Pestilitate* on p. 639; and a second from pp. 81-83 which appears in *De Pestilitate* pp. 639–641. As the passages are quoted in Latin, they rather stand out in *De Pestilitate,* not unlike the Aramaic portions of the Book of Daniel. The *Astronomia Magna* also treats plague, though it considers it more wholly as an astrological phenomenon needing heavenly medicine. See Sudhoff, I/12: 1-444, passim.

14 Sudhoff, I/8: 371-395; Huser, 3: 124–149.

15 Sudhoff, I/9: 565–638 (including fragments); Huser, 3: 150–206.

The text appears largely complete, save for the abrupt third book, although the headings and subsections in the published editions seem rather arbitrary and may have been supplied by later editors. As in the case of *De Pestilitate*, there is no extant autograph of the text; however, its authenticity has not been seriously questioned, and its plague theory is consonant with the Nördlingen treatise. *De Peste* offers a compelling astral-theological vision, akin to Paracelsus' late works such as the *Astronomia magna*, and further articulates the role of the imagination in the spread of plague.

In addition to these two generally acknowledged treatises, Paracelsus also left behind a shorter *Von der Pestilentz, ein Büchlein geschrieben an die Statt Sterzingen* (*On the Pestilence, a Booklet Written for the City of Sterzing*) dated to 1534.[16] This sixteen-page text is unusual in that it is written for a specific community, putatively on behalf of the common man. While the treatise maintains a focus on the heavenly origins of plague, it offers more conventional therapeutic advice than the other authentic treatises. The contained recommendations include bloodletting and a concoction including brandy and theriac to treat the pestilential fever.[17] The relative conservatism of the text may reflect an origin prior to 1530 for some of the content, as the piece is quite unlike his other largely contemporaneous plague treatises. A likely explanation for the book's conservatism may have been its intent to gain a concrete clientele and to offer assistance through means that were ready to hand for this limited audience.

The publication history of these texts is as complicated as the works themselves and proved a substantial barrier to the reception of Paracelsus' genuine plague theory in the late sixteenth century. None of the works were printed in Paracelsus' lifetime, and the first plague treatise to appear under Paracelsus' name was the Pseudo-Paracelsian *Für Pestilentz: Ain seer nützlicher unnd bewerter Tractat* (*For Plague: A Very Useful and Valuable Tract*) which was published in 1554. This non-Paracelsian work is an outlier even among Pseudo-Paracelsica; Sudhoff quipped that the "booklet nowhere breathes a Paracelsian spirit."[18] While there is some possibility that the work may be based on a text owned by Paracelsus and found among his papers – making the ascription to Paracelsus a result of a misunderstanding – it is perhaps more likely that the

16 Sudhoff, 1/9: 454–562; Huser, 3: 109–123.

17 Sudhoff, 1/9, on bloodletting: 549–550; 556; recipe for fever: 552–553. This recommendation of bloodletting is somewhat in conflict with Paracelsus' more general position: see "Vom Aderlaß, Purgiren und Schröpfen," Sudhoff, 1/4: 369–434.

18 Karl Sudhoff, *Bibliographia Paracelsica: Besprechung der unter Theophrast von Hohenheim's Namen 1527–1893 erschienenen Druckschriften* (Berlin, 1894), 50–52 (no. 33); Strein and Telle, "Deutsche Pseudoparacelsica über die Pest," 350–351.

attribution of the work to Paracelsus was simply a marketing ploy to attract a larger readership. Thus, the publication was more a harbinger of the revival of interest in all things Paracelsian than an accurate presentation of Paracelsus' ideas. The reading public first received a fair sense of Paracelsus' own views with the publication of *Zwey Bücher von der Pestilentz* (the Nördlingen treatise) in multiple editions in the 1560s.[19] The more voluminous *De Peste libri tres* appeared in a collection of shorter medical works in 1567.[20] These two works appeared in a joint edition in 1575/6[21] and were belatedly joined by the Sterzing tract in a 1583 edition.[22] Finally, all of the Paracelsian plague texts – including the first published edition of *De Pestilitate*, which was given pride of place in the order of contents – were published in the third volume of Johann Huser's edition of Paracelsus' collected works.[23]

3 Common Features of the "Authentic" Plague Treatises

Before turning to *De Pestilitate* itself, it will be helpful to briefly review the content of the treatises generally acknowledged as Paracelsian, focusing most closely on the *Zwey Bücher von der Pestilentz* and *De Peste libri tres*. These works break noticeably from the conventional wisdom of academic plague treatises and offer an interpretation that is both non-humoral and non-miasmic. A strong animus against Galenic academic medicine is present in these works, and Paracelsus directs a number of caustic barbs toward these physicians. While many late medieval and early modern plague treatises make some reference to celestial conditions as a remote cause of plague epidemics, Paracelsus' treatises emphasize the role of the stars in the generation of plague and employ the analogy of microcosm and macrocosm to explain how this occurs. In a manner generally complementary to this astral primacy, Paracelsus sometimes distinguishes between "natural plague," which may result from the poisoning of food and water, and "supernatural plague" in which astral, divine, or demonic forces may be in play. He may explain this natural corruption by reference to his *tria prima* theory of mercury, salt and sulphur, but this conceptual framework is relatively underdeveloped in the plague treatises. Paracelsus' emphasis on what he would call *astronomia* does not rule out a large role for

19 See Sudhoff, *Bibliographia*, 88–125 (nos. 60, 69, 81).
20 Ibid., 134–138 (no. 87).
21 Ibid., 288–291 (nos. 167–167a).
22 Ibid., 335–336 (no. 193).
23 Huser, 3: 24–206; Sudhoff, *Bibliographia*, 379–382 (no. 218).

human and divine agency in his plague theory. He has a marked biblical vision of the plague as a vehicle for divine retribution upon humanity. Bringing his dual emphases on human culpability and astral causation together, he articulates a theory in which human misdeeds poison the heavens and then cause illnesses such as plague to rain down on humanity.

As previously mentioned, Paracelsus is able to describe the basic symptoms of plague – fever, carbuncles, buboes, headache, and narcotic sleep – well enough to give the impression that he has seen the plague, and he clearly identifies the chief locations of plague buboes or swellings in the throat, armpit, or groin areas. In some texts, he likes to distinguish between the disease itself and some of its side effects or "accidents." In a similar manner, he may differentiate between plague deaths that are targeted, generally by the stars, and those that are accidental, that is, which seem to follow from a larger epidemic. Seen charitably, he is able to align the competing chemical, medical, astrological, and theological dimensions of his theory reasonably successfully. Viewed through a more critical lens, some motifs are quite difficult to reconcile and his schema are redundant, but these inconsistencies likely reflect the different periods in his career when the treatises were composed and/or the different audiences he hoped to address with the works.

4 Background and Provenance of *De Pestilitate*

Naturally, the most basic question to answer in engaging *De Pestilitate* is the issue of its origins. The text includes a few comments about presumably contemporaneous events, mostly in Southern Germany, Switzerland, and Hungary during the 1530s. The last reference is to 1538, making that date the *terminus post quem* of the treatise,[24] although the treatise also incorporates some material from earlier points in Paracelsus' career.[25] While discerning the degree to which the treatise may be considered genuinely Paracelsian is difficult enough, identifying the ghost writer or final editor has proven nearly impossible. As

24 Sudhoff, 1/14: 615: The author notes that: "hat alsbald im Schwabenlant, Schweiz und Bayerlaut eine strenge pestilenz mit zufall einer unerhörten colica geregiret anno 1538"; see Huser, 3: 49. While there are a few hints of plague in the general region ca. 1539, there is little evidence for a widespread epidemic of plague in Upper Germany and Switzerland in 1538. The possible internal references to the *Astronomia Magna*, which Paracelsus composed ca. 1537/38, are also consistent with this chronology. Sudhoff, 1/14: 615, 623, 645; Huser, 3: 49, 60, 92.

25 Including material from the 1527 *Nachschrift aus der Vorlesung über tartarische Krankheiten*, Sudhoff, 1/5: 3.

previously mentioned, a definitive resolution of the authorship question of *De Pestilitate* has not been offered and indeed, has hardly been attempted by prior scholarship. Huser, the diligent editor of the late sixteenth century edition of Paracelsus' collected works, offered little help with his terse comment that this text was "from the manuscript of [Johannes Scultetus] Montanus and of others."[26] Of course, he gave fuller detail when he had access to Paracelsus' own manuscripts, so it is clear that Huser did not possess a genuine autograph of *De Pestilitate* when he assembled his monumental edition. Thus, our text first appeared nearly fifty years after Paracelsus' death, with no claim to be based on any of his original manuscripts and apparently no documented provenance. The intervening four hundred years have done little to illuminate the work's manuscript history.

The question of the origin of *De Pestilitate* is further complicated by the existence of a single-leaf presentation of Paracelsian plague theory in the form of a Ramistic table entitled *Tabula physica, astronomica & medica de pestilitate ...* (hereafter referred to as the *Tabula*). This work was edited in 1578 and published in 1579 by the Silesian Paracelsian Bartolomeus Scultetus, fully ten years before the publication of *De Pestilitate.*[27] Jürgen Strein and Joachim Telle extensively scrutinized the text and have discounted the simple thesis that the *Tabula* was derived directly from the longer, unpublished *De Pestilitate* or from other obvious candidates such as *Zwey Bücher von der Pestilentz*. However, while the *Tabula* does not consist of mere abbreviated excerpts from our treatise, it does share a small measure of material in common with *De Pestilitate*, and one brief passage, concerning the healing power of eels and frogs, is nearly verbatim in both texts.[28] Likewise, both texts claim implicitly or explicitly to be based on

---

26 Huser, 3: fol. Aaa iiiv [table of contents]: "Ex Manuscripto Montani et aliorum." On Huser, see Joachim Telle, "Johann Huser in seinen Briefen: Zum Schlesischen Paracelsismus im 16. Jahrhundert," in Joachim Telle, ed., *Parerga Paracelsica. Paracelsus in Vergangenheit und Gegenwart* (Stuttgart, 1991), 159–248. On Montanus, see Joachim Telle, "Scultetus, Johannes, auch: J.S. Montanus, J. Montanus, Trimontanus," in Wilhelm Kühlmann, ed., *Killy Literaturlexikon,* vol. 10, Ros - Se (Berlin, 2011), 706.

27 The full title is: *Tabula physica, astronomica & medica de pestilitate: In eine kurtze Summ und Inhalt/ auß des Aur. Ph. Th. Par. Buche/ welches er nach dem Jahr 1538 von diese Kranckheit gestellet/ vnd jetzundt erst new herfür an tag komen/ gezogen* (Görlitz: Ambrosius Fritsch, 1586). There is a sufficient evidence that the original 1579 edition was published in a print run of 300 copies: Strein and Telle, "Deutsche Pseudoparacelsica über die Pest," 349–370. Modern scholarship has been based on the extant 1586 edition which purports to be a reprint of the 1579 edition. The 1579 edition was presumed lost, but there is apparently one exemplar in the Universitätsbibliothek Erlangen-Nürnberg (Signatur H61/4 TREW.M 133/141).

28 See Strein and Telle, "Deutsche Pseudoparacelsica über die Pest," 357.

Paracelsus' writings from ca. 1538.[29] The two works are largely independent, with *De Pestilitate* no more a mere expansion of the argument of the *Tabula* than the *Tabula* is a simple abridgement of *De Pestilitate.* While Strein and Telle emphasized the fact that the *Tabula* did not offer verbatim excerpts from *De Pestilitate,* rather than confusing the argument further by proposing the existence of an unknown Q-source, the most economic explanation for the commonality is that some early version of *De Pestilitate* was paraphrased by the compiler of the *Tabula.* The shared material between the *Tabula* and *De Pestilitate* tends to nullify the prospect that *De Pestilitate* was wholly a new forgery of the 1580s. In short, the contention that both works were partially based on common source material, likely at least a fragment of *De Pestilitate* itself, that *claimed* to connect with Paracelsus' own work from ca. 1538, is the most straightforward explanation of the texts' interrelationship. This interpretation does not establish that Paracelsus himself wrote *De Pestilitate,* but it adds some modest corroboration to the theory that both works have a connection to earlier and authentically Paracelsian material.

The title of the work is mildly enigmatic, but this hardly makes it unusual in the Paracelsian corpus. As previously noted, while the work is mostly in German, it has the Latin title *De Pestilitate* (*On Pestilence*). The work also possesses an amply descriptive subtitle: *That is, of the cause and origins of plague, and afterwards the actual and foundational treatment of the same disease, written by the learned Herr Theophrastus von Hohenheim doctor of both medical arts.*[30] While we have no reason to presume that the treatise's title derives directly from Paracelsus, the use of a such a Latinate term in the title would have been a plausible choice for him, given that he had also authored three works which he (or his later editors) all dubbed with the largely equivalent "on plague"/*von der Pestilentz/de peste.* Paracelsus is not known to have used the Latin term *pestilitas* in his oeuvre, and indeed the term is not used within the text of *De Pestilitate* itself. Rosemarie Dilg-Frank – in perhaps the most foundational article published to date on Paracelsus and the plague – excluded the treatise *De Pestilitate* from her linguistic analysis of Paracelsus plague vocabulary due to the treatise's spurious status and in part because Paracelsus never used the term.[31] However, various forms of *pestilitas* are common enough in Latin texts prior to Paracelsus, as a term roughly equivalent to the Latin *pestis*, the German *Pest*, or *Pestilentz*, which is the most common early modern

29 The *Tabula* claims to be based on a Paracelsian book from 1538, and *De Pestilitate* mentions a few historical episodes in the work, the last being an outbreak of plague in 1538.

30 Huser, 3: 24: *De Pestilitate, Das ist, vom Vrsprung vnnd Herkommen Pestis, vnd hernach von derselbigen kranckheit eigentlichen vnd gründlichen Cur.*

31 Dilg-Frank, "Zu Begriff und Bedeutung von '*pestis/pestilentia*'," 55.

German term for plague. Thus, the treatise's title alone clarifies little regarding the work's origin or possible place in Paracelsus' larger corpus.

While *De Pestilitate* is generally regarded as pseudepigraphic, there is currently some disagreement among scholars concerning its relative authenticity. One side sees the work as completely spurious, while the other has generally embraced the text's authentic "Paracelsian" spirit. In the first camp, the mainstream continental position, following Sudhoff, has been to dismiss out of hand the possibility of Paracelsus' direct authorship. As Sudhoff himself concluded: "I cannot regard the book as genuine and have well guessed that perhaps Magister Bartholomäus Scultetus in Görlitz might be the author, … It is only a hunch. The remainder must be left to future scholarship. My standpoint will hardly be able to be shaken."[32] Despite the fact that Huser had given *De Pestilitate* first billing among the plague treatises, Sudhoff relegated *De Pestilitate* to the very last volume of his edition, among the "Spuria for which Huser mostly held to be genuine."[33] Josef Strebel continued this theme, asserting that the work was "an absolutely certain compilation of Silesian Paracelsians, who have done Paracelsus a miserable service with their *Dreckapotheke*."[34] More recently, Joachim Telle agreed with the designation of the work as Pseudo-Paracelsian, but offered the corrective that the humanist Scultetus, the compiler of the aforementioned *Tabula*, should be excluded as a possible ghost author/compiler of the work.[35] Thus, serious German philologists and medical historians have tended to dismiss the work from consideration of what might be considered genuine Paracelsus, and scholars have so far been more successful in eliminating possible authors than in generating plausible alternative candidates.

A second camp, commonly associated with Anglo-American Paracelsus studies, has shown greater interest in the text and has been less quick to dismiss it. The fact that the text was included in all of the important collected editions of Paracelsus, including Huser, Sudhoff, and Peuckert, has led a wide range of scholars to sample its magical worldview. *De Pestilitate* was also uncritically excerpted in the Jungian mélange of Jolande Jacobi, which likely remains the

---

32 Sudhoff, 1/14: XXXIII: "Ich kann das Buch nicht für echt halten und habe wohl vermutet, daß vielleicht Magister Bartholomäus Scultetus in Görlitz der Verfasser sein könne, … Es bleibt also bei der Vermutung. Weiteres muß künftiger Forschung überlassen bleiben. Meinen Standpunkt wird sie kaum erschüttern können."

33 Ibid.: 749.

34 Paracelsus, *Sämtliche Werke in zeitgemässer kurzer Auswahl*, ed. Josef Strebel (St. Gallen, 1944–1949), 6: 344: "Es handelt sich also bei der Schrift De Pestilitate um eine absolut sichere Kompilation schlesischer Paracelsisten, die Paracelsus mit ihrer Dreckapotheke einen miserablen Dienst erwiesen haben."

35 Telle, "Johann Huser in seinen Briefen," 228; see also Strein and Telle, "Deutsche Pseudoparacelsica über die Pest," 356.

most widely distributed edition of Paracelsus in the English-speaking world.[36] However, despite the treatise's relative prominence in the collected editions, most scholars remain reluctant to credit the text wholly to Paracelsus. Charles Webster has given the work the most serious reflection and has asserted that *De Pestilitate* "constitutes a convenient general review of the ideas of *Paracelsus*"[37] and reflects "a case of a competent distillation based on authentic Paracelsus sources."[38] Along these lines, many writers – critically or uncritically – have employed it as a faithful reflection of the Paracelsian worldview, and my comparison of *De Pestilitate* with Paracelsus' acknowledged "authentic" plague treatises will also reveal a fundamental congruence of the work with genuine Paracelsica. While some excerpts of the text are lifted from other Paracelsian works, there are portions of the work which seem to go beyond the boundaries of Paracelsus' corpus. Thus, the work falls in the greyest zone of Pseudo-Paracelsica in exhibiting great coherence with Paracelsus' mature magical worldview, while at the same time revealing a popular or low-magical quality not as commonly seen in his authentic works.

5 Description and Overview of the Treatise

De Pestilitate is both more and less than a plague treatise. While it has often been read as an introduction to the magical Paracelsian cosmos, it is largely about the plague with roughly three quarters of the content dealing with plague in some fashion. An outline of the treatise's contents gives a rough sense of the work's inconsistent organizational schema and topical interests.

Abbreviated Table of Contents of *De Pestilitate*[a]

Chapter Titles/Headings	Location
[*Tractatus Primus*]	Sudhoff, 1/14: 597; Huser, 3: 24
Quid sit homo, hoc est, caro hominis (What is Man and What is the Flesh of Man)	Sudhoff, 1/14: 597; Huser, 3: 25

a Many of the subheadings are omitted (esp. ca. Huser, 3: 80–86; Sudhoff, 1/14: 639–641) as well as the sentences that Huser set in a larger typeface to signify changing topics in the 1589 edition. See below.

36 Paracelsus, *Theophrastus Paracelsus: Lebendiges Erbe,* ed. Jolande Jacobi (Zurich, 1942).

37 Webster, "Medicine as Popular Protest," 69.

38 Charles Webster, *Paracelsus. Medicine, Magic and Mission at the End of Time* (New Haven, CT, 2008), 299 n. 121.

Abbreviated Table of Contents of *De Pestilitat*e (*cont.*)

Chapter Titles/Headings	Location
Vom Ebenbildt des Thierischen Menschen (On the Image of the Animal Human)	Sudhoff, 1/14: 599; Huser, 3: 27
Cabala	Sudhoff, 1/14: 601; Huser, 3: 30
Von dem Mayen der grossen Welt (On the Springtime/Beginning of the Macrocosm)	Sudhoff, 1/14: 604; Huser, 3: 33
[***Tractatus Secundus***]	Sudhoff, 1/14: 630; Huser, 3: 69
Der Ander Theil von der Pestilentz. (The Second Part on the Plague)	
Modus & processus Fiendi (On the Mode and Process of Becoming)	
Von dem natürlichen weg/ durch Speiss und Tranck welcher weg ist *SS nitri* (The Natural Way, through Food and Drink which way is the *Spiritus nitri* [Spirit of Nitre])	Sudhoff, 1/14: 632; Huser, 3: 72
Wie *Pestis* ubernatürlichen im Menschen wird/ und *per Basiliscum Coeli et Aquae* in Menschen kompt, *Cabalisticè per Unarium* (How Plague Comes to Humanity in a Supernatural Manner, and Comes through the Basilisk of Heaven and Water, in Cabalistic Fashion through the Unary)	Sudhoff, 1/14: 633; Huser, 3: 73
Von der Cur und Heilung der natürlichen und ubernatürlichen *Pestis*: und das in 16 Partickel getheilet (On the Cure and Healing of Natural and Supernatural Plague; and this Divided into Sixteen Units)	Sudhoff, 1/14: 639; Huser, 3: 80
Caput Primum, Von der Pyromantischen *Peste* und ihren anhangende zeichen (On Pyromantic Plague and its accompanying Signs)	
Cura Interna (Internal Remedy)	Sudhoff, 1/14: 641; Huser, 3: 82
Cura Externa (External Remedy)	Sudhoff, 1/14: 641; Huser, 3: 83
Caput Primum	Sudhoff, 1/14: 644; Huser, 3: 86
Caput Secundum	Sudhoff, 1/14: 646; Huser, 3: 88
Von der magnetischen kraft der *Mumia* im Menschen (On the Magnetic Power of *Mumia* in Humanity)	Sudhoff, 1/14: 650; Huser, 3: 94

Abbreviated Table of Contents of *De Pestilitat*e (*cont.*)

Chapter Titles/Headings	Location
Tractatus Tertius. *De veneno planetarum, hoc est de igne venenoso* (On the Poison of the Planets, that is on the Poisonous Fire)	Sudhoff, 1/14: 653; Huser, 3: 97
Quomodo nascatur ignis in venenum pestis ex septem planetis (In What Manner the Fire Springs forth in the Plague Poison from the Seven Planets)	
Caput Primum	
Tractatus Quartus, de Peste.	Sudhoff, 1/14: 654; Huser, 3: 99
Caput Primum. *De Quinque Entibus* aller Kranckheiten (On the Five *Entia* [Substances] of All Diseases)	
Caput Secundum. *De Incantatione et Phytonissis Entis Spiritualis* (On Incantation and Prophetesses of the Spiritual *Ens*)	Sudhoff, 1/14: 655; Huser, 3: 99
Caput Tertium	Sudhoff, 1/14: 660; Huser, 3: 107

The outline is not fully representative, as much of the first half of the work is not formally divided into subsections, although Huser flagged the changing topics by setting the opening sentences of new thematic sections in a larger typeface.[39] In a sense, the first two-thirds of the book represents a relatively complete treatise, and then the work's flow and structure is disrupted, on page 80 of the Huser edition, by the insertion of material from *Vorlesung über tartarische Krankheiten*, from which point onwards the organizational scheme never fully recovers. What Huser labels the *Tractatus Secundus* (Huser, 3: 69–96) is a particularly fragmentary jumble including what appears to be the second part of the original treatise (*Der Ander Theil von der Pestilentz*), a new "Caput Primum" which marks the insertion from the works on disease of the tartar, and yet another "Caput Primum," starting six pages later in the Huser edition. The final two "tracts" also seem to stand rather outside of the work's larger structure and are interposed at a brief two and nine pages, respectively, in the Huser edition. A more in-depth analysis of the overall structure and

39 Huser, 3: 33–68; Sudhoff, 1/14: 604–629.

archeology of the work is warranted. More intense study, paired with a stylometric analysis of the German text, may illuminate which parts potentially stem from Paracelsus himself, and from which parts of his career any authentic fragments are derived.

6 Magic and Plague Theory in *De Pestilitate*

As previously mentioned, *De Pestilitate* is a work that is largely concerned with plague although it situates its discussion of plague within the setting of the Paracelsian magical cosmos. This synthesis of plague theory with his magical cosmology is one of the most successful and appealing dimensions of the work. The work opens with a compelling summary of humanity's relationship to the macrocosm:

> Humanity emerged from the first matrix, namely out of the macrocosm (Great World). That is, the macrocosm, formed by God's hand along with all other creatures, gave birth to humanity according to the flesh for a mortal life. For this reason, humanity became earthly and carnal; and humanity received the earthly flesh from earth and water. These two elements constitute the body in its earthly, animal life, which humanity as a natural being received through creation through the hand of God. This animal life is nothing other than the fire and air … . In its earthly life humanity consists of the four elements. Water and earth, of which the body is formed, constitute the dwelling place and body of life. And I am not referring here to the life of the soul, which springs from the breath of God; for my meaning here is not theological but medical, but I mean the life which is animal and transient, the life which is formed from the fire and air. And therefore, the body, thus created from the earth and water, has become the home of life. It is enough to understand that man has two kinds of life – the animal life and the sidereal life.… Hence man has also an animal body and a sidereal body; and both are one, and are not separated.[40]

40 Sudhoff, 1/14: 597–598 (and Huser, 3: 25–26): "Der Mensch erhebt sich also, nemlich aus der ersten matrix das ist aus der großen welt. das ist, die große welt mit und samt allen anderen creaturen durch beschaffung durch die hant gottes hat geboren den menschen, dem fleisch nach zu rechnen zu der sterblikeit. aus solcher ursachen ist der mensch irdisch und fleischlich worden, und dis irdische fleisch hat der mensch empfangen aus der erden und wasser. dise erden und wasser ist nun das corpus des irdischen tierischen

Since so much of human life is integrally related to the macrocosm, a skillful physician must also be an apt "astronomer," artfully discerning the correspondences between the greater and the lesser worlds. A plague physician also needs to be something of a Christian magus, as there are multiple supernatural channels for the plague's transmission.[41]

In common with nearly all of Paracelsus' writings, the author of *De Pestilitate* offers an impressive diatribe against the traditional academic medicine of his day. He insults the leading medical authorities including Persian scholar Avicenna and the Roman encyclopedist Pliny. He especially reviles Galen of Pergamum whom he labels the "arch fool" of medicine, although in another curious passage he claims – from the manuscripts –to have arrived at a better understanding of Galen and Avicenna than the school physicians themselves.[42] He rails against the mercenary nature of academic physicians and asserts that those who studied at universities such as Padua, Paris and Montpellier are not the "Professors of Truth" but rather the "Confessors of Lies."[43] Likewise in keeping with the authentic Paracelsian corpus, he advocates what can be learned from nature, as well as the wisdom of the common folk, over formal academic knowledge. However, he concedes that this folk-wisdom is less effective when employed by the untrained, since "a physician knows how to wield the medicine more artfully than the common man."[44] However, "the physician is a poor

lebens, so der mensch natürlich hat empfangen durch beschaffung durch die hant gotes. dises tierische leben ist an im selber nichts anders dan feur und luft. das ist also zu verstehen: der mensch sovil sein tierisch leben betrift, ist aus den vier elementen, das ist das wasser und die erden, daraus das corpus des menschen beschaffen ist, ist das haus und corpus des lebens. ich verstehe alhier nicht das leben, welchs leben aus der selen, das ist aus dem atem gottes entspringt; dan meine meinung ist an diesem ort nicht theologisch, sonder arzneiisch, sonder ich verstehe das leben, welchs tierisch und zergenglich ist, welchs leben aus feur und luft geschaffen. und also ist das corpus, so aus erden und wasser geschaffen ist, ein haus des lebens worden. Und das ist gnug zu verstehen, wie der mensch zweierlei leben habe, als nemlich das tierische und das siderische leben. … dan der tierische und siderische leib ist ein ding und nicht zwei, und das also." Translation based (with modifications) on Paracelsus, *Selected Writings*, ed. Jolande Jacobi, trans. Norbert Guterman (New York, 1951), 18.

41 Sudhoff, 1/14: 637, 649; Huser, 3: 79, 93.

42 Sudhoff, 1/14: 657, 623; see also 608, 620, 621 and Huser, 3: 102, 80; see Didier Kahn, "*De Pestilitate* and the Paracelsian Cosmology," *Daphnis*, 48 (2020), 65–86.

43 Sudhoff, 1/14: 637; Huser, 3: 79.

44 Sudhoff, 1/14: 606: "sol auch ein doctor mit der arznei wissen zu hantlen kunstlicher dan der gemeine man"; Huser, 3: 38.

creature, if he is only able to equip himself from paper books."[45] He also offers the clever quip that book learning can no more make you a successful physician than reading Livy can make you a successful general.[46] Rather, a skillful doctor must have the esoteric wisdom of the cabala and be able to discern the art of signatures.[47] Such a physician is naturally a Christian, and the author belittles academic physicians whose knowledge is derived only from Pagan authors. [Ps.-]Paracelsus also makes occasional anti-clerical swipes, noting that both "cloisters and brothels"[48] are especially open to demonic agency due to their known lustfulness, and that while a monastery near Braunau possessed a book containing a true commentary on Galen and Avicenna, "the hopeless monks could not understand it."[49]

7 How Plague Strikes

Our author is generally clearer in *De Pestilitate* on the etiology of plague than Paracelsus had been in some of his undisputed treatises. He posits a twofold pathway for the dissemination of plague through natural and supernatural means, which quickly becomes threefold, with the addition of the role of imagination in spreading plague:

> Thus, plague comes naturally through food and drink, and supernaturally via the heavens, therefore this chapter will chiefly treat, as I have noted, the twofold nature of plague, natural and supernatural. For one seed is through food and drink, and the other is caused by the heavens and the stars, and the third way through the imagination, as through fear, horror, etc.[50]

45 Sudhoff, 1/14: 608: "an ein arzt ist ein arme creatur, so er allein aus papirnen büchern sich behelfen will"; Huser, 3: 40.

46 Sudhoff, 1/14: 609; Huser, 3: 40.

47 Sudhoff, 1/14: 607: "das leret dich die kunst signatura in einem ieden dinge; dan also muß cabalia gefüret und gelehret werden von denen, die da wöllen arzt sein"; Huser, 3: 38.

48 Sudhoff, 1/14: 649; Huser, 3: 93.

49 Sudhoff, 1/14: 623; Huser, 3: 60.

50 Sudhoff, 1/14: 632: "Also wird auch pestis durch speis und trank natürlich und durch den himel ubernatürlich. also wird nun dises capitel handeln fürnemlich, und wie ich zweierlei pestes gemeldet, natürlich und ubernatürlich: der eine same ist durch speis und trank, der ander ist durch den himel und sterne verursacht und der dritte weg durch die imagination, als durch furcht, erschrecken, etc."; Huser, 3: 71.

While [Ps-]Paracelsus offers a reasonably straightforward threefold mechanism for the plague to strike, in the book he generally stresses the primacy of astral agency. He connects the spread of plague with many known astrological configurations, and at points, specifically notes the role of the sun, moon, and Mars in sending or "igniting" plague. *De Pestilitate* also includes the theme that Paracelsus developed in *De Peste* that, at times, our evil intentions and imaginings poison the heavens. The heavens then reflect these poisonous intentions back upon the earth with devastating effect.[51] Thus, humans may be "plague shot" (*Sterngeschoß*) as they are infected by the deadly disease solely through the agency of the stars.[52] In other portions of the work he notes that the stars are the "hands" and "rods" of the macrocosm executing the cosmic will.[53]

[Ps.-]Paracelsus' concept of astral infection of plague through heavenly agency connects to his justly famed concept of *imagination*, which has led him to be seen by some as a "harbinger of psychosomatic medicine."[54] As our author comments,

> For when I speak and philosophize of the imagination, I will reveal the secret things, how people through fear, horrors, and sorrow and a powerful imagination fall into illness so easily, and are thrown into sorrowful death. As in the time of the plague, [it happens] to pregnant women and other people, women and even men.[55]

Thus, the power of the imagination, which had likewise been a central theme in *De Peste*, plays a major role in *De Pestilitate* in a manner that connects successfully both to his magical theory of the interrelation of the macrocosm and

51 *Libri tres de Peste*, Sudhoff, 1/9: 571, 579, 592. See also Didier Kahn, "Paracelsus' Ideas on the Heavens, Stars and Comets," in Miguel Angel Granada et al. eds., *Unifying Heaven and Earth: Essays in the history of Early Modern Cosmology* (Barcelona, 2016), 59–116; Frederick R. Jevons, "Paracelsus's Two-Way Astrology," *The British Journal for the History of Science*, 2 (1964), 139–155.

52 Sudhoff, 1/14: 615, 624; Huser, 3: 48, 87. In other passages, the author speaks of "*excrementa stellarum*"; see Sudhoff, 1/14: 615; Huser, 3: 48.

53 Sudhoff, 1/14: 633; Huser, 3: 74.

54 Heinz Schott, "'Invisible Diseases' – Imagination and Magnetism: Paracelsus and the Consequences," in Ole Peter Grell, ed., *Paracelsus: The Man and His Reputation, His Ideas and Their Transformation* (Leiden, 1998), 309–321, here 309.

55 Sudhoff, 1/14: 611: "dan so ich werde von der imagination reden und philosophiren, werde ich die heimlichen dinge offenbaren, warumb so leichtliehen durch furcht, schrecken, trauren und große einbildung so leichtlieh die leute in krankheiten fallen und dem gehen tode underworfen werden. als zu zeiten der pestilenz mit den schwangeren frauen und andern personen, frauen und auch mannen." Huser, 3: 43.

the microcosm but also blends with his theology of human culpability and divine retribution. For example, [Ps.-]Paracelsus states: "When a woman is not mindful of herself or God … in such forgetfulness she imprints upon her fruit an *unform* and a shadow of the image."[56] Thus, the imagination can be a powerful agent for the spread of plague or other maladies. He sometimes treats imagination as a separate category, but at other times views it as a secondary mechanism for supernatural plague alongside astral forces.

While the treatise describes numerous vehicles for the spread of plague, his theory has generous room to encompass divine agency and human culpability. As we have seen above, negative thoughts can effectively poison the heavens leading to infection by plague due to astral influence. A dissolute lifestyle, lived with an excess of food, leads to internal corruption which can also foster the coming of plague.[57] Likewise, the collective sins of humanity can ignite the stars and the sun, prompting a plague. As our author admonishes, "Thus your will and imagination is a mother of this birth of the plague in you yourself, [which you] have brought upon yourself."[58] Focusing again on human agency over the imagination, the author states: "For the world with its imagination is like an ape, for that the human does, the ape will do likewise."[59] However, humans are not without divinely sanctioned aid in a time of plague; faith itself offers some protection against the plague.[60] God both allows humanity to perceive the hidden things of nature and restrains some of the most deadly astral influences.[61] Such an emphasis on plague as the visitation of divine wrath for collective sins was not uncommon in theological works in this period, but [Ps.-]Paracelsus' emphasis on divine agency is rather pronounced for a medical writer on plague.

8 Folklore & Demonology

Another feature of the work which was, no doubt, embarrassing to Paracelsus' modern advocates is *De Pestilitate's* espousal of folkloric cures. While a toad

56 Sudhoff, 1/14: 614: "Die frau wenn die imaginiret, so gedenket sie nimer an sich selbs, vil weniger an got. also auch durch solche ire vergessenheit imprimiret sie irer furcht ein unform und schatten der biltnis"; Huser, 3: 46.

57 Sudhoff, 1/14: 636; Huser, 3: 77.

58 Sudhoff, 1/14: 647: "also ist dein will und imagination eine muter diser geburt der pestilenz in dir selbs, durch dich selbs geursachet"; Huser, 3: 89.

59 Sudhoff, 1/14: 615: "dan die welt mit irer imagination ist gleich einem affen; denn alles so der mensch tut, will ihm der affe nachtun"; Huser, 3: 40.

60 Sudhoff, 1/14: 622; Huser, 3: 59.

61 Sudhoff, 1/14: 634–635; Huser, 3: 75.

cure had also been included in the Sterzing treatise,[62] [Ps.-]Paracelsus' emphasis on possible treatments using eels, toads and frogs is much more extensive in this work. The following quotation offers a reasonable taste of this logic:

> I write here of the plague and its cure: its cure is in the frogs, in the toads, in the fish, [and] in birds. Therefore, everyone should note: A moray eel (*muraena der fisch*) is a cure for plague: if you bind a live [eel] under the armpits before sunrise, it draws out the poison of plague: however, it has no power behind the ears. The frog, whatever the type or color it is, if it is bound to the groin of men, also draws the poison of plague away, if it is bound [there] before sunrise.[63]

Passages like these represent a challenge to the dated monolithic vision of Paracelsus as the progressive experimentalist, repudiating medieval academic medicine. However, even here our author takes pains to develop a rationale for these folk-cures based on the logic of the *tria prima*, within the context of how water serves as a matrix for disease.

De Pestilitate also goes beyond the topical boundaries of the undisputed Paracelsian plague corpus in its speculations on a demonological dimension of the spread of plague. It furthermore embraces much of the contemporary, highly gendered witchcraft paradigm with its acceptance of active demonic agency.[64] In perhaps the most salacious comment of the work, our author writes:

> Thus, the Mumia, that is the fire of man and the magnet of blood, that is the spirit of life, also draws to itself the poison of plague from the poison of the menstruation of women. Then as the basilisk emerges out of the poison of the menstruation of the woman, the plague also emerges in humans from such a menstrual poison. That's why a person is also

62 Sudhoff, 1/9: 552.

63 Sudhoff, 1/14: 606: "welche ire cur in den fröschen hat, in den kröten, in den fischen, in vöglen. darumb merkent alle wol. muraena der fisch ist ein cur pestis under der uchsen, also lebendig vor der sonnen aufgang aufgebunden zeucht aus venenum pestis, aber hinder den oren hat es keine kraft. der frosch, was geschlechts oder farben der ist, auf die schlichten der mannen aufgebunden, zeucht venenum pestis auch hinweg, vor der sonnen aufgang aufgebunden"; Huser, 3: 37. Cf. Hanns Bächtold-Stäubli et al., eds., *Handwörterbuch des deutschen Aberglaubens*, vol. 3 (Berlin/Leipzig, 1930/1931), 124–142.

64 Although he does not engage *De Pestilitate* per se, see Johannes Grabmayer, "Heilige, Heiler und Hexen: Volksmedizin um 1500," in Albrecht Classen, ed., *Paracelsus im Kontext der Wissenschaften seiner Zeit* (Berlin, 2010), 183–200.

> contagious with his eyes, to poison another person, also the moon, stars, and the firmament of heaven.[65]

This spirit of popular credulity is also manifest in [Ps.-]Paracelsus' retelling of an unknown episode of plague in Austria: "It has also happened in St. Veit and Villach that in the time of plague such witches have taken earth and dust from the graves of the dead and prepared them by means of their sorcery, and thus conjured a strong plague and through it many thousands of people have become ill and died." [66] Due to this demonic threat, the authorities would be well advised to employ "pious and god-fearing magicians, who would understand this type of thing."[67]

9 Schema and Comparison

One of the more curious features of *De Pestilitate* are the overlapping organizational schema or lists that our author/compiler deploys over the breadth of the text to explain the origin and course of plague. These include:

- the two and/or three ways in which plague strikes (natural, supernatural and/or imagination)[68]
- the six locations of plague (one for each side for neck, armpit, and groin areas)[69]

---

65 Sudhoff, 1/14: 657: "also zeucht auch mumia, id est ignis hominis et magnes sanguinis, id est spiritus vitae an sich das venenum pestis, aus dem veneno menstrui mulierum. dan wie sich erhebt der basiliscus aus dem gift menstrui des weibes, also erhebt sich auch die pestis im menschen von solchem menstruosischen gift. darum auch der mensch mit seinen augen contagiosisch wird, einen andern menschen zu vergiften, auch den mon stern und das firmamenturn coeli"; Huser, 3: 103.

66 Sudhoff, 1/14: 649: "Es ist auch zu Sanct Veit und zu Villach erfaren, das solche heren haben genomen die erden und staub von den grebem der toten in derzeit der pestilenz und die zugericht mit irer zauberkunst und eine heftige pestilenz damit angerichtet, dadurch vil 1000 menschen inficirt und gestorben seind"; Huser, 3: 92.

67 Sudhoff, 1/14: 649: "solt gut acht gegeben werden auf solche ding, … von der obrigkeit, das sie frome und gotfürchtige magos hetten, welche sich auf solche ding verstünden"; Huser, 3: 93.

68 Sudhoff, 1/14: 632; Huser, 3: 71. Perhaps these could more properly be considered two forms, with the supernatural forms possessing two mechanisms: imagination and the stars.

69 Sudhoff, 1/14: 597; Huser, 3: 24. In *Zwey Bucher* and *De Peste* Paracelsus more commonly speaks of the three locations of plague; see Sudhoff, 1/9: 567 and idem, 1/8: 374.

a fourfold theory of plague that relates to Empedoclean four element theory, which then gets multiplied by each of the elements to become the sixteen forms of plague[70]
two tinctures of humanity (astral and animal) that influence how plague impacts humans[71]
the five *entia* or essences of disease[72]
seven pulses that relate to the seven planets[73]
the three bearers (the heavens, the imagination, and water) of plague.[74]

This list is not definitive and does not reflect the consistent effort through much of the work to integrate plague theory with the concept of the interaction of the *tria prima* of mercury, salt, and sulphur. While these organizational schemes may appear contradictory at the surface level, they are generally reconcilable as one moves from disease etiology to attempted explanations of the generation of chemical toxins, and onward to moral culpability, and finally to divine agency.

Comparison of Paracelsus' Plague Treatises

	Zwey Bücher von der Pestilentz und jhren zufällen	*De peste libri tres*	*Von der Pestilentz [...] an die Statt Sterzingen*	*De Pestilitate*
Humoral Theory	Rejects	Mixed	Employs	Rejects
Anti School-Physicians	Yes	Yes	Implicit but mild	Yes

70 Sudhoff, 1/14: 636: "Dan im menschen sind viererlei oder 4 ignes, 4 aquae, 4 aeres 4 terrae. als ignis per se (id est sol), ignis aquae, ignis aeris, ignis terrae. also aqua per se, aqua ignis, aqua aeris, aqua terrae. Also auch aer per se, aer ignis, aer terrae, aer aquae. also auch terra per se, terra aeris, terra ignis und terra aquae. Dieser verstaut, weil er cabalistisch ist und euch etwas schwer mag dünken, wisset also zu underscheiden"; Huser, 3: 78. See also Sudhoff, 1/14: 616–617; Huser, 3: 49–50.

71 Sudhoff, 1/14: 600; Huser, 3: 29.

72 Sudhoff, 1/14: 645, 654; Huser, 3: 87, 99.

73 Sudhoff, 1/14: 625; Huser, 3: 62–63.

74 Sudhoff, 1/14: 612–613: "das element wasser aber das ist der acker; die imaginatio hominis ist der same. der himel und imagination microcosmi ist der koch und bereiter des samens in sein ultimam materiam zu bringen, welcher same aus der imagination der menschen geboren wird: die drei ding geberen die pesten"; Huser, 3: 45.

(*cont.*)

	Comparison of Paracelsus' Plague Treatises			
	Zwey Bücher von der Pestilentz und jhren zufällen	***De peste libri tres***	***Von der Pestilentz [...] an die Statt Sterzingen***	***De Pestilitate***
Primacy of Astral Causation	Yes	Yes	Yes	Yes
Reverse Astral Causation	Yes	Yes	No	Yes
Imagination	Yes	Yes	No	Yes
Basilisk or Astral shot	Yes	Yes	No	Yes
Plague as Divine Wrath	Yes	Yes	No	Yes
3 Locations of Plague	Yes	Yes	Yes	Yes
Tria Prima	Yes	Yes	No	Yes
Microcosm-Macrocosm	Yes	Yes	Mild	Yes
Bloodletting	No	No	Yes	No
Folkloric Cures	No	No	Yes (Toads, etc.)	Yes – extreme
Demonic Agency	No	Limited/Mild	No	Yes

The comparative table of Paracelsian plague treatises gives at-a-glance confirmation of the contention that the plague theory of *De Pestilitate* is generally consistent with the genuine works of Paracelsus. *De Pestilitate* is, in fact, much more in harmony with the *Zwey Bücher von der Pestilentz* and *De peste* than is *Von der Pestilentz, ein Büchlein geschrieben an die Statt Sterzingen.*

Conclusion

While this focused study cannot offer a definitive assessment of the Paracelsian qualities of the work, the comparison of *De Pestilitate* with Paracelsus' mature plague treatises reveals a general consonance with Paracelsus' corpus. The plague theory is genuinely Paracelsian, although the work's extreme qualities seem held less in check by Paracelsus' consistent Biblicism or the mature astrological theory seen in the *Zwey Bücher von der Pestilentz* and *De Peste.* Ironically, the Paracelsus plague text that we can situate most clearly within a particular episode of Paracelsus' career, the Sterzing treatise (*Von der Pestilentz … an die Statt Sterzingen*), is less fully Paracelsian in its communication of Paracelsus' plague theory than is *De Pestilitate.* Viewed from the perspective of its plague theory alone, *De Pestilitate* is not the outlier among Paracelsian plague treatises. While it goes beyond the other plague treatises in its folkloric content, speculations concerning demonic agency, and general credulity, it is not overburdened by excessively hermetic or neo-platonic theory throughout.[75] It is arguably more driven by the Christian magical-astronomical worldview of the *Astronomia Magna* than the conventions of Renaissance Hermeticism.

While the ambiguous status of *De Pestilitate* in Paracelsus' canon remains unchanged, this study has shown that its Paracelsian qualities are of a higher order than many Pseudo-Paracelsian texts. It shares some commonality with the *Tabula physica, astronomica & medica de pestilitate* and other treatises, and its fragmentary structure and its competing organizing schemes hint at authentic Paracelsian sources or even a reworking of genuine Paracelsica.[76] Ultimately, the text is simply too strange to think that someone would have invented it from whole cloth in the late 1500s. All told, it might be considered a "quasi-Paracelsian" quilt, albeit of an uneven quality. Despite its defects and its lack of appeal to Paracelsus' twentieth-century champions, the work enjoyed a wide and influential readership in the seventeenth century, including Oswald Croll, Jacob Boehme, and Jan Baptist Van Helmont.[77] *De Pestilitate* offers an accessible introduction to the Paracelsian magical worldview, and it likewise serves as a generally reliable guide to Paracelsian plague theory.

---

75 There are certainly Platonic echoes in part one of the treatise, where our author speaks of an unknown work entitled "*Volumen de Anima & Spiritu Mundi*," Sudhoff, 1/14: 613 and Huser, 3: 44; see Kahn, "*De Pestilitate* and the Paracelsian Cosmology," 66.

76 Kahn has discerned many more parallels with other Paracelsian texts including *De matrice* and *De meteoris*; see idem, "*De Pestilitate* and the Paracelsian Cosmology."

77 See most recently Jo Hedesan, *An Alchemical Quest for Universal Knowledge: The "Christian Philosophy" of Jan Baptist Van Helmont (1579–1644)* (London, 2016), 75–76, 95. Even the inquisitor Johannes Faber engaged the work: see Lyke de Vries and Leen Spruit, "Paracelsus and Roman censorship – Johannes Faber's 1616 report in context," *Intellectual History Review*, 28 (2018), 225–254.

The *Astronomia Olympi novi* and the *Theologia Cabalistica*: Two Pseudo-Paracelsian Works of the *Philosophia Mystica* (1618)

Martin Žemla
Palacky University Olomouc
zemlam@gmail.com

Abstract

The *Astronomia Olympi novi* and the *Theologia Cabalistica* were published as part of the *Philosophia Mystica* (1618). This influential collection of Paracelsian and Weigelian texts was among the first to include a publication of the *theologica* of Paracelsus. Both of these short pseudo-Paracelsian works were written by Adam Haslmayr (1560–1630), the propagator of the "Theophrastia Sancta," a philosophical theology of Paracelsus mixed with Weigelian and alchemical influences. These works, taken in the *Philosophia Mystica*, are among the very few of his texts that appeared in print in his lifetime. In this paper, the content of both pseudepigrapha is analyzed and related to ideas found in both the authentic and the pseudepigraphic works of Paracelsus, as well as in relevant works of some Paracelsians (Khunrath, Figulus, Weigel, Croll, Siebmacher and Mögling). Specifically, the notions of the "*Olympus novus*" and the "three cabalistic principles" are discussed in a wider context.

Keywords

Paracelsus – Valentin Weigel – Adam Haslmayr – Johann Siebmacher – Benedictus Figulus – *Theophrastia Sancta* – cabala – alchemy – universal reform – Rosicrucianism

1 The Pseudo-Paracelsian Works of the *Philosophia Mystica*

The *Astronomia Olympi novi* and the *Theologia Cabalistica*, allegedly written by "Theophrastus," were published in 1618 in "Newstat," i.e., Frankfurt,[1] as part

1 For the identification of "Newstat" as Frankfurt, see Hermann Geyer, *Verborgene Weisheit. Johann Arndts "Vier Bücher vom Wahren Christentum" als Programm einer spiritualistisch-hermetischen Theologie*, 3 vols. (Berlin-New York, 2001), 1: 34–39.

 | DOI:10.1163/9789004503380_008

of the *Philosophia Mystica*, an influential compendium of Paracelsian and Weigelian texts.[2,3] It was one of the numerous publications on medical and spiritual alchemy prepared by the Frankfurt engraver, publisher and bookseller Lucas Jennis between 1616 and 1630.[4] However, in its composition, scope, extent and influence, it was quite unique.

The *Philosophia Mystica* is divided into two parts, consisting of fifteen texts.[5] In addition to the two pseudo-Paracelsian works, the first half includes the *De poenitentiis*, *De iustitia* and *De confessione, poenitentia et remissione*, as well as the *Commentarius in Danielem Prophetam* by Paracelsus, plus a "Life and teachings of the hermit Nicolaus in Switzerland," the latter being actually an epitome of the Eckhartian teaching on the *abegescheidenheit* ("detachment"). The other part consists of five works ascribed to Valentin Weigel (1533–1588), most of them authentic: 1) the *Einführung in teutsche Theologiam*, 2) *Scholasterium Christianum, Vom himmlischen Jerusalem*, and 3) the *Betrachtung vom Leben Christi*; 4) the *Philosophia theologica, daß nemblich Gott allein gut sey* was written by Weigel's deacon Benedict Biedermann, who penned much

2 *Philosophica Mystica, Darinn begriffen Eilff unterschidene Theologico-Philosophische / doch teutsche Tractätlein / zum theil auß Theophrasti Paracelsi, zum theil auch M. Valentini Weigelii* [...] *bißhero verborgenen manuscriptis der Theosophischen Warheit liebhabern* [...] *beyde Liechter / der Gnaden und der Natur / in uns zuerwecken* [...] "Gedruckt zur Newstadt / und zu finden bey Lucas Jennis / Buchhändler", 1618 (hereafter: "PM"). Cf. Horst Pfefferl, "Einleitung," in Valentin Weigel, *Sämtliche Schriften. Neue Edition*, ed. Horst Pfefferl (Stuttgart-Bad Cannstatt, 1996–2015), 7: xxxvii and xii, with relevant notes.

3 This study is a result of research funded by the Czech Science Foundation as the project GA ČR 19–07439S "Philosophy of Renaissance Medicine in the Czech Lands (Jessenius – Hájek – Khunrath – Paracelsus). New Approaches and Contexts." I would like to thank Didier Kahn for his insightful revision of the text and his useful suggestions and emendations.

4 For the (incomplete) list, see <www.alchemywebsite.com/printer_jennis.html>, last accessed 10 October 2021. In fact, the title-page announces Jennis only as the seller of the book ("Buchhändler"), and Pfefferl mentions the possibility that Heinrich Starck from Neustadt an der Hard was the printer of the book (Pfefferl, 10: xix; cf. the "Verzeichnis der im deutschen Sprachraum erschienenen Drucke des 17. Jahrhunderts" [VD17], position 3:604297L, <http://www.vd17.de>, last accessed 10 October 2021). Together with Johann Karl Unckel, Starck bought the printing office of Nikolaus Schramm in Neustadt in 1614 (see <https://thesaurus.cerl.org/record/cni00031389>, last accessed 10 October 2021). He printed, for example, the *Psalter und Psalmen Davids* (1619), bearing his and Unckel's name on the title-page ("Gedruckt zur Newstadt and der Hard / durch Heinrich Starck / In Verlegung Johann Carl Unckels / Buchhändlers zu Franckfurt am Mayn"). On Jennis, see Wilhelm Kühlmann, "Poeta, Chymicus, Mathematicus. Das Stammbuch des böhmischen Paracelsisten Daniel Stoltzius von Stoltzenberg," in Joachim Telle, ed., *Parerga Paracelsica. Paracelsus in Vergangenheit und Gegenwart* (Stuttgart, 1991), 277–300, here 278–279. For the notion of "spiritual alchemy", see Mike A. Zuber, *Spiritual Alchemy from the Age of Jacob Boehme to Mary Anne Atwood, 1600–1900*, PhD thesis, University of Amsterdam, 2017, 34–51.

5 Cf. Karl Sudhoff, *Bibliographia Paracelsica* (Berlin, 1894), 513–516.

of the pseudo-Weigeliana;[6] 5) the last and second longest text, the *Introductio hominis, oder kurtze anleitung zu einem christlichen Leben*, was published by an "anonymous" author. It was followed by a short *Appendix*, still under the heading of the "Kurtze Anleitung", signed by "Huldrich Bachsmeier von Regenbrun" – an anagram of Johannes Siebmacher, the author of the renowned theoalchemical work *Wasserstein der Weisen* (1619).[7] The final page is dedicated to the *Querela Authoris*, lamenting over the blindness of his contemporaries who do not recognize the precious "Stone."

The works by Paracelsus published in the *Philosophia Mystica* deal with genuinely theological matters; only the *Astronomia Olympi novi* and the *Theologia Cabalistica* use some vague astrological motifs. Paracelsus' *Commentarius in Danielem* added an apocalyptic tone suggesting that the last days were coming, bringing an ever deeper and clearer understanding of things and of the words of Christ, although the exact date of the end of time remained unknown.[8] Finally, alchemical elements were admixed, mostly in the *Introductio hominis* and the *Appendix*.

The *Philosophia Mystica* was one of the earliest printed publications of the theological works of Paracelsus; it was named "the most important forerunner of the theological edition" of Paracelsus.[9] The theological texts by Paracelsus

6 Horst Pfefferl, *Die Überlieferung der Schriften Valentin Weigels* (Marburg/Lahn, 1991), Kap. B, 381.

7 For the identification of Siebmacher, see Carlos Gilly, "*Theophrastia Sancta.* Der Paracelsismus als Religion im Streit mit den offiziellen Kirchen," in Joachim Telle, ed., *Analecta Paracelsica* (Stuttgart, 1994), 425–488, on 456 n. 69 (cf. the somewhat differing English version in Ole P. Grell, ed., *Paracelsus: The Man and His Reputation, His Ideas and Their Transformation* [Leiden, Boston and Cologne, 1998], 151–185 and the updated version, archived here: <https://web.archive.org/web/20190824152935/> <https://www.ritmanlibrary.com/collection/comparative-religion/theophrastia-sancta-paracelsianism-as-a-religion-in-conflict-with-the-established-churches/>, note 49.); Carlos Gilly, *Adam Haslmayr. Der erste Verkünder der Manifeste der Rosenkreuzer* (Amsterdam, 1994), 184; Zuber, *Spiritual Alchemy*, 54 and 74–95. A manuscript dated 1607 proves that the *Introductio* was written by the same author, "Huldrich Bachsmeier von Regenbrun," i.e., Johannes Siebmacher von Nürnberg, as the *Wasserstein der Weisen*, published in 1619 in Frankfurt and repeatedly reedited (1661, etc.), even under the title *Das goldene Vlies, oder Das allerhöchste, edelste, kunstreichste Kleinod und der urälteste verborgene Schatz der Weisen* (referred to in the PM, 269, but published under this title only in 1736); see the updated version of Gilly's *Theophrastia Sancta*, note 49 (at the archive URL above).

8 Paracelsus, *Commentarius in Danielem*, PM, 79. The critical edition was published in Paracelsus, *Sämtliche Werke*, 2. Abt.: *Theologische und religionsphilosophische Schriften*, ed. Kurt Goldammer, vols. 2–7 (Wiesbaden, 1955–1973) (hereafter: Goldammer, II/2, II/3, etc.), here II/7.

9 Paracelsus, *Theologische Werke 1: Vita Beata – Vom seligen Leben*, ed. Urs Leo Gantenbein et al. (Berlin-New York, 2008) (hereafter: Gantenbein), 1: 38.

published here and known, then, only in manuscripts, together with the texts accompanying them, could – and, obviously, *should* – substantiate the picture of Paracelsus as a theologian and, as the *Theologia Cabalistica* remarks, the author of the *sancta Theophrastia*.[10]

Under this title, Adam Haslmayr (ca. 1560–after 1630), the prolific Tyrolean Paracelsian who penned many spiritual, alchemical, cabalistic, and magical works, preferred to propagate the philosophical religion of Paracelsus as he understood it, i.e., as a combination of Paracelsus' theology and natural philosophy with Weigelian theology as well as with alchemical, magical, and cabalistic motifs. Not much of Haslmayr's work was printed during his life. In fact, according to Carlos Gilly, besides his two "Gesangbücher" and his *Antwort an die lobwürdige Brüderschafft der Theosophen von RosenCreutz* (1612), published again two years later together with the first edition of the *Fama fraternitatis*, only two other works were published in his lifetime: the *Astronomia Olympi novi* and the *Theologia Cabalistica*.[11] The style, terminology and content of these two brief pseudo-Paracelsian works from the *Philosophia Mystica* indicate that their author was one and the same person. Besides the *sancta Theophrastia* mentioned in the *Theologia Cabalistica*, there is also a reference in the *Astronomia Olympi novi* to the "Signatstern," also a Paracelsian term preferred by Haslmayr.[12] And most obviously, both texts refer to the "three cabalistic principles," another of his favored motifs.

How did the two works find their way to the publishing house of Lucas Jennis? According to Carlos Gilly, Haslmayr himself was appointed to manage a secret printing press for August von Anhalt from 1611 onwards and to publish Paracelsus' and Weigel's theological works.[13] The aim was to spread the new religion of the *Theophrastia sancta*. But because of Haslmayr's imprisonment (he was sentenced to the galleys for four and a half years in 1612 for publishing his answer to the *Fama Fraternitatis*), most manuscripts must have been given over to other publishers, among others to Jennis. This way, the *Astronomia Olympi novi* and the *Theologia Cabalistica*, too, might have appeared in Jennis'

10 *Theologia Cabalistica*, PM, 49.

11 Gilly, *Haslmayr*, 184 and 202–203; idem, "*Theophrastia Sancta*," 450–451 and 456 (see also the English version, 171). Haslmayr's sign, his "Character Cabalisticus," appeared in the Latin and German versions of the *Nucleus sophicus* (Frankfurt, 1623), published by Jennis. An *Astronomia fidelium Paracelsica IVDICIVM d.i. 1613 Jahr* (i.e., perhaps the *Astronomia Olympi novi*) appears as No. 90 in the *Nuncius Olympicus*, reprinted by Gilly, *Haslmayr*; as for the *Theologia Cabalistica*, I could not identify it in the *Nuncius*.

12 For the "*stella signata*", see Gilly, *Haslmayr*, 188 and 199 n. 2.

13 Idem, "*Theophrastia Sancta*," 467–468.

workshop, together perhaps with the idea of joining them up with some more Paracelsian and Weigelian *theologica*. For Halsmayr, Weigel was a highly learned, pious and "enlightened" man, second only to Paracelsus, so that a plan to unite the "*theologia Paracelsi et Weigelii*" seemed quite natural for him.[14] Similarly, the German Mysticism which formed the core of the teachings of "brother Nicolaus ... the hermit in Switzerland", was also favored by Haslmayr.[15] He liked to call Paracelsus a "hermit:"[16] he was from Switzerland, and from "Einsiedeln", literally a "hermitage."

We know that Jennis collaborated with Johannes Siebmacher at that time. It was just one year after the publication of the *Philosophia Mystica* that Siebmacher's *Wasserstein der Weisen* left his printing press (the title chosen, possibly by the publisher, while the author might have had another preference).[17] Carlos Gilly even suggested in 1994 that Siebmacher might have been the editor of the whole *Philosophia Mystica*.[18] Given that his interests largely corresponded with those of Haslmayr, Siebmacher would surely be tempted to contribute to the volume, and to supplement alchemical motifs somewhat lacking here. But, as Gilly later noticed, some copies of the *Philosophia Mystica* contained an admonition to the reader written by the alchemist Benedictus Figulus (1567–after 1619); here he reported his acquisition of some texts by his friend Haslmayr which he appended to the volume.[19] Although it is impossible to say if the editorial plan for the *Philosophia Mystica* was his, it seems entirely plausible that Figulus, as a friend of Haslmayr and an experienced editor of Paracelsian works, might have adopted and adapted an existing project of this kind. It would have served his purposes, given his avowed aim to fight against rigid orthodoxy with the "cabalistic and theological books" of Paracelsus and to promote the "true christian philosophy" and Christ as the "highest spagyrus".[20]

14 See titles No. 120, 124 and 132 in the *Nuncius Olympicus*, in idem, *Haslmayr*, 267–269, see also 125–126.

15 Idem, "*Theophrastia Sancta*," 213, 216, 274.

16 See *Nuncius Olympicus*, No. 89, 92, 102, 177, in idem, *Haslmayr*, 262–278. On this appellation used by Paracelsus himself since 1527, see Eduard Schubert and Karl Sudhoff, *Paracelsus-Forschungen*, 2 (1889), 105.

17 Zuber, *Spiritual Alchemy*, 80.

18 Gilly, "*Theophrastia Sancta*," 456; idem, *Haslmayr*, 184.

19 CP 2: 229; CP 3: 1070; see also Gilly, *Haslmayr*, 184 and 202–203.

20 CP 3: 1075.

2 *Astronomia Olympi novi*

The *Astronomia Olympi novi* consists of seven short "aphorisms," presented as a genuine work "Theophrasti."[21] The author speaks of "our 4 books of *Meteors*,"[22] thus appropriating the *De meteoris* by Paracelsus, not to leave any doubts about the *Astronomia Olympi novi* authorship.[23]

Indeed, we recognize here a number of typically Paracelsian, but also Weigelian, terms and motifs: the "*magnalia* of God," rejection of all "sects," the "Aniadus"[24] and "Signatstern." But we find also allusions to the *Theologia Deutsch*: the teaching that there is nothing in hell but free will;[25] or the accentuation of the "obedience" to God – a frequent motif in the *Philosophia Mystica*.[26]

In its title, the *Astronomia Olympi novi* obviously refers to the lacking third part, or the third "*astronomia*," of the *Astronomia magna*, announced by Paracelsus on the "first figure."[27] He distinguishes "four astronomies" here: *naturalis, supera, olympi novi*, and *inferorum*. The natural astronomy deals with the firmament, the light of nature, man as microcosm and as a natural being made from the *limus terrae*. The *supera* astronomy "has its essential dwelling in heaven"; it is given to those who are "reborn" by Christ. The "astronomy of the new Olympus" stems from the "faith," and, as the title claims, it deals with "the powers of the new Olympus, or of the faith" (*Kräften Olympi novi seu fidei*). Whether the book really existed, or had never been written, we do not know. Huser (and others)[28] believed that it would appear one day; in the mean-

21 *Astronomia Olympi novi, Das ist, Die Gestirnkunst deß newen Himmels / welche allein auß dem Glauben entspringet / darauß der Mensch alle Magnalia Gottes und der Natur / die den glaubigen seind zuwissen / sehen und erlernen mag.*

22 *Astronomia Olympi novi*, PM, 34.

23 The reference may be to the two-volumes *Opera* (1616), 2: 106–125, a reedition of the four-volume 1603–1605 edition made after the quarto Huser edition of 1589–1591. The part Haslmayr had in mind is perhaps ibid., 107; see also ibid., 108: "wir anfahen unser Philosophiam der 4. Bücher *Meteororum* [...]" Haslmayr's interest in the *De meteoris* is mentioned by Gilly, *Haslmayr*, 182–183.

24 *Astronomia Olympi novi*, PM, 33 and 35.

25 In fact, the *Theologia Deutsch* speaks of the *own* will, see *Theologia Deutsch*, chap. 47; *Astronomia Olympi novi*, PM, 37; cf. *Scholasterium Christianum*, PM, 172.

26 *Astronomia Olympi novi*, PM, 38; *Das Leben und Lehrpuncten*, PM, 104; *Vom himmlischen Jerusalem*, PM, 191. Cf. *Theologia Deutsch*, chap. 14 and 15.

27 *Astronomia magna*, in Paracelsus, *Bücher und Schrifften*, ed. Johannes Huser, 11 vols. (Basel, 1589–1591) (hereafter: Huser), 10: 66 and the "Catalogus" at the beginning of vol. 10. Sudhoff, *Bibliographia*, 514–515, assumed the *Astronomia Olympi novi* could be an abridged version or a draft of *Astronomia magna*.

28 For example, Michael Toxites; see CP 2: 228–229.

time, this "deficiency" should be recompensed with other writings in which Paracelsus dealt with faith and its powers. Huser also notes that this astronomy is practiced by the believers, "who can bring their natural faith and imagination to such a power and exaltation that they can do with it everything that otherwise happens by the natural astronomy of heaven."[29]

What does the title "*Astronomia Olympi novi*" really mean? *Astronomia*, of course, is to be understood in the Paracelsian way. As Paracelsus explains in the foreword to the *Astronomia magna*, "God has given to man all arts and everything natural through the firmament, so that the firmament is the natural light and man has all nature from the firmament."[30] That is why astronomy, dealing with invisible powers in things and in man, is the "mother of all other arts."[31] And as Paracelsus continually reminds his reader in the *Astronomia magna*, he is writing in the "natural light" only. Thus, there remains space for the "supernatural." According to Huser's recommendation, the reader can use other writings of Paracelsus to fill in the gap, to find this higher *astronomia*, a "divine" *astronomia*.

Why "*Olympi novi*"? The intimation of the third book of the *Astronomia magna* was not the only place where Paracelsus used the terms "Olympus," "Olympic." For example, the *impressiones Olympi* refer to the effects coming from the sky, i.e., rain, dew, etc.[32] The *Astronomia magna* names *methridatum Olympi* among the *siderische arcana*, next to *thyriaca coelestis*, thus connecting it with heaven and the stars;[33] a similar analogy is obvious from the pairs *panis firmamenti* / *panis agri* and *vinum olympi* / *vinum terrae*;[34] or from formulations like "*der inclination schuld von dem olympischen sidus*."[35] In the *Bücher von den unsichtbaren Krankheiten*, Paracelsus mentions the "*olimpische Geister*" in the context of imagination and calls them the "*gestirn* in man."[36] The pseudo-Paracelsian *Archidoxa magica* even claims that the "*gestirn* in man [can] be led by the *olympischen geist* and brought into another one," and such a

29 Huser, "Ad lectorem," in idem, 10: 364. Cf. the *addenda* ("Weiteres") to the *Astronomia magna*, in Paracelsus, *Sämtliche Werke*, 1. Abt.: *Medizinische, naturwissenschaftliche und philosophische Schriften*, ed. Karl Sudhoff, 14 vols. (Munich, then Munich-Berlin, 1922–1933) (hereafter: Sudhoff, 1/1, 1/2, etc.), here 1/12: 448.

30 Ibid., 1/12: 3. Similarly, the *Astrum* is, according to Gerhard Dorn's *Dictionarium Theophrasti Paracelsi* (1584), 20: "virtus & potentia rerum ex praeparationibus acquisita [...]."

31 *Astronomia magna*, Sudhoff, 1/12: 5.

32 *Das Buch von den tartarischen Krankheiten*, Sudhoff, 1/11: 116.

33 *Astronomia magna*, Sudhoff, 1/12: 98.

34 Ibid.: 222.

35 Ibid.: 230.

36 *Bücher von den unsichtbaren Krankheiten*, Sudhoff, 1/9: 298–299. Cf. e.g., the *Archidoxa magica*, idem, 1/14: 488.

transfer is possible also for the transmutation of metals.[37] According to Oswald Croll, "the inner heaven of man, which is the *Spiritus Olympicus*, contains all stars in particular."[38] The "*spiritus Olympi*" were also the subject of a distinctive discussion, in the context of the "*Olympica magica*," in an influential work on magic, the *Arbatel* (Basel, 1575) – one of the first early modern books to use the term "*theosophia*." According to the *Arbatel*, the "Olympic spirits" are inhabitants of the firmament and its stars; their duty is to administer destiny; they are powers but are, at the same time, also real, separate "spirits" which can be magically controlled.[39]

Thus, for Paracelsus, "Olympus" is related to the powers of both the starry heavens and the "*gestirn* in man."[40] But here, the "*new* heavens" are advocated, accessible *only* to the believers, in contrast to the "natural" heavens: "the astronomy of the new heavens stemming only from faith, through which man can see and learn all the *magnalia* of God and of nature which are to be known to the believers."[41] The new heavens are Christ himself; in him, believers can see everything "according to the vision of the veritable, holy, unspoiled faith."[42]

It seems, however, that later authors were not systematic in distinguishing between the "new" and the "old" Olympus, and the standard meaning of the Olympus as the "*divine* Olympus" became established. This is the case with the *Nuncius Olympicus* (1625), the important catalogue of "some secret books and treatises," collected by a "famous theosopher and physician" (i.e., Adam Haslmayr) and dealing with "*Theosophia, Cabala, Magia, Chemia, Medicina und Pilologia* [sic]."[43] The introductory admonition "*Ad Momum*" already indicates the "Olympic" character of the listed works – Momus was a mocker and

---

37 *Archidoxa magica*, Sudhoff, I/14: 488: "das gestirn im menschen [kan] [...] durch den olympischen geist gefürt und hingetragen werden in ein anderen [...]"; cf. idem, I/10: 644: "gestirn ist das obriste geschöpf der zerstörlikeit und ist im Olympo [...]."

38 Oswald Croll, *Basilica chymica* (Prague, 1609), 30.

39 *Arbatel. De magia veterum* (Basel, 1575), 22 and 35–37. For *theosophy*, see ibid., 86; the earlier mention is in Balthasar Flöter's dedication in *Philosophia magna* (Cologne, 1567), fol. Aiir. On the early modern notion of "theosophy," see Carlos Gilly, "Khunrath und das Entstehen der frühneuzeitlichen Theosophie," in Heinrich Khunrath, *Amphitheatrum Sapientiae Aeternae*, ed. Carlos Gilly, Anja Hallacker, Hanns-Peter Neumann and Wilhelm Schmidt-Biggemann (Stuttgart-Bad Cannstatt, 2014), 9–21.

40 The relationship between Olympian gods and celestial bodies is, of course, defined by their very names. The connection between the inner *astrum* and imagination is clearly stated at the end of the *De imaginibus idolatriae*, hinted at in *De iustitia*, PM, 19.

41 See note 18.

42 *Astronomia Olympi novi*, PM, 37.

43 See the reprint of the *Nuncius Olympicus*, in Gilly, *Haslmayr*, 243–290, and the list of Haslmayr's works ibid., 239.

pedant who was exiled from the divine Olympus for his criticism. As a sharp critic and scoffer, he appears also in Khunrath's *Amphitheatrum*.[44]

If we look for other "Olympic" publications of that time, we can find three: the *Rosarium novum olympicum et benedictum* (Basel, 1608), the *Thesaurinella Olympica aurea tripartita* (Frankfurt, 1609), and appended to the latter, the *Hortulus Olympicus aureolus*. All these anthologies of alchemical texts were put together by Benedictus Figulus, an author in personal and epistolary contact with Haslmayr,[45] and, as we have seen, a possible editor of the *Philosophia Mystica*. Figulus copied Haslmayr's manuscripts and shared his ideas among his friends.[46] He not only borrowed the "Olympic" attribute from him, but was also the first to mention Haslmayr's concept of the *Theophrastia sancta* in print in 1608.[47]

The motif of the "*Olympus novus*" was tackled also under the heading of the "theologizing of astrology" in other works.[48] Only one year before the *Philosophia Mystica*, the *Astrologia Theologizata* was published in Frankfurt.[49] Although a later English translation ascribed this work to Weigel,[50] the Latin original gives only an "anonymous theosopher" as its author. But there is another work, published under Weigel's name in 1618, which mentioned *Astrologia theologizata* in its subtitle: the second, pseudepigraphic part of the *Gnothi seauton*.[51] And the foreword of the pseudo-Weigelian *Studium*

44 See the long title as well as the admonition after the 365th commentary in Heinrich Khunrath's *Amphitheatrum sapientiae aeternae* (1609).

45 Gilly, *Haslmayr*, 33. It was perhaps Figulus who brought the *Fama fraternitatis* from Haslmayr to the publisher at Kassel; see Carlos Gilly, "Die Rosenkreuzer als europäisches Phänomen im 17. Jahrhundert und die verschlungenen Pfade der Forschung," in Carlos Gilly and Friedrich Niewöhner, eds., *Rosenkreuz als europäisches Phänomen* (Amsterdam 2002), 19–57, here 40–41.

46 Gilly, "*Theophrastia Sancta*," 457 n. 71; idem, *Haslmayr*, 103 n. 19.

47 Benedictus Figulus, *Pandora magnalium naturalium aurea et benedicta* (Strasbourg, 1608), 8r; see Joachim Telle, "Benedictus Figulus. Zu Leben und Werk eines deutschen Paracelsisten," *Medizinhistorisches Journal*, 22 (1987), 303–326, here 323–324.

48 In the *Astronomia magna*, "astrology" is one of the nine "*membra*" belonging to each of the four "astronomies." Here, it is used rather in the Weigelian sense, where astrology pertains to the nature of the *homo externus*.

49 *Astrologia Theologizata, hoc est: quod externus homo cum omnibus operibus, quantumvis lumine naturae in omni scientiarum genere splendibus, deponi, abnegari & plane emori: internus autem per lumen gratiae assumi* [...] *debeat* (Frankfurt, 1617).

50 *Astrologie Theologized: Wherein is set forth what Astrologie and the Light of Nature is* (London, 1649; new edition London, 1886).

51 [Pseudo]-Valentin Weigel, *Gnothi seauton. Nosce te ipsum. Erkenne dich selber O Mensch: Heisset Astrologia Theologizata, Auff den Andern Theil des Menschen die Seel nemblichen/ so auß dem firmament formieret/ geführet vnnd gestellet* ("Neustadt," 1618). Some scholars suggest that Johann Arndt was the author of both works, connecting them with ideas he expressed in his *Iudicium* on Khunrath's *Amphitheatrum sapientiae aeternae* (1st edition,

universale (1618) suggests "*Astrologia theologizata*" as an alternative title for the work.[52] Weigel himself debates the "use and worth of astronomy" as well as the relationship between theology and astronomy in a chapter of his *Gnothi seauton*,[53] which might have served as the point of departure for the speculations in the pseudonymous *Astrologia theologizata*. Put simply, the idea was founded on an old saying that "the wise man will dominate the stars" (*sapiens dominabitur astris*); this adage occurs not only in the subtitle of the anonymous *Astrologia theologizata*, but also, for example, in the *Astronomia magna* by Paracelsus and in the (at least partly) pseudonymous *De natura rerum*,[54] as well as in Haslmayr's *Theologia Cabalistica*, in its "*particulae*".

For the author of the *Astronomia Olympi novi*, not following the new "light of heaven" and its stars, i.e., the Apostles, is a "Necromantic vision," because here the saints are "dead."[55] In this perspective, lacking the light of the "*Olympus novus*," all sciences are necessarily false. The author calls this the "cabalistic teaching;"[56] indeed, similarly, a reform of sciences through the "religion of cabala" was proclaimed by Paracelsus in his *De religione perpetua*.[57] Both pseudo-Paracelsian works claim that such a teaching is based on "the three *cabalistica et prima principia*": "ask, seek, knock";[58] which is a motif typical of Haslmayr.[59]

1595), appended to Khunrath's *De igne magorum* (1608); see Jole Shackelford, "Western Esotericism and the history of European science and medicine in the early modern period," *Scripta Instituti Donneriani Aboensis* [S.l.], 20 (2014), 173–203, here 185.

52 [Pseudo]-Valentin Weigel, *Studium universale* (I am quoting the later edition Frankfurt-Leipzig, 1700), fol.):(a: "Ich schreibe jetzund dieses Buch und nenne es Studium universale, oder: Encyclopaediam Omnium Scientiarum, oder: Astrologiam theologizatam [...]."

53 Weigel, *Gnothi seauton* I, chap. 14, ed. Pfefferl, 3: 89–91 (in the manuscript, this is the last chapter); cf. the *astrologia coelestis* in *Astronomia magna*, Sudhoff, I/12: 382–386.

54 *Astronomia magna*, Sudhoff, I/12: 41; *De natura rerum*, Sudhoff, I/11: 378. For the authenticity of the *De natura rerum*, see Urs Leo Gantenbein, "Real or Fake? New Light on the Paracelsian *De natura rerum*," *Ambix*, 67 (2020), 4–29.

55 Although the "*nectromantia*" of Paracelsus relates to visions, it does not allude to anything dead (nor does his "*nigromantia*"): for example, the "*nectromantia coelestis*" is a "*Vision in Christo*," and "*nectromantia*" alludes to Elias and Moses appearing in front of Christ; see *Astronomia magna*, Sudhoff, I/12: 148–161. The "new light of heaven," should here be understood as the Apostles, as opposed to the sidereal "*gubernatores Olympi*" in the *Arbatel*, 23–24 (which refers also to an *Astronomia gratiae*).

56 *Astronomia Olympi novi*, PM, 37 and 34.

57 *De religione perpetua*, in Paracelsus, *Sämtliche Werke. II. Abteilung. Die theologischen und religionsphilosophischen Schriften*, vol. I, ed. Wilhelm Matthießen (Munich, 1923) (hereafter: Matthiessen), II/1: 100 ff; Gantenbein, 1: 299–320.

58 *Astronomia Olympi novi*, PM, 38; *Theologia Cabalistica*, PM, 42.

59 Gilly, *Haslmayr*, 188; cf. Gilly, "*Theophrastia Sancta*," 456, and Haslmayr's manuscript *Oratio revelatoria*, ibid., 454–455: "bitten, suchen vnd anklopfen [...] wie D[octor] Paracelsus Magnus erklert, der gwaltig Cabalista."

Generally speaking, such a method was nothing new to the Paracelsian audience. The "three principles" are prefigured in Paracelsus' *Astronomia magna*, without, however, being called "cabalistic."[60] Adam von Bodenstein dedicates his publication of the *Metamorphosis Doctoris Theophrasti* (1572) to the "knocking and searching sons of Wisdom" (*den anklopfenden und suchenden Filiis sapientiae*). The first part of this anthology consists of nothing else but the first seven books of the renowned *De natura rerum* ascribed to Paracelsus. In the ninth book, called *De signatura rerum*, which appeared in the 1584 edition, the author draws on the "three principles" from the *Astronomia magna* – this time connecting them explicitly with the "magical and cabalistic art."[61] These "three cabalistic principles" can be found later in Croll's *Basilica chymica*,[62] and repeatedly in the *Rosarium* and the *Thesaurinella* by Figulus. For him – as for Haslmayr – Paracelsus is a "highly enlightened magus and cabalist."[63]

3 Theologia Cabalistica

Now let us turn our attention to the longest of the two pseudo-Paracelsian treatises written by Haslmayr, the *Theologia Cabalistica*.[64] Again, this treatise claims to be an authentic work by Paracelsus, though the three appended "*particulae*," as the editor acknowledges, "are not the work of Theophrastus Paracelsus"; their writer even refers to "philosophical treatises by Paracelsus … which will be brought to light by a disciple of the *sancta Theophrastia*." Nonetheless, the editor says, because they were not "inconsistent" (*zuwider*) with the rest of the treatises and because there was no time to alter anything, they were left in place.[65] And rightly so, because, contrary to most of the text, they were extracted from an authentic work of Paracelsus.

If there were no explicit references to the *Astronomia magna* in the *Astronomia Olympi novi*, the *Theologia Cabalistica* quite straightforwardly appropriates it as "our great astronomy." It adduces its teaching that the "lunatic children of man led by the drunken stars (*Gestirn*) as brutes" must be

60 *Astronomia magna*, Sudhoff, 1/12: 185: "die <ding> ale zu erforschen alein in dreien punkten stehet, als in biten, suchen und anklopfen [...] die drei ding der hauptpunct ist [...]."

61 *De natura rerum*, idem, 1/11: 394: "bitten, suchen anklopfen im glauben alles erlangen. in disen dreien hauptpunkten stehet al unser grunt der magischen und cabalistischen kunst."

62 Croll, *Basilica chymica*, 46; see Gilly, "*Theophrastia Sancta*," 456.

63 Figulus, *Rosarium novum olympicum* (Basel, 1608), 1: fol. a3r-v; idem, *Thesaurinella Olympica aurea tripartita* (Frankfurt, 1609), fol. 1r; see Gilly, *Haslmayr*, 199 n. 5.

64 *Theologia Cabalistica, De perfecto homine in Christo Iesu, et contra de perdito animali homine in Adam, qui Lunaticus dicitur.*

65 *Theologia Cabalistica*, PM, 49 and 47.

considered "possessed Antichrists."[66] This notion of the "lunatics" is so crucial for the *Theologia cabalistica* that it even appears in its subtitle.

Although the "lunatics" are treated *in extenso* in Paracelsus' *De lunaticis*, which was a part of the so-called *Philosophia magna*, the same concept is presented in the *Astronomia magna*. There Paracelsus discusses the "drunken sidereal men" who drank the *wrong* "Olympic cider" (*olympischen most*), instead of the "cider of the Holy Spirit".[67] Similarly, according to the *Theologia Cabalistica*, such a "sidereal wisdom" stems from the "spirit of the Limbus" (*Geist deß Limbi*), and therefore is "brutish natural" (*viehisch natürlich*), belonging to those who "have not learned to *read themselves*."[68] In contrast to the "lunatics," it is necessary to have one's "mind drowned and immersed fully in God" to have the "wisdom" which the "ancient wise naturalists" called "*Nosce teipsum*."[69] This ancient motif was apparently popular at that time – not only favored by Weigel, but widespread among "enthusiasts,"[70] and also in medicine.[71] A passage similar to the one in the *Theologia Cabalistica* can be found in Paracelsus' *Liber de imaginibus*. Here, Paracelsus connects self-knowledge with the *exaltation* of mind which is "drowned and immersed in itself," so that God's secrets are revealed to it.[72]

We have said that the "three cabalistic points," mentioned in the *Theologia Cabalistica*, were advocated by Figulus. Interestingly, in the foreword to his *Rosarium*, he also ascribed a "*theologia cabalistica*" to Paracelsus, the *artium Monarcha*, when claiming that he had found in him a fundamental elucidation and exposition of the "light of nature" in "*Astronomia, Philosophia, Magia & Theologia Cabalistica*".[73] Similarly, the *Astronomia Olympi novi* proclaims that

66 Ibid., 45: "Diese lunatischen vnnd von trunckenen Gestirn viehisch geführten Menschenkind [...] bessene Antichristische [...]."

67 *Astronomia magna*, Sudhoff, 1/12: 224–225.

68 *Theologia Cabalistica*, PM, 45. This "reading" is related to the idea of the "three divine books" (the Bible, world, man or human conscience), well established in the Paracelsian tradition. Cf. the idea that Christ must be "writing" in us and that he is a "book in our heart," in the *Scholasterium christianum*, PM, 155; *Von Betrachtung deß Lebens Christi*, PM, 15–16, 4–5; *Introductio hominis*, PM, 234.

69 *Theologia Cabalistica*, PM, 42 and 44.

70 See Ehregott Daniel Colberg, *Das Platonisch-Hermetische Christenthum* (Leipzig, 1710), 2: 143–144.

71 The appeal to "know oneself" was used to justify anatomical investigations, including human dissections.

72 *Liber de imaginibus*, Sudhoff, 1/13: 382.

73 Figulus, *Rosarium novum olympicum*, fol. a2v and a3r; Gottfried Arnold's *Unparteyische Kirchen- und Ketzer-Historien* (Schaffhausen, 1742), 3: 4, with reference to Weigel's *Studium universale*. But besides being a later pseudepigraph, it says only: *wie man in Theologia Cabalistice beweisen kan* (*Studium universale*, chap. 2, [fol. A6r]).

our salvation depends solely on the "sacred knowledge" of Christ as the "new heaven," of the eternal Wisdom and of ourselves, and "no one has ever so clarified it on the cabalistic ground until this our monarchy," i.e., until the coming of Paracelsus and his teachings.[74] Besides this, the *Theologia Cabalistica* also speaks of four "principles (*Hauptpuncten*) of our Christianity": 1) To love our enemies; 2) to leave everything one owns behind; 3) to suffer patiently any contempt; and 4) and to renounce all offered honors.[75] These principles were also described by Haslmayr in many of his works, as Carlos Gilly remarks, and we find them also in the foreword to Figulus' *Thesaurinella*.[76]

Those who search for God primarily are "cabalists," according to both the *Astronomia Olympi novi* and the *Theologia Cabalistica*. So, who was a "cabalist"? The "cabala", closely related to magic and alchemy, was, more than anything, quite fashionable in the period. The Christian cabala was, in general, vague and, compared to the Jewish kabbalah, simplified to some basic concepts, ideas and hermeneutical techniques. Despite his relatively rare and rather idiosyncratic use of cabala, Paracelsus was soon taken for a great cabalist, Adam Haslmayr being among the first to bestow this noble title upon him.[77] In his *Astronomia magna*, Paracelsus mentions cabala under the heading "magica" as its sixth and last part.[78] This *caballa, caballia, caballistica* or *gaba nala* is a pagan (i.e., ancient) invention which the Hebrews only corrupted, having no understanding of nature.[79] The cabala, Paracelsus argues, deals with the "spiritual body".[80] Or, as he put it in his *Von den unsichtbaren Krankheiten*, with the "*gestirn* in man" or the "Olympic spirits," related to the imagination.[81] According to the *De natura rerum*, the "sidereal body" is the substance of the specters appearing occasionally on graves, and it is exactly this which the *caballia* or *kunst*

---

74 *Astronomia Olympi novi*, PM, 35 and 38.

75 *Theologia Cabalistica*, PM, 42.

76 Gilly, *Haslmayr*, 188 and 199 n. 5; idem, "*Theophrastia Sancta*," 457.

77 See Hartmut Rudolph, "Die Kabbala im Werk des Paracelsus," in Wilhelm Schmidt-Biggemann, ed., *Christliche Kabbala* (Ostfildern, 2003), 109–120, on 109, 113, 120.

78 *Astronomia magna*, Sudhoff, 1/12: 78.

79 Ibid., 158; see also *Spitalbuch*, Sudhoff, 1/1: 398 with its distinction between the authentic, magical and empirical *gabalia* and the erroneous *gabalistica*, i.e., *gabala* put on paper.

80 *Astronomia magna*, Sudhoff, 1/12: 156 and 127–128: "so das der elementirt leib kan und vermag, wievil mer der spiritualisch leib in diser arte cabalistica den elementirten leib fürtrift [...]."

81 *Von den unsichtbaren Krankheiten*, Sudhoff, 1/9: 298: "in dem Olimpischen geist ligt die kunst gabalistica mit iren annexis, welche kunst bewert, das der inbildung noch vil mer müglich ist in dem, so die zusamenfügung der olimpischen geister sich verfügen mögen, dan zu gleicher weis wie die sichtbaren corpora können zusamen komen, das können auch die Olimpischen geister der schöpfung, welche das gestirn sind in dem menschen, die ding werden in die bücher der gabalia geschriben."

cabalistica is concerned with[82] – although, elsewhere, as we have seen, the *De natura rerum* conjoins cabala with searching and knocking.[83] Paracelsus mentions *die große gabal* also in the context of illnesses of the "innate spirit" (*angeboren geist*).[84] The *gabal* is the natural spirit, containing all arts and wisdom in itself; that is also why Adam and Moses are called the *gabalischen* because they found everything in the light of nature in themselves.

So, in the medical and natural philosophical works, Paracelsus takes *cabala* or *gabal* for a part of the natural magic;[85] besides, it is attested that the "Olympic spirit" is clearly natural, not supernatural. The *gabalia* is a knowledge of invisible powers hidden in visible things.[86]

But turning to Paracelsus' *theologica*, the picture seems to change.[87] In *De religione perpetua* Paracelsus claims that the Devil is a *magus* because he knows all the natural course of the heavens and all natural magic.[88] Yet contrary to this, "*die religion der gabalien*" is firmly rooted in God, and its essence are the biblical words: "seek ye first the kingdom of God, and all things shall be added unto you" (Matthew 6:33). This *gabalia* is the "rock" upon which Samson and Moses were building. Such a "religion of cabala" differs from the "religion of nature" dealing with natural "signatures," although a "cabalistic fundament" is also needed in the study of nature. Again, this fundament is nothing other than the search for the hidden "divine Wisdom" that can change the ungodly arts and sciences into a source of wisdom and of the "blessed life" (*seliglich zu dem seligen leben gezogen werden*).[89]

Such a notion of cabala differs from the cabala as a part of magic in the natural philosophical texts. The pseudo-Paracelsian *Aurora philosophorum*

---

82 *De natura rerum*, Sudhoff, I/11: 361.

83 Together with this, the author, contrary to Paracelsus, confuses the "sidereal" (mortal) and the "celestial" (sacramental, immortal) body.

84 *Vorläufige Ausarbeitungen zu* […] *5 Bücher von den Unsichtbaren Krankheiten*, ed. Sudhoff, I/9: 359–360.

85 See Urs Leo Gantenbein, "Converging Magical Legends: Faustus, Paracelsus, and Trithemius," in *The Faustian Century. German Literature and Culture in the Age of the Enlightenment*, eds., James M. van der Laan and Andrew Weeks (Rochester, NY, 2013), 93–123, here 99–103.

86 Rudolph, "Die Kabbala," 110–111, 113.

87 Ibid., 110; Rudolph omitted this part of the Paracelsian *oeuvre* in his study, because, as he remarked, the notion of *gabalia* was "hardly present in the theological work" of Paracelsus.

88 *De religione perpetua*, Matthiessen, II/1: 98–99 (Gantenbein, 1: 308). All mantic arts are "abwürflig von der gabalia" (Matthiessen, II/1:101), i.e., they are deduced from, or rebellious against the cabala; cf. Gantenbein, 1: 313, note, and *Deutsches Wörterbuch von Jacob Grimm und Wilhelm Grimm* (Leipzig, 1854), 1: 152, s.v. "*abwerf*".

89 *De religione perpetua*, Matthiessen, II/1: 100; Gantenbein, 1: 313.

follows this pattern, making a clear distinction between the "true magic", as an instrument of the "knowledge of elements," hidden powers in nature and "natural secrets," and the "cabala" which is the way to God and the "divine mysteries."[90] The "cabala" in the *Theologia Cabalistica*, where it pertains to the "*Olympus novus*," is the same; indeed, its author might have had in mind the *De religione perpetua*, as the specific use of the term "religion" suggests, although it occurs also in other works.[91] Such a concept of cabala was, in fact, predominant among the authors drawing on Paracelsian and Weigelian ideas,[92] and it fits well also in the context of the early Rosicrucianism. For example, Julius Sperber (c. 1540–1616), closely connected with Rosicrucianism, says in his Biblical florilegium *Kabalisticae praecationes* that cabala is not taken from nature and human books, but is "spread divinely" (*divinitus promulgata*) into souls.[93] Such cabala, inspired by Pico,[94] Reuchlin and Agrippa, was promulgated by the Paracelsian, alchemist and theosopher Heinrich Khunrath (c. 1560–1605) in his influential *Amphitheatrum sapientiae aeternae* (1609).[95] Indeed, we cannot but suppose that his ideas on theosophy, cabala and alchemy had some bearing both on Haslmayr and on the general plan of the *Philosophia Mystica*.[96] We know, for example, about Haslmayr's acquaintance with Daniel Mögling (alias Theophilus Schweighardt), another important promulgator of Rosicrucianism (with whom Jennis collaborated),[97] whose *Speculum sophicum rhodo-stauroticum* (1618) contains recognizable

90 [Pseudo]-Paracelsus, *Aurora thesaurusque philosophorum* (Basel, 1577), 14. Here "cabala" is connected also with the prophetic spirit with reference to Daniel 5:26 (ibid., 9), but this idea remains undeveloped by contrast with the natural realm of "cabala."

91 *Theologia Cabalistica*, PM, 45. Cf. also *Labyrinthus medicorum errantium*, "Vorrede," Sudhoff, 1/11: 166; *Astronomia magna*, Sudhoff, 1/12: 3 ("religion aller creaturen").

92 Although not on Weigel's authentic texts. The word "cabala" is not especially favored by Weigel, and when he occasionally makes use of it, he does so in the context of the *natural* "sidereal spirit" which is called *gabalis*, and from here the "noble art *cabalistica*" is born and perfected (see *Gnothi seauton* 1, chap. 2; cf. chap. 7, taking up the fragments relating to Paracelsus' *De fundamento sapientiae*, Huser, 9: 457).

93 Julius Sperber, *Kabalisticae praecationes* (Magdeburg, 1600), fol. 4r.

94 According to Giovanni Pico della Mirandola, *Apologia*, in *Opera omnia* (Basel, 1557), 1: 178 and 180, cabala gives the most subtle and divine interpretation of the Bible, bringing man from worldly matters to heavenly and spiritual things. Etymologically, Kabbalah means "reception."

95 This is the second, substantially expanded edition of the book that appeared in 1595, in a limited print run.

96 For the influence of Khunrath's *Amphitheatrum* on Siebmacher, see Zuber, *Spiritual Alchemy*, 90.

97 See Ulrich Neumann, "*Olim, da die Rosen Creutzerey noch florirt, Theophilus Schweighart genant*: Wilhelm Schickards Freund und Briefpartner Daniel Mögling (1596–1635)," in Friedrich Seck, ed., *Zum 400. Geburtstag von Wilhelm Schickard. Zweites Tübinger*

Khunrathian elements.[98] Haslmayr knew and commented on parts of the *Speculum*,[99] and he partly copied Mögling's *Rosa florescens* (1617), an apologia for Rosicrucianism, which he included in his own works.[100] Mögling's *Colloquium rhodo-stauroticum* (1621), in turn, listed the *Philosophia Mystica* among the books recommended for further reading,[101] together with other Paracelsian, Weigelian and Rosicrucian works.

Let us now focus on the "particles" appended as a separate part to the body of the *Theologia Cabalistica*. Having pronounced doubts about the authorship of the "particles," the author of the "*praemonitio ad lectorem*" explains their content: namely, that the inclinations of stars create a "cagastric" life and "lunatic" people, as well as the "cacosophers"[102] who do not know how to "dominate over stars" (i.e., we could say, do not know how to "theologize the astrology"). The "short foreword to a prince" (*praefatiuncula ad principem quondam*) is obviously addressed to a real person as the salutation "E.F.D." ("Eure Fürstliche Durchlaucht," "Your Princely Highness") indicates. According to Carlos Gilly, it is Haslmayr's letter to the archduke Maximilian of Tyrol.[103] He is addressed here as the *amator sapientiae aeternae*,[104] an appellation

Schickard-Symposion 25. bis 27. Juni 1992 (Sigmaringen, 1995; *Contubernium. Tübinger Beiträge zur Universitäts- und Wissenschaftsgeschichte*, Bd. 41), 93–115, on 105.

98 This is clear, not only when the author calls himself an "indigne proclaimer of the divine-magical, physical-chymical, tertriune-catholic philosophy," but is echoed also in words like "Zoilorum [...] calumni[ae]" (Mögling, *Speculum*, 5), "Thraso [...] Ardelio [...] Autophilus" (ibid., 7, falsely printed as "6"), "Theosophi" (ibid., 10, 12), "Harpocrate[s]" (ibid., 13), "Philauti" (ibid., 14, 16; cf. Khunrath, *Amphitheatrum*, com. 274), "Labyrinth" (ibid., 16), "sacratissima monas triade ligate" (ibid., 16, 18 [falsely "13"]; cf. Khunrath, *Amphitheatrum*, "Proclamation of Theodorus Aquilinus," com. 1, 5, 157, etc.), "particularia and Archetyp[us]" (ibid., 17 [falsely "13"]), or neologisms like "Philopansophus" (ibid., 14). The fourth engraving from the *Amphitheatrum* was also taken in an adapted version into the "Ergon and Parergon" image in the *Speculum*. The final "Arbor Pansophiae" with its use of Hebrew letters, the *typus / Archetypus* relation and its general form are again strongly reminiscent of Khunrath's style. Cf. Gilly, "*Theophrastia Sancta*," 460 and n. 75, quoting a manuscript by Haslmayr echoing terminology of the *Amphitheatrum* ("die wahre Weißheit von Gott durch das Gebett, labore, arte et Nature zue erlangen"; "Felsen Spei et Fidei sub Urim et Thumim"; "Petite et dabitur vobis in Oratorio Theosophico"; etc.).

99 Gilly, "*Theophrastia Sancta*," 460 n. 74. Mögling himself was in contact with Haslmayr's friends Karl Widemann and August von Anhalt; see Gilly, *Haslmayr*, 111 and 185 n. 12.

100 Gilly, *Haslmayr*, 182. Here, nevertheless, nothing seems specific to Khunrath.

101 *Colloquium rhodo-stauroticum* (no place, 1621), 108.

102 Cf. *Arbatel*, 86, according to which the *kakosophia* and *cacodaemonia* belong to the *scientia mali*, while *theosophia* and *antroposophia* are parts of the *scientia boni*. The term "cagastrum" appears in the pseudo-Paracelsian *Azoth*, Huser, 10, Appendix: 6.

103 Gilly, *Haslmayr*, 186 n. 21; in fact, Gilly takes the whole first *particula* for the letter.

104 *Theologia Cabalistica*, PM, 48.

reminiscent to the *amator theosophiae* and *amator sapientiae verae*, the names Khunrath uses for himself and for the "disciples of the Wisdom."[105] Haslmayr recommends the study of "philosophical works" of Paracelsus, namely his *De ordinatione Dei vel Domini*; the work was promised to be published soon by a "disciple of the *sancta Theophrastia*," but it never appeared in print.[106] This text addresses the neighbor's love and one's own poverty,[107] reminds the reader that often man is no "monarch" but a *knecht*,[108] and warns against false apostles seducing the ordinary man.[109]

After the "*praemonitio ad lectorem*," three short "particles" follow, developing one and the same motif. First, two contrary spirits of man are distinguished, the eternal and the mortal, the latter coming from the "limbus" and ruling in the "lunatics." A perfect theologian, astronomer and physician, however, must know both spirits well.[110] His teacher must be God, not the heavens nor the "walled schools."[111] Then he will be free of the sidereal influences because they reach the "brutish man" only.[112] To get rid of this "illness," we must follow the "life of Christ," know ourselves and God – a teaching ascribed to "Bernhardus in *Canticis*" – and find the kingdom of God in ourselves.[113] This is also one of the key messages of the whole *Philosophia Mystica*.

For the distinction between "lunatics" and "fools," the text refers to the *De generatione stultorum* by Paracelsus.[114] This work is a part of the Paracelsian *Philosophia de divinis operibus*,[115] and it directly follows the *De lunaticis* (where it is also cited). However, the "particles" are nothing but an abbreviated paraphrasis, sometimes even a literal copy, of the first part of the *De lunaticis*. In the 1567 edition (not so in the 1591 edition by Huser), this text itself was organized into four "*particulae*," of which the *Theologia Cabalistica* adapted only the first one, thus dividing the text in a specific way. It is worth noting that in the text following the *Theologia Cabalistica* in the *Philosophia Mystica*, the

105 Khunrath, *Amphitheatrum*, com. 65.
106 See *De ordine doni*, Goldammer, II/2: 49–69; Gantenbein, 1: 627–634.
107 Ibid.: 117–118.
108 Ibid.: 119.
109 Ibid.: 129.
110 *Theologia Cabalistica*, PM, 50.
111 *Theologia Cabalistica*, PM, 51. Cf. the radical spiritualists' (Paracelsus, Franck, Weigel…) hostility against the "walled church" (*Mauerkirche*).
112 *Theologia Cabalistica*, PM, 51.
113 Ibid., 52: "weil kein Mensch selig sol werden / wie Bernhardus in Canticis sagt / der sich selbsten nicht erkennet nach dem Geist Gottes zu leben."
114 Ibid., 49.
115 *Philosophia de divinis operibus*, Sudhoff, I/14: 43–72 (*De lunaticis*) and 73–92 (*De generatione stultorum*); see also the *Philosophia magna* (Cologne, 1567), 118–138 and 139–155.

"Commentary on Daniel," a motif is adopted from the third particle of the *De lunaticis*: it is as if the editor became inspired to dedicate not just a "particle," but a whole treatise to the topic of the "false prophets," an important theme at the time of the Rosicrucian enthusiasm.[116]

4 Conclusion: Paracelsian Theology, Channeled and Contextualized

We have seen that the *Astronomia Olympi novi* and the *Theologia Cabalistica* helped to spread and establish some concepts coined or adapted by Haslmayr, though they might have appeared earlier in the works of Figulus or Haslmayr himself (in his *Responsio*). Nevertheless, to understand their possible role and impact, we must understand them in their larger context. It has been said that the *Philosophia Mystica* helped to establish Paracelsus as an eminently theological thinker. Though the existence of Paracelsus' theological writings was not unknown to the interested public,[117] very few of them were published before the *Philosophia Mystica*.[118] It presented a deliberately concocted mélange of Paracelsian-Weigelian thoughts, flavored with the mysticism of Eckhart, Tauler and the *Theologia Deutsch*, with just a pinch of alchemy. While the (more or less distant) echoes of German Mysticism are spread all over the volume, the presence of alchemy is delicate, yet perceptible, mostly in the penultimate *Introductio hominis* and the *Appendix* by Siebmacher,[119] which can be seen as a kind of summary of the *Philosophia Mystica*. Besides the famous concepts

116 Cf. Will-Erich Peuckert, "Nachwort," in *Philosophia Mystika (Nachdruck der Ausgabe 1618)*, ed. idem (Berlin, 1943), 135.

117 Gilly, "*Theophrastia Sancta*," 430–437.

118 *Philosophia magna*, ed. Balthasar Flöter (1567), containing 17 treatises by Paracelsus, among others the *De lunaticis*; *De secretis creationis*, ed. Michael Toxites (Strasburg, 1575), discussing also theological matters; *Kleine Handt-Bibel*, ed. Johann Thölde (Lych, 1605), a "Paracelsian" biblical florilegium with a foreword allegedly by Paracelsus, but drawing upon Weigel's *Dialogus de Christianismo* (see Ute Gause, "Theophrasti Paracelsi Kleine Handbibel," *Nova Acta Paracelsica*, N.F. 19 [2005], 85–104, on 91); *Philosophia de limbo aeterno*, ed. Johannes Staricius (Magdeburg, 1618), a collection of "Abendmahlschriften" (treatises on the Lord's Supper). Furthermore, Jennis also published other *theologica* by Paracelsus between 1618 and 1619: *Sermo de purgatorio* (1618, printed at the end of the anonymous *Libellus theosophiae*, together with Weigel's *Deus non potest se negare*); *Commentatio in Epistolam S. Judae Apostoli* (1618); *Liber Sermonum in Antichristos, et Pseudoprophetas Veteris et Novi Testamenti* (1619, biblical commentaries); and *Außlegung oder Bericht uber die Wort: Sursum Corda* (1619); the last work reappeared in one volume as *Drey underscheydene Tractätlein* in 1619.

119 Cf. *Commentarius in Danielem*, PM, 90, using alchemical signs and claiming that the just must be examined like gold is examined seven times by fire.

of German Mysticism (*Gelassenheit, Abgescheidenheit*, etc.)[120] and Weigelian-Paracelsian motifs, the *Introductio hominis* discusses remarkable practices and methods recently invented in the world, blaming them, however – like Khunrath in his *Amphitheatre*[121] – for being "impious folly" because of their erroneous orientation towards worldly life. Among these inventions, *Chymia* or the "Art of Stone" is named, which was being practiced without regard to its foremost goal: *self-knowledge*.[122] In spite of this, it is the highest, most hidden, and best-founded art because it gives perfect knowledge of natural and divine things. Indeed, it is described not as a matter of some prescriptions, but of the Divine Mercy and Goodness; that is why it is a "divine and heavenly art"[123] (thus we could infer: an "Olympic" or "cabalistic" art). In the true knowledge, the humble, (self-)knowing and *Gottgelassene* person is finally assimilated to the known "object" and unified with it.[124]

In Weigelian terms, what the *Introductio* is talking about here is the highest cognition through the "eye of mind," where God is active in seeing and knowing Himself through the passive human "eye." It is a mystical cognition, also advocated, among others, by Khunrath in his concept of the "divinization" of the theosopher.[125] Similarly to his *Amphitheatrum*, Siebmacher's *Wasserstein* or the pseudo-Weigelian *Azoth et ignis* (printed in 1702, but written before 1600),[126] Christ is called a "heavenly Corner Stone," with a clear allusion (by analogy) to the "Philosopher's Stone."[127] Even if there is not much alchemy and natural philosophy in either the Weigelian or the Paracelsian parts of the *Philosophia Mystica*,[128] the reader can have no doubts that the "mystery of the Philosopher's Stone"[129] is in perfect concordance with Weigel's ideas, not to speak of those of Paracelsus. Perhaps the *Introductio*, signed only at the end of the *Appendix*, was supposed to be understood as a Weigelian text. In that time,

120 Esp. *Introductio hominis*, PM, 232, 235, 264–265.

121 Khunrath, *Amphitheatrum*, com. 10, 101 and 251; see also above, note 58, for Paracelsus.

122 *Introductio hominis*, PM, 231, 234 and 268–269.

123 Ibid., 231–232.

124 Ibid., 232 and 235. The Pseudo-Weigelian *Studium universale*, fol. B3r, speaks of "*Transplantatio, Renascentia, Regeneratio*".

125 Khunrath, *Amphitheatrum*, com. 132.

126 For the *Azoth et Ignis* (written before 1600) as the primary text of "spiritual alchemy", see Zuber, *Spiritual Alchemy*, 57–67.

127 *Introductio hominis*, PM, 248.

128 Besides the works by Haslmayr and Siebmacher, mainly the *Scholasterium Christianum* and the pseudo-Weigelian *Philosophia theologica* include some natural philosophical topics.

129 *Appendix*, PM, 269.

after all, Weigel could easily fit into the alchemical context, having already been acclaimed openly by alchemists like Khunrath and Croll.[130]

The timing and the formula of the *Philosophia Mystica* were ideal. From 1614 onwards, the Rosicrucian manifestos were attracting considerable attention. Weigel had just left the (not so narrow, in fact) circle of his manuscript readers a couple of years before, and his works (both authentic and pseudepigraphic) were coming off printing presses in large numbers: there were at least eighteen Weigelian and pseudo-Weigelian publications in 1618 alone.[131] Paracelsus' theological opinions had been under discussion since the 1560s;[132] Weigel himself was among the first to praise him as a great theologian,[133] and the theological aspects of his work were also important when the *Fama fraternitatis* celebrated Paracelsus' name.

In fact, the whole orientation of the *Philosophia Mystica* was rather Weigelian. There are typical Weigelian motifs here: priority of the inner essence against the outer form; the uselessness of external forms in arts, sciences and cult; the importance of the "essential" faith and self-knowledge; the three "divine books" (God, nature, and man); the two divine "lights"; the threefold *corpus* of man (the body, the spirit, the soul), etc. These motifs also fit well with most of the Paracelsian part of the volume, which is no coincidence but, obviously, the goal of the editor. Thus, if the congruities between Paracelsus and Weigel (and between theology and alchemy) had not been obvious to the public before, now they were made clear. And the message was not overlooked. For the orthodox Lutheran theologian Nicolaus Hunnius, the Weigelian theology was just a development of the "fanatical theology" spawned by Paracelsus, and the most dangerous heresy of Weigel was fully grounded in Paracelsus.[134] Similarly to Rosicrucianism, Weigel's "new theology" "set the school of Paracelsus

130 Khunrath, *Amphitheatrum*, com. 66; Croll, *Basilica chymica*, 70. Croll tacitly copied almost a whole chapter of Weigel's *De vita beata* (printed 1609) in his introduction to *De signaturis internis rerum* (Frankfurt, 1609), *Praefatio*, 31–35.

131 See e.g., Arnold, *Unpartheyische Kirchen- und Ketzer-Historie* (1742), 2: 1091–1092.

132 Gilly, "*Theophrastia Sancta*," 430–437.

133 Weigel copied Paracelsus' interpretation of the first five chapters of Matthew's Gospel; see Winfried Zeller, "Naturmystik und spiritualistische Theologie bei Valentin Weigel," in Antoine Faivre and Rolf Ch. Zimmermann, eds., *Epochen der Naturmystik. Hermetische Tradition im wissenschaftlichen Fortschritt* (Berlin, 1979), 105–125, esp. 116–118. In his *Kirchen- oder Hauspostille*, Pfefferl, 12/2: 582 (see also 252 and 413), Weigel regrets that the "*theologica scripta*" of Paracelsus had not yet been printed.

134 Nicolaus Hunnius, *Principia Theologiae Fanaticae, quam Theophrastus Paracelsus genuit, Weigelius interpolavit* (Wittenberg, 1619); idem, *Christliche Betrachtung der neuen Paracelsischen, und Weigelischen Theologie* (Wittenberg, 1622). In the *Principia Theologiae Fanaticae*, cap. I, § IX, the *Philosophia mystica* is mentioned.

in motion".[135] Another critic, Andreas Merck, remarks that "Weigelianism" is "called Paracelsianism by some."[136] And Johannes Schelhammer in his *Widerlegung der vermeynten Postill Valentini Weigelii* speaks of a "Weigelian and Paracelsian foolish theology (*Tollogia*)."[137] E.D. Colberg later (1710) notes that this new theology was initiated by Paracelsus out of a desire to reform not only Galen but also Luther,[138] and its "foolishness" consisted of his unorthodox mixing of alchemical terms with theology which caused obscuration and perversion of theology.[139] The Rosicrucians, for Colberg, combined the worst excesses of Paracelsians and Weigelians; in Rosicrucianism, he saw little more than "a confraternity of Paracelsians, enthusiasts and chiliasts" who fancied a "Golden Age" was imminent, and were aiming at a universal reformation on magical and cabalistic grounds.[140]

Colberg's bias notwithstanding, this is exactly what the *Philosophia Mystica* was aiming at. Its goal was not only to present Paracelsian theological ideas and show Paracelsus as a theologian, but, primarily, to give an impetus for practical orientation.

The two works by Haslmayr, the *Astronomia Olympi novi* and the *Theologia Cabalistica*, stress the importance of the "cabalistic" return to God, the spiritual conquest of the astrally conditioned human nature, and the creation of the "new Olympus." This, however, is not the last and only message of Haslmayr nor that of the *Philosophia Mystica* and its editors (be it Siebmacher and/or Figulus, or perhaps someone else). To avoid being "enthusiasts," we must return from the quiet "Sabbath" of contemplation, as the Weigelian texts put it, to our own occupation in the world.[141] As the *Astronomia Olympi novi* remarks, taking up a motif from Paracelsus' *theologica*,[142] the light of the "*Olympus novus*" is a necessary precondition for bringing all arts and sciences to their true form and goal.[143] Through faith, one can know all "*magnalia*" of God and nature, but it is works that "make the faith" – *die Werck machen den Glauben*.[144] The relevance

135 Hunnius, Christliche Betrachtung, fol. 111b.

136 Andreas Merck, *Trewhertzige Warnung fürm Weigelianismo* (Hall in Sachsen, 1620) [not pag.].

137 Johannes Schelhammer, *Widerlegung der vermeynten Postill Valentini Weigelii* (1621), 1: 12.

138 Colberg, *Christenthum*, 1: 187.

139 Ibid., 2: "Vorrede," [**7r]. For Colberg's view, see Christine Maillard, "Alchimie et hétérodoxie: critiques et mises en cause du 'christianisme chymique' dans l'espace germanique au XVIIe siècle," *Aries*, 3/1 (2003), 1–24.

140 Colberg, *Christenthum*, 1: 269–274, esp. 274 and 278–279.

141 *Vom himmlischen Jerusalem in uns*, PM, 190.

142 *De religione perpetua*, Matthiessen, II/1: 100–112; Gantenbein, 1: 299–319.

143 *Astronomia Olympi novi*, PM, 34.

144 Ibid., 33 and 35; cf. *De poenitentiis*, PM, 16.

of this idea for the whole volume is underlined by the call for the reorientation of new inventions in arts and sciences, particularly, but not exclusively, in alchemy, to their true spiritual goal, expressed in the final *Introductio hominis* by Siebmacher.[145] Within a context where calls for universal reform were regularly heard, this was certainly a timely topic. Such an appeal for the reform of sciences, connected with the most important "reform," the spiritual restoration ("rebirth") of man, and finally with the eschatological renovation of the whole world, was already part of the "theosophy" conceived by Heinrich Khunrath,[146] hardly unknown to Haslmayr and Siebmacher. And it is, of course, much more explicitly a key part of the Rosicrucian ethos. It is in this way that the readers of both pseudo-Paracelsian texts, as well as of the whole *Philosophia Mystica*, were being directed to take the path which the "*monarcha*" Paracelsus himself had followed.

145 *Introductio hominis*, PM, 231–232 and 268–269.
146 See Khunrath, *Amphitheatrum, com. 136–137, as well as the introductory text:* ibid., (*I*) 5.

The Development of the Basil Valentine Corpus and Biography: Pseudepigraphic Corpora and Paracelsian Ideas

Lawrence M. Principe
Johns Hopkins University, Dept. of History of Science and Technology
Lmafp@jhu.edu

Abstract

Early modern alchemical literature is full of pseudonymous corpora. One of the most famous of these is connected with the name Basil Valentine, a supposed Benedictine monk and master of both medicinal and transmutational chymistry. Accreted over a period of nearly a century, the Valentine corpus is complex and heterogeneous. This paper endeavors to organize and recount the construction of the corpus by an array of authors, editors, publishers, and bibliographers, to sort out some of its strata, and to trace the origins and modifications of some of its texts. This exercise will be useful not only for further investigations of Basil Valentine and other pseudonymous chymical corpora, but also for broader studies of forgery and book history.

Keywords

Basil Valentine – Johann Thoelde – pseudepigraphy – alchemy – decipherment – Paracelsus – antimony

Introduction

The mysterious Basil Valentine is one of the most celebrated and prized of all early modern alchemical writers. He also presents a difficult but intriguing case of pseudonymity with relevance for the study of both the real and the pseudo-Paracelsus. The writings that bear the name of Basil Valentine and their supposed author(s) have already been the subject of extensive scholarly work, both in recent times and in the seventeenth century. Much of the recent work focusses on the question of who lies behind the pseudonym. Rather than attempting to add directly to that ongoing discussion, this paper examines

 | DOI:10.1163/9789004503380_009

instead the complex details of the gradual construction of the texts comprising the Valentine corpus, and the accretion of a biography for its supposed author. A study of how the Valentine corpus and biography grew – rather than of the equally important questions of authorship and influence – promises to contribute to the studies of forgery, pseudonymity, and book history that have recently become important and lively areas of scholarship. The investigation of the Basil Valentine corpus can also provide a useful point of comparison for the larger and more complex corpus of the pseudo-Paracelsus, and for pseudonymous corpora more broadly. Given the prevalence of pseudepigraphic corpora in the history of alchemy, a better understanding of how and why they agglomerate is important for the field as a whole.

This paper therefore examines three aspects of the Basil Valentine corpus: first and foremost, the gradual emergence and evolution of a large body of writings under his supposed authorship over the course of nearly a century; secondly, the related construction of a constantly expanding pseudobiography for their supposed author – purportedly a Benedictine monk; and thirdly, the eventual clash that emerged due to the false dating of his corpus, resulting in the conclusion that Paracelsus was Valentine's student and/or his plagiarist. In a peculiar and unexpected sense therefore, the writings of Basil Valentine might be considered an ancillary part of the pseudo-Paracelsian literature.

1 Emergence of Corpus A, 1599–1604

Basil Valentine first appeared in print in a slim volume published at Eisleben in 1599 and entitled *Von dem grossen Stein der Uralten*. This text is often referred to as the *Zwölff Schlüssel*, which is (strictly speaking) the subtitle for the second half of the book. This alternate title refers to the twelve allegorical chapters, labelled as "keys," that supposedly encrypt the means of making the philosophers' stone. A second edition appeared in 1602, which provided an illustrative and emblematic woodcut for each key.[1] Much of the book's fame both in its own day and in the subsequent history of chymistry is due to these engaging graphic images. The 1602 second edition also added a *Kurtzer Anhang und Klare Repetition* (pp. 104–160) that expounds some of the allegories and concepts in terse but somewhat clearer terms, and includes a lengthy (yet still largely

1 Many bibliographic entries list this second edition as published at Zerbst; it was indeed printed there as stated in the colophon, but the publisher was Jacob Apel, who was based in Leipzig (as also stated in the colophon), even though neither appears on the title-page.

allusive) description of chymical substances, their secrets, and their products. These first two texts are entirely devoted to the topic of metallic transmutation. They have been shown to contain very specific technical material that reveals a talented and highly-experienced laboratory practitioner. Although in modern times the text has often been thought devoid of practical meaning, I have successfully deciphered the first four keys into workable chemical processes and reproduced them in the laboratory. It is therefore possible to say with confidence that this early text hides some rather sophisticated chemistry.[2]

The second edition included three further works: *De microcosmo* (pp. 161–184), *Von der grossen Heimligkeit der Welt und ihrer Artzney* (pp. 185–226), and *Von der Meisterschafft der sieben Planeten* (pp. 227–240). Two of these short treatises represent a striking contrast with the first two tracts in the volume. While the first two tracts (*Von dem grossen Stein* and *Kurzer Anhang*) deal exclusively with the philosophers' stone and transmutation, two of the new additions are predominantly medical. *De microcosmo* deals with the origin and functioning of the human body, its composition and relation to the *tria prima* (body-salt, soul-sulphur, spirit-mercury), and correct medical practices (e.g., 'like cures like'). *Von der grossen Heimligkeit* also treats predominantly medical topics, such as the attributes of a true physician, the role of seeds and the *tria prima* in health and disease, and two universal medicines called *Asa* and *Phalaia*, the former for external illnesses and the latter for internal illnesses. No specifics are given about their preparation, only the suggestion that they are both prepared from the same substance by "returning it to its first beginning." The prose text is followed by a series of short rhymed verses about chymical substances. This text was later given the title *De macrocosmo* by other printers, at once pairing it with, and distinguishing it from, the preceding *De microcosmo*.[3] Finally, the brief *Von der Meisterschafft* returns to metallic transmutation, giving allusive treatments of each of the planets/metals in rhymed verse, with emphasis on their "hidden secrets and wonders." Unlike the first two treatises, none of the latter three includes the name Basil Valentine in its text, nor do they make any reference to each other or to other works under Valentine's name.

---

2 Lawrence M. Principe, *The Secrets of Alchemy* (Chicago, IL, 2013), 133–158.

3 The earliest titling as *De macrocosmo* appears in the collected edition of Basil Valentine, *Chymische Schrifften*, 2 vols. (Hamburg, 1677), 1: 129. On *Von der Meisterschafft der sieben Planeten*, see Joachim Telle, "Astrologie et alchimie au XVIe siècle: à propos des poèmes astro-alchimiques de Christoph von Hirschenberg et de Basile Valentin," *Chrysopoeia*, 3 (1989), 163–192, esp. 182–188.

In 1603, two further texts appeared under the name of Basil Valentine. The first, *Von der heimlichen Wundergeburt der sieben Planeten und Metallen*, also called *De occulta philosophia*, is an odd work composed of extravagant fables, personifications, and allegories in mythological form, describing the origin of the planetary metals/gods. It is unclear what practical alchemical purpose this slim volume might serve, save as an entertaining allegorical *Gedicht*. Only the last page contains some vague instructions about making the philosophers' stone. The second publication of 1603 is *Von den natürlichen unnd übernatürlichen Dingen* which covers the metals, their composition, and their potential for medicine and transmutation in a more literal although still cryptic way. The book ends with the author's promise of "a particular little book I shall write later on antimony, vitriol, sulphur, [and the] magnet."[4] Part of this plan seemed to be fulfilled in the following year when the most celebrated member of the Valentine corpus was published under the title *Triumph-Wagen antimonii*. This book presents a fairly comprehensive study of antimonial preparations, particularly for pharmaceuticals. It does not neglect transmutation entirely, although that is not its main concern; an appendix deals with the *Stein ignis*, a transmuting agent (specifically, a *particularium*) made from antimonial materials but with weaker potency than the philosophers' stone.[5]

Together, these seven works (published 1599 to 1604) comprise what I call Corpus A.[6] All of them indicate Johann Thoelde (ca. 1563–ca. 1614), *Rathcamerarius* and *Pfannenherr* of the saltworks in Franckenhausen, as their editor. All but the first were published and sold by Jacob Apel of Leipzig. Thoelde claims that they were printed from "old manuscripts" that came into his possession, and it is now widely believed that Thoelde in fact wrote them.7 Thoelde's

4 Basil Valentine, *Von den natürlichen unnd übernatürlichen Dingen* (Leipzig: Jacob Apel, 1603), 122: "das andere von der heimlichen Verborgenheit der *Mineralien* sol gesparet werden zu meinem Vornehmen / da ich hernacher schreiben werde / in einem sonderlichen *Libell* von dem *Antimonio*, Victriol / Schwebel / Magneten / und welcher insonderheit für andern begabet sein."

5 On particularia, see Principe, *Secrets*, 112–113.

6 I originally suggested this division into three corpora in "Apparatus and Reproducibility in Alchemy," in Frederic L. Holmes and Trevor Levere, eds., *Instruments and Experimentation in the History of Chemistry* (Cambridge, MA, 2000), 55–74. Karl Sudhoff, "Die Schriften des sogenannten Basilius Valentinus," *Philobiblion*, 6 (1933), 163–170, likewise divided off what I call Corpus A from the rest of the writings under Valentine's name.

7 The literature is considerable on this topic; see, inter alia, Joachim Telle, "Basilius Valentinus," in Wilhelm Kühlmann, ed., *Killy Literaturlexikon*, 12 vols. (Berlin, 2008), 1: 348–350; essays by Hans Gerhard Lenz and Oliver Humberg in *Triumphwagen des Antimons: Basilius Valentinus, Kerckring, Kirchweger*, ed. Hans Gerhard Lenz (Elberfeld, 2004); Claus Priesner, "Johann Thoelde und die Schriften des Basilius Valentinus," in Christoph Meinel, ed., *Die Alchemie in der europäischen Kultur- und Wissenschaftsgeschichte, Wolfenbütteler Forschungen* 32

authorship was first suggested in print by Vincent Placcius in 1674; one of his colleagues believed that they were the work of "someone in the family of Johann Thoelde."[8] The seven works are however quite heterogeneous both in content and in style, and Thoelde's authorship of all of them has not been convincingly demonstrated, even though it is clear that he had much to do with the *Triumph-Wagen*, and that his role generally is more than that of an editor. I am not convinced, for example, that *Von dem grossen Stein* is entirely by the same author as the other works. One piece of evidence for this assertion is that the references to *Von dem grossen Stein* in *Von den natürlichen unnd übernatürlichen Dingen* do not appear to indicate a correct understanding of its hidden chemical content. In one place, the later text describes a *spiritus mercurii* as the "chief key of my second key [*Haupt-Schlüssel meiner anderen Schlüssel*]," and then provides a preparation of this material from the ores of mercury and gold using corrosive sublimate and spirit of wine. This *spiritus mercurii* is described as a material "without any admixture of earthly aquosity [*ohne einige vermischung der irrdischen aquositet*]" that can dissolve all metals and resolve them into their *prima materia* "without any corrosive [*ohn einig corrosiv*]."[9] In contrast, while the second key of *Von dem grossen Stein* does describe a liquid it calls "mercurial," that product is a highly-corrosive aqueous acid, prepared by the codistillation of sal ammoniac (eagle) and niter (dragon). Thus, *Von den natürlichen unnd übernatürlichen Dingen* appears to misinterpret the second key of *Von dem grossen Stein*, strongly suggesting that they do not share the same author. Further insights into multiple authorship within Corpus A might be gleaned from analyzing the frequency with which certain words are used, although that endeavor is beyond the scope of the current paper.

2 Emergence of Corpus B, 1622–1625

A second cluster of writings published under Valentine's name, which I will call Corpus B, began to appear in print only about twenty years after Corpus A was complete, and roughly a decade after Thoelde's death. The first of this new set of publications was a short volume published by Johann Birckner in Erfurt

(Wiesbaden, 1986), 107–118; and Hans-Henning Walter, ed., *Johann Thölde: Um 1565 – um 1614* (Freiburg, 2011).

8 Vincent Placcius, *De scriptis et scriptoribus anonymis atque pseudonymis syntagma*, (Hamburg, 1674), 159.

9 Basil Valentine, *Von den natürlichen unnd übernatürlichen Dingen*, 36–37 and 45–47; and Principe, *Secrets*, 146–149.

in 1622 and entitled *Conclusiones oder Schlussreden Fratris Basilii Valentini aller seiner Schrifften und Tractaten von Schwefel, Vitriol und Magneten*. The text on sulphur, vitriol, and the magnet could be seen as fulfilling the author's promise of such a title made in *Von den natürlichen unnd übernatürlichen Dingen*, or as a work written by someone else using the otherwise unfulfilled promise as a point of departure. The publisher's preface claims that manuscripts of this and other unpublished Valentine texts were being offered and sold at exorbitant prices, and that he chose to publish this volume in order to make the Benedictine's final (and, one might hope, explanatory) teachings more easily available.[10] Birckner promised to publish another manuscript called *Offenbahrung der verborgenen Handgriffe*, and hinted at an unpublished "explication and clarification of Basil's parabolic writings" that might follow thereafter.[11]

The publisher kept his word. The *Offenbahrung* appeared, along with a second printing of the *Schlussreden*, in 1624. The *Offenbahrung* contains a much fuller and more workable exposition of the *Zwölff Schlüssel* than had been provided by the 1602 *Klare Repetition*. As its one-page preface promises, "herein you will find something more than in the *Twelve Keys*." It names exact materials, apparatus, and manipulations for several – but not all – of the keys. The publisher rounded out the volume with the *Tractatus von Natur der Elementen*, a German translation of a work by Cornelis Drebbel (1572–1633) originally published in 1621 in Dutch.[12]

These first two works of Corpus B were republished quickly in a noticeably altered format, and accompanied by two additional texts. In 1625, Lucas Jennis (1590–1630) – renowned for his publications of the works of Michael Maier – published the *Dyas chemica tripartita*, a collection of alchemical treatises. The editor is listed only as "H. C. D." on the title-pages, but expanded as "Hermann

10 Benedictus Figulus also reported that high prices were being paid for copies of the *Schlussreden*, including 300 florins paid to the brother of the deceased Johann Thölde; see Wilhelm Kühlmann and Joachim Telle, eds. *Corpus Paracelsisticum: Dokumente frühneuzeitlicher Naturphilosophie in Deutschland*, vol. 3, part 2: *Der Frühparacelsismus* (Berlin, 2013), 1172. Such prices obviously encouraged further productions and publications under Valentine's name.

11 Basil Valentine, *Conclusiones oder Schlussreden* (Erfurt: Johann Birckner, 1622), "Ad lectorem." The hint about the third possible treatise disappeared from the "Ad lectorem" in the reprint of the *Schlussreden* published with the 1624 *Offenbahrung* (see next note), 26.

12 Basil Valentine, *Offenbahrung der verborgenen Handgriffe* [...] *item Conclusiones oder Schlussreden* (Erfurt: Johann Birckner, 1624). The name of an editor (or author?) may be hidden in the phrase "Medium Tenuere BeatI" which occurs at the end of both the preface and the book, and appears also in the 1622 publication; the capitals M.T.B.I. may represent initials (Iohann Birckner?).

Condeesyanus" at the end of the dedicatory epistle; this figure has been identified as Johann Rhenanus, physician to Moritz of Hessen-Kassel.[13] The "medieval section" of his *Dyas* contains *Vier Tractätlein Basilii Valentini*.[14] The first of these "four little treatises" is the *Offenbahrung*, published for the first time just the year before, but here it is altered in two significant ways. First, the terse recipe-like text of the 1624 Erfurt version has been expanded, generally in order to give fuller, more explicit instructions for the practical work. The gist and basic wording remain, although at one point the impossible instruction to sublime a material with tartar is replaced with the more reasonable instruction to do so with sal ammoniac.[15] Second, and more dramatically, this new version interpolates the allegorical images from the *Zwölff Schlüssel*, using the redesigned and more elegant engravings produced for Michael Maier's Latin translation – *Practica cum duodecim clavibus* – published in 1618. The result is a monstrous hybrid of reworked images and a second-generation text to produce a "new" Basil Valentine treatise.

The *Vier Tractätlein* also contains a third printing of the *Schlussreden*, whose text, relative to the 1624 Erfurt edition, is again slightly expanded throughout. It is further augmented with a short *Concordantz* by an unknown author who, surprisingly enough, does *not* claim to be Basil Valentine.

The *Vier Tractätlein* also put forth two new titles. The first is the *Handgriffe der Artzney*, a compendium of fifty-one recipes for various products, not all of them in fact medicinal.[16] Georg Wolfgang Wedel (1645–1721) claimed in 1704 to own a manuscript entitled *Handgriff und Bereitung seiner vornehmen Artzeney*, which is unquestionably the same work, and which he considered to be a yet-unpublished work by Basil Valentine. He dates his manuscript to "around 1620,"

13 Condeesyanus has often been identified as Johann Grasse (or Grasshoff), but this identification was refuted by Thomas Lederer, *Der Kölner Kurfürst Herzog Ernst von Bayern (1554–1612) und sein Rat Johann Grasse (um 1560–1618) als Alchemiker der Frühen Neuzeit: Ein Beitrag zur Geschichte des Paracelsismus* (Inaug. diss., Ruprecht-Karls-Universität Heidelberg, 1992), 207–210; I thank Prof. William Newman for reference to this thesis. A contemporaneous inscription in Condeesyanus' *Harmoniae inperscrutabilis chymico-philosophicae* (Frankfurt, 1625) identifies him as Rhenanus, as noted by Carlos Gilly, *Adam Haslmayr. Der erste Verkünder der Manifeste der Rosenkreuzer* (Amsterdam, 1994), 150 n. 32.

14 H.C.D. (Hermann Condeesyanus [Johann Rhenanus]), ed., *Dyas chemica tripartita* [...] *Sechs herrliche teutsche philosophischeTractätlein* (Frankfurt: Lucas Jennis, 1625).

15 Cf. *Offenbahrung*, 15 and *Dyas*, 16. The *Dyas* also omits two recipes, for *spiritus vini* and *fewriges Wein-Geist*, printed in 1624 at the end of the *Offenbahrung*, 21–24.

16 This text was republished in *Basilius Innovatus* [...] *Chymische Schrifften*, ed. Benedikt Nikolaus Petraeus, 3 vols. (Hamburg, 1717), 3: 1057–1096, but not in other collected works of Basil Valentine, such as the 1677 Hamburg *Chymische Schrifften*.

thus placing it in the temporal vicinity of the 1625 *Vier Tractätlein*, although he was clearly unaware of that publication. Wedel likewise recognizes that not everything in that manuscript was actually from Basil Valentine.[17] The last text of the *Vier Tractätlein* is another new title, the *Zugab oder Supplementum*, composed of various recipes for chymical products. This text would later be reprinted in the collected works of Basil Valentine, although sometimes prefaced with the notice that it is a "doubtful" work.[18] Both of these new works are clearly recipe collections assembled from various sources of various dates. I therefore prefer to place them in a third category, Corpus C, composed of works more distantly (or not at all) related to Corpus A.

The works of Corpus B that appeared from 1622 to 1625 differ markedly from those of Corpus A, not only in terms of when they were published, but also in style. This difference was noted in 1728 by Friedrich Roth-Scholtz, who dismissed both the *Offenbahrung* and the *Schlussreden* from the canon of "authentic" Valentine writings:

> The *Conclusiones, oder Schlussreden* was compiled by somebody else, and yet can pass. The *Offenbahrung der verborgenen Handgriffe*, however, is not worthy to be counted among the writings of Basil Valentine – it does not agree in the least with the rest of his other writings.[19]

I have suggested elsewhere that the *Offenbahrung* is the production of a later practitioner who interpreted the allegories of *Von dem grossen Stein* and followed them successfully – up to a point – in the laboratory.[20] What I had not noticed previously is that the *Offenbahrung*'s "Erste Arbeit" (corresponding to the first four keys of the *Zwölff Schüssel*) closely paraphrases – with some expansions – the method of making *sal Solis* given in Johann Thoelde's

---

17 Georg Wolfgang Wedel, *Propemticon inaugurale de Basilio Valentino* (Jena, 1704), 7–8. The identity of Wedel's manuscript with the text printed in the *Vier Tractätlein* is proven by the fact that Wedel mentions that his manuscript contains the "menstruum Zwelfferi ad solvendas perlas, e spiritu viridis aeris crystallini," and the making of "nitrum vitriolatum Mynsichti" both of which are likewise in the *Dyas*, 44–46, although without the names of the authors to whom Wedel attributes them.

18 For example, *Chymische Schrifften*, 2 vols. (1677), 2: 383.

19 Friedrich Roth-Scholtz, ed. *Deutsches Theatrum Chemicum*, 3 vols. (Nuremberg, 1728), 1: 663 (foreword to *Triumph-Wagen antimonii*): "Die *Conclusiones*, oder Schlussreden / sind von einem andern zusammen getragen werden / und können allenfalls noch *passi*ren. Die Offenbahrung der verborgenen Handgriffe aber / sind nicht wehrt / unter die Schrifften Basili Valentini gezehlet werden / die auch mit seinen übrigen Schrifften im geringsten nicht überein kommet."

20 Principe, "Apparatus," 60 and idem, *Secrets*, 241 n. 28.

Haliographia of 1603.[21] The texts are close enough that either a later author of the *Offenbahrung* copied the description virtually directly from the *Haliographia*, or, perhaps less likely, that Thoelde wrote the *Offenbahrung* himself and it circulated in manuscript after his death.

Strikingly, the author of the *Offenbahrung* was not the only practitioner to write a practical decipherment of the first several keys, which indicates a vigorous and widespread response to Corpus A, particularly to *Von dem grossen Stein*. Clearly, multiple practitioners were interested in translating its allegories into practical laboratory terms. One of these early treatments is Lorenz Meisner's 1608 *Gemma gemmarum alchimistarum*, which offers practical explanations for each of the twelve keys of *Von dem grossen Stein*, and for many of the verse enigmas in *Von der grossen Heimligkeit der Welt*. It is likely that *Gemma gemmarum* was the text that Birckner, the publisher of the 1622 *Schlussreden*, had in mind to (re)publish as a new Valentine text, since Meisner's subtitle, *Erleuterung der Parabolischen und Philosophischen Schrifften Basilii*, is strikingly close to Birckner's promised "Erlauterung der Parabolischen Schrifften Basilii."[22]

An interpretation by Konrad Schüler of *Von dem grossen Stein*'s concluding poem about the *materia prima* was published along with Meisner's work.[23] Meisner's and Schüler's books also came from the shop of Jacob Apel, which published most of Corpus A. In fact, the *Gemma gemmarum* reuses several of the woodcut images that were originally printed in the 1602 *Klare Repetition*. Both of these books, despite having their authors' names printed clearly on their title-pages, are occasionally encountered in less-than-critical bibliographies as works by Valentine himself – and such works also can be classed in Corpus C.

Yet another explication of the *Zwölff Schlüssel* appeared in Hans Christof Reinhart's 1611 *Gülden Gesundbrunnen*. Many aspects of his interpretation differ from both the earlier *Gemma gemmarum* and the later *Offenbahrung*,

---

21 Compare Johann Thoelde, *Haliographia* (Leipzig: Jacob Apel, 1603), 168–171 with *Offenbahrung*, 3–8.

22 Lorenz Meisner, *Gemma gemmarum alchimistarum* (Eisleben: Jacob Apel, 1608) and *Schlussreden* (1622), fol. Aiir-v.

23 Conrad Schüler, *Gründliche Ausslegung und warhafftige Erklerung der Rythmorum Basilii Valentini* (Eisleben: Jacob Apel, 1608). The poem appears first at the end of the 1599 *Von dem grossen Stein*. Schüler's explanation is itself in verse, and is interpolated within the lines of the original Valentine poem; the two authorships are distinguished only by a moderately different typeface, suggesting one way that pseudocorpora can expand. On the original poem, as well as on Schüler and Meisner, see Joachim Telle, "'De prima materia lapidis philosophici': Zu einer deutschen Lehrdichtung im Basilius-Valentinus-Alchemicacorpus," *Zeitschrift für Germanistik*, N.F. 1 (2009), 12–40.

indicating that these three decipherments of the same text were arrived at independently.[24] Previously, in 1608, Reinhart had published *Liecht der Natur*, a text devoted to transmutational alchemy and based predominantly on the ideas expressed in Corpus A. He stressed that his understanding of the Valentine texts came not merely from literary interpretation, but from experiments done with his own hands as a means of getting at the correct meaning.[25] (*Liecht der Natur* in older bibliographies also occasionally shows up as a product of Basil Valentine, so it too fits into Corpus C.) These four instances, by three different authors – Meisner, Schüler, and Reinhart – of resolving the cryptic parts of Corpus A into practical experimental processes indicates that there was a substantial amount of such activity taking place immediately after the emergence of Corpus A. It is to this milieu that we undoubtedly owe some of texts that later appeared as part of Corpus B.

3 Corpus B Continues: The *Letztes Testament* 1626–1680

In 1626, another work under the name of Basil Valentine, the *Letztes Testament*, appeared from a printer in Jena. This member of Corpus B has the most complex history, yet its origins can be traced the most fully and confidently. As such, it provides an intriguing and illustrative example of how a pseudonymous corpus can accrete. The 1626 title-page claims that this new Valentine text came from a recently discovered manuscript that had long lain hidden within an altar. Oddly, even though its editor and publisher, Georgius Claromontanus (Georg Hellberger), advertised on the title-page that the *Letztes Testament* contained five parts and published the full titles of all five, he printed only the first two.[26] He never published the remaining three parts. Those further three parts appeared only twenty years later from a completely different press,

24 Hans Christof Reinhart, *Der Gülden Gesundbrunnen* (Halle: Joachim Krüsicken, 1611). It is worth mentioning that, of the early editions of *Von dem grossen Stein* that I have examined, a significant proportion contain marginalia indicating that their readers seriously endeavored to interpret the allegories into practical laboratory operations; see, for example, Principe, *Secrets*, 149 (figure 6–5) and 154 (figure 6–7). Equally interesting is the wide diversity of decipherments proposed.

25 Hans Christof Reinhart, *Liecht der Natur* [...] *der warhafftigen Kunst alchimiae höchstes Geheimniss* (Halle: Joachim Krüsicken, 1608), fol. Biiv: "nicht allein nach dem aussern Buchstaben ansehen / sondern durch [...] meiner eignen Händen richtiger Arbeit befindung."

26 Basil Valentine, *Letztes Testament und Offenbahrung der himmlischen und irrdischen Geheimnüß so in einem Altar gefunden / in fünff Bücher abgetheilet* [...], ed. Georgius Claromontanus (Jena: Hans Eyring, 1626).

publisher, and city – namely, from Caspar Dietzel in Strasbourg in 1645.[27] But even this publication does not represent the final stage of the *Letztes Testament*'s growth. Thirty-five years later, Johann Cardilucius (1630–1697) pointed out inconsistences in the fifth (and shortest) book of the work, in particular, cross-references to topics that were nowhere to be found. Owing to his "manifold travels and acquaintance with many eminent people," as he claims, he was at last given a manuscript that contained much more text for the fifth book. He published this fuller version of the fifth book, more than *five times* the length of the 1645 version, in 1680.[28] Where did all these texts of the *Letztes Testament* originate?

In 1674, Daniel Georg Morhof (1639–1691) made an intriguing but not entirely clear reference to the real author of an unspecified text attributed to Basil Valentine. He claimed that "*On the Increase and Decrease of Metals* by Andreas de Solea [...] is commonly attributed to Basil Valentine and is customarily published with his *Twelve Keys*, even though Basil preceded Solea by a whole century."[29] Morhof repeated this assertion in 1688, writing that a friend showed him Solea's book in quarto with the title *De incrementis et decrementis metallorum*, and further suggested that Solea may be a partial anagram for Johann Thoelde (because the letters of "de Solea" are contained in the name Johannes Thoelde, even though "the remaining letters cannot make a first name"). In the posthumously published last part of his *Polyhistor* (1708), Morhof referred a third time to "Andreas Solea, praised by Maier, whose German book *Uhrsprung der Metallen* was published under the name of Basil Valentine."[30] Nowhere did

---

27 Basil Valentine, *Geheime Bücher oder letztes Testament* [...] *Auß dem Original / so in dem hohen Altar zu Erffurt unter einem Marmorsteinen Täfflein gefunden* [...] (Strasbourg: Caspar Dietzel, 1645). In the same year, Dietzel also republished all the works in Corpus A with the exception of the *Triumph-Wagen*.

28 *Basilii Valentini Fünfften Theil seines Letzten Testaments*, in Johann Hiskia Cardilucius, ed., *Magnalia medico-chymica continuata, oder Fortsetzung der hohen Artzney und feuerkunstigen Geheimnüssen* (Nuremburg, 1680), 711–797.

29 Daniel Georg Morhof, *Ad Langelottum epistola*, in J.J. Manget, ed., *Bibliotheca chemica curiosa*, 2 vols. (Geneva, 1702), 1: 169: "*Andreae de Solea* Liber de Incremento et Decremento Metallorum, qui Basilio Valentino vulgo adscribitur, ac cum eius XII clavibus edi solet, quanquam integro seculo hunc ille superet." On Morhof and his *Epistola*, see Lawrence M. Principe, "D.G. Morhof's Analysis and Defence of Transmutational Alchemy," in Françoise Wacquet, ed., *Mapping the World of Learning: The Polyhistor of Daniel Georg Morhof*, *Wolfenbüttler Forschungen* 91 (Wiesbaden, 2000), 138–153.

30 Daniel Georg Morhof, *Polyhistor* (Lübeck, 1688), 85–86 and idem, 3 vols. (Lübeck, 1708), 2: 434. The reference in the quotation is to Michael Maier, *Symbola aureae mensae duodecim nationum* (Frankfurt, 1617), 51, where that author makes an argument regarding transmutation "ex incremento & decremento metallorum, veluti *And. Solea demonstrat, in his rebus artifex peritissimus*."

Morhof make clear *which* Valentine work was supposedly by Solea; neither the German nor the Latin title he gives corresponds clearly with any book published under the name of either Valentine or Solea.

In 1717, however, Benedikt Nikolaus Petraeus explicitly identified the Solea work in the foreword to his new edition of Valentine's collected works. Petraeus observed that the text of the first book of the *Letztes Testament* had been published in 1600 under the name of Nicolas (not Andreas) Solea, a Bohemian, with the title *Büchlein vom Bergwergk*.[31] This connection was rediscovered by Felix Fritz in 1925.[32] Fritz also suggested – correctly, in my opinion – that, based on style and content, the second part of *Letztes Testament* was by the same author, a fact seemingly assumed by Petraeus who always refers to the two parts together as Valentine's *Berg-Buch*, as does the third part of *Letztes Testament* which refers to them as the "first and second parts of the *Berg-Sachen*."[33] What has not been noted previously is that part of the subtitle of the second part of the *Letztes Testament* – "wie auch numehr die *Metallen generirt*, geborn / und zu Tage gebracht werden" – harmonizes well with the title given by Morhof, *Uhrsprung der Metallen*.[34] The origin and authorship of the 1624 version of the *Letztes Testament* can therefore be clearly explained: part one comes almost directly from Solea's 1600 book (or a manuscript thereof) and part two very probably from another, but currently unidentified Solea book or manuscript. Of course, this is a modern historian's conclusion; Petraeus himself had another explanation to which I shall return in part 6, below.

But under what circumstances was Solea's text reassigned to Basil Valentine? A newly-discovered manuscript indicates that this transmutation of authorship had already been carried out by 1612.[35] A manuscript bearing that date and preserved in Leipzig presents Solea's text as the first two books of Valentine's *Letztes Testament*, and prefaces them with a "foreword and introduction by Basil Valentine" that claims the material as the supposed Benedictine's own. This foreword was printed with only minor alterations in the 1626 *Letztes*

31 *Basilius Innovatus*, 1: fol. f[7]; Nicolas Solea, *Ein Büchlein von dem Bergwergk*, ed. Elias Montanus (Zerbst, 1600); reprinted without Solea's name as *Bergwerckschatz* (Frankfurt, 1618).

32 Felix Fritz, "Basilius Valentinus," *Zeitschrift für angewandte Chemie*, 38 (1925), 325–329.

33 Basil Valentine, *Letztes Testament* (1645), 177.

34 The subtitle to the first part, "[…] Bergwerge / derer Ursprung / Natur und Eigenschafft," also has some resonance with Morhof's title.

35 Gerhard Görmar, "'Fratris Basilii Valentini Benedicter Ordens V Letzte Bücher, welche sindt sein letztes Testament' – Ein bisher unbeachtetes Manuskript in der Universitätsbibliothek Leipzig," *Sudhoffs Archiv* 103 (2019): 26–54. I have not yet seen this new manuscript myself.

Testament. Again like the printed version, the foreword describes *five* books that comprise the *Letztes Testament*, but the manuscript provides only *three* of them: the two published in 1626 and the "dritter Theil," not published until 1645. The manuscript itself, which was owned successively by two apothecaries prior to 1626, is a very high-quality production with elegant, ornate capitals and a professional book hand, thus suggesting that it was originally intended as a presentation copy rather than as something destined immediately for the press. It does not provide evidence of who compiled it or for whom it was intended, but this extremely important discovery does push the origins of the *Letztes Testament* back to at least fourteen years before its appearance in print.

The latter three parts of the *Letztes Testament*, published in 1645, do not come from Solea. The third book, a highly-annotated version of which appears already in the Leipzig manuscript, is yet another exposition of the *Zwölff Schlüssel*, but with several interesting features. First of all, its practical directions differ significantly from those of the *Offenbahrung*, as well as from the conclusions of Meisner and Reinhart. Thus, it is clearly a composition by yet *another* practitioner largely or entirely unrelated to these previous publications. Again unlike the *Offenbahrung*, the author of the third book of the *Letztes Testament* takes great pains to pass himself off as Basil Valentine. The *Offenbahrung* contains only a perfunctory reference to the Benedictine at the start ("Ich, Fr. Basilius Valentinus, ein Closter Diener des Herrn [...]") before becoming a clear and matter-of-fact description of laboratory operations – essentially, a lengthy recipe without theoretical content, literary formulations or embellishments. The third book of the *Letztes Testament*, on the other hand, includes frequent references to "my other writings" (in Corpus A); it adds new allegories and lengthy theoretical digressions, and employs throughout many of the same characteristic turns of phrase that occur in the *Triumph-Wagen*. Despite being part of a "last testament," the text still refers to the "book on antimony" as something the author will soon write, clearly in reference to the 1604 *Triumph-Wagen*.[36] This third part is thus a very conscious attempt, made by unknown author before 1612, to write a seemingly "authentic" Basil Valentine text.

The fourth and fifth parts, however, are the most interesting and revealing of how the Valentine corpus continued to develop. In 1842, Ferdinand Hoefer noted tersely (and very imprecisely) that the *Letztes Testament* was identical with *Haliographia*, a work on salts first published in 1603 and written by

36 Basil Valentine, *Letztes Testament* (1645), 169: "Wie du nachmals / als ich insonderheit allein vom *Antimonio* schreiben werde / weitern Bericht vernemmen wirst."

Johann Thoelde.[37] In fact, only part of the *Letztes Testament* is related to the *Haliographia*, and not directly. The fourth book of the *Letztes Testament* does draw heavily upon the fourth book of the *Haliographia*, but the original material is heavily edited, partly rewritten, and smoothly interpolated with new text that doubles its length. Some of the added material is clearly intended to link the text to the earlier Valentine corpus by means of frequent references to works in Corpus A, as well as to the *Offenbahrung*. Most strikingly, the interpolations and new headings completely subvert Thoelde's original text and intent. His original *Haliographia* text dealt entirely with metallic salts and their medicinal uses, but in the *Letztes Testament* this material has been completely reworked to deal almost entirely with transmutation. The recipes that originally yielded medicinal salts, now promise to provide transmuting agents. Thoelde's original title for this part of his *Haliographia* – "how to extract and prepare salts from all metals" – becomes, in the *Letztes Testament*, "On the *particularia* [transmuting agents] of all the metals." The unknown reworker also deleted everything about medicinal virtues and the lengthy sections on mineral, animal, and vegetable salts. He thus transformed Thoelde's straightforward set of iatrochemical recipes into a new and cryptic Basil Valentine treatise about transmutational agents.

The 1651 edition goes a little further; it also tacks the 1624 *Offenbahrung* onto this fourth part of the *Testament*. This addition appears to have been a last-minute decision at the press, since the printer paginated it using letters rather than numbers so that it could be inserted in the middle of already typeset and paginated (and perhaps already printed) text. This late insertion becomes (mostly) seamless in the later collected editions.[38]

The fifth part of the *Letztes Testament* provides yet another interesting case. As published in 1645 it is very short – only twelve pages in length – and deals predominantly with making *aurum potabile*.[39] The much longer version published by Cardilucius in 1680, on the other hand, gives an exhaustive treatment of how to prepare a wide variety of salts. This longer version preserves the first eight pages of the 1645 version, but then continues by tacking on, in rearranged order, all the sections of the fourth book of Thoelde's *Haliographia* that were not previously used to create the fourth part of the *Letztes Testament*![40] The

37 Ferdinand Hoefer, *Histoire de la chimie*, 2 vols. (Paris, 1842), 1: 456.

38 The collected edition *Chymische Schrifften*, 2 vols. (1677) rather inexplicably still marks the first two pages as "a" and "b."

39 Basil Valentine, *Letztes Testament* (1645), 252–260.

40 Since this connection has not, to my knowledge, been noticed before, I include here a concise analysis of the origins of Cardilucius' 1680 fifth part: *Magnalia*, 715–723 (= 1645 *Letztes Testament*, 252–260); 723–745 (= 1603 *Haliographia*, 187–225); 745–750 (= *Haliographia*,

Haliographia text is repeated verbatim here, just as Thoelde wrote it, without the reworkings and interpolations made for the fourth part of the *Letztes Testament*, save for one brief paragraph that refers back to the preceding part.

The *Letztes Testament* thus showcases the accretion of a pseudonymous text over the course of more than half a century. By 1612, two texts by Solea, one of which was already published under its true author's name, had been reassigned to Basil Valentine in the Leipzig manuscript. These two texts were then published as Valentine's in 1626. That publication, like the newly-discovered manuscript upon which it ultimately depended, mentioned three further texts, but did not print them. Only one of those three exists in the manuscript now in Leipzig; did the other two exist in some form in other manuscripts at this time, or were only their titles written? These three promised texts were finally published in 1645: one, a very conscious effort to mimic Corpus A; another, a massively reworked and redirected section of Thoelde's *Haliographia*; and the last, a brief, obviously fragmentary text, mostly about potable gold. Finally, in 1680, this last fragment was massively augmented by text taken verbatim from the *Haliographia*, but with care being taken not to duplicate any of the material already pilfered from that book to create the foregoing fourth part of the *Letztes Testament*.

4 Corpus C: More Distantly Related, Reassigned, and Untraceable Works

Corpus A contains the works published from 1599 to 1604 in connection with Johann Thoelde, and Corpus B contains the works published later, as purported continuations of Corpus A. This leaves Corpus C, a miscellany of other texts attributed in one way or another to Basil Valentine and more distantly, or not at all, related to Corpus A. Some of these attributions are the result of publishers cashing in on the popularity of Basil Valentine, such as a Latin translation of the entire fourth book of Thoelde's *Haliographia* – without any of the alterations found in the *Letztes Testament* – published in Bologna in 1644 under the name of Basil Valentine and supposedly from his original manuscripts.[41]

300–309); 751–765 (= *Haliographia*, 274–300); 766–785 (= *Haliographia*, 238–273); 786–795 (= *Haliographia*, 226–237); 795–796 (= 1645 *Letztes Testament*, 264); 796–797 (new concluding text). There is also brief new text included on 723 and 750–751, designed to make the transitions to the following topics smoother.

41 Basil Valentine, *Haliographia, de praeparatione, usu, ac virtutibus omnium salium, mineralium, animalium, & vegetabilium, ex manuscriptis, & originalibus Fratris Basilii Valentini*, ed. Andreas Salmincius (Bologna: Niccolo Tebaldini, 1644).

I mentioned previously the two recipe collections *Handgriffe der Artzeney* and *Zugab oder Supplementum*, published in 1625 as parts of the *Vier Tractätlein*.

Other texts gained membership in Corpus C owing to three centuries of sloppy bibliographic work. I have already mentioned the early interpretations of Corpus A, by Meisner (*Gemma gemmarum*), Schüler (*Erklerung der Rythmorum Basilii*), and Reinhart (*Liecht der Natur*). But the most commonly encountered example in this category is *Azoth, sive aureliae occultae philosophorum*, which seems to have been attributed to Valentine for two reasons, both very inadequate. First, it contains a Latin prose version of the German poem "De prima materia" from *Von dem grossen Stein*. Valentine is even mentioned on the title-page as the author of some of the contained text, but his name is garbled there as "Basilius Vicentinus." Second, the book is a Latin translation of a German original, issued by the same printer in the same year, and whose title, *Occulta philosophia von den verborgenen philosophischen Geheimnussen*, is superficially similar to the 1603 *De occulta philosophia* of Corpus A.[42] The link between Valentine and *Azoth* may have been strengthened by the loose French translation of the *Zwölff Schlüssel* published in 1624 which promises on its title-page to include "Azoth, ou le moyen de faire l'Or caché des Philosophes," but this latter is nothing more than a French versification of "De prima materia."[43] The puzzling *Dialogus fratris Alberti cum spiritu*, cited by Pierre Borel (1620–1671) in his list of Basil Valentine's writings, and dutifully but uncritically copied by bibliographers until the twentieth century, also derives from this French publication which includes a "Colloque de l'esprit de Mercure à frère Albert." This composition is itself a translation, with alterations, of a German work with the fuller original title *Ein Colloquium oder Gesprech Spiritus Mercurii & Monachi fratris Alberti Beieri*. This *Gesprech* appeared first in a collection of alchemical treatises published in 1604 by an editor hidden under the initials I.P.S.M.S.[44]

---

42 *Azoth, sive Aureliae occultae philosophorum*, trans. Georgius Beatus (Frankfurt: Johann Bringer, 1613); "Symbolum Basilii Valentini" [Latin version of "De prima materia"], 53–54; *Occulta philosophia von den verborgenen philosophischen Geheimnussen der heimlichen Goldblumen und Lapidis Philosophorum* (Frankfurt: Johann Bringer, 1613). This publication is also mentioned by Petraeus as a work potentially by Valentine, although he had not examined it; see *Basilius innovatus*, fol. f4v.

43 *Les douze clefs de philosophie de Frere Basile Valentin* (Paris: Jeremie and Christophle Perier, 1624), 131–133.

44 Pierre Borel, *Bibliotheca chimica* (Heidelberg, 1656), 212; *Douze clefs*, 151–169; *Alchimia vera, das ist, der wahren und von Gott hochbenedeyten Naturgemessenen edlen Kunst Alchimia wahre Beschreibung* (s.l., [Erfurt? Heinrich Birnstiel?], 1604), 61–74; see Joachim Telle, "*Alchimia vera* (1604)," in Hans-Gert Roloff, ed., *Die Deutsche Literatur. Biographisches und bibliographisches Lexikon*, Reihe II: *Die Deutsche Literatur zwischen 1450 und 1620*, Abt. A: *Autorenlexikon*, vol. 2 (Bern, 1991), 84–86, on 85b. The *Colloquium spiritus mercurii* was

Corpus C also contains the *Opus praeclarum ad utrumque* that appeared in the *Theatrum chemicum*; it is attributed to a *magister Valentinus*, but is unrelated to Basil Valentine.[45] Petraeus mentions an unpublished Valentine manuscript entitled *Via veritatis*, supposedly held by the Canons Regular at Thurnstein near Krems-an-der-Donau, but nothing is known of this suppositious work.[46] Finally, there is also the *Apocalypsis chymica*, reportedly published in Erfurt in 1624, but which I have been unable to locate.[47]

5 Basil Valentine's "Biography"

The texts of Corpora A and B provide several details about the life of their supposed author. *Von dem grossen Stein* begins by mentioning Valentine's profession as a Benedictine monk. The "author" then recounts how after some years, he found that he had free time after celebrating the Office and his duties. So that "idleness might not give cause for [his] thoughts to fall into evil things," he decided to set about a "dissection of nature." In the monastery library he found many books on alchemy "written long before my time by the wise masters," and thereafter spent all his free time discovering the secrets of transmutational and medical chymistry, curing his sick brethren and eventually succeeding in making the philosophers' stone.[48] The *Triumph-Wagen* gives a bit more information, namely that Basil Valentine was born in "the upper Rhineland, in high Germany," that he travelled in Belgium and England while young, and in later years made a "difficult journey" as a pilgrim to Santiago di Compostella.[49]

---

supposedly written by the fictitious Carmelite monk Albert Beyer, who was said to have met with a spirit (namely the spirit of mercury) in 1560 in the cloister of Maria Magdalena della Stella Nova; my thanks to Didier Kahn for this information. The French version omits the concluding autobiographical material about Beyer and appends an "Explication de l'esprit" regarding the qualities of the *première matière*. It also introduces bizarre obfuscation by rendering the spirit's straightforward instruction to obtain "etliche lb des besten Orientalischen Gold / wie es ohn Fewr aus seiner Mutter der Erden kömpt" as "dix livres du meilleur 98756 Æ 5Æ [sic] d'Orient, tout ainsi qu'il vient du ventre de sa mere sans feu."

45 *Theatrum chemicum*, 6 vols. (Strasbourg, 1659–1661), 4: 941–954. On this medieval text, see Lynn Thorndike, *A History of Magic and Experimental Science*, vol. 3 (New York, 1934), 182–190.

46 *Basilius innovatus*, fol. f4v.

47 Listed as one of Valentine's works by Magnoaldo Ziegelbauer, *Historia rei literariae Ordinis S. Benedicti*, 4 vols. (Vienna, 1754), 3: 191. Birckner published several alchemical titles in Erfurt in 1624, but I have found no record of this *Apocalypsis*.

48 *Von dem grossen Stein* (1599), fols. A[viii]v–Biir.

49 *Triumph-Wagen antimonii* (1604), 46–47 and 211.

Regarding Valentine's dates, nothing explicit is offered. The foreword by Thoelde to the 1599 *Von dem grossen Stein* remarks only that the manuscripts of the work were in "a very old script," making them difficult to decipher.[50] The *Triumph-Wagen* however frequently mentions "the new war-disease" that had recently been brought into Germany by the French, presumably a reference to the French disease, or syphilis, which was thought to have appeared in Central Europe only after the 1494 French invasion of Italy. Mention of this disease thus makes the texts appear to be from the late fifteenth or very early sixteenth century.[51]

The first biographical claims made outside of the corpus itself come from Michael Maier in 1617, who claimed that Valentine was born in Alsace (probably just a further specification of Valentine's own mention of the Upper Rhineland). Maier described Valentine as having "flourished in the time of our great grandparents" – again, situating him around 1500 – and as having lived in "the metropolis of Thuringia," a reference to Erfurt.[52] Maier does not give a source for this last claim but the connection to Erfurt continued to grow for the rest of the century (and has recently been shown to have an unexpected basis in reality).[53] The 1645 *Letztes Testament* expands on the claim made in the 1624 edition that the manuscript was found in an altar, specifying that it was found in the "high altar in Erfurt." Olaus Borrichius tells a more embellished story that the manuscripts were discovered due to "a bolt of lightning which struck a column in the church at Erfurt, in the shattered middle of which his writings had lain hidden until then."[54] Johann Moritz Guden (1639–1688), chief magistrate, counsellor, and university professor at Erfurt provides yet

50 *Von dem grossen Stein* (1599), Vorrede, fol. Av-v.

51 *Triumph-Wagen antimonii* (1604), 44–45.

52 Maier, *Symbola aureae mensae*, 257–258.

53 Although Maier's unexpected mention of Erfurt is puzzling, there now seems to be some truth to it. Lenz determined that Thoelde was a student in Erfurt, and at least one recipe in the manuscript *Proces Buch* that Thoelde gave to Moritz of Hessen in 1594 is cited as having been "copied down from an old book in the cloister on Petersberg in Erfurt" – exactly the locale later writers would say Basil Valentine had inhabited; see Lenz, *Triumphwagen*, 211 (text from the *Proces Buch*) and 287, and Priesner, "Johann Thoelde," 113–115. Might there have been contact between Thoelde and Maier? Both moved in the circle around Moritz where Basil Valentine's writings were highly prized; see Bruce Moran, *The Alchemical World of the German Court: Occult Philosophy and Chemical Medicine in the Circle of Moritz of Hessen (1572–1632)*, *Sudhoffs Archiv*, Beiheft 29 (Stuttgart, 1991).

54 Olaus Borrichius, *Conspectus scriptorum chemicorum illustriorum* (Copenhagen, 1697), 30: "per ictum fulminis columnâ Templi Erffurtensis, in cujus medio diffracto scriptum ejus hactenus delituerat." The Leipzig manuscript (see note 35) bears a general title that also cites the discovery of the text in the high altar at Erfurt, but this inscription is written by a different and later hand on a separate sheet, and so is possibly copied from the 1645 (or a later) publication.

more precise information in his 1675 *Historia Erfurtensis*, clarifying that "Basil Valentine, marvelously skilled in the medical art and in the investigation of natural things, lived in St. Peter's monastery" in 1413.[55] The publication of this exact date, even though it threw Valentine's dates back yet another hundred years, provoked the emergence of "corroborating evidence" to support it. Thus in 1717, Petraeus published a supposed manuscript of Basil Valentine that includes, rather suspiciously, the date 1417 – certainly chosen by the unknown author to correlate with Guden's date of 1413. Petraeus also published a text, detailing how to make the medicine called Phalaia (mentioned in *Von der grossen Heimligkeit*), which states it was written down in 1401 by one Johann von Ebertz after having received the process from Basil Valentine's "own hands in the cloister at Erfurt."[56]

More than one researcher went to the Benedictine abbey in Erfurt to search for more information. Valentin Andreas Möllenbrock (1623–1675), professor of medicine at Erfurt, received answer from the monastery's abbot that no Basil Valentine had ever lived there. But Georg Wolfgang Wedel tried again in 1704, at which time a different abbot acknowledged that Valentine had indeed been a monk there, and that the monastic library once held his original manuscripts. These manuscripts, the abbot said (according to Wedel), together with a box full of a golden-yellow powder, were found hidden in a wall under the refectory. Almost all of the original manuscripts, however, were looted by the Swedes during the Thirty Years War (1618–1648) and brought to Queen Christina. Only two manuscripts were spared; but one was sent, by command of the Elector of Mainz, to the Elector of Köln, and the other – on the quintessence – was lent to a prior of the Carthusians, and never returned.[57]

6 Anachronisms and Paracelsus as Plagiarist

Situating Basil Valentine in the early fifteenth century raised two problems. The first is that he mentions such things as America, the printing press, and tobacco. This problem was "solved" however in 1730, when Johann Conrad

55 Johann Moritz Guden, *Historia Erfurtensis* (Dudenstadt, 1675), 129.

56 *Basilius innovatus*, 3: 1097 and 1109–1111. The alternate suggestion that Basil Valentine lived instead at the cloister of Walkenried was made by Samuel Reyher in 1692, on the authority of one "Dr. Waitz." On this topic, see Gerhard Görmar, "Jakob Waitz und Basilius Valentinus im Kloster Walkenried: Legende und Wirklichkeit," *Mitteilungen der Gesellschaft deutscher Chemiker*, 23 (2013), 31–51.

57 Valentin Möllenbrock, *De varis seu arthritide* (Leipzig, 1672), 153 (thanks to Noah Millstone for this reference); *Basilius innovatus*, "Neue Vorrede," sig. f3v; Wedel, *Propemticon de Basilio*, 7.

Creiling reported that he saw a manuscript by one Johannes Macarius regarding the Stone of Fire – a transmutational agent described in the *Triumph-Wagen* – which contained a note that read: "This Macarius was the student of Brother Basil Valentine, from whom he learned this art. He, Basil, lived in 1386, cured about three thousand persons of great sicknesses, lived to 136 years of age, and even longer, and had a quiet old age."[58] Such longevity would mean that Basil survived into the sixteenth century, giving him time enough to witness the printing press, the discovery of America, and the importation of tobacco.

A second problem was potentially more problematic, but turned out to be even easier to solve. The Valentine texts are obviously Paracelsian, but their supposed date of composition is prior to Paracelsus (1493–1541). *Von dem grossen Stein* and almost all of Corpus A deploys the *tria prima* of mercury, sulphur, and salt as the principles of mixed bodies. *De microcosmo* refers to the coming of the "enlightened Elias." *Von der grossen Heimlichkeit* uses the term "spagyrical" and emphasizes the operation of chymical *Scheidung*. *Von den natürlichen unnd übernatürlichen Dingen* cites the *archeus* and defends the weapon salve. Corpus A in fact becomes increasingly Paracelsian with time – many pages of the *Triumph-Wagen* contain vitriolic denunciations of Galenic physicians, perfectly in the style of Paracelsus; such critiques do not occur in the earlier texts. One might readily conclude that these Paracelsian elements would immediately throw the pre-Paracelsian dating of the Valentine corpus into doubt, but very few contemporaneous authors drew that conclusion. The dedicatory poem for the 1602 edition of *Von dem grossen Stein* by Herman Kirckner, professor at Marburg, already confidently expresses what would become the standard reading. Noting that "the writings of Valentine were hidden for many ages," Kirckner praises him as "the most learned author of the Theophrastian teaching."[59] Thus, the apparent anachronisms are nothing of the kind; rather, they are simply earlier expressions of the correct natural philosophy, later picked up by Paracelsus.

Thoelde was himself avowedly Paracelsian. He praises Paracelsus by name as a discover of new medicines in the foreword to the 1602 *Von dem grossen*

58 Johann Conrad Creiling, *Die edelgeborne Jungfer Alchymia* (Tübingen, 1730), 297: "Dieser *Macarius* ist des *Fratris Basilii Val. Discipulus* gewesen, den er diese Kunst gelehret. Er der *Basilius* hat gelebet *An.* 1386 hat bey 3000 Menschen von grossen Kranckheiten *curi*ret, ist 136 Jahr alt gewesen, hat auch noch länger gelebet, und hat noch ein ruhiges Alter gehabt." Like Reyher (note 54), Creiling also claims that Valentine lived at the cloister of Walkenried in the Harz, rather than at the Peterkloster in Erfurt, ibid., 296.

59 *Von dem grossen Stein* (1602), fol. A2r: "Basilium Valentinum Theophrasticae disciplinae auctorem eruditissimum."

Stein.[60] He even added to the pseudo-Paracelsus corpus by publishing the *Kleine Handt-Bibel und Einführung oder Lehr zum ewigen Leben* in 1605.[61] If, as is virtually certain, he is the real author of at least some of Corpus A, one wonders why he inserted (what are to us) such obvious anachronisms. Perhaps he never considered the ramifications of implying that the manuscripts dated from before Paracelsus. Perhaps he was simply ahistorical. Or perhaps, as a Paracelsian himself, he considered the inclusion of such ideas as a means of making Paracelsian doctrines older and more venerable, perhaps even more religiously orthodox, if they really originated from a Benedictine monk.

What Thoelde probably did not foresee was how these inclusions could be read in negative way – not against the dating of the texts, but against Paracelsus himself. The first such negative reading that I have located comes from Joan Baptista Van Helmont (1579–1644). In the tract "Tria prima chymicorum principia" of his *Ortus medicinae*, Van Helmont writes that Paracelsus,

> taught by Basil Valentine that water, oil, and salt are separated from many bodies by distillation, began not only to call these three the universal first beginnings of corporeal entities, but also to introduce them into diseases and the means of curing them, so that he referred all diseases directly to one of these three principles […]. Theophrastus Paracelsus, who was more than a century Basil's junior, then by marvellous industry turned these things he took from Basil into the entire principles of bodies. Having suppressed the author's name, he stole Basil's teaching for his own […][62]

In this reading, Paracelsus is nothing other than a plagiarist and thief who could not even use correctly the ideas he stole from his Benedictine master.

Other early modern claims supported this observation. As noted above, the first part of *Letztes Testament* first appeared under the name of Nicolas Solea, and the second part is very likely his composition as well, but some

60 Ibid., fol. Br.

61 Karl Sudhoff, *Versuch einer Kritik der Echtheit der Paracelsischen Schriften*, vol. 1: *Bibliographia Paracelsica. Die unter Hohenheim's Namen erschienenen Druckschriften* (Berlin, 1894), 465 and 647–648.

62 Joan Baptista Van Helmont, "Tria prima chymicorum principia," in *Ortus medicinae* (Amsterdam, 1648), 399, #3 and 6–7: "Hic vir ergo à Basileo Valentino didicerat, per distillationem à plerisque corporibus separari aquam, oleum & salem, coepit haec tria vocare, nedum entium corporalium primia initia universalia: verùm etiam intra morbos & medendi necessitates, sic introduxit, ut omnes morbos immediatè in aliquod è tribus illis primis referret […] Quae sic à Basilio mutuata, deinceps in cuncta corporum principia, mirifica indagine, transtulit Theophrastus Paracelsus, uno plus seculo, Basilio junior. Cujus doctrinam, suppresso Auctoris nomine, in se rapuit […]."

contemporaneous authors read this parallel differently than we would today. Petraeus noted that Elias Montanus, the editor of Solea's work, claimed that the manuscript he worked from was in a handwriting strikingly similar to that of Paracelsus. Petraeus then observed that the references to Paracelsus in Solea's work, as published in 1600, were absent from the 1626 *Letztes Testament*. The disappearance of these references from the *Letztes Testament* seems reasonable enough to us because the text was being passed off as pre-Paracelsian. But Petraeus came to a different conclusion, namely, that Montanus' manuscript was *in fact* in Paracelsus' hand; Paracelsus had produced it by coopting Valentine's original text and then interpolating his own name and ideas into it. The 1626 *Letztes Testament*, therefore, represents an older more correct version of the text, free from later Paracelsian intrusions.[63]

Conclusion

The writings of Basil Valentine form a heterogenous mass agglomerated over a long period of time at the hands of multiple authors, editors, and publishers. In addition to original works, several already published texts were folded into the Valentine corpus, with varying degrees of retailoring. Yet neither the identification of these previously printed works, nor the anachronisms in the texts themselves, sufficed to call the authenticity of the corpus seriously into question during the early modern period. Instead, there was always another explanation that could be adduced, such as the plagiarism of Paracelsus, or the discovery of a new source corroborating Valentine's biography. The Benedictine's biography accreted information and details in step with the expansion of his corpus of writings. Due to the primary interest of this paper in organizing and presenting the complexities and connections within the corpora, I have here only gestured towards what is perhaps the most significant feature of the growth of a pseudonymous corpus – the incentives and motives of authors, editors, publishers, and printers, and how they benefitted from 'muddying the waters' of authorship. The difficult historical issues of intentionality and belief run in parallel with such questions. To what extent did the various builders of the Valentine corpus and legend believe what they were claiming? Was Cardilucius really given a better manuscript of the *Letztes Testament*, and did he really believe it was by Basil Valentine? Did Claromontanus know he was publishing Solea's

63 *Basilius Innovatus*, 1: f[vii]r–v.

text as Valentine's? Similar questions will arise during the analysis of any of the many other pseudonymous corpora within chymistry, as well as in other fields, and certainly in regard to the pseudo-Paracelsus materials. Despite the difficulty, or in some cases the impossibility, of resolving some of these questions, one thing is abundantly clear: *habent sua fata libelli.*

APPENDIX

A Catalogue raisonné of Pseudo-Paracelsian Writings: Texts Attributed to Paracelsus and Paracelsian Writings of Doubtful Authenticity

Julian Paulus

Contents

 | DOI:10.1163/9789004503380_010

Introduction

For centuries, the question of the authenticity of the Paracelsian writings was of little interest, except for occasional statements by various editors such as Adam von Bodenstein, Michael Toxites or Johannes Huser. In the 18th century, historiography was primarily concerned with establishing a reasonably reliable biography of Theophrastus von Hohenheim, commonly called Paracelsus, for which increasingly autobiographical statements and also newly discovered documents were used. Although efforts were made in the 18th century to systematically record the various printings of the Paracelsian works, Huser's edition (or the Latin Geneva edition of 1658) was essentially considered authoritative and the question of authenticity was not raised: the works published by Huser were seen to be both authentic and complete. Thus, on the one hand, the entire theological oeuvre that remained in manuscript was left out of consideration; on the other hand, a number of writings that are certainly considered spurious from today's perspective were considered part of the corpus of authentic Paracelsian writings.

The question of authenticity and inauthenticity was probably first raised in a systematic way by the medical historian Karl Friedrich Heinrich Marx (1796–1877),[1] who saw the multitude of writings under Paracelsus's name as suspicious: "The writings allegedly originating from him are so numerous that, in order to write them, a long life and an undisturbed activity and state of mind would have been necessary."[2]

On the problem of the distinction between authentic and spurious, Marx wrote:

> An assessment based on writing style possesses great difficulties, because the style changes astonishingly in the course of the years, and many a treatise written in the first youth hardly resembles the one written in later maturity. Only a certain archetype reveals itself, and not easily, without the biographical backstory providing some explanation, may a complete contradiction be found in the same.
>
> In this uncertainty in the discernment between the authentic and inauthentic writings, it is undoubtedly far more advisable to attribute too few than too many [texts] to the author. Some external criteria also come to our aid here, which, though not sufficient in themselves, nevertheless, because they usually coincide with the content, provide a starting point for the selection.[3]

1 Karl Friedrich Heinrich Marx, *Zur Würdigung des Theophrast von Hohenheim* (Göttingen, 1842), 18–24.

2 *Ibid.*, 19.

3 *Ibid.*, 20.

Marx formulated the following criteria: authentic works were those that (1) had a "dedication indicating the place and time where and when they were written down" and (2) also bore the "signature of his true name Theophrastus von Hohenheim".[4] These rigorous criteria, however, led to the fact that only a total of ten writings could be regarded as authentic, all of them with a medical, surgical or natural philosophical content, including *De natura rerum*.

The question of the authenticity was further investigated by Friedrich Mook (1844–1880), a young Egyptologist, physician and traveler, in his *Theophrastus Paracelsus: Eine kritische Studie* (1876) where he undertook two things: first, a bibliography of Paracelsus's printed works, based on autopsy, as well as a list of some manuscripts; second, the setting up of new criteria to prove authenticity.

Mook's approach was a new one: (1) All writings that were available in manuscripts written by Paracelsus himself were considered authentic. (2) All writings that appeared in print during Paracelsus's lifetime were considered authentic. (3) Based on that, one could decide, according to Mook, if Huser was a reliable editor who included all authentic writings that were established under criteria (1) and (2) and also described them as authentic. (4a) If Huser proved to be a reliable editor, then those writings of Hohenheim should also be considered authentic which Huser had edited from Paracelsus's autograph manuscripts according to Huser's own statements. (4b) Furthermore, all writings that were quoted in texts previously established as authentic (according to the stated criteria) should be considered authentic. (5) Finally, texts that corresponded in form and content to the previously examined writings should be considered authentic.[5]

Mook reached the following conclusions: (1) No original manuscripts were known. (2) During Paracelsus's lifetime several texts were published; Mook counted a total of eleven different ones, six prognostications and five medical writings. The criteria of authenticity established by Marx (dedication, naming) were not suitable according to Mook, since many of the writings published during Paracelsus's lifetime have no dedication. (3) According to Mook, Huser proved to be a conscientious editor who carefully reproduced the writings published before 1541 and also named his sources and informants for all other writings. Two writings printed before 1541 are not to be found in Huser's edition, which according to Mook could be explained by their extraordinary rarity. Furthermore, Mook (4) made no statements about the works cited in Paracelsus's texts, and (5) he preferred not to answer the last point, since it "gives the most room for arbitrariness."[6]

4 *Ibid.*, 20–21.

5 Friedrich Mook, Theophrastus Paracelsus. Eine kritische Studie (Würzburg, 1876), 21.

6 *Ibid.*, 133.

In a review published shortly after Mook's work appeared, John Ferguson (1838–1916) described Mook's criteria of authenticity as "doubtless sufficient".[7] The bibliography compiled by Mook, on the other hand, did not find favor with Ferguson, because in his eyes it was incomplete and inaccurate as well as distorted by numerous errors and inconsistencies.

In a series of altogether six publications, which were printed from 1877 to 1896 (however, only in a very limited edition of 50 to 100 copies) under the title *Bibliographia Paracelsica*, Ferguson endeavored to record the printed works of Hohenheim bibliographically as reliably as possible. On the other hand, Ferguson hardly commented on the question of the authenticity of individual writings.

Mook's criteria of authenticity were rejected in extraordinarily sharp words by the medical historian Heinrich Rohlfs (1827–1898).[8] According to Rohlfs, the authenticity of writings could be proven exclusively from their content, not by criteria such as those established by Mook. Rohlfs's aim was clearly to characterize Paracelsus as a solely scientific-medical writer; consequently all texts that contained speculative, prognostic, mantic, or theological thoughts proved themselves to be inauthentic. In another essay Rohlfs claimed to have discovered proof that Paracelsus "wrote only 6 books prior to the year 1536" and thus to have finally solved the "three hundred years old controversy concerning the number of genuine writings of Theophrastus von Hohenheim."[9]

In response to Rohlfs's sharp criticism of Mook, in 1887 the physicians Eduard Schubert (1822–1892) and Karl Sudhoff (1853–1938) published the first issue of their *Paracelsus-Forschungen* under the title *Inwiefern ist unser Wissen über Theophrastus von Hohenheim durch Friedrich Mook und seinen Kritiker Heinrich Rohlfs gefördert worden?* (The degree to which our knowledge of Theophrastus von Hohenheim has been furthered by Friedrich Mook and his critic Heinrich Rohlfs?).

Schubert and Sudhoff first emphasized – quite in agreement with Mook – that the most important task was to check the reliability of Huser's edition.[10] For this purpose, the knowledge and examination of all Paracelsus editions published before Huser was necessary.

However, Schubert and Sudhoff explained: "Mook has not, in any respect, examined in depth the bibliographical material that he has found; his material is still absolutely

7 John Ferguson, Bibliographia Paracelsica. An Examination of Dr. Friedrich Mook's "Theophrastus Paracelsus. Eine kritische Studie" (Glasgow, 1877), 388.

8 Heinrich Rohlfs, review of Ludwig Graf Ütterodt zu Scharffenberg, *Zur Geschichte der Heilkunde*, in *Schmidt's Jahrbücher der in- und ausländischen gesammten Medicin*, 176 (1877), 111–112, on 112; Heinrich Rohlfs, review of F. Mook, *Theophrastus Paracelsus*, in *Deutsches Archiv für Geschichte der Medicin*, 2 (1882), 213–242, on 228–239.

9 Rohlfs, 1882, 473–474. For this "discovery" see Schubert and Sudhoff, *Paracelsus-Forschungen*, 1, 50–56.

10 Eduard Schubert and Karl Sudhoff, *Paracelsus-Forschungen*, 1 (Frankfurt a.M., 1887), IV.

unripe to draw conclusions from. And so, in spite of the so much increased and established material [...] the question of the authenticity of Hohenheim's writings is only marginally advanced by Mook."[11] And further: "The so-called criteria of authenticity statued by Mook are however more than contestable."[12]

As far as the question of "Paracelsus's own manuscripts" is concerned, which are said to have been available to the various early editors "we are even more skeptical than Rohlfs, because these [editors], with the only exception of Huser, rarely or never, as we have checked ourselves, say anything sufficient about these manuscripts. From this, however, one should not draw the conclusion that in this early era after Paracelsus no original manuscripts existed and all these men were impostors or deceivers."[13] In summary, Schubert and Sudhoff state, "Our many years of acquaintance with Huser's Basel and Strasbourg editions have made us realize more and more clearly the manifold unreliability of Huser in detail. However, we have no reason to make a total *tabula rasa* with his edition."[14]

A few years later – Eduard Schubert had died in the meantime, and Schubert's extensive Paracelsus collection including various manuscripts and the original manuscripts of Friedrich Mook had been bought *in toto* by John Ferguson – Karl Sudhoff's *Bibliographia Paracelsica* was published. The medical historian Julius Pagel (1851–1912) judged in a review in 1895:

> After the author had introduced us to his intellectual workplace in the sensational two issues of *Paracelsus-Forschungen* [...] and offered a foretaste of the fruits that we owe to his quiet and diligent, decades-long sowing and collecting, he has now completed the basement of a three-story monumental building in the present work, the contemplation of which already today gives reason to hope that this work, when completed, will not only tower over everything that has been achieved by others in the field of Paracelsus research, but will also be one of the best works that have been produced on the terrain of medical literary history in the recent past.[15]

Highly impressed by Sudhoff's achievement, Pagel writes: "It is precisely this bibliographical patterning, reaching the summit of completeness, as it has been undertaken here [...] by Sudhoff, that already provides the uninitiated with proof of the

11 *Ibid.*, 36.
12 *Ibid.*, 38.
13 *Ibid.*, 73.
14 *Ibid.*, 79.
15 Julius Pagel, review of Karl Sudhoff, *Bibliographia Paracelsica*, in *Göttingische gelehrte Anzeigen*, 1895, 8–11, on 9–10.

possibility of arriving at the most definitive and certain solution possible to the question of authenticity along the path taken by the author."[16]

Sudhoff's *Bibliographia Paracelsica* was the first part of a three-part *Versuch einer Kritik der Echtheit der Paracelsischen Schriften* (*Attempt of a Critique of the Authenticity of the Paracelsian Writings*). The second part was published in 1898/99 under the title *Paracelsus-Handschriften gesammelt und besprochen*. The third part "will then discuss in a coherent presentation the authenticity of the individual writings of Hohenheim on the basis of the offered material, whereby the course of Hohenheim's thinking and work in the various periods of his life will emerge by itself in broad outlines."[17] Unfortunately, the third part never appeared and was probably never written. Neither the preface nor the introduction to the *Bibliographia Paracelsica* goes into more detail about criteria of authenticity, even though both the *Bibliographia Paracelsica* and the *Paracelsus-Handschriften* frequently make judgments about authenticity and inauthenticity.

When in 1919 Karl Sudhoff began to prepare his large edition of the medical, scientific and philosophical works of Paracelsus for publication – the first volume (vol. VI) was not published until 1922 as a result of the First World War – he had largely ceased to raise questions of authenticity. The *Bibliographia Paracelsica* and the *Paracelsus-Handschriften* had previously used Huser's edition as a reference point, and thus the *Sämtliche Werke* might be seen as a re-ordered and linguistically unified reprint of Huser with a few additions. For Sudhoff, the question of authenticity seems to have been solved, even if he never actually asked. Through decades of study of the Paracelsian writings, Sudhoff had become so immersed in the world of Paracelsus that he believed he could distinguish authentic from spurious with great certainty – and he was probably right.

Alternatively, for Will-Erich Peuckert (1895–1969), the most interesting part of the Paracelsian writings was always that of the so-called "spurious" works. Already in the *Pansophie* (1929) Peuckert had thoroughly utilized the countless excerpts Sudhoff had made in the *Bibliographia Paracelsica* and in the *Paracelsus-Handschriften*. Peuckert's five-volume edition of Paracelsus (1967–1968) devoted an entire volume (V) to the "pansophic, magical, and gabalic writings." In the preface to this volume, Peuckert stated: "This fifth volume brings a body of texts long attributed to Hohenheim, but today more often rejected. These texts, long attributed to him, have at the same time had a stronger impact than his own assured statements."[18] From the language of dip-

16 *Ibid.*, 10.

17 Sudhoff, *Bibliographia Paracelsica* (1894), V–VI.

18 Will-Erich Peuckert, "Vorwort", in Theophrastus Paracelsus, *Werke*, ed. Peuckert, vol. 5 (Darmstadt, 1968), vii. The term "verunechtet" is also frequently used by Peuckert in his *Gabalia* (1967); see Peuckert, *Gabalia*, 161, 162, 177, 193, 203, 470.

lomatics, Peuckert adopted the term "Verunechtung," which in historiography refers to the subsequent alteration of a genuine document, as opposed to a forgery, which is inauthentic from the outset.

In contrast, however, for Peuckert the expression "verunechet" did not necessarily mean that the respective text was based on a genuine writing of Paracelsus, but he was rather concerned with a "Verunechtung" as a continuation of genuine thoughts – in his view, the so-called spurious writings conveyed the genuine thoughts of Paracelsus perhaps better than the so-called authentic ones.[19]

Walter Pagel (1898–1983) took a similar stance, but without explicitly acknowledging it. His much acclaimed essays on Paracelsus were based to a considerable extent on spurious Paracelsus works such as *Liber Azoth*, *De natura rerum*, *Philosophia ad Athenienses*, *De secretis creationis*, *De tinctura physicorum*, which implied no contradiction for Pagel, since he obviously but tacidly considered the spurious writings to be expressions of genuine Paracelsian thoughts.

Actual examples of a "Verunechtung" in the historiographical sense came to light in 1973 with the publication of the so-called "Kurzfassungen" (abridged versions) of various theological works of Paracelsus by Kurt Goldammer (1916–1997). In detailed investigations Goldammer and his co-workers proved what Karl Sudhoff had already noted in his *Paracelsus-Handschriften* – that the "abridged versions" were later adaptations, which were based on genuine works and actually conveyed the concepts of the genuine writings largely unchanged.

•••

The present catalogue is an attempt to give an overview of the totality of the non-authentic writings under the name of Paracelsus.

Listed are in the first place all the pseudo-Paracelsian texts that Sudhoff describes in the *Bibliographia Paracelsica* (printed books) and in the *Paracelsus-Handschriften* (manuscripts). This basic inventory was supplemented by entries in the catalogue of Kassel manuscripts by Hartmut Broszinski as well as in many other manuscript catalogs. In addition, I was able to draw on my own research in the late 1980s and early 1990s, when I systematically reviewed and excerpted numerous manuscripts in Hamburg, Vienna, Wolfenbüttel, Dresden, Lindau, Kassel, Bamberg, Heidelberg, Erlangen, Göttingen, London, Überlingen, and Hannover. I have now been able to access many of these and other manuscripts again through scans provided by the libraries. I have created working editions for most of the texts which for the first time

19 The process of "Verunechtung" began for Peuckert already with Huser, who, however, already recognized that they were "Theophrasti Sachen nicht vngemeß" (Huser, 10, appendix, 66; quoted by Peuckert, *Gabalia*, 203).

have made it possible to compare different versions as well as gather information about their size and contents.

For easier orientation, the various texts have been divided into groups according to their main content: theology, natural philosophy, medicine and surgery, alchemy, magic, prophecies and treasures, alchemical processes, collections of recipes, processes and experimenta, single medical recipes and experimenta. In addition, there is a group of miscellaneous texts, which is further subdivided according to formal criteria (poems, dialogues, prayers, others) and the "Kurzfassungen" of theological texts are included in an appendix. It goes without saying that a clear assignment was not always possible. However, the respective texts can be easily found through the various indexes.

About the individual entries.

The heading gives a normalized title of the work in question. Other titles, as found in the various prints and manuscripts, are listed below. All ascertainable information on the texts has been compiled in seven sections.

The first section (*Printing History, Manuscripts*) provides concise information about the first printed editions and their editors and gives an overview of the manuscript tradition. Reprints, translations, and modern editions have generally not been listed, because otherwise the size of the present catalogue would have grown immeasurably.

The second section (*Editions*) lists only the complete editions by Huser, Sudhoff, Goldammer, and Gantenbein.

The third and fourth sections (*Relationship between different versions*; *Structure, genre/form, perspective, style*) essentially record new findings that emerged from the preparation of working editions during the compilation of the catalogue.

The fifth section (*Relationship to other texts*) deals primarily with source citations and quotations in the cataloged texts.

The sixth section (*Authenticity, authorship*) brings together the judgments of the scholarly literature on authenticity and inauthenticity and attribution to specific authors.

The seventh section (*Time of writing*) attempts, as far as possible, to place the texts approximately in time on the basis of the printed and manuscript tradition as well as the scholarly literature. In most cases this classification is *highly* speculative, but nevertheless perhaps helpful as a starting point for ongoing research.

On supplementary bibliographical references.

Word Count: This information allows a comparative impression of the size of the texts. If the text was printed by Huser, the information is usually based on this edition. In very few cases, no or no complete working edition could be prepared. In these cases, the information is missing or an estimate was made based on a partial working edition.

Incipit: The incipit allows the identification of texts without a title or with a different title.

Manuscripts: As far as possible, all German-language manuscripts are listed. Manuscript copies from prints are only partially recorded, translations into other languages not at all (with a few exceptions).

First Printed: Only the first printing is recorded, not later editions or print runs. If different versions or versions from different manuscripts were published, these are recorded. Translations are not usually recorded unless they preceded the first German edition.

Historical Manuscript Catalogues: This is a list of (1) references in what Kühlmann and Telle called "Paracelsicaverzeichnisse"[20] or "Manuscripta-Paracelsica-Verzeichnisse":[21] for the most part 16th century lists of Paracelsian texts, some of them library catalogues or lending lists, some of them search lists or offerings of manuscripts to a prince; (2) references to Paracelsian manuscripts in 18th century sales catalogues.

Essential Bibliography: Here references to the following publications have been listed: Sudhoff, *Bibliographia Paracelsica*; Sudhoff, *Paracelsus-Handschriften*; Sudhoff, prefaces to Paracelsus, *Sämtliche Werke*; Kühlmann-Telle, *Corpus Paracelsisticum.*

•••

The idea for this catalogue was born at a meeting of the Fondation des Treilles in 2019, where Didier Kahn presented a preliminary catalogue of printed pseudo-Paracelsica that he had prepared. It is my pleasant duty to thank various people. First and foremost, my thanks go to Didier Kahn, without whom this catalogue would not have been started, written, or ever finished. Carlos Gilly has always generously gifted me with his books over the years; this catalogue owes a great deal to his publications. Kathrin Pfister provided materials for the "Lion from Midnight." To Stefan Guzy I owe valuable materials on Karl Widemann. Urs Leo Gantenbein contributed "a few remarks". Chad Gunnoe and Kathrin Pfister were kind enough to improve the English of my introduction. Sabine Wenzel of the Haldensleben Library uncomplicatedly prepared photographs for me from the manuscripts there, as did Katja Lorenz of the Klassik Stiftung Weimar. Various libraries and archives (Hamburg, Greifswald, Bamberg, Göttingen, Basel, Hannover, Nürnberg, Stuttgart, Dresden) provided me with scans from their manuscripts.

20 CP 1 (2001): 451, 593, 596, 664; CP 2 (2004): 546.

21 CP 2 (2004): 242–243, 244.

Abbreviations

Abbreviations used in "Historical Manuscript Catalogues" (in chronological order)

Toxites, *Zugesagte Bücher* (Dresden) Michael Toxites, *Mir seindt zugesagt von namhafftigen leutten, vil buecher* ["Notable people promised me the following books"], letter to Kurfürst August von Sachsen (Strasburg, 14 July 1571); manuscript in Dresden (Hauptstaatsarchiv, 10024 Geheimer Rat, Loc. 04417/16); ed. in CP 2: 232–249, n° 49, on 234–235 (commentary on 242–247).

Catalogus Osek (Prague) 1: *Catalogus librorum Theophrastj*; 2: *Catalogus oder kurtze Verzaichnuss aller Theophrastischen Paracelsischen Büecher, die Ich bey meinen Hand, vnd noch nitt Imm truckh khommen, sonder noch Inn gehaim vnd verborgen sein*; 3: *Verzaichnuss ettlicher Theophrastischer büecher, so bey mir Just vnd klar gefunden werden*; manuscript in Prague (Národní knihovna, Ms. Osek 43), ca. 1571(?); ed. (with omissions) by Sudhoff, *Paracelsus-Handschriften*, 29–31, 31–37, 37–39. Unnumbered; quoted after Sudhoff with our own numbering.

Libri Theophrasti (Dresden) *Libri Theophrasti*; manuscript in Dresden (Sächsische Landes- und Universitätsbibliothek, Mscr.Dresd.C.453, fol. 1r–7r); probably written in the 1590s. Unpublished.

Catalogus (Stuttgart) 1: *Catalogus Continens libros chÿmicos, Magicos, Medicos, Astrologicos et Cabalisticos, cum uarijs alijs materijs et experimentis, insertis*; 2: *Alius quidam Catalogus rerum chÿmicarum et Medicinalium*; manuscript in Stuttgart (Landesarchiv Baden-Württemberg, A47 Bü 3/5, Nr. 4 & 5); written by Johannes Andreae according to Carlos Gilly, but more probably written by Johann Marcell Hess, as shown by comparison with an autograph letter of Hess (Königsbronn, 12 October 1598) accompanying the catalogue. Unpublished; unfoliated and unnumbered.

Catalogus Adelwert (Leiden) *Catalogus Librorum et variarum rerum et artificiorum, quae omnia habentur apud doctorem Jonam Atelwertinum Canonicum nouj Monasterij herbipolensis*; manuscript in Leiden (Cod. Voss. Chym. Q.56, 42–56), ca. 1600; quoted after an unpublished edition prepared by Carlos Gilly in 1998 with Gilly's numbering; extracts in Sudhoff, *Paracelsus-Handschriften*, 28–29.

Scripta Theophrasti (Wien) *Scripta Theophrasti*, manuscript in Vienna (Österreichische Nationalbibliothek, Cod. 11133), dated 30 July 1604; ed. by Sudhoff, *Paracelsus-Handschriften*, 187–188.

Widemann, *Unausgangene Bücher* (Hannover) *Vnausgangene Büecher Thëophrastj & aliorum denen nachzuefragen, die Ich zum thail hab*; manuscript in Hannover (Gottfried Wilhelm Leibniz Bibliothek/Niedersächsische Landesbibibliothek, Ms. IV,341, 538e–538h); written by Karl Widemann probably in the 1610s; unpublished.

Widemann, *Verzeichnisse* (Kassel) Ten lists of manuscripts in the library of Karl Widemann; manuscript in Kassel (Landesbibliothek, 2° Ms. chem. 7), mostly written in the early 1620s; extracts ed. by Sudhoff, *Paracelsus-Handschriften*, 747–764; quoted after our own unpublished edition using the numbering introduced by Carlos Gilly.

Herzog Johann Friedrich von Sachsen-Weimar (1627) *Vorzeichnuß Was von Magischen v[nd] andern verdechtigen sachen in der custodirten F[ürsten] person gemach zu Ichtershausen befunden worden*, manuscript (location unknown), prepared after the duke's arrest on 1 June 1627; ed. in Bernhard Röse, *Johann Friedrich der Sechste, Herzog zu Sachsen, Ernestinischer Linie. Ein biographischer Versuch* (Neustadt a.d. Orla: Wagner, 1827), 216–218.

Catalogus Pfeffer (1680) "Catalogus Der Bücher des Philosophi E: P: J: H:", in Franciscus Rottman, *Treuhertzige Vermahnung In diesen itzigen gefährlichen/ doch wunderbahrlichen Zeiten/ an alle und jede Sucher des sonst heiligen und gerechten Arcani Arcanorum* (Hamburg: Georg Wolff, 1680), sig. c4r–c6v.

Catalogus Pfeffer (1683) "Catalogus der Schrifften des erleuchteten Magi, Cabalisten und Philosophi E.P.J.H. welche alle ich nach dem Todt seines neuen in Gott ruhenden Discipels erlanget", in Georg Ernst Aurelius Reger, *Gründlicher Bericht Auff einige Fragen* [...]. Worbey angefüget der Catalogus vieler raren und sonderlichen Manuscripten des neulichsten Philosophi E.P.I.H. &c. (Hamburg: Georg Wolff, 1683), 121–137.

Thomasius (1699) "Habes heic manuscriptorum codicum qui a me asservantur brevem atque succinctum indicem," in Rudolf Martin Meelführer, *Accessiones ad celeber. viri, Theodori Jansonii ab Almeloveen* [...] *Bibliothecam promissam et latentem* (Nuremberg-Leipzig: Andreas Otto, 1699), 131–144. – A list of manuscripts in the library of Gottfried Thomasius (1660–1746).

Schröer (1710) *Catalogus rariorum manuscriptorum*, 4 printed pages, no place or date; facsimile in Daniel Bellingradt and Bernd-Christian Otto, *Magical Manuscripts in Early Modern Europe. The Clandestine Trade In Illegal Book Collection. Including a critical edition of a 'catalogus rariorum manuscriptorum' from 1710* (Cham, 2017), 153–156. – Sale catalogue of part of the library of Samuel Schröer (1669–1716).

Uffenbach (1720) *Bibliotheca Uffenbachiana MSSta seu catalogus et recensio MSStorum codicum qui in bibliotheca Zachariae Conradi ab Uffenbach* [...] *adservantur* (Halle, 1720).

Arpe (1726) "De Theophrasto Paracelso," in Peter Friedrich Arpe, *Feriae aestivales. Sive, scriptorum suorum historia* (Hamburg: Johann Christoph Kisner, 1726), 120–123.

Schröer-Roth-Scholtz (1732) *Catalogus rariorum librorum et manuscriptorum magico-cabbalistico-chymicorum*, in Friedrich Roth-Scholtz, *Veterum sophorum sigilla et imagines magicae* (Herrenstadt: Samuel Roth-Scholtz, 1732), Appendix, 16 pages. – A reprint of the 1710 *Catalogus* with a few additions.

Crusius (1735) "Catalogus. Derjenigen Manuscriptorum, so Augustinus Crusius, Buchhändler in Erffurth drucken zu lassen gesonnen ist," in *R. Abrahami Eleazaris*

Uraltes Chymisches Werck, ed. Julius Gervasius (Erfurt: Augustinus Crusius, 1735), two unnumbered pages, (sig. 6 r–v).

Jahn (1754–1758) *Verzeichnis der Bücher so gesamlet Johan Christian Gottfried Jahn* (Frankfurt-Leipzig: Johann Samuel Heinsius Erben, 1754–1758), 3 vols. in 6 parts.

von Linden (1786) *Catalogus Manuscriptorum chemico-alchemico-magico-cabalistico-medico-physico-curiosorum*, anonymously ed. by Max Joseph Freiherr von Linden (no place [Wien: Graeffer], 1786). – According to a review in *Allgemeine deutsche Bibliothek*, 85/1 (1789), 474, this catalogue was not published by the well-known Vienna bookseller Rudolph Graeffer, but by his brother August.

Gräffer (1786) *Anhang von unterschiedlichen chymischen, alchymischen, und andern Handschriften, welche bey einem vornehmen Kunstverständigen in Wien, aufbehalten waren* (no place, 1786). – Printed as an appendix to von Linden (1786).

Linck (1787) "Manuscripta magica et alia," in *Index Musaei Linckiani, oder kurzes systematisches Verzeichniß der vornehmsten Stücke der Linckischen Naturaliensammlung zu Leipzig* (Leipzig: Beygang, 1787), 3 vols., here 3: 242–244.

von Linden (1788) *Catalogus Manuscriptorum chemico-alchemico-magico-cabalistico-medico-physico-curiosorum* (no place, 1788). – A revised and considerably expanded version of the 1786 catalogue. According to *Reise des Herrn von Bretschneider nach London und Paris*, ed. L.F.J. von Göckingk (Berlin-Stettin: Nicolai, 1817), 310, it was edited by von Linden ("den ich gut kenne") and published by the Vienna bookseller Johann Georg Binz; a copy of this extremely rare catalogue is in London, Wellcome Library.

Binz (1791) *Verzeichniß von chymischen, alchymischen, physikalischen, theosophischen, kabalistischen und Kunstbüchern, welche bey Johann Georg Binz, Buchhändler in Wien, im Zwettelhof, verkauft werden* (Vienna, 1791). – A sales catalogue of ca. 2000 printed works and 60 manuscripts.

Ponickau (1791) "Bücher so zu verkaufen," in *Intelligenzblatt der Allgem*[*einen*] *Literatur-Zeitung*, n° 149 (21 December 1791), col. 1227–1233. – A sales list of some magical and alchemical printed works and manuscripts sold by H.F.W. von Ponickau in Zeitz.

Dost (1791–1798) (1) "Manuscripte, so zu verkaufen," in *Intelligenzblatt der Allgem* [*einen*] *Literatur-Zeitung*, n° 30 (5 March 1791), col. 240. – (2) "Manuscripte so zu verkaufen," in *Intelligenzblatt der Allgem*[*einen*] *Literatur-Zeitung*, n° 74 (20 June 1792), col. 595–597. – (3) "Folgende alchymistische und magische Manuskripte sollen aus dem Nachlaß eines bekannten Adepten um beigesetzte Preise in Sächs. Münzsorten verkauft werden," in *Allgemeiner Litterarischer Anzeiger*, Beilage zu n° 37 (28 March 1797), col. 394–397. – (4) "Verkauf einer kostbaren und seltenen Sammlung alchymistischer und magischer Manuskripte," in *Allgemeiner Litterarischer Anzeiger*, Beilage zu n° 30 (20 February 1798), col. 305–310. – The manuscripts were first offered by Johann Friedrich Dost in Halle; after his death (probably 1795 or 1796), they were offered anonymously, probably by his heirs. The number of manuscripts varies in the different offers. All these publications also appeared in other journals and papers like *Der Anzeiger*, *Kaiserlich privilegirter Reichs-Anzeiger* or *Der Verkündiger*.

Abbreviations Used in the Whole Catalogue

Bellingradt and Otto, *Magical Manuscripts*
Daniel Bellingradt and Bernd-Christian Otto, Magical Manuscripts in Early Modern Europe. The Clandestine Trade In Illegal Book Collection. Including a critical edition of a 'catalogus rariorum manuscriptorum' from 1710 (Cham, 2017).

Boeren, *Codices Vossiani Chymici*
Petrus Cornelis Boeren, *Codices Vossiani Chymici* (Leiden, 1975).

Broszinski, *Manuscripta chemica in quarto*
Hartmut Broszinski, *Manuscripta chemica in quarto* (*Die Handschriften der Universitätsbibliothek Kassel, Landesbibliothek und Murhardsche Bibliothek der Stadt Kassel*, vol. 3,2,2) (Wiesbaden, 2011).

Büttner, *Die pseudoparacelsischen "Secreta Secretorum"*
Stephanie Büttner, *Die pseudoparacelsischen "Secreta Secretorum". Untersuchungen und Texte zur frühneuzeitlichen Chemiatrie* (Heidelberg, 2001).

CP 1, CP 2, CP 3
Wilhelm Kühlmann and Joachim Telle, *Corpus Paracelsisticum. Dokumente frühneuzeitlichern Naturphilosophie in Deutschland*, herausgegeben und erläutert von Wilhelm Kühlmann und Joachim Telle, 3 vols.
vol. 1: *Der Frühparacelsismus. Erster Teil* (Tübingen, 2001)
vol. 2: *Der Frühparacelsismus. Zweiter Teil* (Tübingen, 2004)
vol. 3: *Der Frühparacelsismus. Dritter Teil* (Berlin-Boston, 2013)

Czifra, *Paracelsus' Abendmahlschriften*
Nikolaus Czifra, *Paracelsus' Abendmahlschriften. Überlieferung und Kontext*, Diss. phil. (Salzburg, 2014).

Gantenbein, *Neue Paracelsus-Edition*
Neue Paracelsus-Edition, vol. 1: *Theologische Werke 1: Vita Beata – Vom seligen Leben*, ed. Urs Leo Gantenbein (Berlin, 2008).

Gilly, "Zwischen Erfahrung und Spekulation" (1977)
Carlos Gilly, "Zwischen Erfahrung und Spekulation. Theodor Zwinger und die religiöse und kulturelle Krise seiner Zeit," *Basler Zeitschrift für Geschichte und Altertumskunde*, 77 (1977), 57–137.

Gilly, "Zwischen Erfahrung und Spekulation" (1979)

Carlos Gilly, "Zwischen Erfahrung und Spekulation. Theodor Zwinger und die religiöse und kulturelle Krise seiner Zeit, 2. Teil: Zwingers Philosophie," *Basler Zeitschrift für Geschichte und Altertumskunde*, 79 (1979), 125–223.

Gilly, *Paracelsus in der BPH*

Carlos Gilly, *Paracelsus in der Bibliotheca Philosophica Hermetica* (Amsterdam, 1993).

Gilly, *Adam Haslmayr*

Carlos Gilly, *Adam Haslmayr: Der erste Verkünder der Manifeste der Rosenkreuzer* (Amsterdam, 1994).

Gilly, "Theophrastia Sancta"

Carlos Gilly, "*Theophrastia Sancta.* Der Paracelsismus als Religion im Streit mit den offiziellen Kirchen," in *Analecta Paracelsica*, ed. Joachim Telle (Stuttgart, 1994), 425–488.

Carlos Gilly, "*Theophrastia Sancta* – Paracelsianism as a Religion, in Conflict with the Established Churches," in Ole Peter Grell, ed., *Paracelsus. The Man and His Reputation. His Ideas and Their Transformation* (Leiden, 1998), 151–185.

Gilly, "Das Bekenntnis zur Gnosis"

Carlos Gilly, "Das Bekenntnis zur Gnosis von Paracelsus bis auf die Schüler Jacob Böhmes", in Roelof van den Broek and Cis van Heertum, eds., From Poimandres to Jacob Böhme: Gnosis, Hermetism and the Christian Tradition (Amsterdam, 2000), 385–425.

Gilly, "Un bel trattato ermetico del paracelsismo"

Carlos Gilly, "Un bel trattato ermetico del paracelsismo: Il 'De tribus facultatibus' di Alexander von Suchten" / "Paracelsism brings forth a fine Hermetical treatise: Suchten's 'De tribus facultatibus'", in Carlos Gilly and Cis van Heertum, eds., *Magia, alchimia, scienza dal '400 al '700. L'influsso di Ermete Trismegisto* (Venice, 2002), 185–198.

Gilly, "Il primo prontuario di magia bianca"

Carlos Gilly, "Il primo prontuario di magia bianca in Germania" / "The first book of white magic in Germany," in Carlos Gilly and Cis van Heertum, eds., *Magia, alchimia, scienza dal '400 al '700. L'influsso di Ermete Trismegisto* (Venice, 2002), 199–217 and 179–180 (color plates).

Gilly, "Vom ägyptischen Hermes zum Trismegistus Germanus"

Carlos Gilly, "Vom ägyptischen Hermes zum Trismegistus Germanus. Wandlungen des Hermetismus in der paracelsischen und rosenkreuzerischen Literatur," in Peter-André Alt and Volkhard Wels, eds., *Konzepte des Hermetismus in der Literatur der Frühen Neuzeit* (Göttingen, 2010), 71–131.

Goldammer, *Der göttliche Magier*

Kurt Goldammer, *Der göttliche Magier und die Magierin Natur. Religion, Naturmagie und die Anfänge der Naturwissenschaft vom Spätmittelalter bis zur Renaissance. Mit Beiträgen zum Magie-Verständnis des Paracelsus* (Stuttgart, 1991).

Huser
Paracelsus, *Bücher und Schrifften*, ed. Johann Huser (Basel: Konrad Waldkirch, 1589–1591), 10 vols.

Huser, *Chir.* 1605 (or 1618)
Paracelsus, *Chirurgische Bücher vnd Schriften*, ed. Johann Huser (Strasburg: Zetzner, 1605; re-ed. 1618 with a different page-setting).

Ps.-Huser, *Opera Bücher vnd Schrifften*
Paracelsus, *Opera Bücher vnd Schrifften* (Strasburg: Zetzner, 1603), 2 vols., 2: 668–691 (spurious treatises not edited by Huser himself, but added by the editors to this careless re-edition of Huser).

Kahn, "Le retour de Gérard Dorn"
Didier Kahn, "Le retour de Gérard Dorn sur la scène éditoriale de Bâle (1577–1578)," *Nova Acta Paracelsica*, N.F. 24/25 (2010–2011), 125–156.

Pagel, "The Paracelsian Elias Artista"
Walter Pagel, "The Paracelsian Elias Artista and the Alchemical Tradition," *Medizinhistorisches Journal*, 16 (1981), 6–19; also published in Rosemarie Dilg-Frank, ed., *Kreatur und Kosmos. Internationale Beiträge zur Paracelsusforschung* (Stuttgart, 1981), 6–19.

Paracelsus, *Sämtliche Werke*
Theophrast von Hohenheim gen. Paracelsus, *Sämtliche Werke*
1. Abteilung: *Medizinische, naturwissenschaftliche und philosophische Schriften*, ed. Karl Sudhoff, 14 vols. (Munich, 1922–1925; Munich-Berlin, 1928–1933)
2. Abteilung: *Theologische und religionsphilosophische Schriften*, ed. Kurt Goldammer, vols. 2–7 and Supplement (Wiesbaden, 1955–1986)

Peuckert, *Pansophie*
Will-Erich Peuckert, *Pansophie* (Stuttgart, 1936).
Will-Erich Peuckert, *Pansophie*, vol. 1 (Berlin, 1956) [revised 2nd ed.]

Peuckert, *Gabalia*
Will-Erich Peuckert, *Pansophie*, vol. 2: *Gabalia* (Berlin, 1967).

Peuckert, *Rosenkreutz*
Will-Erich Peuckert, *Pansophie*, vol. 3: *Das Rosenkreutz* (Berlin, 1973).

Sudhoff, *Bibliographia Paracelsica*
Karl Sudhoff, *Bibliographia Paracelsica. Besprechung der unter Theophrast von Hohenheim's Namen 1527–1893 erschienenen Druckschriften* (*Versuch einer Kritik der Echtheit der Paracelsischen Schriften*, 1) (Berlin, 1894).

Sudhoff, *Paracelsus-Handschriften*
Karl Sudhoff, *Paracelsus-Handschriften gesammelt und besprochen* (*Versuch einer Kritik der Echtheit der Paracelsischen Schriften*, 2) (Berlin, 1899).

Telle, "Paracelsus in pseudoparacelsischen Briefen"
Joachim Telle, "Paracelsus in pseudoparacelsischen Briefen," *Berichte zur Wissenschaftsgeschichte*, 27 (2004), 149–160; reprinted in *Nova Acta Paracelsica*, N.F. 20 (2006–2007), 147–164.

Telle, "'Wo sind meine Vers'?'"

Joachim Telle, "'Wo sind meine Vers'?' – Paracelsus als Poet," in Joachim Telle, *Paracelsus im Gedicht. Theophrastus von Hohenheim in der Poesie des 16. bis 21. Jahrhunderts* (Hürtgenwald, 2008), 333–336.

Reprinted in Telle, *Alchemie und Poesie* (2013), 889–895.

Telle, *Alchemie und Poesie*

Joachim Telle, *Alchemie und Poesie. Deutsche Alchemikerdichtungen des 15. bis 17. Jahrhunderts* (Berlin-Boston, 2013).

VD16

Verzeichnis der im deutschen Sprachbereich erschienenen Drucke des 16. Jahrhunderts (*VD 16*), http://www.vd16.de

VD17

Verzeichnis der im deutschen Sprachraum erschienenen Drucke des 17. Jahrhunderts (*VD 17*), http://www.vd17.de

VD18

Das Verzeichnis Deutscher Drucke des 18. Jahrhunderts (*VD 18*), http://www.vd18.de

1 Theology

1.1 *Alia explicatio uberior in Magnificat Eremitae*

1. *Printing History, Manuscripts.* Not printed. Two manuscripts, one Latin translation.
2. *Editions.* Not edited by Huser or Sudhoff.
3. *Relationship between different versions.* "This fragment, found in one version only among excerpts, [other] fragments and texts not found anywhere else is obviously part of a larger text since it ends abruptly in the middle of a sentence."[22] There exists a Latin translation which shows "that in earlier times the fragment was considered genuine and worthy of translation along with other important theological treatises written by Paracelsus."[23]
4. *Structure, genre/form, perspective, style.*
5. *Relationship to other texts.* According to Biegger there were four different explications of the "Magnificat" written by Paracelsus, all of them written as part of larger texts.[24] The Latin text of the London manuscript represents, according to

22 Goldammer, "Einleitung," in Paracelsus, *Sämtliche Werke*, II/7: XXIX.

23 Ibid.

24 Biegger, "De invocatione Beatae Mariae Virginis" (1990), 212–214.

Biegger, the same text as the German explication found in the Leiden and Fulda manuscripts.[25] Further study required.

6. *Authenticity, authorship.* The authenticity of the *Explicatio* was not questioned by Sudhoff. According to Goldammer, "there might be some concerns in regard to the authorship of Paracelsus although the vocabulary is essentially the common Paracelsian, and also the imagery and the forms of thought correspond to those of Paracelsus."[26] The text is found "amidst texts of doubtless authenticity" while, on the other hand, the text is not directly attributed to Paracelsus.[27]
7. *Time of writing.*

Word count:

Incipit: "Die seel Maria, das ist die tugent Maria macht groß Gott den Vatter …" [Leiden manuscript]; "Hoc canticum composuit Maria, in quo comprehendat omnem doctrinam suam …" [London manuscript]

Manuscripts:

German

- Fulda, *Hochschul- und Landesbibliothek*: Cod. 22a (1), 2°, f. 158r–162v
- Leiden, *Universiteitsbibliotheek*: Voss. Chym. F.24, f. 373r–376r

Latin

- London, *British Library*: Harley ms 516, f. 89v–93v

First printed: not printed

Essential bibliography: Sudhoff, *Paracelsus-Handschriften*, 369, 644; Goldammer, "Einleitung," in Paracelsus, *Sämtliche Werke*, ii/7: xxix–xxx.

Further bibliographical references:

Katharina Biegger, "Wie gelangten theologische Paracelsusschriften nach London?", *Nova Acta Paracelsica*, n.f. 4 (1989), 24–37.

Katharina Biegger, *"De invocatione Beatae Mariae Virginis". Paracelsus und die Marienverehrung* (Stuttgart, 1990), 134, 212.

25 *Ibid.*, 134 ("third explication"), 212 ("first explication").

26 Goldammer, "Einleitung," in Paracelsus, *Sämtliche Werke*, ii/7: xxx.

27 Ibid.

1.2 *Astronomia Olympi Novi* *also: Gestirnkunst des neuen Himmels*

1. *Printing History, Manuscripts.* First printed in 1618 along with *Theologica Cabalistica* (§ 1.23) in a collection of Paracelsian, Pseudo-Paracelsian and Weigelian texts called *Philosophia Mystica*. No manuscripts known.
2. *Editions.* Not edited by Huser or Sudhoff.
3. *Relationship between different versions.* Only one known version.
4. *Structure, genre/form, perspective, style.*
5. *Relationship to other texts.* The title *Astronomia Olympi Novi* is inspired by the division of astronomy in Paracelsus's *Astronomia magna*.[28] Like the *Arbatel* (§ 5.6), the *Astronomia Olympi Novi* may have been a kind of replacement for the missing third book of the *Astronomia magna*.[29]
6. *Authenticity, authorship.* In a "Praemonitio" printed after both *Astronomia Olympi Novi* and *Theologia Cabalistica*, the anonymous editor of the *Philosophia Mystica* explained that the texts that were printed thereafter were not by Paracelsus, as opposed (he implied) to those printed before (i.e. our two treatises).[30] Thus the editor of the volume either considered these two treatises as authentic, or at least presented them this way.[31] – In at least one copy of the *Philosophia Mystica*, discovered by Carlos Gilly (but that we have not been able to locate), the "Praemonitio" was replaced by a "Vermahnung [...] an den Leser" by Benedictus Figulus. According to Gilly, Figulus stated that he had received some "Paracelsian" texts written by Adam Haslmayr and had them printed there.[32] This would refer to the *Astronomia Olympi Novi*, the *Theologia Cabalistica* and two other texts. – According to Gilly, the *Astronomia Olympi Novi* and the *Theologia Cabalistica* can be ascribed to Adam Haslmayr with reasonable certainty: "There are some sentences and phrases like 'elementi verbique arcanum' present in the *Astronomia Olympi Novi* that Haslmayr used in all his writings almost like a ritornello."[33]

28 Huser, 10: 67: "Das seind die 4. Astronomey [...] Naturalis. Supera. Olympi novi. Satanistae." See Gilly, *Paracelsus in der BPH* (1993), 41; Pörksen, "Was sagt Paracelsus, wenn er 'Mensch' sagt?" (2002), 6; CP 2 (2004): 229.

29 Gilly, "Zwischen Erfahrung und Spekulation" (1979), 185.

30 "Praemonitio ad Lectorem", *Philosophia Mystica*, 47.

31 Gilly, "*Theophrastia Sancta*" (1994), 456–457.

32 CP 3 (2013), 1070; for this reason Žemla, "The *Astronomia Olympi novi* and the *Theologia Cabalistica*" (2019), 531 (above, 119), suggests that Figulus "might have adopted and adapted an existing project" such as the whole collection of the *Philosophia Mystica*.

33 Carlos Gilly, private communication to Joachim Telle, quoted in CP 2 (2004): 229.

Also in the *Theologica Cabalistica* there are concepts and ideas typical of Adam Haslmayr.[34]

7. *Time of writing*. Probably written by Adam Haslmayr in the 1610s.

Word count: ca. 2.100 words

Incipit: "Dieweil wir wissen daß der meiste theil der menschenkind noch nicht verstehen ..."

Manuscripts: no manuscripts known

First printed:

- 1618 ("Astronomia Olympi Novi, Das ist: Die Gestirnkunst deß newen Himmels/ welche allein auß dem Glauben entspringet/ darauß der Mensch alle Magnalia Gottes vnd der Natur/ die den glaubigen seind zuwissen/ sehen vnd erlernen mag", in *Philosophia Mystica, Darinn begriffen Eilff vnterschidene Theologico-Philosophische/ doch teutsche Tractätlein/ zum theil auß Theophrasti Paracelsi, zum theil auch M. Valentini Weigelii* (Newstadt [Frankfurt a.M.]: Lucas Jennis, 1618); VD17 3:604297L; Sudhoff, *Bibliographia Paracelsica*, 513–516 n° 306)

Historical Manuscript Catalogues: *Catalogus* Pfeffer (1683), 137

Essential bibliography: Sudhoff, *Bibliographia Paracelsica*, 696, n° 36; Sudhoff, "Kritischer Anhang", in Paracelsus, *Sämtliche Werke*, I/12: 603; CP 2: 229; CP 3: 1070.

Further bibliographical references:

Peuckert, *Pansophie* (1956), 1: 350–352, 471, 484.

Peuckert, *Gabalia* (1967), 135.

Gilly, *Adam Haslmayr* (1994), 184, 199 n. 5, 202–203 n. 21.

Gilly, "Theophrastia Sancta" (1998), 171.

Gilly, "Il primo prontuario di magia bianca" (2002), 202, 211–212.

Gunhild Pörksen, "Was sagt Paracelsus, wenn er 'Mensch' sagt?," *Nova Acta Paracelsica*, N.F. 16 (2002), 3–18.

Martin Žemla, "The *Astronomia Olympi novi* and the *Theologia Cabalistica*: Two Pseudo-Paracelsian Works of the Philosophia Mystica (1618)," *Early Science and Medicine*, 24 (2019), 527–548 (in the present volume, 115–136).

---

34 Gilly, "*Theophrastia Sancta*" (1994), 457, 466; Gilly, *Adam Haslmayr* (1994), 188, 199 n. 5; Gilly, "Vom ägyptischen Hermes zum Trismegistus Germanus" (2010), 89.

Martin Žemla, "From Paracelsus to Universal Reform. The (Pseudo-) Paracelsian-Weigelian *Philosophia Mystica* (1618)," *Daphnis*, 48 (2020), 184–213.

1.3 *Auslegung des 91. Psalms*

1. *Printing History, Manuscripts.* One manuscript, written by Karl Widemann.
2. *Editions.* Not edited by Huser or Sudhoff or Goldammer.
3. *Relationship between different versions.* No information about different versions.
4. *Structure, genre/form, perspective, style.*
5. *Relationship to other texts.*
6. *Authenticity, authorship.* Not authentic according to Sudhoff: "The [biblical] text is always given in Luther's German translation, which already speaks against Paracelsus as the author."[35] This is further supported by Goldammer: "The eschatology presented here is unparacelsian in this sharpness; the actualistic genuine interpretation of the verse in the commentary on Psalms proceeds quite differently."[36] – Based on a sentence in this text ("Gott wirt auch ein Neapolin dz ist ein Neustatt bauen vnd Halberstatt einreißen"), Winfried Zeller identified the printing place "Newstatt", where several Paracelsian and Weigelian texts were issued, as a "Paracelsistic *deckname* for Magdeburg".[37]
7. *Time of writing.* Probably written in the 1590s or later.

Word count:

Incipit: "Diser Psalm ist der fürnembsten ainer der da weisßagung jnn sich heldtt vnd eröffnet …"

Manuscripts:

– Wolfenbüttel, *Herzog August Bibliothek*: Cod. Guelf. 37.2 Aug. 2°, f. 1r–6r

First printed: not printed

Historical Manuscript Catalogues: Widemann, *Verzeichnisse* (Kassel), n° v, 876; x, 46.

35 Sudhoff, *Paracelsus-Handschriften*, 572.

36 Goldammer, "Einleitung," in Paracelsus, *Sämtliche Werke*, II/3: LIV.

37 See Horst Pfefferl, "Valentin Weigel und Paracelsus", in *Paracelsus und sein dämonengläubiges Jahrhundert* (Wien, 1988) (Salzburger Beiträge zur Paracelsusforschung, 26), 77–95, on 82.

ESSENTIAL BIBLIOGRAPHY: Sudhoff, *Paracelsus-Handschriften*, 571–572, 791 n° 86; Goldammer, "Einleitendes zum Psalmenkommentar," in Paracelsus, *Sämtliche Werke*, II/4: XXVIII–IL, on XXXIX.

1.4 *Brief an die Wittenberger Theologen* (*Luther, Bugenhagen, Melanchthon*) also: *Brief an Luther, Bugenhagen und Melanchthon*

1. *Printing History, Manuscripts.* Edited by Johannes Staricius in the collection *Philosophia de limbo* (1618) where it was followed by a compilation of different Paracelsian writings on the Lord's Supper. Five manuscripts (the Gotha manuscript in Valentin Weigel's hand).
2. *Editions.* Not edited by Huser, Sudhoff or Goldammer.
3. *Relationship between different versions.* There seems to be only one version.
4. *Structure, genre/form, perspective, style.* Written in the first person in the form of a letter, directly addressing the recipients.
5. *Relationship to other texts.* Written to accompany and recommend the author's (i.e. Paracelsus's) interpretation of Matthew 1–5.
6. *Authenticity, authorship.* The authenticity of this letter of Paracelsus to Martin Luther, Johannes Bugenhagen and Philipp Melanchthon accompanying Paracelsus's interpretation of Matthew 1–5 was and is still disputed, especially among historians of Paracelsus. It has been included in the critical editions of the letters of both Luther (1933) and Melanchthon (1995). Sudhoff considered it spurious[38] or at least questionable.[39] Kurt Goldammer was uncertain about the date 1525,[40] Hartmut Rudolph contributed arguments for its authenticity,[41] Stefan Rhein voted against authenticity[42] while Ute Gause tended to consider

38 Sudhoff, *Bibliographia Paracelsica*, 510: "dieser Brief scheint mir untergeschoben."

39 Sudhoff, *Paracelsus-Handschriften*, 237: "über die Echtheit des Briefes habe ich doch meine grossen Bedenken."

40 Kurt Goldammer, "Neues zur Lebensgeschichte und Persönlichkeit des Theophrastus Paracelsus," *Theologische Zeitschrift*, 3 (1947), 191–221, on 199.

41 Hartmut Rudolph, "Einige Gesichtspunkte zum Thema 'Paracelsus und Luther'," *Archiv für Reformationsgeschichte*, 72 (1981), 34–54; reprinted in *Von Paracelsus zu Goethe und Wilhelm von Humboldt* (Vienna, 1981) (*Salzburger Beiträge zur Paracelsusforschung*, 22), 9–26.

42 Stefan Rhein, "Vergil oder die 'Königskerze': War Paracelsus Humanist?," *Nova Acta Paracelsica*, N.F. 7 (1993), 45–71, on 54: "fingiert, um Paracelsus aus der häretischen Ecke in die protestantische Legitimität zu rücken."

it authentic,[43] and Kühlmann and Telle rather doubted its authenticity.[44] Dane Daniel, however, sees "little reason to doubt" it, based on the theology implied in the letter.[45]

7. *Time of writing.* Written ca. 1525 (if authentic) or the early 1560s (if spurious).

Word count: ca. 500 words

Incipit: "Bruderliche Liebhaber der Wahrheit in Christo [...] / Christliche vnnd Ehrsame Hochgelehrte Herrn vnnd Brüder ..."

Manuscripts:
- Görlitz, *Oberlausitzische Bibliothek der Wissenschaften*: Th VI 146, 167–168, 249
- Gotha, *Forschungsbibliothek*: Chart. B 908, f. III^{r}, IV^{r}–VI^{v}, 250–251
- Kopenhagen, *Det Kongelige Bibliotek*: Thott 35 folio, f. 1r–1v
- Kopenhagen, *Det Kongelige Bibliotek*: Thott 119 kvart, f. 1r–1v
- Wolfenbüttel, *Herzog August Bibliothek*: Cod. Guelf. 194.3 Extrav., f. 1r–2r

First printed:
- 1618 (in: *Philosophia de limbo [...] Doctoris Aurelij Philippi Theophrasti Paracelsi ab Hohenhaim*, ed. Johannes Staricius (Magdeburg: Johann Francke, 1618); VD17 23:263455M; Sudhoff, *Bibliographia Paracelsica*, 508–511 n° 303).

Critical edition:
- *D. Martin Luthers Werke. Kritische Gesamtausgabe* (*Weimarer Ausgabe*), Briefwechsel vol. 3 (Weimar, 1933), 465–467 n° 850 [based on the Gotha manuscript]
- *Melanchthons Briefwechsel*, vol. T2, ed. Richard Wetzel (Stuttgart, 1995), 269–273 n° 385 [based on the Görlitz manuscript].

Historical Manuscript Catalogues: *Catalogus* Osek (Prague) 3, n° 48; *Catalogus* (Stuttgart), I, f. 4v n° [155]; *Catalogus* Adelwert (Leiden), n° [106]; Widemann, *Verzeichnisse* (Kassel), n° V, 778; *Catalogus* Pfeffer (1683), 137

Essential bibliography: Sudhoff, *Bibliographia Paracelsica*, 510, 696 n° 26; Sudhoff, *Paracelsus-Handschriften*, 237, 788 n° 63; CP 1 (2001): 556; CP 3 (2013): 24, 313.

---

43 Ute Gause, *Paracelsus (1493–1541). Genese und Entfaltung seiner frühen Theologie* (Tübingen, 1993), 163.

44 CP 1 (2001): 556: "die Echtheit [...] ist zweifelhaft"; CP 3 (2013): 24: "wohl eher unecht[...]"; CP 3: 313: "vielleicht unecht[...]."

45 Daniel, "The Authenticity" (2019), 428 (above, 16).

Further bibliographical references:

Stefan Rhein, "Melanchthon und Paracelsus," in Joachim Telle, ed., *Parerga Paracelsica. Paracelsus in Vergangenheit und Gegenwart* (Stuttgart, 1991), 57–73.

Kathrin Pfister, "Paracelsus und Luther," *Nova Acta Paracelsica*, 28 (2018), 139–158, on 149, 150.

Dane T. Daniel, "The Authenticity of Paracelsus' *Astronomia Magna* and *Brief an die Wittenberger Theologen*: Towards a Diagnostic Rubric Clarifying Authentic and Spurious Elements in Paracelsus' Oeuvre on the Basis of Theological Motifs," *Early Science and Medicine*, 24 (2019), 419–438 (in the present volume, 7–26).

1.5 *Brief an Johannes Oekolampad*

1. *Printing History, Manuscripts.* Discovered ca. 1980, transcribed and prepared for an edition by Hartmut Rudolph in 1983 (unpublished draft). First published in 2014 by Nikolaus Czifra.[46] One single manuscript.
2. *Editions.* Not edited by Huser, Sudhoff or Goldammer.
3. *Relationship between different versions.* Only one known version.
4. *Structure, genre/form, perspective, style.* Written in the first person in the form of a letter, directly addressing the recipient Johannes Oecolampadius (1482–1531), the Basel humanist and reformator. The letter comprises *salutatio* and *narratio*, but is missing *datum* and *subscriptio* in the manuscript.
5. *Relationship to other texts.*
6. *Authenticity, authorship.* While Goldammer was uncertain about its authenticity,[47] Czifra considered it to be the Latin translation of an authentic letter;[48] to him there seems to be no significant argument against its authenticity.[49] Urs Leo Gantenbein considers the letter to be "rather inauthentic."[50]
7. *Time of writing.* Probably written ca. 1528 (if authentic).[51] The manuscript was dated by Goldammer from the end of the sixteenth century.[52]

Word count: ca. 2.000 words

Incipit: "O Oecolampi, tuum scriptum multo et duro labore …"

46 Czifra, *Paracelsus' Abendmahlschriften* (2014), 274–288, transcription and modern German translation.

47 Goldammer, "Aufgaben der Paracelsusforschung" (1991), 13: "ein […] nicht sicher echter Brief".

48 Czifra, *Paracelsus' Abendmahlschriften* (2014), 283.

49 *Ibid.*, 288.

50 Urs Leo Gantenbein, private communication, 6 May 2021.

51 *Ibid.*, 288.

52 Goldammer, "Einleitung," in Paracelsus, *Sämtliche Werke*, II/2: XXIII.

MANUSCRIPTS:
- LONDON, *British Library*: Harley MS 508, f. 2r–6v

FIRST PRINTED: not printed before Czifra (2014)

ESSENTIAL BIBLIOGRAPHY: Czifra, *Paracelsus' Abendmahlschriften* (2014), 37–39, 274–288.

FURTHER BIBLIOGRAPHICAL REFERENCES:
Katharina Biegger, "Wie gelangten theologische Paracelsusschriften nach London?," *Nova Acta Paracelsica*, N.F. 4 (1989), 24–37, on 33–35.
Kurt Goldammer, "Aufgaben der Paracelsusforschung," in Joachim Telle, ed., *Parerga Paracelsica* (Stuttgart, 1991), 1–26, on 13.

1.6 *Brief an einen Gelehrten*

1. *Printing History, Manuscripts.* The letter is found in a single manuscript volume of authentic theological writings of Paracelsus, dated 1597. The copyist of the volume, Karl Widemann, obtained the letter from Lorenz Dehn or Thenn (1522–1599), a Paracelsist and goldsmith at Augsburg.
2. *Editions.* First published by Sudhoff in 1899. Not edited by Huser, or in the *Sämtliche Werke* by Sudhoff or Goldammer. Identified and critically edited by Kühlmann and Telle in CP 3 (2013), 300–322, n° 108.
3. *Relationship between different versions.* The author of the letter was not Paracelsus but Sebastian Franck, as discovered by Kühlmann and Telle.[53] The original letter (1531) was written in Latin and addressed to Johannes Campanus (ca. 1500–after 1574), an antrinitarian and anabaptist. While only the first part of the original letter survived in a Vienna manuscript, a German translation of the full text was printed in 1563, not addressed to Campanus, but to "N." The anonymous Paracelsist who copied it and ascribed it to Paracelsus took up about the first sixth of this printed letter. The heading was changed from "Sebastian Franck von Wörd/ an einen Gelehrten" to "Theophrastj Paracelsj vonn hohenhaimb Schreiben ahn einen glertten". Apart from this, the Pseudo-Paracelsian letter is an almost word-by-word copy of the printed text with only a few different words on the last two pages.
4. *Structure, genre/form, perspective, style.*
5. *Relationship to other texts.*
6. *Authenticity, authorship.*

53 CP 3: 300.

7. *Time of writing.* Written in 1531 by Sebastian Franck. Attributed to Paracelsus probably in the mid-1590s.

Word count: ca. 700 words [Pseudo-Paracelsus letter]; ca. 4.000 words [complete letter]

Incipit: "Gnad sey dir vom vatter des liechtes durch den herren Christum …"

Manuscripts:

- Wolfenbüttel, *Herzog August Bibliothek*: Cod. Guelf. 28.4 Aug. 4°, f. 96v–98v

First printed:

- **1563** (in: *Von dem Kindertauff/ Bestendiger vnd klarer Gegenbericht/* […] *Item zween Sendbrieff Sebstiani Francken*, n.p. 1563; VD16 E 4697; not in Sudhoff, *Bibliographia Paracelsica*) [not attributed to Paracelsus, but to the actual author, Sebastian Franck]

Essential bibliography: Sudhoff, *Paracelsus-Handschriften*, 556–558, 788 n° 62; CP 3: 300–322, n° 108.

Further bibliographical references:

Bruno Becker, "Fragment van Francks latijnse brief aan Campanus," *Nederlands archief voor kerkgeschiedenis*, 46 (1964–1965), 197–205.

Melanchthons Briefwechsel, vol. 11, ed. Heinz Scheible: *Personen A-E* (Stuttgart, 2003), 261, s.v. Campanus.

Telle, "Paracelsus in pseudoparacelsischen Briefen" (2004), reprinted (2006/2007), esp. 158 n. 40.

1.7 *De diebus Aegyptiacis*

1. *Printing History, Manuscripts.* Not printed, no manuscripts known. A short fragment on imagination possibly associated with *De diebus Aegyptiacis* is found in one manuscript.
2. *Editions.* The fragment was edited by Goldammer in Paracelsus, *Sämtliche Werke*, II/3: 289.[54]
3. *Relationship between different versions.* No information about different versions.
4. *Structure, genre/form, perspective, style.*

54 Due to a printer's error, in vol. 3 of Goldammer's edition the pages 289 and 297 were interchanged.

5. *Relationship to other texts.* The first word of the short fragment is "Diebsstrick" (reading uncertain, "thief's rope") a combination of terms also found in the *Liber de Superstitionibus et Ceremoniis* and the *Fragmentum Libri de Sagis et earum Operibus* as well as in the 4th book of the *Astronomia magna* (Huser, 9: 237, 245; Huser, 10: 378). Mention of the "dies aegyptiacas" can be found in an astrological passage of *De caducis* (1530), Sudhoff, VIII, 288.
6. *Authenticity, authorship.* A text on the "Egyptian days" was mentioned in the Osek *Catalogus*: "Etliche teill von den Egiptischen tägen".[55] No Paracelsian text with this title is known, except for a very short fragment where "De diebus Aegyptiacis" was written at the end, possibly as a source or title. It is the first of a series of short fragments found only in one single manuscript. "The authorship of Paracelsus is probable in all cases, even if the wording that has come down to us is often unclear and the context of the text is often disturbed," according to Goldammer, who suggested to compare it with *De virtute imaginativa* (Huser, 9: 298–307).[56]
7. *Time of writing.* Probably originally written by Paracelsus in the 1530s, according to Goldammer.

WORD COUNT: ca. 65 words

INCIPIT: "Diebsstrick hindersich wieder gohn zu tode beten fasten …"

MANUSCRIPTS:
- WROCŁAW/BRESLAU, *Biblioteka Uniwersytecka*: R 334 [probably lost]

FIRST PRINTED: not printed before Goldammer (1986)

HISTORICAL MANUSCRIPT CATALOGUES: *Catalogus* Osek (Prague) 3, n° 40

ESSENTIAL BIBLIOGRAPHY: Sudhoff, *Paracelsus-Handschriften*, 527.

1.8 *De dignitate sanctae Mariae* also: *De dignitate Mariae*

1. *Printing History, Manuscripts.* Not printed. One Latin manuscript.
2. *Editions.* Not edited so far.
3. *Relationship between different versions.* Only one version known.
4. *Structure, genre/form, perspective, style.*

55 Sudhoff, Paracelsus-Handschriften, 39.
56 Goldammer, "Einleitung," in Paracelsus, *Sämtliche Werke*, II/3: LIV.

5. *Relationship to other texts.*
6. *Authenticity, authorship.* Goldammer was already skeptical in regard to the authenticity ("Echtheit noch eingehender zu prüfen").[57] Biegger states that *De dignitate sanctae Mariae* "was probably not [written] by Paracelsus; in any case, a German text with a corresponding title is not known."[58] However, the text seems to have been translated directly from German.[59]
7. *Time of writing.*

WORD COUNT:

INCIPIT: "Ave Maria, tu es sine vae, plena gratia ..." / "Maria in statu gloriae suae tria habet officia distincta ..."

MANUSCRIPTS:
- LONDON, *British Library*: Harley MS 516, f. 15v–19v

FIRST PRINTED: not printed

ESSENTIAL BIBLIOGRAPHY: Sudhoff, *Paracelsus-Handschriften*, 641–642, 796 n° 128; Goldammer, "Einleitung," in Paracelsus, *Sämtliche Werke*, II/6: XXIX.

FURTHER BIBLIOGRAPHICAL REFERENCES:
Katharina Biegger, "Wie gelangten theologische Paracelsusschriften nach London?", *Nova Acta Paracelsica*, N.F. 4 (1989), 24–37.
Katharina Biegger, *"De invocatione Beatae Mariae Virginis". Paracelsus und die Marienverehrung* (Stuttgart, 1990), 133.

1.9 *De iustitia*

1. *Printing History, Manuscripts.* First printed in 1618 in a collection of Paracelsian, Pseudo-Paracelsian and Weigelian texts called *Philosophia Mystica*. Three manuscripts. – Karl Widemann was in the possession of several copies as shown by his *Verzeichnisse*. One of these was written by himself (I, 19).
2. *Editions.* Edited by Goldammer in Paracelsus, *Sämtliche Werke*, II/2: 151–163.
3. *Relationship between different versions.* No information about different versions.

---

57 Goldammer, "Einleitung," in Paracelsus, *Sämtliche Werke*, II/6: XXIX.
58 Biegger, "De invocatione Beatae Mariae Virginis" (1990), 133.
59 Biegger, "Wie gelangten theologische Paracelsusschriften nach London?" (1989), 29 note 20.

4. *Structure, genre/form, perspective, style.* Written in the third person, directly addressing the reader as "du".
5. *Relationship to other texts.* Refers to *De imaginibus idolatriae* ("als jhr de imaginibus Idolatriae hören werdet"), an authentic treatise of Paracelsus (ed. by Goldammer in 1986 in Paracelsus, *Sämtliche Werke*, II/3: 275–286). – *De iustitia* is the first text of a four-part section in the Leiden manuscript comprising "Librj. 1 De Iustitia. 2 De Penetentiæ [sic]. 3 De Imaginibus Idolatrie. 4 De Sursum corda." (f. 424r).
6. *Authenticity, authorship.* Not mentioned in any of the early lists of Paracelsian writings,[60] although the address to the reader by the otherwise unknown Valentius de Retiis (written before 1562) mentions that Paracelsus "wrote splendid works on philosophy, medicine and mathematics, as well as on the State and justice" (*atque de Republica & Iustitia*).[61] Considered to be "probably" authentic by Goldammer.[62] Cited in 1569/70 by Lambert Wacker in his preface to the Pseudo-Paracelsian *Kurze Auslegung der zehn Gebote Gottes* (§ 1.18), although his quotation (in Latin) is not found in any of the known manuscripts of *De iustitia*.[63] In fact, the quotation is not from *De iustitia* but from Adam Schröter's Latin translation of the *Archidoxa* (Cracow, 1569)[64] where the source is given as "Liber Monarchiae".[65]
7. *Time of writing.* Probably written ca. 1525 (if authentic) or in the 1560s (if spurious).

Word count: ca. 3.600 words

Incipit: "So der grundt der gerechtigkeit betrachtet vnd gehalten sol werden ..."

60 Goldammer, "Einleitung," in Paracelsus, *Sämtliche Werke*, II/2: XLI.

61 Huser, 6: 99: "Apud Germanos nunc Vir Adolescens existit, cui par Orbis non fert, qui adeo excellenter in Philosophia, Medica & Mathematica arte, atque de Republica & Iustitia scripsit."

62 *Ibid.*

63 CP 3 (2013): 201.

64 *Archidoxae Philippi Theophrasti Paracelsi Magni* [...] Nunc primum studio et diligentia Adami Schröteri [...] è Germanico in Latinum translati (Cravoviae, 1569), 4. – For this translation see Sudhoff, *Bibliographia Paracelsica*, 170–173.

65 For the "Liber Monarchiae" see Udo Benzenhöfer, *Studien zum Frühwerk des Paracelsus im Bereich Medizin und Naturkunde* (Münster, 2005), 131, 139, 158. – Goldammer, Paracelsus in neuen Horizonten. Gesammelte Aufsätze (Vienna, 1986) (Salzburger Beiträge zur Paracelsusforschung, 24), 147 note 37: "Paracelsus plante – vielleicht in Ergänzung zu Dantes De monarchia? – ein Werk de »monarchia dei« (III, 100), das nicht erhalten oder nicht geschrieben ist, oder sich unter einem anderen paracelsischen Buchtitel verbirgt."

MANUSCRIPTS:
- LEIDEN, *Universiteitsbibliotheek*: Voss. Chym. F.24, f. 425r–430r
- WOLFENBÜTTEL, *Herzog August Bibliothek*: Cod. Guelf. 50 Aug. 4°, f. 33r–46v
- BEROMÜNSTER, *Haus zum Dolder*: B.MPa.31, f. 23r–30r

FIRST PRINTED:
- 1618 (in: *Philosophia Mystica, Darinn begriffen Eilff vnterschidene Theologico-Philosophische/ doch teutsche Tractätlein/ zum theil auß Theophrasti Paracelsi, zum theil auch M. Valentini Weigelii* (Newstadt [Frankfurt a.M.]: Lucas Jennis, 1618), 13–21; VD17 3:604297L; Sudhoff, *Bibliographia Paracelsica*, 513–516 n° 306)

HISTORICAL MANUSCRIPT CATALOGUES: Widemann, *Unausgangene Bücher* (Hannover), n° 106; Widemann, *Verzeichnisse* (Kassel), n° I, 19 / V, 284, 775 / IX, 100.

ESSENTIAL BIBLIOGRAPHY: Sudhoff, *Bibliographia Paracelsica*, 514, 696 n° 34; Sudhoff, *Paracelsus-Handschriften*, 785 n° 31; Goldammer, "Einleitung," in Paracelsus, *Sämtliche Werke*, II/2: XLI; CP 1 (2001), 597; CP 3 (2013), 180, 201–202.

FURTHER BIBLIOGRAPHICAL REFERENCES:
Mirjam Reich, "Der 'Liber de iustitia' des Theophrast von Hohenheim genannt Paracelsus," *Nova Acta Paracelsica*, N.F. 12 (1998), 117–146.

1.10 *De praedestinatione et libera voluntate*

1. *Printing History, Manuscripts.* Three manuscripts. Not printed before Goldammer.
2. *Editions.* Edited by Goldammer in Paracelsus, *Sämtliche Werke*, II/2: 109–118.
3. *Relationship between different versions.* The text of the Munich manuscript is part of a compilated commentary on Matthew; it slightly differs from the text of the other two manuscripts and has some additional paragraphs.
4. *Structure, genre/form, perspective, style.*
5. *Relationship to other texts.*
6. *Authenticity, authorship.* Goldammer had "slight doubts" ("leise Zweifel") with regard to the authenticity of *De praedestinatione et libera voluntate*, but due to the "originality of this little text that may have suffered from some textual corruptions" he regarded it as probably authentic.[66] *De praedestinatione et libera voluntate* treats "higher" and "lower gods", which for an authentic text seems

66 Goldammer, "Einleitung," in Paracelsus, *Sämtliche Werke*, II/2: XXXVIII.

strange to Goldammer, who sees a parallel in *De fundamento scientiarum sapientiaeque* (§ 5.4), however.[67]

7. *Time of writing.*

Word count: ca. 2.700 words

Incipit: "So haben wir von den göttern ein freien willen, den sie uns geben haben …"

Manuscripts:
- Fulda, *Hochschul- und Landesbibliothek*: Cod. 22a (2), 2°, f. 185v–189v
- Leiden, *Universiteitsbibliotheek*: Voss. Chym. F.24, f. 418v–421v
- München, *Bayerische Staatsbibliothek*: cgm 949, 111–123

First printed: Not printed before Goldammer (1965).

Essential bibliography: Sudhoff, *Paracelsus-Handschriften*, 381–382, 497, 784 n° 25; Goldammer, "Einleitung," in Paracelsus, *Sämtliche Werke*, II/2: XXVI, XXVIII, XXXVII–XXXVIII.

Further bibliographical references:

Kurt Goldammer, "Humor bei Paracelsus", *Nova Acta Paracelsica*, 9 (1977), 44–68, on 67; reprinted in Goldammer, *Paracelsus in neuen Horizonten. Gesammelte Aufsätze* (Vienna, 1986) (*Salzburger Beiträge zur Paracelsusforschung*, 24), 376–394, on 394.

Jean-Pierre Brach, "Quelques aspects de la doctrine de la prédestination chez Paracelse", in *Paracelse et les siens* (Paris, 1995), 20–25.

Hartmut Rudolph, "Prädestination und 'seliges Leben'. Ein Beitrag zur Verhältnisbestimmung von Weltbild und Theologie bei Paracelsus", in Volker Zimmermann, ed., *Paracelsus. Das Werk – die Rezeption* (Stuttgart, 1995), 85–98, on 95.

Maximiliam Bergengruen, *Nachfolge Christi – Nachahmung der Natur. Himmlische und natürliche Magie bei Paracelsus, im Paracelsismus und in der Barockliteratur (Scheffler, Zesen, Grimmelshausen)* (Hamburg, 2007), 79.

1.11 *De sanctis*

1. *Printing History, Manuscripts.* Not printed before Gantenbein. Two manuscripts.
2. *Editions.* Edited by Gantenbein in *Neue Paracelsus-Edition* 1: 541–545, 843–845
3. *Relationship between different versions.* Only one version known.
4. *Structure, genre/form, perspective, style.*

67 *Ibid.* See also Kurt Goldammer, "Humor bei Paracelsus" (1977), 67.

5. *Relationship to other texts.* In both manuscripts, *De sanctis* is – together with an explication of psalms 149 and 150, otherwise unknown – appended to the authentic *De venerandis sanctis.*
6. *Authenticity, authorship.* According to Gantenbein "doubts must be raised as to its authenticity since the uncritical and affirmative approach to the veneration of the saints is rather untypical for Paracelsus."[68] Otherwise *De sanctis* may only be one of the early theological writings of Paracelsus.[69]
7. *Time of writing.* Further study required.

WORD COUNT: ca. 1400 words

INCIPIT: "Es ist ein solliche meinung, dass ein jeglicher mensch ein sondere natur an ihm hat im leib …"

MANUSCRIPTS:
- FULDA, *Hochschul- und Landesbibliothek*: Cod. 22a (2), 2°, f. 201v–204r
- LEIDEN, *Universiteitsbibliotheek*: Voss. Chym. F.24, f. 456v–458v

FIRST PRINTED: not printed before Gantenbein (2008)

ESSENTIAL BIBLIOGRAPHY: Sudhoff, *Paracelsus-Handschriften*, 387.

1.12 *De secretis creationis*

1. *Printing History, Manuscripts.* First published by Michael Toxites as a separate work in 1575. A Latin translation of the last chapter, from a different source, was made by Georg Forberger in 1574 and published in 1575. Three manuscripts. – According to Toxites (dedicatory epistle, dated 1st January 1575), this work will repudiate the accusation that Paracelsus denied the resurrection of the body. Toxites reprinted it in *Drey Tractat Ph. Theophrasti Paracelsi* in 1577. – The text of the Heidelberg manuscript is part of a collection of alchemical texts. The 1701 edition in *Das Geheimnüß der Schöpfung*, a collection of mostly (pseudo-) Weigelian texts printed in Amsterdam,[70] has a few notes of the editor that refer to an alchemical context as well.
2. *Editions.* Edited by Huser, *Chir.* 1605, Appendix, 102–115. Not edited by Sudhoff.

---

68 Gantenbein, "Einleitung," in *Neue Paracelsus-Edition* (2008), 83, 107.

69 *Ibid.* 83.

70 Horst Pfefferl, "Die Rezeption des Paracelsischen Schrifttums bei Valentin Weigel," in Peter Dilg and Hartmut Rudolph, eds., *Neue Beiträge zur Paracelsus-Forschung* (Stuttgart, 1995), 151–168, on 161 n. 27.

3. *Relationship between different versions.* The original version contains a diagram with text in 14 circles. This diagram is preserved only in the 1701 edition and (scattered over four pages) in the Heidelberg manuscript. The earlier printed versions, including Huser's edition, preserve only the text of the diagram.
4. *Structure, genre/form, perspective, style.*
5. *Relationship to other texts.*
6. *Authenticity, authorship.* Georg Forberger doubted the authenticity of the work. In 1574, obviously not knowing the 1575 edition (as Sudhoff surmised), he translated its last chapter into Latin from a manuscript, and wrote in the margin: "Videtur esse fragmentum commentarii in Genesim: dubito tamen an Theophrasti sit."[71] Rejected as spurious by Huser[72] and Sudhoff.[73] – Fritz Lieb tried to show that Valentin Weigel's deacon, Benedikt Biedermann, was the author of *De secretis creationis*;[74] this, however, is not generally accepted. – The Heidelberg manuscript (dated 1574) is not attributed to Paracelsus.
7. *Time of writing.* In Toxites's edition (1575), the text is dated 1570 on the last page ("Gott allein die Ehr. M.D.LXX."). However, this date may refer either to the original writing of the text, or to the writing of the manuscript copy used by Toxites.[75] Probably written in the 1560s – or in 1570.

Word count: ca. 15.000 words

Incipit: "In dem namen der heiligen dryfaltigkeit Gott des vatters [...] / Seid der zeit da Gott der allmechtig Adam schůff biß auff dise zeit ..."

Manuscripts:
- Heidelberg, *Universitätsbibliothek*: Cod. Pal. germ. 598, f. 1r–41r [anonymous]
- Karlsruhe, *Badische Landesbibliothek*: Hs. Bernus 148, f. 10r–46v
- Leiden, *Universiteitsbibliotheek*: Voss. Chym. Q.59, f. 121r–135v

First printed:
- 1575 (*De secretis creationis. Von Heimligkeiten der Schöpffung aller dingen. Philipp. Theophrastus Paracelsus*, ed. Michael Toxites (Strasburg: Christian Müller, 1575); VD16 P 561; Sudhoff, *Bibliographia Paracelsica*, 277–278 n° 161)

71 Paracelsus, *Operum Latine redditorum Tomus II* (Basel: Pietro Perna, 1575), 753.
72 "gar nicht Theophrasti/ [wird] doch von vielen darfür gehalten": Huser, *Chir.* 1605, Appendix, sig. Q1v (Huser, *Chir.* 1618, 681).
73 "untergeschoben[es] Werk[]": Sudhoff, *Bibliographia Paracelsica*, 278.
74 Lieb, *Valentin Weigels Kommentar zur Schöpfungsgeschichte* (1962), 21–23, 131–134.
75 Sudhoff, *Bibliographia Paracelsica*, 287–288: "wer will das heute noch entscheiden?"

- 1575 (Latin fragment; in *Aureoli Theophrasti Paracelsi eremitae, philosophi summi Operum latine redditorum tomus I*, ed. Georg Forberger (Basel: Pietro Perna, 1575); VD16 P 382; Sudhoff, *Bibliographia Paracelsica*, 284–288 n° 166)
- 1577 (in: *Drey Tractat Ph. Theophrasti Paracelsi*, ed. Michael Toxites (Strasburg: Christian Müller, 1577); VD16 P 446; Sudhoff, *Bibliographia Paracelsica*, 299–300 n° 175)
- 1701 (in: *Das Geheimnüß Der Schöpffung/ Nach ihren sichtbaren und unsichtbaren Wundern* (Amsterdam: Heinrich Betke, 1701); VD18, 11639903)

Historical Manuscript Catalogues: Thomasius (1699), 141.

Essential bibliography: Sudhoff, *Bibliographia Paracelsica*, 287, 463; CP 2: 463–476 n° 66, 669; CP 3: 529, 1189.

Further bibliographical references:

Fritz Lieb, *Valentin Weigels Kommentar zur Schöpfungsgeschichte und das Schrifttum seines Schülers Benedikt Biedermann. Eine literarkritische Untersuchung zur mystischen Theologie des 16. Jahrhunderts* (Zurich, 1962), 21–23, 38, 47, 131–134, 151.

Walter Pagel, *Das medizinische Weltbild des Paracelsus. Seine Zusammenhänge mit Neuplatonismus und Gnosis* (Wiesbaden, 1962), 108.

Boeren, *Codices Vossiani Chymici* (1975), 235.

Gilly, *Paracelsus in der BPH* (1993), 62.

Carlos Gilly, in *500 лет гнозиса в Европе. гностическая традиция в печатных и рукописных книгах* (Amsterdam, 1993), 116 n° 20a.

Hildegard Elisabeth Keller, "'zwo welt in einer haut'. Paracelsische 'Augenlehr' am Beispiel der Frau," *Nova Acta Paracelsica*, N.F. 14 (2000), 41–78, on 49–59.

Gilly, "Vom ägyptischen Hermes," 82, 102, 109, 110, 113.

Pamela Kalnig, Matthias Miller and Karin Zimmermann, Die *Codices Palatini germanici in der Universitätsbibliothek Heidelberg* (*Cod. Pal. germ. 496–670*) (Wiesbaden, 2014), 260–261.

Einleitung der heilligen geheimen Kunst und Weisheit der Propheten: see *Theophrastiae cabalisticae isagoge* (§ 1.25)

1.13 *Erklärung über Lukas 1* also: *Declaratio super dictum Lucae cap. 1 v. 35/Explicatio salutationis angelicae/Declaratio spiritus sanctus superveniet in te*

1. *Printing History, Manuscripts.* First edited by Johannes Staricius in 1618. Three manuscripts.
2. *Editions.* Not edited by Huser or Sudhoff.

3. *Relationship between different versions.* Only one version known.
4. *Structure, genre/form, perspective, style.*
5. *Relationship to other texts.*
6. *Authenticity, authorship.* According to Gantenbein "possibly inauthentic, having been handed down in none of the early manuscripts. [...] Quite untypical for authentic Paracelsian theology is the consistent use of alchemical terminology."[76] Sudhoff noted: "It is worth noting that this allegorical writing is completely absent from all the older manuscripts and we encounter it only in relatively late manuscripts."[77] "Another dubious writing [...] most likely written by late Paracelsists," according to Czifra.[78]
7. *Time of writing.* Possibly written in the 1590s.

WORD COUNT:

INCIPIT: "Spiritus sanctus superveniet in te ..." / "Ein mensch kan mit der vernunfft nimmermehr fassen ..." [ed. 1618] / "Khein mensch khan mit der vernunfft nimmermehr fassen ..." [MS Kopenhagen]

MANUSCRIPTS:
- KOPENHAGEN, *Det Kongelige Bibliotek*: GKS 1396 kvart, f. 215v–223r
- ROMA, *Biblioteca Apostolica Vaticana*: Reg. lat. 1344, f. 119r–122v
- SALZBURG, *Salzburg Museum*: Hs 840 (*olim* Paracelsiana 160), f. 93*v–98*v
- WROCŁAW/BRESLAU, *Biblioteka Uniwersytecka*: R 2316, f. 103v [fragment copied from the 1618 edition]

FIRST PRINTED:
- **1618** (in: *Philosophia de limbo* [...] *Doctoris Aurelij Philippi Theophrasti Paracelsi ab Hohenhaim*, ed. Johannes Staricius (Magdeburg: Johann Francke, 1618); VD17 23:263455M; Sudhoff, *Bibliographia Paracelsica*, 508–511 n° 303).

ESSENTIAL BIBLIOGRAPHY: Sudhoff, *Bibliographia Paracelsica*, 511; Sudhoff, *Paracelsus-Handschriften*, 607, 614, 618, 665, 796 n° 127.

FURTHER BIBLIOGRAPHICAL REFERENCES:
Nikolaus Czifra, *Paracelsus' Abendmahlschriften. Überlieferung und Kontext*, PhD thesis (Salzburg, 2014), 30, 42, 48, 126.

76 Gantenbein, "The New Adam" (2019), 193.
77 Sudhoff, Paracelsus-Handschriften, 614.
78 Czifra, Paracelsus' Abendmahlschriften (2014), 126.

Urs Leo Gantenbein, "The New Adam: Jacob Böhme and the Theology of Paracelsus (1493/94–1541)," in *Jacob Böhme and His World*, ed. Bo Andersson et al. (Leiden, 2019), 166–196.

1.14 *Erklärung über Matthäus 10*

1. *Printing History, Manuscripts.* Not printed before Weimann (1975). One manuscript.
2. *Editions.* Not edited by Huser or Sudhoff.
3. *Relationship between different versions.* Only one version known.
4. *Structure, genre/form, perspective, style.* The second part directly addresses the reader twice.
5. *Relationship to other texts.* Refers once to the *Apocalypsis Hermetis* (§ 4.1).
6. *Authenticity, authorship.* The text is not taken from any of the known authentic commentaries on Matthew. According to Weimann, syntax and semantics are typically Paracelsian and as a whole the *Erklärung* seems to be authentic; on the other hand there are strong indications of a later Paracelsian or spiritualist revision of the original text, e.g. the mention of "den von Hohenheim Eremiten."
7. *Time of writing.* Possibly written in the 1530s (if authentic), probably revised in the 1560s or later.

Word count: ca. 1.800 words

Incipit: "Es ist vill leichter selig werden weder verdambt …"

Manuscripts:
- Hannover, *Gottfried Wilhelm Leibniz Bibliothek/Niedersächsische Landesbibliothek*: Ms. I 22; 4 pages

First printed: not printed before Weimann (1975).

Essential bibliography: "Neu entdeckte Paracelsus-Handschriften," in Paracelsus, *Sämtliche Werke*, II/3: XXV.

Further bibliographical references:

Karl-Heinz Weimann, "Eine neu aufgefundene Paracelsus-Handschrift," in Sepp Domandl, ed., *Paracelsus. Werk und Wirkung* (Vienna, 1975), 353–361.

1.15 *Es sind auf der Welt viel wunderbarlicher Sachen beschehen* also: *Closing part of 'De invocatione beatae Mariae virginis'*

1. *Printing History, Manuscripts.* One manuscript. First edited by Katharina Biegger (1990).[79]
2. *Editions.* Not edited by Huser or Sudhoff.
3. *Relationship between different versions.* Only one version known.
4. *Structure, genre/form, perspective, style.*
5. *Relationship to other texts.*
6. *Authenticity, authorship.* The text is found in only one manuscript, as the closing part of the authentic *De invocatione beatae Mariae virginis*. According to Biegger, "the authenticity of this section is doubtful, but it appears to have been seamlessly connected in the transcript (the same scribe in the same ink, with no spatial separation)."[80]
7. *Time of writing.*

WORD COUNT: ca. 400 words

INCIPIT: "Es sind auf der Welt viel wunderbarlicher Sachen beschehen ..."

MANUSCRIPTS:
- DRESDEN, *Sächsische Landesbibliothek*: Mscr.Dresd.J.469, f. 1–13

FIRST PRINTED: not printed before Biegger (1990)

FURTHER BIBLIOGRAPHICAL REFERENCES:
Katharina Biegger, *"De invocatione Beatae Mariae Virginis". Paracelsus und die Marienverehrung* (Stuttgart, 1990), 124, 189–191.

1.16 *Es wird gefragt, ob auch Maria sei in Sünden empfangen und geboren*

1. *Printing History, Manuscripts.* Preserved in a manuscript in Wolfenbüttel.
2. *Editions.* Not edited so far.
3. *Relationship between different versions.* Only one known version.
4. *Structure, genre/form, perspective, style.*
5. *Relationship to other texts.*
6. *Authenticity, authorship.* According to Weimann the text "is obviously spurious [...]. Style and content are completely unparacelsian as can already be seen from

79 Biegger, "De invocatione Beatae Mariae Virginis" (1990), 189–191.
80 *Ibid.*, 189.

cursory reading. Additionally, this text is only found in one single manuscript and even there it is not denoted as Paracelsian. It is merely found in the vicinity of other Paracelsian works contained in this manuscript."[81]

7. *Time of writing.* Further study required.

Word count:

Incipit: "Da soll ein vnderscheid zweyerley Samen verstanden werden …"

Manuscripts:

- Wolfenbüttel, *Herzog August Bibliothek*: Cod. Guelf. 51.3 Aug. 4°, f. 66r–72v

First printed: not printed

Essential bibliography: Sudhoff, *Paracelsus-Handschriften*, 796 n° 123.

Further bibliographical references:

Karl-Heinz Weimann, "Einige paracelsische Werke weniger. Untersuchungen zum überlieferten Bestand an echten Paracelsischen Werken," *Nova Acta Paracelsica*, 8 (1957), 118–125.

Extractus et Theophrastiae Cabalisticae Isagogen: see *Theophrastiae cabalisticae isagoge* (§ 1.25)

Isagoge in artem cabalam: see *Theophrastiae cabalisticae isagoge* (§ 1.25)

1.17 *Kleine Handbibel und Einführung* also: *Kleine Hand- und Denk-Bibel*

1. *Printing History, Manuscripts.* First printed in 1605 in Lich (near Gießen and Marburg), edited by Johann Thoelde and dedicated to Count Gottfried von Oettingen. One manuscript in Freiburg im Breisgau (attributed to Jacob Böhme).
2. *Editions.* Not edited by Huser or Sudhoff.
3. *Relationship between different versions.* No information about different versions.
4. *Structure, genre/form, perspective, style.* A collection of quotations from the Bible.

81 Weimann, "Einige paracelsische Werke weniger" (1957), 123.

5. *Relationship to other texts.* Strangely enough, in some of the later editions of the *Kleine Handbibel* is appended the alchemical *Zehnte Buch der Archidoxen* (§ 4.25), which was put into circulation by Adam Haslmayr.[82]
6. *Authenticity, authorship.* According to Sudhoff it is "devoid of the faintest trace of Paracelsian spirit [...] Apart from the chapter headings and some short headings [...] the whole book consists solely of passages from the Old and New Testament. It is spurious without any doubt."[83] Gause points out that a closer look at the Biblical quotations reveal an author influenced by both Paracelsus and Weigel[84] with a profound knowledge of the Bible,[85] and with an approach similar to that which can be observed in the circles of Johann Arndt.[86] According to Gilly, the text was published in 1605 by Thoelde to defend Paracelsus against the widespread accusations of atheism.[87] According to Peuckert, Thoelde seems to be the author,[88] which is most probably not true.
7. *Time of writing.* According to Gause not written before 1584.[89]

Word count:

Incipit: "Es schreyet Christus der schöpffer, der neuwen creatur [...] / Der weise spricht, ich bin auch ein sterblicher mensch ..."

Manuscripts:
- Freiburg i.Br., *Universitätsbibliothek*: Hs. 1301; 325 fols.

First printed:
- **1605** (*Aureoli Theophrasti Paracelsi à Bombast, in Hohenheim* [...] *Kleine Handt-Bibel vnd Einführung*, ed. Johann Thoelde (Lich: Wolfgang Ketzel, 1605); VD17 39:151845N; Sudhoff, *Bibliographia Paracelsica*, 464–465 n° 268)
- **1684** (in: *Aureoli Theophrasti Paracelsi kleine Hand- und Denck-Bibel* (Frankfurt a.M.-Leipzig: Andreas Luppius, 1684); VD17 12:103261E; Sudhoff, *Bibliographia Paracelsica*, 623–626 n° 418)

---

82 Gilly, *Paracelsus in der BPH* (1993), 60.
83 Sudhoff, *Bibliographia Paracelsica*, 465.
84 Gause, "Theophrasti Paracelsi Kleine Handbibel" (2005), 93.
85 *Ibid.*, 95.
86 *Ibid.*, 103.
87 Gilly, *Paracelsus in der BPH* (1993), 60.
88 Peuckert, *Pansophie* (1956), 483.
89 Gause, "Theophrasti Paracelsi Kleine Handbibel" (2005), 90: "Im Verlauf der Untersuchung fand sich zudem noch ein schlagendes Argument, warum der 'Terminus ante quem non' der Entstehung der Schrift nicht vor 1584 liegen kann. Darauf wird später noch einzugehen sein." This argument, however, is missing in Gause's publication.

Essential bibliography: Sudhoff, *Bibliographia Paracelsica*, 464–465, 696 n° 16; CP 3: 984–997 n° 153, 998–1002 n° 154.

Further bibliographical references:

Peuckert, *Pansophie* (1956), 380, 483, 491.

Peuckert, *Rosenkreutz* (1973), 229.

Gilly, *Paracelsus in der BPH* (1993), 14, 18, 60–61.

Ute Gause, "Theophrasti Paracelsi Kleine Handbibel. Ein Beitrag zur Wirkungsgeschichte Valentin Weigels im 17. Jahrhundert," *Nova Acta Paracelsica*, N.F. 19 (2005), 85–104.

1.18 *Kurze Auslegung der zehn Gebote Gottes*

1. *Printing History, Manuscripts.* Two manuscripts in Berlin and Rostock.
2. *Editions.* First edited by Goldammer in 1973 in Paracelsus, *Sämtliche Werke*, II/ Suppl.: 167–203.
3. *Relationship between different versions.* The Berlin manuscript has a dedication letter to Joachim, Markgraf von Brandenburg (1505–1571) by Lambert Wacker, dated Berlin 7 May 1569 and asking for support to bring this text into print. Wacker reports this text to be taken from an autograph of Paracelsus ("aus des autoris eigen handschrift"). The Rostock manuscript has a dedication by Wacker to Johann Albrecht, Herzog von Mecklenburg (1525–1576), dated Prague 5 April 1570 and asking essentially the same. According to Kühlmann and Telle,[90] who edited Wacker's dedication in *Corpus Paracelsisticum*, there is no positive or negative indication whether the Berlin and Rostock manuscripts are autograph copies by Wacker or not.
4. *Structure, genre/form, perspective, style.* The text consists of a dedication written and signed by Lambert Wacker (see above), a preface on Paracelsus not signed by Wacker and written either by him or by the author of the following text.[91] The dedication is written in the first person, directly addressing the recipient. The preface is partly written in the third person, partly in the first person. The main text is designated as "Liber primus expositionum", although there is no second book. The Ten Commandments are treated in five chapters.
5. *Relationship to other texts.* The dedication opens with a citation of Sebastian Franck's well-known sentences on Paracelsus (without Franck's name). It further cites excerpts from "Velentinus Rhetius" and the (Pseudo?)-Paracelsian *De iustitia* (§ 1.9).
6. *Authenticity, authorship.* Weimann seriously doubts the authenticity of this "Short Commentary on the Ten Commandments" that has nothing in common

90 CP 3: 180.

91 Attributed to Wacker by Kühlmann-Telle, CP 1: 617.

with the authentic Paracelsian "Commentary on the Ten Commandments" and that is found in none of the known manuscript collections of Paracelsian theological works.[92] The *Kurze Auslegung* is considered spurious by Kühlmann and Telle.[93]

7. *Time of writing.* Probably written in the 1560s.

Word count: ca. 300 words (Wacker's dedication letter), ca. 2.600 words (Wacker's *Praefatio*), ca. 8.300 words (*Kurze auslegung*).

Incipit: "Es schreiben etliche in ihren chronicis [...] / Do hat uns gott geben anfenglich zwei gebott ..."

Manuscripts:

- Berlin, *Staatsbibliothek*: Ms. germ. quart. 50; 64 fols.
- Rostock, *Universitätsbibliothek*: Mss. theol. 76

First printed: not printed before Goldammer (1973).

Essential bibliography: Sudhoff, *Paracelsus-Handschriften*, 791 n° 82; CP 3: 175–218 n° 101.

Further bibliographical references:

Kurt Goldammer, "Paracelsus, Osiander and Theological Paracelsism in the Middle of the 16th Century. Some remarks concerning a Paracelsian pseudepigraphon on the Ten Commandments," in Allen G. Debus, ed., *Science, Medicine and Society in the Renaissance: Essays to honor Walter Pagel* (London, 1972), 2 vols., 1: 105–120.

Karl-Heinz Weimann, "Einleitendes zur Text- und Überlieferungsgeschichte", in *Religiöse und sozialphilosophische Schriften in Kurfassungen*, Paracelsus, *Sämtliche Werke*, II/Supplement (1973): XIX–XXVII.

Gilly, "Theophrastia Sancta" (1994), 425.

Gilly, "Theophrastia Sancta" (1998), 151.

Gilly, "Vom ägyptischen Hermes zum Trismegistus Germanus" (2010), 78.

1.19 *Maria ist aus Anna geboren, aber aus Joachim nicht*

1. *Printing History, Manuscripts.* Preserved in one manuscript in Wrocław/Breslau.
2. *Editions.* Not edited so far.

92 Weimann, "Einige paracelsische Werke weniger" (1957), XXVI–XXVII.

93 CP 3: 212.

3. *Relationship between different versions.* Only one known version.
4. *Structure, genre/form, perspective, style.*
5. *Relationship to other texts.* Not a separate work, but a "very concise excerpt" (Weimann)[94] from the main version of the authentic *Libellus de virgine sancta theotoca.* According to Weimann the shortening "is so radical that whole chapters are reduced to two or three sentences. On the other hand there is nothing in this excerpt that is not found in the main version. And most sentences are more or less taken word-by-word from the main version."[95] – Contrary to Weimann, Gantenbein considers *Maria ist aus Anna geboren* to be a work on its own, possibly "fragmentary working notes by Paracelsus when he was about to revise the *Book on the Virgin Mary.*"[96]
6. *Authenticity, authorship.*
7. *Time of writing.* Further study required.

Word count:

Incipit: "Eine Jungfrau soll gebehren ..."

Manuscripts:
– Wrocław/Breslau, *Biblioteka Uniwersytecka*: R 333, f. 129v–133r

First printed: not printed

Essential bibliography: Sudhoff, *Paracelsus-Handschriften*, 540–542, 796 n° 122.

Further bibliographical references:

Karl-Heinz Weimann, "Einige paracelsische Werke weniger. Untersuchungen zum überlieferten Bestand an echten Paracelsischen Werken," *Nova Acta Paracelsica*, 8 (1957), 118–125, on 118, 121–123.

Angelika Dörfler-Dierken, *Die Verehrung der heiligen Anna in Spätmittelalter und früher Neuzeit* (Göttingen, 1992), 50.

Urs Leo Gantenbein, "The Virgin Mary and the Universal Reformation of Paracelsus," in *Daphnis* 48 (2020), 4–37, on 9–10.

94 Weimann, "Einige paracelsische Werke weniger" (1957), 121.
95 *Ibid.*
96 Gantenbein, "The Virgin Mary" (2020), 10.

1.20 *Parasarchum*
also: *Theophrastia,* book two

1. *Printing History, Manuscripts.* Not printed, no known manuscripts.[97] *Parasarchum* is the second of three works composing the book *Theophrastia*, according to Valentius de Retiis (see § 1.24). Like the *Carboantes* (§ 4.22), it is apparently spurious, although nowhere to be found.
2. *Editions.* Not edited by Huser or Sudhoff. No text available.
3. *Relationship between different versions.* It is not certain that the text existed at all.
4. *Structure, genre/form, perspective, style.* Valentius de Retiis described it thus: "The other [is] *Parasarchum,* in which he [Paracelsus] deals with the *summum bonum* in eternity."[98] Karl Widemann characterized the text as follows: "The *Parasarchum* of Theophrastus treats theological and cabalistical matters much like the *Paragranum.*"[99] The Dresden *Libri Theophrasti* summarizes the text: "Parasarchum id est Summum Bonum et Aeternitas. De Sapientia Lib. 3. Philosophiae Lib. 4."
5. *Relationship to other texts.*
6. *Authenticity, authorship.* Both *Parasarchus* and *Carboantes* are mentioned in various 16th and 17th century lists of works written by Paracelsus.[100] Gerhard Dorn[101] and Michael Toxites both believed in the existence of *Theophrastia*, Sudhoff denotes it as "more than legendary,"[102] Gilly doubts its existence[103] while Kühlmann and Telle reject any such doubts as "unjustified".[104]
7. *Time of writing.* Further study required.

Word count:

Manuscripts: no manuscripts known

Historical Manuscript Catalogues: *Libri Theophrasti* (Dresden), fol. 5vb; Widemann, *Unausgangene Bücher* (Hannover), n° 34

97 Adam Haslmayr wrote in 1611 in a letter to Karl Widemann: "Das Buech Carboantes und Parasarchum hab ich nicht gesehen. Ich hör, es habs der Thurneisser gehabtt." (Gilly, *Adam Haslmayr*, 1994, 97).

98 "secundum Parasarchum dictum, in quo de summo bono tractat in aeternitate," edited in CP 1: 590.

99 "Parasarchum Theophrasti, tractat theologica et cabalistica sicut etiam Paragranum," quoted in Gilly, *Adam Haslmayr* (1994), 103 n. 20.

100 CP 1: 596.

101 Gilly, *Paracelsus in der BPH* (1993), 37.

102 Sudhoff, *Bibliographia Paracelsica*, 338.

103 Gilly, *Adam Haslmayr* (1994), 103 n. 20.

104 CP 1: 595.

ESSENTIAL BIBLIOGRAPHY: Sudhoff, *Bibliographia Paracelsica*, 142; Sudhoff, *Paracelsus-Handschriften*, 700; CP 1: 585, 590, 591, 593, 595–596, 617; CP 2: 165.

FURTHER BIBLIOGRAPHICAL REFERENCES:
Peuckert, *Pansophie* (1956), 460.
Gilly, *Paracelsus in der BPH* (1993), 37, 38.
Gilly, *Adam Haslmayr* (1994), 97, 103.

Selige Leben (Das): see *Summum bonum sive vita sancta* (§ 1.22)

1.21 *Sermo ad filios*
also: *Gmeine philosophei Paracelsi Magni*

1. *Printing History, Manuscripts.* One manuscript in Hannover.
2. *Editions.* Not edited so far.
3. *Relationship between different versions.* Only one known version.
4. *Structure, genre/form, perspective, style.*
5. *Relationship to other texts.*
6. *Authenticity, authorship.* The *Sermo ad filios* was first pointed out by Weimann as a work of Paracelsus (Weimann, "Der pseudoparacelsische *Sermo ad filios*," 1963). However, the term "philosophei Paracelsi Magni" in the heading is not to be read as "philosophy by Paracelsus" but "philosophy of Paracelsus." The manuscript is in the hand of Adam Haslmayr, who is also seems to be the author.[105]
7. *Time of writing.* Further study required.

WORD COUNT:

INCIPIT: "Liebe kinder der verdambten lehr …"

MANUSCRIPTS:
- HANNOVER, *Gottfried Wilhelm Leibniz Bibliothek/Niedersächsische Landesbibliothek*: Ms. IV 370; 16 pages

FIRST PRINTED: not printed

HISTORICAL MANUSCRIPT CATALOGUES: Widemann, *Verzeichnisse* (Kassel), n° IX, 67

105 Gilly, "*Theophrastia Sancta*" (1994), 459 n. 73; Gilly, *Adam Haslmayr* (1994), 206.

Essential bibliography: Sudhoff, "Einleitendes," in Paracelsus, *Sämtliche Werke*, I/1: XXII.

Further bibliographical references:

Karl-Heinz Weimann, "Der pseudoparacelsische *Sermo ad filios*," *Archiv für Reformationsgeschichte*, 54 (1963), 234–237.

Carlos Gilly, "Iter Rosicrucianum. Auf der Suche nach unbekannten Quellen der frühen Rosenkreuzer", in *Das Erbe des Christian Rosenkreuz* (Amsterdam, 1988), 63–89, on 87 n. 22.

Gilly, "Theophrastia Sancta" (1994), 459 n.73, 483.

Gilly, "Theophrastia Sancta" (1998), 175 n. 52.

Gilly, *Adam Haslmayr* (1994), 14, 199 n. 7, 200 n. 9, 206.

1.22 *Summum bonum sive vita sancta*
also: *Das selige Leben; Von dem höchsten Gut ins ewige Leben*

1. *Printing History, Manuscripts.* Not printed. One manuscript.
2. *Editions.* Not edited by Huser or Sudhoff.
3. *Relationship between different versions.* Only one version known.
4. *Structure, genre/form, perspective, style.* Written in the first person, directly addressing the reader.
5. *Relationship to other texts.* Refers to "our book Azoth". A reference to *Liber Azoth* (§ 2.5)?
6. *Authenticity, authorship.* The anonymous author is Adam Haslmayr, according to Gilly.
7. *Time of writing.* Probably written in the 1610s or 1620s.

Word count: ca. 2.900 words

Incipit: "Seitemall vill schreyer vnnd schreiber in aller wellt seindt [...] / Der mensch soll erstlich vor allen dingen wissen ..."

Manuscripts:

– Weimar, *Herzogin Anna Amalia Bibliothek*: Q 286/20, f. 14r–22r

First printed: not printed

Essential bibliography: Sudhoff: *Paracelsus-Handschriften*, 638–638, 797 n° III/132; Matthießen, "Tabellarische Aufstellung der Handschriften," in Paracelsus, *Sämtliche Werke*, II/1 (1923): 23–45, on 44.

FURTHER BIBLIOGRAPHICAL REFERENCES:
Gilly, *Adam Haslmayr* (1994), 209–210.

Super Nosce te ipsum: see *Vom höchsten Gut und seligen heiligen Leben* (§ 1.26)

1.23 *Theologia cabalistica* also: *De perfecto homine in Christo Iesu/Von dem vollkommenen Menschen in Christo Jesu*

See *Astronomia Olympi Novi* (§ 1.2).

WORD COUNT: ca. 2.300 words

INCIPIT: "Demnach vnser monarchey das erlebet daß seid der apostel abschiedt …"

MANUSCRIPTS: no manuscripts known.

FIRST PRINTED:
- 1618 ("Theologia Cabalistica De perfecto homine in Christo Iesu, et contra de perdito animali homine in Adam, qui Lunaticus dicitur. Das ist: Von dem vollkommenen Menschen in Christo Jesu/ vnd hergegen von dem verdorbenen thierischen Menschen in Adam/ welcher ein Mondsüchtiger genennet wirdt," in *Philosophia Mystica, Darinn begriffen Eilff vnterschidene Theologico-Philosophische/ doch teutsche Tractätlein/ zum theil auß Theophrasti Paracelsi, zum theil auch M. Valentini Weigelii* (Newstadt [Frankfurt a.M.]: Lucas Jennis, 1618); VD17 3:604297L; Sudhoff, *Bibliographia Paracelsica*, 513–516 n° 306).

ESSENTIAL BIBLIOGRAPHY: Sudhoff, *Bibliographia Paracelsica*, 696, n° 37.

1.24 *Theophrastia* (*De archidoxis, Parasarchum, Carboantes*)

1. *Printing History, Manuscripts.* Not printed, no manuscripts known.[106] The *Theophrastia* may or may not have existed at some time. *Theophrastia* was the title of a "book" (*librum*) mentioned by Valentius de Retiis (in a text of uncertain dating, first published in 1562).[107] It consisted of three works: *De archidoxis* (§ 3.4), *Parasarchum* (§ 1.20), and *Carboantes* (§ 4.22).

---

106 Adam Haslmayr wrote in 1611 in a letter to Karl Widemann: "Das Buech Carboantes und Parasarchum hab ich nicht gesehen. Ich hör, es habs der Thurneisser gehabtt." (Gilly, *Adam Haslmayr*, 1994, 97).

107 Sudhoff, *Bibliographia Paracelsica*, 189; CP 1: 590: "et tria opera simul in unum composuit librum qui Theophrastia nuncupatur."

2. *Editions.* Not edited by Huser or Sudhoff.
3. *Relationship between different versions.* It is not certain if the book existed at all.
4. *Structure, genre/form, perspective, style.* A "book" consisting of three "works."
5. *Relationship to other texts.* See *De archidoxis* (§ 3.4).
6. *Authenticity, authorship.* Gerhard Dorn[108] and Michael Toxites both believed in the existence of *Theophrastia*, Sudhoff denotes it as "more than legendary,"[109] Gilly doubts its existence[110] while Kühlmann and Telle reject any such doubts as "unjustified."[111]
7. *Time of writing.* Further study required.

Word count:

Manuscripts: no manuscripts known

First printed: not printed

Essential bibliography: Sudhoff, *Bibliographia Paracelsica*, 142, 171, 172, 189–194, 198–199, 271, 338, 390–391; Sudhoff, *Paracelsus-Handschriften*, 635; CP 1: 585, 587, 590–596, 617; CP 2: 152, 156, 165, 984, 1003; CP 3: 116, 121, 129, 170, 388.

Historical Manuscript Catalogues: *Libri Theophrasti* (Dresden), fol. 5vb

Further bibliographical references:
Peuckert, *Pansophie* (1956), 325.
Gilly, *Paracelsus in der BPH* (1993), 37, 38.
Gilly, *Adam Haslmayr* (1994), 97, 103.

1.25 *Theophrastiae cabalisticae isagoge* also: *Extractus et Theophrastiae Cabalisticae Isagogen; Theophrastica Cabalistica; Einleitung der heilligen geheimen Kunst und Weisheit der Propheten; Isagoge in artem cabalam*

1. *Printing History, Manuscripts.* Not printed. One manuscript.
2. *Editions.* Not edited by Huser or Sudhoff.
3. *Relationship between different versions.* Only one version known.
4. *Structure, genre/form, perspective, style.* Partially written in the first person.

108 Gilly, *Paracelsus in der BPH* (1993), 37.
109 Sudhoff, *Bibliographia Paracelsica*, 338.
110 Gilly, *Adam Haslmayr* (1994), 103 n. 20.
111 CP 1: 595.

5. *Relationship to other texts.*
6. *Authenticity, authorship.* The anonymous author is Adam Haslmayr, according to Gilly.
7. *Time of writing.* Probably written in the 1610s or 1620s.

Word count: ca. 1.500 words

Incipit: "Die cabala ist der olimpische geist oder der sacramentalische leib …"

Manuscripts:
- Weimar, *Herzogin Anna Amalia Bibliothek*: Q 286/20, f. 1r–5r

First printed: not printed

Historical Manuscript Catalogues: Widemann, *Verzeichnisse* (Kassel), n° v, 89

Essential bibliography: Sudhoff: *Paracelsus-Handschriften*, 635–636, 796 n° III/130; Matthießen, "Tabellarische Aufstellung der Handschriften," in Paracelsus, *Sämtliche Werke* II/1 (1923): 23–45, on 44.

Further bibliographical references:

Gilly, *Adam Haslmayr* (1994), 187, 199 n. 2, 209, 214.

Gilly, "Theophrastia Sancta" (1994), 453, 455 n. 66, 480.

Gilly, "Theophrastia Sancta" (1998), 170, 173 n. 50.

Gilly, "Vom ägyptischen Hermes zum Trismegistus Germanus" (2010), 88 n. 58.

Pierre Béhar, "Von der Mystik zur Magie. Die Verwandlung der christlichen Kabbala von Reuchlin zu Agrippa von Nettesheim," in Wilhelm Schmidt-Biggemann, ed., *Christliche Kabbala* (Ostfildern, 2003), 101–108, on 104.

Hartmut Rudolph, "Die Kabbala im Werk des Paracelsus," in Wilhelm Schmidt-Biggemann, ed., *Christliche Kabbala* (Ostfildern, 2003), 109–120, on 109.

1.26 *Vom höchsten Gut und seligen heiligen Leben* also: *Super Nosce te ipsum*

1. *Printing History, Manuscripts.* Not printed. One manuscript.
2. *Editions.* Not edited by Huser or Sudhoff.
3. *Relationship between different versions.* Only one version known.
4. *Structure, genre/form, perspective, style.* Written in the first person, directly addressing the reader. Divided into a preface and eight "paragraphs" (chapters).

5. *Relationship to other texts.* Refers to "our archidoxic and philosophical christian writings."
6. *Authenticity, authorship.* Found in a manuscript with three texts, the first (*Theophrastiae cabalisticae isagoge*) and third (*Summum bonum*) attributed by Gilly to Adam Haslmayr. Not authentic, according to Sudhoff.[112] Widemann's *Verzeichnisse* list a German text "Philosophia Paracelsica christiana de fundamento sapientiae aeternae et inventione artium christianum vera super Nosce teipsum" written by Adam Haslmayr in 1625, but divided into ten chapters ("In .X. Text abgetaildtt a°. 1625").
7. *Time of writing.* Probably written in the 1610s or 1620s.

Word count: ca. 2.800 words

Incipit: "So wir nun vernünfftiger lieber leser mit khurtzen ainfaltigen verstandt [...] / Weill Gott alle seine werckh vnd jhre mysteria will vnter des menschen khindern ..."

Manuscripts:
– Weimar, *Herzogin Anna Amalia Bibliothek*: Q 286/20, f. 5v–13r

First printed: not printed

Historical Manuscript Catalogues: cf. Widemann, *Verzeichnisse* (Kassel), n° IX, 69

Essential bibliography: Sudhoff, *Paracelsus-Handschriften*, 636–638, 796 n° III/131.

2 Natural Philosophy

2.1 *De mumia*

1. *Printing History, Manuscripts.* Not printed before Sudhoff. Two manuscripts known.
2. *Editions.* First published by Karl Sudhoff in 1933 in Paracelsus, *Sämtliche Werke*, I/14: 305–308.

112 Sudhoff, *Paracelsus-Handschriften*, 638: "Wenn in diesem zweiten Theile also auch nicht geradezu in der dritten Person von Hohenheim gesprochen wird, so ist er doch ebenso sicher als untergeschoben zu bezeichnen, wie der erste."

3. *Relationship between different versions. De mumia* is found in two manuscript collections of authentic and semi-authentic theological texts. Both manuscripts present essentially the same text, the Leiden manuscript missing many sentences found in the Heidelberg manuscript. However, the order of paragraphs in the Heidelberg manuscript is distorted, making the text completely unintelligible.
4. *Structure, genre/form, perspective, style.*
5. *Relationship to other texts. De mumia* is a short text or fragment pertaining to the *Volumen primum suae Philosophiae, de Diuinis Operibus & factis, & de Secretis Naturae.*[113]
6. *Authenticity, authorship.* Both manuscripts are drawn from the same source, which consisted obviously of single slips of paper marked with numbers and signs. This source was also used by Huser for editing some texts in vol. IX.[114] However, Huser did not publish *De mumia* in his edition of the *Bücher vnd Schrifften* of Paracelsus, apparently because these specific sheets were unavailable to him at the time (see Huser, 9: 328, #14): "De Mumia. *Desideratur totus.*" – Held by Sudhoff to be authentic.
7. *Time of writing.* Probably written in the 1530s (if authentic).

Word count: ca. 1.400 words (Heidelberg ms.), ca. 900 words (Leiden ms.)

Incipit: "Darauff weither so wissen, das hie furgenommen ist ..."

Manuscripts:
- Heidelberg, *Universitätsbibliothek*: Cod. Pal. germ. 476, f. 251r–253v, 254v–255v
- Leiden, *Universiteitsbibliotheek*: Voss. Chym. F.24, f. 351r–352v

First printed: not printed before Sudhoff (1933).

Historical Manuscript Catalogues: *Libri Theophrasti* (Dresden), fol. 6r; *Catalogus* (Stuttgart), I, f. 4v n° [156]; *Catalogus* Adelwert (Leiden), n° [115]

Essential bibliography: Sudhoff, *Paracelsus-Handschriften*, 778 n° I/7; Sudhoff, "Vorwort," in Paracelsus, *Sämtliche Werke*, I/14: VI.

113 See Huser, 9: sig. A4r and 328 (#14).
114 Sudhoff, *Paracelsus-Handschriften*, 463.

FURTHER BIBLIOGRAPHICAL REFERENCES:

Josef Strebel, "Zur Echtheitsfrage der paracelsischen Geheimphilosophie (*Ph. Occulta*) und die *Philosophia ad Athenienses*," *Nova Acta Paracelsica*, 4 (1947), 48–54, on 50 [questionable].

Peuckert, *Gabalia* (1967), 250, 552.

Hartmut Rudolph, "Fragen zum sogenannten 'Vita beata'-Schrifttum des Paracelsus," *Nova Acta Paracelsica*, N.F. 9 (1977), 193–204, on 200.

2.2 *De natura rerum*

1. *Printing History, Manuscripts.* Books 1–7 first printed in 1572, edited by Adam von Bodenstein. Books 8 and 9 first printed in 1584 by Lucas Bathodius.[115] Manuscripts in Freiberg (Sachsen), Leiden, Kopenhagen, Wolfenbüttel, Hamburg, some of them containing only certain books. The oldest manuscript (Freiberg) is dated 1571.
2. *Editions.* Edited by Huser, 6 (1590): 255–362. Edited by Sudhoff in Paracelsus, *Sämtliche Werke*, I/11: 307–403. The dedicatory preface only edited by Kühlmann and Telle in CP 3: 266–278 n° 106.
3. *Relationship between different versions.* Further study required.
4. *Structure, genre/form, perspective, style. De natura rerum* consists of nine books and a dedicatory preface to Johann Winckelsteiner dated 1537.
5. *Relationship to other texts.*
6. *Authenticity, authorship.* Sudhoff deemed the whole work spurious,[116] but probably reworked on the basis of possibly authentic fragments or drafts.[117] The different views of Sudhoff and other specialists on its authenticity have been summarized by Kühlmann-Telle.[118] According to Gantenbein *De natura rerum* was probably written by at least three different authors, three chapters probably by Paracelsus himself:[119] in Gantenbein's view, the preface and books 1 and 2 are "rather crude forgeries;" books 3 and 8 are "properly fake, their content [being] not typical for Paracelsus" and "probably both [are] written by another author, a craftsman who possessed great expertise;" as for books 4 and 7, "there are no strong arguments against their authenticity. However, they are probably fake in terms of their structure;" books 5, 6 and 9 "represent reasonably authentic writings of Paracelsus."

---

115 Hirai, "Into the Forger's Library" (2019), 485–486 (in the present volume, above, on 82).

116 "So wie es uns vorliegt, ist das Werk nicht aus Hohenheims Feder hervorgegangen:" Sudhoff, "Vorwort," in Paracelsus, *Sämtliche Werke*, I/11: XXXIII.

117 "vermutlich überarbeitet:" *ibid.*; "doch möglicherweise zum Teil Hohenheimsche Ausarbeitungen oder Entwürfe:" *ibid.*

118 CP 3: 266–269.

119 Gantenbein, "Real or Fake?" (2020), 4.

7. *Time of writing*. Allegedly written in 1537 (date of the dedicatory letter). The non-authentic parts were probably written in the 1560s.

Word count: ca. 36.000 words

Incipit: "Es ist billich, lieber vertrawter freund vnd bruder …" [dedication] / "Die generation aller natürlichen dingen ist zweyerley …" [1. *De generatione rerum naturalium*] / "Es ist gnůgsam offenbar vnnd meniglich in wissens …" [2. *De crescentibus rerum naturalium*] / "Ein ding zůerhalten vnd vor schaden zůbewaren …" [3. *De conservatione rerum naturalium*] / "Es kan niemands leugnen das der lufft …" [4. *De vita rerum naturalium*] / "Der todt aller natürlichen dingen …" [5. *De morte rerum naturalium*] / "Die resuscitation vnnd wiederbringung aller natürlichen dingen …" [6. *De resuscitatione rerum naturalium*] / "Zu schreiben von der transmutation aller natürlichen dingen …" [7. *De transmutationibus rerum naturalium*] / "In schöpffung der Welt hat die Erste Separation an den vier Elementen angefangen …" [8. *De separationibus rerum naturalium*] / "Soll nun hie in diesem bůch de signatura rerum philosophiert werden …" [9. *De signatura rerum naturalium*]

Manuscripts:
- Freiberg, *TU Bergakademie – Universitätsbibliothek*: Wissenschaftl. Altbestand, XVII 57 8; 75 fols.
- Hamburg, *Staats- und Universitätsbibliothek*: Cod. alchim. 192, f. 161r–197v
- Kopenhagen, *Det Kongelige Bibliotek*: GKS 1756 kvart, f. 1r–69r
- Leiden, *Universiteitsbibliotheek*: Voss. Chym. Q.4, f. 1r–3v
- Wolfenbüttel, *Herzog August Bibliothek*: Cod. Guelf. 60.6 Aug. 8°, f. 229–232, 233r–237v, 254r–255
- *olim* Tovačov, *Zámecká knihovna*: R 328[120]

First printed:
- 1572 (in: *Metamorphosis. Doctoris Theophrasti von Hohenheim/ der zerstörten gůten küsnten vnnd artzney/ restauratoris/ gewaltigs vnnd nutzlichs schreiben*, ed. Adam von Bodenstein (no place [Basel: Samuel Apiarius], 1572); VD16 P 694; Sudhoff, *Bibliographia Paracelsica*, 229–231 n° 137)

120 The Tovačov Castle Library is currently stored in the Olomouc branch of the Provincial Archive in Opava, see Eva Beňová, Theatrum chemicum, praecipuos selectorum avctorum tractatus de chemiae et lapidis philosophici antiqvitate, veritate, jure, praestantia, & operationibus, continens. Küenburská knižní sbírka 1532–1815, master thesis (Olomouc 2009), online (https://theses.cz/id/na62ww, accessed 18 January 2021), 13.

– 1584 (in: *De Natura Rerum, IX Bücher. Ph. Theophrasti von Hohenheim genant Paracelsi*, ed. Lucas Bathodius (Strasburg: Bernhard Jobin, 1584); VD16 P 698; Sudhoff, *Bibliographia Paracelsica*, 345–346 n° 199)

Historical Manuscript Catalogues: Toxites, *Zugesagte Bücher* (Dresden), n° 3, 4, 5, 11; Widemann, *Verzeichnisse* (Kassel), n° v, 427; *Scripta Theophrasti* (Wien); Thomasius (1699), 140; Binz (1791), 139 [*Metamorphosis*]; Ponickau (1791), col. 1232 n° Quarto 6

Essential bibliography: Sudhoff, *Bibliographia Paracelsica*, 392; Sudhoff, *Paracelsus-Handschriften*, 773; Sudhoff, "Vorwort," in Paracelsus, *Sämtliche Werke*, I/11: XXXI–XXXIII; CP 3: 266–278 n° 106, 757–772 n° 141, 1025.

Further bibliographical references:

Ernst Darmstaedter, "Paracelsus, *De Natura Rerum*. Eine kritische Studie," *Janus*, 37 (1933), 1–18, 48–62, 109–115, 323–324.

Peuckert, *Pansophie* (1956), 452.

Will-Erich Peuckert, "Paracelsische Zauberei," *Nova Acta Paracelsica*, 8 (1957), 71–94, on 89, 91.

Peuckert, *Gabalia* (1967), 159–162, 177, 180, 193.

Boeren, *Codices Vossiani Chymici* (1975), 119.

Sepp Domandl, "Hohenheims *De natura rerum*. Eine Einführung in die geistige Welt des Paracelsus," in *Kunst und Wissenschaft um Paracelsus* (Vienna, 1984) (*Salzburger Beiträge zur Paracelsusforschung*, 23), 61–88.

Gilly, "Vom ägyptischen Hermes zum Trismegistus Germanus" (2010), 102.

Hiro Hirai, "Into the Forger's Library: The Genesis of *De natura rerum* in Publication History," *Early Science and Medicine*, 24 (2019), 485–503 (also published in this volume, 73–91).

Urs Leo Gantenbein, "Real or Fake? New Light on the Paracelsian *De natura rerum*," *Ambix*, 67 (2020), 4–29.

William R. Newman, "Bad Chemistry: Basilisks and Women in Paracelsus and pseudo-Paracelsus," *Ambix*, 67 (2020), 30–46.

2.3 *De summis naturae mysteriis*

1. *Printing History, Manuscripts.* First printed in Latin in 1570, translated (and edited) by Gerhard Dorn.
2. *Editions.*
3. *Relationship between different versions.*

4. *Structure, genre/form, perspective, style.* A collection of translations of pseudo-Paracelsian texts: *De spiritibus metallorum* (§ 4.29), *De occulta philosophia* (§ 5.7), *Archidoxis magica* (§ 5.1).
5. *Relationship to other texts.* One of Widemann's *Verzeichnisse* lists an "Auslegung des Bueches Theophrasti de summis naturae mysteriis von ime intituliertt."
6. *Authenticity, authorship.*
7. *Time of writing.* Translated and edited by Gerhard Dorn in 1570.

Word count:

Incipit: "Invocato prius nomine domini Iesu Christi nostri servatoris ..."

Manuscripts:
- Kassel, *Landesbibliothek*: 4° Ms. chem. 60[3,2, f. 16v–28r [excerpts]

First printed:
- 1570 (*Aureoli Philippi Theophrasti Paracelsi Philosophorum atque Medicorum excellentissimi, De summis Naturae Mysterijs Libri tres*, trans. Gerhard Dorn (Basel: Pietro Perna, 1570); Sudhoff, *Bibliographia Paracelsica*, 209–211 n° 125; VD16 P 565)

Historical Manuscript Catalogues: cf. Widemann, *Verzeichnisse* (Kassel), n° IV, 28

Essential bibliography: Sudhoff, *Bibliographia Paracelsica*, 209–211.

2.4 *De tribus facultatibus* also: *Liber de tribus facultatibus et magica inventione theologiae, astronomiae et medicinae/Kurzer Bericht von der Wahrheit und Sophisterei dreier der fürnehmsten Fakultäten, nämlich Theologie, Astronomie und Medizin*

1. *Printing History, Manuscripts.* First printed in 1608, edited by Benedictus Figulus in *Pandora Magnalium Naturalium Aurea et Benedicta* and attributed to Alexander von Suchten (who is the real author). Two manuscripts with attribution to Paracelsus.
2. *Editions.* Not edited by Huser or Sudhoff.
3. *Relationship between different versions.* Further study required.
4. *Structure, genre/form, perspective, style.*
5. *Relationship to other texts.*

6. *Authenticity, authorship.* Written by Alexander von Suchten.
7. *Time of writing.* The manuscripts attributed to Paracelsus are dated 1542 and both written in the late 1590s. The text itself was written in the early 1560s according to Carlos Gilly.[121]

WORD COUNT:

INCIPIT: "Nuhn das ich zur sache komme fahe jch an vnd sage ..."

MANUSCRIPTS:
- AMSTERDAM, *Bibliotheca Philosophica Hermetica*: M382 [formerly in the library of Walter Pagel][122]
- GREIFSWALD, *Universitätsbibliothek*: Ms. 1002 (*olim* theol. quart. 8), part 3; I + 36 fols.

FIRST PRINTED:
- **1608** ("Ex Libro de Tribus Facultatibus Alexandri à Suchten", in *Pandora Magnalium Naturalium Aurea Benedicta, De Benedicto Lapidis Philosoph. Mysterio*, ed. Benedictus Figulus (Strasburg: Lazarus Zetzner, 1608); VD17 23:277709W; Sudhoff, *Bibliographia Paracelsica*, 480–482 n° 282).
- **1680** (in: *Alexandri von Suchten/ Eines wahren Philosophi und der Artzneyen Doctoris Chymische Schrifften Alle*, ed. Ulr. C. von Dagitza, Johann Görlin (Hamburg) für Georg Wolff (Frankfurt a.M.), 1680; VD17 23:239279L; Sudhoff, *Bibliographia Paracelsica*, 618–619 n° 411; probably reprinted from Figulus's edition with many small variants)

HISTORICAL MANUSCRIPT CATALOGUES: cf. Widemann, *Verzeichnisse* (Kassel), n° III, 22 / IV, 48 / V, 402

ESSENTIAL BIBLIOGRAPHY: CP 1: 547.

FURTHER BIBLIOGRAPHICAL REFERENCES:
Peuckert, *Pansophie* (1956), 462, 470.
Walter Pagel, *The Smiling Spleen. Paracelsianism in Storm and Stress* (Basel, 1984), 13–17, 164–165.

121 Gilly, "Zwischen Erfahrung und Spekulation" (1977), 77; Gilly, *Johann Valentin Andreae* (1987), 31: "um das Jahr 1563 verfasst;" Gilly, "Das Sprichwort 'Die Gelehrten die Verkehrten'" (1991), 344 n. 284: "unmittelbar nach 1560 geschrieben;" Gilly, "Vom ägyptischen Hermes zum Trismegistus Germanus" (2010), 103: "von ca. 1563–1569."

122 Fac-simile of the first page in Pagel, *The Smiling Spleen* (1984), 16, and Gilly, "Un bel trattato ermetico del paracelsismo" (2002), 179 (fac-simile of four pages).

Gilly, "Zwischen Erfahrung und Spekulation" (1977), 77–82.

Carlos Gilly, *Johann Valentin Andreae 1586–1986. Die Manifeste der Rosenkreuzerbruderschaft*, 2nd ed. (Amsterdam, 1987), 31.

Carlos Gilly, "Das Sprichwort 'Die Gelehrten die Verkehrten' oder der Verrat der Intellektuellen im Zeitalter der Glaubensspaltung," in Antonio Rotondò, ed., *Forme e destinazione del messaggio religioso. Aspetti della propaganda religiosa nel Cinquecento* (Florence, 1991), 229–375, on 344.

Gilly, "Theophrastia Sancta" (1994), 437–439.

Gilly, "Theophrastia Sancta" (1998), 159.

Gilly, "Das Bekenntnis zur Gnosis" (2000), 398.

Ralf Bröer, "Friedenspolitik durch Verketzerung Johannes Crato (1519–1585) und die Denunziation der Paracelsisten als Arianer," *Medizinhistorisches Journal*, 37 (2002), 139–182, on 161–166.

Carlos Gilly, "Un bel trattato ermetico del paracelsismo: Il 'De tribus facultatibus' di Alexander von Suchten" / "Paracelsism brings forth a fine Hermetical treatise: Suchten's *De tribus facultatibus*," in Carlos Gilly and Cis van Heertum, eds., *Magia, alchimia, scienza dal '400 al '700. L'influsso di Ermete Trismegisto* (Venice, 2002), 185–198.

Carlos Gilly, "Hermes oder Luther. Der philosophische Hintergrund von Johann Arndts Frühschrift *De antiqua philosophia et divina veterum Magorum Sapientia recuperanda*," in Hans Otte and Hans Schneider, eds., *Frömmigkeit oder Theologie. Johann Arndt und die 'Vier Bücher vom wahren Christentum'* (Göttingen, 2007), 163–199, on 174.

Gilly, "Vom ägyptischen Hermes zum Trismegistus Germanus" (2010), 103, 107.

Volkhard Wels, "Die Tradierung alchemischen Wissens bei Michael Maier, Andreas Libavius und Oswald Croll," in Thorsten Burkhard, ed., *Natur – Religion – Medien. Transformationen frühneuzeitlichen Wissens* (Berlin, 2013), 63–85, on 81–82.

Volkhard Wels, *Manifestationen des Geistes. Frömmigkeit, Spiritualismus und Dichtung in der Frühen Neuzeit* (Göttingen, 2014), 135, 164, 220–222, 255.

Decretum magicum: see *Secretum magicum* (§ 2.7)

2.5 *Liber Azoth*

1. *Printing History, Manuscripts.* First published by Huser in 1590. No manuscripts known. Huser published the *Liber Azoth* "ex manuscriptis aliorum" ("from manuscripts of other people," i.e. neither Paracelsus nor Johannes Montanus). There are no further editions of this text in the 16th to 18th century apart from re-editions or translations of Huser's edition.
2. *Editions.* Edited by Huser, 10 (1590): Appendix, 3–66. Edited by Sudhoff in Paracelsus, *Sämtliche Werke*, 1/14: 547–595.

3. *Relationship between different versions.* Only one known version.
4. *Structure, genre/form, perspective, style.*
5. *Relationship to other texts.*
6. *Authenticity, authorship.* Sudhoff did not even comment on the authenticity of the text, he included it in his *Sämtliche Werke* "only not to withhold anything from the reader of this present edition that Huser considered authentic in 1590/91."[123]
7. *Time of writing.* Probably written in the 1580s.

Word count: ca. 22.000 words

Incipit: "Welcher die Geheimnuß allerley verborgenen dingen zuwissen begeret …"

Manuscripts: no manuscripts known

First printed: not printed before Huser (1590).

Essential bibliography: Sudhoff, *Bibliographia Paracelsica*, 406–407; Sudhoff, "Vorwort," in Paracelsus, *Sämtliche Werke*, I/14: XXXII–XXXIII; CP 3: 1105–1106.

Further bibliographical references:

Will-Erich Peuckert, "Paracelsische Zauberei," *Nova Acta Paracelsica*, 8 (1957), 71–94, on 76.

Peuckert, *Gabalia* (1967), 131, 463–464.

Peuckert, *Das Rosenkreutz* (1973), 30, 124, 325.

Pagel, "The Paracelsian Elias Artista," 8.

Carlos Gilly, "Der Paracelsus-Einblattdruck des Matthis Quad von 1606. Kritische Überlegungen über die sogenannten Rosenkreuzer-Bildnisse des Paracelsus," in Gilly, *Paracelsus in der BPH* (1993), 13–27, on 14, 16, 17.

Goldammer, *Der göttliche Magier* (1991), 87, 106.

Ute Frietsch, *Häresie und Wissenschaft. Eine Genealogie der paracelsischen Alchemie* (Paderborn, 2013), 355–369.

Didier Kahn, "Quintessence and the Prolongation of Life in the Works of Paracelsus," *Micrologus*, 26 (2018), 183–225, on 218.

2.6 *Philosophia ad Athenienses*

1. *Printing History, Manuscripts.* First printed in the Köln printing office of the heirs of Arnold Birckmann in 1564. No manuscripts known.

123 Sudhoff, "Vorwort," in Paracelsus, *Sämtliche Werke*, I/14: XXXII.

2. *Editions.* Edited by Huser, 8 (1590): 1–47. Edited by Sudhoff in Paracelsus, *Sämtliche Werke*, 1/13: 387–423.
3. *Relationship between different versions.* Only one known version.
4. *Structure, genre/form, perspective, style.* The text consists of three "books," the last one incomplete, containing respectively 24, 23 and only 6 chapters (called "Texts"). Each book has a preface and a conclusion summarizing its contents, except the unfinished third book, with no conclusion. The prefaces and conclusions describe what the "Fürst Theophrastus" (Prince Theophrastus) has explained, which implies that they are the work of an editor, not the author himself, although their summaries do not always match the actual content of the work and mention topics not discussed in the text.

   In the preface to book 1, the whole work is introduced as "the second Paras [*sic*], containing three books, of the first volume of the philosophy of Paracelsus," as if a first "Paras" (i.e. another work) should have preceded the *Philosophia ad Athenienses*. However, "the word 'Paras' is used in all other conclusions and prefaces of the work as if the 'ander Paras' referred only to the first book of the *Philosophia ad Athenienses*, instead of to the whole work."[124] Thus the second book is introduced as "dritt Paras," and the third book as "vierdt Paras," leading the reader to conclude that the very first book of the *Philosophia ad Athenienses* is missing. But the publisher of the volume "had understood it differently, for he chose 'Das erst Buch,' 'Das ander Buch,' and 'Das dritt Buch' as running titles, which is consistent with the preface to the first book."[125]
5. *Relationship to other texts.* The text has close similarities with the *Philosophia de generationibus et fructibus quatuor elementorum* and mentions a number of elemental beings, which implies a knowledge of the *Liber de nymphis*. Other authentic texts possibly known to the author may be *De meteoris*, the *Archidoxis*, and the section *De ente astrorum* in the *Volumen medicinae Paramirum*. Knowledge of any of these texts imply that the author had access to the manuscripts of Paracelsus.
6. *Authenticity, authorship.* According to Gilly the early editors had not the least doubt about the authenticity of the *Philosophia ad Athenienses* and the anti-Paracelsians recognised it at once as the embodiment of Paracelsus's heresies.[126]

124 Kahn, "The *Philosophia ad Athenienses*" (2019), 448; in the present volume, 36.

125 *Ibid.*

126 Gilly, "Das Bekenntnis zur Gnosis" (2000), 393, 395–396. – In 1570 the physician Bartholomäus Reußner published a resolute refutation of the *Philosophia ad Athenienses*: Bartholomäus Reußner, *Ein kurtze Erklerung vnd Christliche widerlegung/ Der vnerhörten Gotteslesterungen vnd Lügen/ welche Paracelsus in den dreyen Büchern Philosophiae ad Athenienses hat wider Gott/ sein Wort/ vnd die löbliche Kunst der Artzney außgeschüttet* (Görlitz 1570); VD16 R 1339.

As "the most accurate account of Paracelsus's gnostic views" (Carlos Gilly)[127] it was central to Walter Pagel, who did rarely differentiate between authentic and non-authentic writings of Paracelsus. Sudhoff doubted its authenticity – a "rather suspicious writing"[128] –, but he saw the possibility of an authentic core: "it will be difficult to estimate how far it is a revision of a fragmentary work – if at all we can assume an authentic core".[129] Kahn rejects its authenticity "both in style and content:" "the unknown author wrote this text and ascribed it to Paracelsus in order to produce a more Platonist Paracelsian cosmology than that of the genuine treatises of Paracelsus, which apparently disappointed the unknown author in this regard."[130]

7. *Time of writing.* Probably written in the late 1550s or early 1560s.

Word count: ca. 15.000 words

Incipit: "Hierinnen beschreibt der fürst Theophrastus […] / Aller geschaffnen dingen die da in zergencklichem wesen stehen …"

Manuscripts: no manuscripts known

First printed:

- **1564** (*Des Hocherfarnen vnd Hochgelehrten Herrn Theophrasti Paracelsi von Hohenheim* […] *Philosophiae ad Athenienses drey Bücher* (Köln: Arnold Birckmanns Erben, 1564); VD16 P 528; Sudhoff, *Bibliographia Paracelsica*, 101–104 n° 65)

Historical Manuscript Catalogues: *Catalogus* Osek (Prague) 1, n° 3

Essential bibliography: Sudhoff, *Bibliographia Paracelsica*, 102–103, 399; Sudhoff, "Vorwort," in Paracelsus, *Sämtliche Werke*, I/13: XI–XIII; CP 2: 17, 236–237, 898–899.

Further bibliographical references:

Josef Strebel, "Zur Echtheitsfrage der paracelsischen Geheimphilosophie (*Ph. Occulta*) und die *Philosophia ad Athenienses*," *Nova Acta Paracelsica*, 4 (1947), 48–54 [questionable].

Peuckert, *Pansophie* (1956), 218, 454.

127 Gilly, "Iter gnostico-russicum" (1993), 54.

128 Sudhoff, *Bibliographia Paracelsica*, 399.

129 *Ibid.* 102–103; the same view in Sudhoff, "Vorwort," in Paracelsus, *Sämtliche Werke*, I/13: XII.

130 Kahn, "The *Philosophia ad Athenienses*" (2019), 439; in the present volume, 27.

Carlos Gilly, "Iter gnostico-russicum," in *500 лет гнозиса в Европе. гностическая традиция в печатных и рукописных книгах* (Amsterdam, 1993), 54–58, 114 n° 19a.
Gilly, *Paracelsus in der BPH* (1993), 36.
Gilly, "Theophrastia Sancta" (1994), 436.
Gilly, "Theophrastia Sancta" (1998), 156, 158.
Gilly, "Das Bekenntnis zur Gnosis" (2000), 392–393, 395–396, 404, 408, 416.
Gilly, "Vom ägyptischen Hermes zum Trismegistus Germanus" (2010), 77–78, 104.
Katharina Dück, "Transformationen des paracelsischen 'Prima-Materia'-Begriffs in der 'Philosophia ad Athenienses'", *Sprachreport*, 25 (2009), n° 3, 12–16.
Ute Frietsch, *Häresie und Wissenschaft. Eine Genealogie der paracelsischen Alchemie* (Paderborn, 2013), 346–355.
Didier Kahn, "The *Philosophia ad Athenienses* in the Light of Genuine Paracelsian Cosmology," *Early Science and Medicine*, 24 (2019), 439–472; also published in this volume, 27–60.
Katharina Dück, *Materia prima. Zur Semantik des Begriffs in naturkundlichen Sachschriften des 16. Jahrhunderts* (Wiesbaden, 2020), 149–153.
Didier Kahn, "La Création *ex nihilo* et la notion d'*increatum* chez Paracelse", in Edouard Mehl and Isabelle Pantin, eds., *"De mundi recentioribus phœnomenis." Cosmologie et science dans l'Europe des Temps modernes, XVe–XVIIe siècles. Essais en l'honneur de Miguel Ángel Granada* (Turnhout: Brepols, forthcoming).

2.7 *Secretum magicum* also: *Decretum magicum/Von dreien gebenedeiten magischen Steinen*

1. *Printing History, Manuscripts.* First printed in 1603, edited by Ps.-Johann Huser (with many misreadings). Printed again in 1608, edited by Benedictus Figulus from a different source. Four manuscripts. Another, hitherto not located, manuscript was in the library of Johann Christian Gottfried Jahn ("Secretum Magicum de Lapide Philosophorum vel Apocalypsis Hermetis ab illustrissimo viro Aureolo Helvetio, qui fuit Hermes II.").
2. *Editions.* Edited by Ps.-Huser in *Opera Bücher vnd Schrifften* (1603), 2: 671–682.
3. *Relationship between different versions.*
4. *Structure, genre/form, perspective, style.*
5. *Relationship to other texts.* There is another *Secretum magicum*, associated with the name of Johann Christan von Wildeck, in a manuscript that contains some pseudo-Paracelsian texts (WIEN, *Österreichische Nationalbibliothek*: Cod. 11259).[131]
6. *Authenticity, authorship.* Huser's text was printed allegedly from the autograph of Paracelsus ("ex manuscripto exemplari Theophrasti"). The Kopenhagen

131 See Sudhoff, *Paracelsus-Handschriften*, 198–199.

manuscript is associated with Philipp Jakob Gustenhofer (dead before 1607)[132] whom Carlos Gilly assumes to be the author.[133] The original title seems to have been *Decretum magicum*.[134]

7. *Time of writing.* Probably written in the 1590s or early 1600s.

Word count: ca. 10.000 words

Incipit: "Ob wol von den gar alten philosophis durch langwürige mühe vnnd nachforschung ..."

Manuscripts:

- Dresden, *Sächsische Landesbibliothek*: Mscr.Dresd.N.54, f. 1–13
- Kassel, *Landesbibliothek*: 4° Ms. chem. 60[5,3, f. 106v (Fragment)

---

132 According to a genealogical notice by Franz J. Müller, "Aus Offenburg: Name Gustenhofer, 1510 bis 1828" (https://forum.ortenau.eu/showthread.php?tid=2006, accessed 12 December 2020), Gustenhofer was a member of an old and important Offenburg family. – On Gustenhofer see Johann Jacob Heilmann, "Praefatio ad lectorem," in *Theatrum Chemicum*, vol. 6 (Strasburg: Eberhard Zetzner's Heirs, 1661), sig. *5v–*8v, on sig. *7v/*8r; Johann Conrad Creiling, *Die Edelgeborne Junger Alchymia* (Tübingen: Cotta, 1730), 80–82; John Ferguson, *Bibliotheca Chemica* (Glasgow, 1906), 2 vols., 1: 372, 2: 375, 387; Victor Klarwill, ed., *Fugger-Zeitungen. Ungedruckte Briefe an das Haus Fugger aus den Jahren 1568–1605* (Vienna, 1923), 237 n° 233, 277; Gerhard Eis, *Vor und nach Paracelsus* (Stuttgart, 1965), 146; François Secret, "Littérature et alchimie," *Bibliothèque d'Humanisme et Renaissance*, 35 (1973), 499–531, on 517; Marc Lienhard, "Gustenhover (Gustenhoffer, Gossenhauer) Philippe," in *Nouveau dictionnaire de biographie alsacienne*, vol. 14 (1989), 1337; Julian Paulus, "Seton, Alexander," in C. Priesner and K. Figala, eds., *Alchemie. Lexikon einer Hermetischen Wissenschaft* (Munich, 1998), 335; Didier Kahn, "King Henry IV, Alchemy, and Paracelsianism in France (1589–1610)," in Lawrence M. Principe, ed., *Chymists and Chymistry. Studies in the History of Alchemy and Early Modern Chemistry* (Sagamore Beach, 2007), 1–11, on 5; Rafał T. Prinke and Mike A. Zuber, "'Learn to Restrain Your Mouth': Alchemical Rumours and their Historiographical Afterlives", *Early Science and Medicine*, 25 (2020), 413–452, on 425, 426. 447. – See also München, *Bayerische Staatsbibliothek*, cgm 4120: an alchemical process by Johann Justenhofer von Offenburg, Strasburg, 1587; Praha, *Strahovská knihovna*, Cod. DF IV 14: a "tinctura" by Ph. J. Justenhoven; Wien, *Haus-, Hof- und Staatsarchiv*, Hofakten des Ministeriums des Innern, Karton 12, Fasz. 1 E 4, f. 172–174: "Interrogation" of the Baden Chancellor [Johann] Aschmann "in der Güsterhoferischen [sic] sachen", 6 May 1602; more texts in Göttingen, *Niedersächsische Staats- und Universitätsbibliothek*, 8 Ms. hist. nat. 77 and Wien, *Österreichische Nationalbibliothek*, Cod. 11290 and Cod. 11329.

133 Gilly, *Adam Haslmayr* (1994), 177, n. 11: "Aus dieser Kopie erfahren wir übrigens den Namen des wirklichen Verfassers dieses pseudoparacelsischen Traktats: Es war der bekannte Alchemist Philipp Jakob Gustenhoffer aus Strassburg."

134 *Ibid.*, 166.

- KOPENHAGEN, *Det Kongelige Bibliotek*: GKS 1769 kvart, f. 1r–41v
- PRAHA, *Strahovská knihovna*: DF IV 59, f. 284r–306v
- TARTU, *Ülikooli Raamatukogu* [*University Library*]: Mscr. 655, item n° 11

FIRST PRINTED: not printed before Ps.-Huser (1603)
- 1603 (in: Ps.-Huser, *Opera Bücher vnd Schrifften* (1603); VD17 12:168390P)
- 1608 (in: *Thesaurinella Olympica aurea tripartita. Das ist: Ein himmlisch güldenes Schatzkämmerlein*, ed. Benedictus Figulus, Frankfurt a.M.: Wolfgang Richter for Nicolaus Stein 1608; VD17 3:302600E; Sudhoff, *Bibliographia Paracelsica*, 477–479 n° 280)
- 1700 ("Theophrasti Paracelsi Secretum Magicum," in *Gottfried Arnolds Fortsetzung und Erläuterung Oder Dritter und Vierdter Theil der unpartheyischen Kirchen und Ketzer-Historie* (Frankfurt a.M.: Thomas Fritsch, 1700), 148–159)

HISTORICAL MANUSCRIPT CATALOGUES: Jahn (1754–1758), vol. 2, 166 n° 3042

ESSENTIAL BIBLIOGRAPHY: Sudhoff, *Bibliographia Paracelsica*, 445; CP 1: 450; CP 3: 1086–1087.

FURTHER BIBLIOGRAPHICAL REFERENCES:

Sten Lindroth, *Paracelsismen i Sverige till 1600-talets mitt* (Uppsala: Almqvist & Wiksells, 1943), 119, 120, 124.

Peuckert, *Pansophie* (1956), 316–322.

Sten Lindroth, review of Alexandre Koyré, *Mystiques, spirituels, alchimistes*, in *Lychnos*, 1956, 259.

Walter Pagel, "Paracelsus and the Neoplatonic and Gnostic Tradition," *Ambix*, 8 (1960), 125–166, on 155.

Walter Pagel, *Das medizinische Weltbild des Paracelsus. Seine Zusammenhänge mit Neuplatonismus und Gnosis* (Wiesbaden, 1962), 92, 109.

Karl-Heinz Weimann, "Der pseudoparacelsische *Sermo ad filios*," *Archiv für Reformationsgeschichte*, 54 (1963), 234–237, on 236 n.11.

Hans Biedermann, *Materia Prima. Eine Bildersammlung zur Ideengeschichte der Alchemie* (Graz, 1973), 30–31.

Walter Pagel, *Paracelsus. An Introduction to Philosophical Medicine in the Era of the Renaissance*, 2nd rev. ed. (Basel, 1982), 113 n. 301, 114 n. 305, 205 n. 6, 360.

Joachim Telle, "Paracelsus im Gedicht. Materialien zur Wirkungsgeschichte Theophrasts von Hohenheim im 16. und 17. Jahrhundert," in Gundolf Keil, ed., *Fachprosa-Studien. Beiträge zur mittelalterlichen Wissenschafts- und Geistesgeschichte* (Berlin, 1982), 552–573, on 557.

Gilly, *Adam Haslmayr* (1994), 166, 177 n. 11.

Roland Edighoffer, "La gnoséologie rosicrucienne d'après la *Fama Fraternitatis* (1614). Position du problème," in Ilana Zinguer and Heinz Schott, eds., *Systèmes de pensée précartésiens* (Paris, 1998), 313–337, on 321, 329.

Rudolf Schlögl, "Hermetismus als Sprache der 'unsichtbaren Kirche': Luther, Paracelsus und die Neutralisten in der Kirchen- und Ketzerhistorie Gottfried Arnolds," in Anne-Charlott Trapp and Hartmut Lehmann, eds., *Antike Weisheit und kulturelle Praxis in der Frühen Neuzeit* (Göttingen, 2001), 165–188, on 181–182 n. 69.

Didier Kahn, *La Messe alchimique attribuée à Melchior de Sibiu* (Paris, 2015), 105.

Von dreien gebenedeiten magischen Steinen: see *Secretum magicum*

3 Medicine and Surgery

3.1 *Anatomia corporum adhuc viventium* also: *Anatomia vivorum/Anatomie, das ist Zerlegung der lebendigen Körper*

1. *Printing History, Manuscripts.* First printed 1577, edited by Gerhard Dorn, who appended the text to his translation of pseudo-Paracelsian treatises: *Aurora thesaurusque philosophorum Theophrasti Paracelsi.* No manuscripts known.

2. *Editions.* Edited by Huser, *Chir.* 1605, Appendix, 58–70 (German version, translated from Dorn's Latin text). Not edited by Sudhoff.

3. *Relationship between different versions.* Only one known version.

4. *Structure, genre/form, perspective, style.* The *Anatomia* is an instruction on the analysis of human urine.[135] It is written entirely in the third person.

5. *Relationship to other texts.*

6. *Authenticity, authorship.* The authenticity of the *Anatomia* was heavily questioned already by Huser, who in 1605 described the text as "attributed" to Paracelsus ("Theophrasto Paracelso zugeeignet"). It was probably written in Latin by Gerhard Dorn,[136] who states, however, that he has just appended his own remarks to the doctrine of Paracelsus (ed. 1577, 131). Dorn had promised to his new patron, Prince François of Valois (1555–1584), Duke of Anjou and Alençon, the French King's brother, to bring to light new writings of Paracelsus. However, he had seriously fallen out with both Pietro Perna and Adam von Bodenstein since 1573 and had been excluded from the Basel Paracelsian publishing movement, so that no new Paracelsian text was available to him. All he could do was

135 "The term 'anatomia' as used in the *Paramirum* has nothing in common with Vesalius's concept of 'anatomia'" (CP 1: 281).

136 Kahn, "Les débuts de Gérard Dorn" (1994), 109; Sudhoff, *Bibliographia Paracelsica*, 304.

editing spurious texts like the *Aurora* (§ 4.5) and *Thesaurus thesaurorum* (§ 4.52), write himself treatises on Paracelsianism like his *Monarchia Triadis in Unitate* and produce forgeries like the *Anatomia corporum adhuc viventium*, all of which filled his 1577 volume dedicated to François of Valois.[137]

7. *Time of writing.* Probably written in the mid 1570s.

Word count: ca. 11.000 words

Incipit: "Ehe vnd zuvor wir etwas von der anatomi vnd zerlegung des menschlichen corpers tractiren …"

Manuscripts: no manuscripts known

First printed:

- 1577 (Latin; in *Aurora thesaurusque philosophorum Theophrasti Paracelsi*, ed. Gerhard Dorn (Basel: [Thomas Guarin], 1577); VD16 P 405; Sudhoff, *Bibliographia Paracelsica*, 302–304 n° 177)

Essential bibliography: Sudhoff, *Bibliographia Paracelsica*, 303–304, 462–463.

Further bibliographical references:

Didier Kahn, "Les débuts de Gérard Dorn," in Joachim Telle, ed., *Analecta Paracelsica* (Stuttgart, 1994), 59–126, on 109, 110.

Kahn, "Le retour de Gérard Dorn" (2010–2011), 135–136, 138.

3.2 *Clavis chymiae Philippi Theophrasti* also: *Thesaurus chymiatricus*

1. *Printing History, Manuscripts.* First printed in 1623 as a work of Guido Magnus de Monte, edited by Hermannus Condeesyanus [i.e. Johannes Rhenanus]. Five manuscripts, some of them anonymous, some attributed to Guido Magnus de Monte.
2. *Editions.* Not edited by Huser or Sudhoff.
3. *Relationship between different versions.* Further study required.
4. *Structure, genre/form, perspective, style.*
5. *Relationship to other texts.*
6. *Authenticity, authorship.* Identified by Broszinski as the work printed in 1623 under the title *Tesaurus* [sic] *chymiatricus*, ascribed to "Guido Magnus de

137 Kahn, "Le retour de Gérard Dorn" (2010–2011), 132–137.

Monte."[138] The New Haven manuscript is explicitly attributed to this author in the explicit.

7. *Time of writing.* Possibly written in the 1610s.

WORD COUNT:

INCIPIT: "Recipe einen guten starcken rothen wein welcher besser ist denn der weisse ..."

MANUSCRIPTS:

- KASSEL, *Landesbibliothek*: 4° Ms. chem. 8, f. 87r–125r
- KASSEL, *Landesbibliothek*: 4° Ms. chem. 60[6,5, f. 32r–40v
- KASSEL, *Landesbibliothek*: 4° Ms. chem. 74, f. 7r–50v
- KASSEL, *Landesbibliothek*: 4° Ms. chem. 75, 1–59
- NEW HAVEN, *Beinecke Library*: Beinecke MS 309, f. 134r–172r

FIRST PRINTED:

- 1623 (*Tesaurus Chymiatricus, Das ist/ Lang verborgener Schatz der hochberühmten vnd herrlichsten Kunst Chymiae* [...] *Von dem hocherleuchten Philosopho Guidone Magno de Monte*, ed. Hermannus Condeisyanus [*sic*; i.e. Johannes Rhenanus] (Halle: Peter Schmied for Caspar Klosemann, 1623); VD17 23:294776N; not in Sudhoff, *Bibliographia Paracelsica*)

ESSENTIAL BIBLIOGRAPHY: Sudhoff, *Paracelsus-Handschriften*, 718–720, 779 n° 1/23.

FURTHER BIBLIOGRAPHICAL REFERENCES:
Broszinski, *Manuscripta chemica in quarto* (2011), 23, 96, 161, 213, 251, 263, 299, 335, 338.

3.3 *Consilium für Bernhard Rechlinger*

1. *Printing History, Manuscripts.* One manuscript.
2. *Editions.* Not edited by Huser. Edited by Sudhoff in *Paracelsus-Handschriften* (1899), 47–48, and in Paracelsus, *Sämtliche Werke*, I/10: appendix, 681–681.
3. *Relationship between different versions.* Only one version known.

---

138 Guido Magnus de Monte is often identified with a 14th or 15th century alchemical writer Guido de Montanor, see John Ferguson, *Bibliotheca chemica* (Glasgow, 1906), 2: 100. Contrary to this assumption, Guido Magnus de Monte's chemiatrical works have been published by Benedictus Figulus and Johannes Rhenanus in the first decades of the 17th century (sometimes disguised as translations). A manuscript in HAMBURG (*Staats- und Universitätsbibliothek*, Cod. alchim. 614) is dated Fulda, 28 October 1601, another in KASSEL (*Landesbibliothek*, 4° Ms. chem. 74) is dated Halle, 14 July 1607.

4. *Structure, genre/form, perspective, style.* A medical advice for the Augsburg citizen Bernhard Rechlinger, directly addressing the recipient. This may refer to Bernhard Rehlinger (ca. 1477–1537) or his son Bernhard Rehlinger (ca. 1502–1572).[139]
5. *Relationship to other texts.*
6. *Authenticity, authorship.* While Sudhoff in 1899 still thought that it was "very well possible"[140] that the Consilium originated from Paracelsus, he stated in 1928: "whether it may really be attributed to Paracelsus has become quite doubtful to me."[141] Benzenhöfer accepted Sudhoff's judgment characterizing Sudhoff's reasons, however, as "unclear".
7. *Time of writing.* Probably written in 1536/37 while Paracelsus stayed in Augsburg (if authentic).

Word count: ca. 450 words

Incipit: "Fester Junckher, mein dienst sind euch alzeit bereit [...]" / "Zu extinguiren die coleram sol also geschehen ..."

Manuscripts:
– Wien, *Österreichische Nationalbibliothek*: Cod. 11200, f. 240v–241v

First printed: not printed before Sudhoff (1899, 1928)

Essential bibliography: Sudhoff, *Paracelsus-Handschriften*, 47; Sudhoff, "Vorwort," in Paracelsus, *Sämtliche Werke*, I/10: XLII.

Further bibliographical references:
Udo Benzenhöfer, *Paracelsus* (Reinbek: Rowohl 1997), 102.

3.4 *De archidoxis* also: *Theophrastia*, book one

1. *Printing History, Manuscripts.* Not printed. No manuscripts known.
2. *Editions.* Not edited by Huser or Sudhoff.
3. *Relationship between different versions.* Text not preserved.
4. *Structure, genre/form, perspective, style.*

139 Wolfgang Reinhard, ed., Augsburger Eliten des 16. Jahrhunderts. Prosopographie wirtschaftlicher und politischer Führungsgruppen 1500–1620 (Berlin: Akademie-Verlag 1996), 653–654.
140 Sudhoff, *Paracelsus-Handschriften*, 47.
141 Sudhoff, "Vorwort," in Paracelsus, *Sämtliche Werke*, I/10: XLII.

5. *Relationship to other texts.*
6. *Authenticity, authorship. De archidoxis* is the first of three works composing the book *Theophrastia*, according to Valentius de Retiis (see § 1.24). Since the other two works, *Parasarchum* (§ 1.20) and *Carboantes* (§ 4.22) are apparently spurious (and nowhere to be found), one cannot exclude that this *De archidoxis* was fictitious, too. However, the most plausible hypothesis is that it refers to the authentic work of Paracelsus entitled *Archidoxis*, for two reasons: firstly, Valentius de Retiis describes *De archidoxis* as a text "in which [Paracelsus] explains how to extract the virtues and separate them from what is devoid of any strength:"[142] a fitting description for the authentic *Archidoxis*. Secondly, the authentic *Archidoxis* is expressly said to be taken "*ex Theophrastia Paracelsi Magni*", as the most reliable manuscripts and editions have it.[143]
7. *Time of writing.*

Word count:

Manuscripts: no manuscripts known

First printed: not printed

Historical Manuscript Catalogues: *Libri Theophrasti* (Dresden), fol. 5vb

Essential bibliography: Sudhoff, *Bibliographia Paracelsica*, 142, 171–172, 190–194, 198–199, 391; Sudhoff, *Paracelsus-Handschriften*, 87–89, 93, 98, 102, 710; CP 1: 585, 590–595; CP 2: 165, 984, 1003; CP 3: 116, 121, 129, 170.

Further bibliographical references:

Peuckert, *Pansophie* (1956), 460.
Gilly, *Paracelsus in der BPH* (1993), 37, 38.
Gilly, *Adam Haslmayr* (1994), 97, 103.
Kurt Goldammer, "Zur philosophischen und religiösen Sinngebung von Heilung und Heilmittel bei Paracelsus," in Kurt Goldammer, *Paracelsus in neuen Horizonten. Gesammelte Aufsätze* (Vienna, 1986) (*Salzburger Beiträge zur Paracelsusforschung*, 24), 343–357, on 347.

142 "primum opus de Archidoxis dictum, in quo declarat extractiones et separationes virtutum ex invalido," edited in CP 1: 590.

143 Sudhoff, *Bibliographia Paracelsica*, 172, 190–194, 198; Sudhoff, *Paracelsus-Handschriften*, 87, 88, 98, 102, 710; Gilly, *Paracelsus in der BPH* (1993), 38.

Udo Benzenhöfer, "Zur 'Archidoxis'-Schrift des Paracelsus," *Nova Acta Paracelsica*, N.F. 19 (2005), 105–124, on 109.

Didier Kahn, "Quintessence and the Prolongation of Life in the Works of Paracelsus," *Micrologus*, 26 (2018), 183–225.

3.5 *De pestilitate*
also: *Von Ursprung und Herkommen der Pest und ihrer Kur*

1. *Printing History, Manuscripts.* First printed in 1589 in Huser's edition together with all the authentic treatises on the plague. Both *De pestilitate* and the plague treatise addressed by Paracelsus to the city of Sterzingen were taken, according to Huser, "ex manuscripto D. Johan. Montani et aliorum" ("from the manuscript of Johannes Montanus and other people"). Thus, contrarily to Sudhoff's statement,[144] the text of *De pestilitate* does not necessarily stem from a Montanus manuscript. No known manuscripts.
2. *Editions.* Edited by Huser, 3 (1589): 24–107. Edited by Sudhoff in Paracelsus, *Sämtliche Werke*, 1/14: 597–660.
3. *Relationship between different versions.* Only one known version of the original text. – *De pestilitate* was summarized by Bartholomäus Scultetus in December 1578 in a so-called *Tabula physica, astronomica et medica de pestilitate*, a broadsheet published in 1579 and again in 1586.[145] – Two similar tabular summaries of different texts, probably also prepared by Bartholomäus Scultetus, are found in a Lübeck manuscript (Ms. math. 4° 9).
4. *Structure, genre/form, perspective, style. De pestilitate* consists of four parts ("tractatus"). The sections in part one are not numbered in Huser's edition. Scultetus however, in his *Tabula*, refers to numbered chapters, the last one being "cap. 13." His quotation corresponds with a sentence in the last section of the first part. Both the first and the second part are written in the first person and directed to the reader. The third and fourth part are rather short and consist of one single chapter (part 3) and three chapters (part 4). The four parts are quite different both in style and content.
5. *Relationship to other texts.* According to Gunnoe, "the plague theory of *De pestilitate* is generally consistent with the genuine works of Paracelsus," especially the Nördlingen treatise (*Zwey Bücher von der Pestilentz*) and *De peste libri tres*, despite obvious differences.[146] The author also knew the doctrine of the "fünf Entien" presented in the *Volumen medicinæ Paramirum*, a work published for

144 Sudhoff, "Vorwort," in Paracelsus, *Sämtliche Werke*, 1/14: XXXIII.

145 Copies of both broadsheets are preserved in Erlangen, Universitätsbibliothek.

146 See Gunnoe, "Paracelsus, the Plague, and *De Pestilitate*" (2019), 525.

the first time in 1575. Other authentic works used in *De pestilitate* include *De matrice* and *De meteoris*. Spurious texts were equally known to the author, like the *Liber Principiorum* (§ 5.18), *De natura rerum* (§ 2.2), book 1 – if it can be confirmed that he borrowed his notion of the basilisk from this latter source – and the *Philosophia ad Athenienses* (§ 2.6).[147]

6. *Authenticity, authorship.* Huser did not have the slightest doubt about the authenticity of the text (see Huser, 3: 108). However, mentions of the *anima mundi*, references taken from the spurious *Liber Principiorum* (§ 5.18), or far-fetched mentions of alleged manuscripts in various libraries (including so-called autographs of Galen and Avicenna) are only some of the evidence pointing to an entire forgery.
7. *Time of writing.* The first part was written allegedly "after the year 1538" according to Scultetus. However, the whole text was written or composed rather in the 1570s.[148]

Word count: ca. 28.000 words

Incipit: "Zu philosophiren waher pestis komme …"

Manuscripts: no manuscripts known

First printed: not printed before Huser (1589)
- 1579 (broadsheet; tabular summary: *Tabula Physica, Astronomica & Medica De pestilitate: In eine kurtze Summ vnd Inhalt/ ausz des Aur. Ph. Th. Par. Buche/ welchs er nach dem Jahr 1538. von dieser Kranckheit gestellet/ vnd jetzund erst nw an tag komen/ gezogen. An. 1578*, ed. Bartholomäus Scultetus (Görlitz: Ambrosius Fritsch, 1579); not in VD16; not in Sudhoff, *Bibliographia Paracelsica*)
- 1589 (in Huser, 3)

Essential bibliography: Sudhoff, *Bibliographia Paracelsica*, 381; Sudhoff, "Vorwort," in Paracelsus, *Sämtliche Werke*, I/14: XXXIII; CP 2: 726–727; CP 3: 814–815.

---

147 Kahn, "*De Pestilitate* and Paracelsian Cosmology" (2020), 82–84; Newman, "Bad Chemistry" (2020).

148 *De pestilitate* is partly inspired by the "five entia" theory as outlined in Paracelsus's *Volumen medicinae Paramirum* (first published in 1575), which does not, however, necessarily give a *terminus post quem*.

Further bibliographical references:

Ernst von Frisch, "Hans Baumann, der erste Buchdrucker, in Salzburg 1550–1557", in *Gutenberg-Jahrbuch* 1927, 68–71, on 70.

Edwin Scheidegger, "Paracelsus und Rademacher," *Nova Acta Paracelsica*, 6 (1952), 34–46, on 40.

Peuckert, *Pansophie* (1956), 457.

Will-Erich Peuckert, "Paracelsische Zauberei," *Nova Acta Paracelsica*, 8 (1957), 71–94, on 76.

Johannes Steudel, "Paracelsisches Gut bei Goethe," *Nova Acta Paracelsica*, 8 (1957), 95–101, on 100.

Felix F. Strauss, "Herzog Ernst von Bayern und Hofbuchdrucker Hans Baumann", in *Salzburger Museum Carolino Augusteum, Jahresschrift* 5 (1959), 193–203, on 201.

Peuckert, *Gabalia* (1967), 131, 175, 350, 362, 463–465, 479, 486.

Pagel, "The Paracelsian Elias Artista," 8.

Goldammer, *Der göttliche Magier* (1991), 106.

Jürgen Strein and Joachim Telle, "Deutsche Pseudoparacelsica über die Pest: Ein 'Begriff' zur Pestdiagnose (1553) und die 'Tabula de pestilitate' von Bartholomäus Scultetus (1578)," in Dominik Groß and Monika Reininger, eds., *Medizin in Geschichte, Philologie und Ethnologie: Festschrift für Gundolf Keil* (Würzburg, 2003), 349–370.

Maximiliam Bergengruen, *Nachfolge Christi – Nachahmung der Natur. Himmlische und natürliche Magie bei Paracelsus, im Paracelsismus und in der Barockliteratur* (*Scheffler, Zesen, Grimmelshausen*) (Hamburg, 2007), 153, 162, 168.

Joachim Telle, "Theophrastus von Hohenheim – Irrlehrer oder Leitgestalt einer Alternativmedizin des 21. Jahrhunderts?," *Nova Acta Paracelsica*, 24/25 (2010–2011), 17–62, on 41.

Charles D. Gunnoe, Jr., "Paracelsus, the Plague, and *De Pestilitate*," *Early Science and Medicine*, 24 (2019), 504–526; also published in this volume, 92–114.

Didier Kahn, "*De Pestilitate* and Paracelsian Cosmology," *Daphnis*, 48 (2020), 65–86.

William R. Newman, "Bad Chemistry: Basilisks and Women in Paracelsus and pseudo-Paracelsus," *Ambix*, 67 (2020), 30–46.

3.6 *De quinta essentia* also: *Libellus de quinta essentia*

1. *Printing History, Manuscripts.* First published in Latin in 1582 along with the pseudo-Paracelsian *Manualia* (§ 8.9) by the Basel printer Pietro Perna in a volume called *Aureoli Theophrasti Paracelsi Archidoxorum seu de Secretis Naturae mysteriis libri decem.* First published in German by Huser in 1605. There are two German manuscripts in Kassel and Leiden.

2. *Editions.* Edited by Huser in *Chirurgische Bücher vnd Schriften* (1605), Appendix: 92–98. Not edited by Sudhoff.
3. *Relationship between different versions.* Contrary to Sudhoff's assumption,[149] the German text published by Huser is not a translation of the Latin text printed in 1582. On the contrary, the Latin text was a translation from German by a French-speaking person (who may or may not have been Gerhard Dorn).[150] The manuscripts in Leiden and Kassel preserve the same original German text printed by Huser, although both of them have fewer sections and present them in a different order.
4. *Structure, genre/form, perspective, style. De quinta essentia* consists of a theory-oriented introduction written in the first person, and a practical part discussing the preparation of oils (or quintessences) and salts of different metals, minerals and plants.
5. *Relationship to other texts.* Further study required.
6. *Authenticity, authorship.* The authenticity of *De quinta essentia* was already denied by Huser.[151] Sudhoff considers the text "obviously spurious."[152] Paracelsus had frequently referred to a text by himself entitled *De quinta essentia* in his *Siebente Buch in der Arznei von den Krankheiten, die der Vernunft berauben* and *Das neunte Buch in der Arznei. Von Ursachen und Kuren der Kontrakturen und Lähme.*[153] This *De quinta essentia* quoted by Paracelsus himself was most probably the corresponding chapter in the authentic *Archidoxis*.[154]
7. *Time of writing.* Possibly written in the 1570s to replace the *De quinta essentia* quoted by Paracelsus and supposed to be missing.

Word count: ca. 5.000 words [German text, Huser]

149 Sudhoff, *Bibliographia Paracelsica*, 333: "Sie erscheinen hier lateinisch, Huser hat sie ins Deutsche hiernach übersetzen lassen."

150 *Aureoli Theophrasti Paracelsi Archidoxorum* (Basel 1582), 190: "haec omnes lepras sanat, omnes fistulas, Schorbock (puto esse *les veroles*) omnes denique eius generis incurabiles morbos." 'Schorbock' is set in German type.

151 Huser to the Reader, in Huser, *Chir.* 1605, Appendix, sig. Qqq 1v: "In diesem Appendice/ günstiger Leser/ sind zusamen geordnet etliche Bücher vnd Tractätlein/ welche bißher von andern vnter dem nahmen Theophrasti Paracelsi, in truck gegeben/ deren er doch kein Author ist/ sondern etwa von andern hochverständigen/ gelerthen Leuthen gemacht vnnd componirt worden. [...] Also auch die zwey Büchlein/ das Eine de Quinta essentia/ das ander/ de Mercuriis Metallorum, sind nicht zuverwerffen: doch seind sie für Theophrasti werck vnd arbeyten nicht zuhalten."

152 Sudhoff, *Paracelsus-Handschriften*, 156: "Im Einzelnen finden sich zahlreiche Abweichungen zwischen Huser und der [Leidener] Handschrift, auf welche ich bei diesem sicher untergeschobenen Buche nicht weiter eingehen mag".

153 See Benzenhöfer, *Studien zum Frühwerk des Paracelsus* (2005), 96, 106–109, 113.

154 *Ibid.*, 133, 207.

INCIPIT: "Scripserunt multi de quinto esse rerum earum ..." [ed. 1582] / "Ihrer viel haben geschrieben vom funfften wesen der dingen ..." [MS Kassel] / "Viel haben geschrieben von dem fünfften wesen der dingen ..." [Huser]

MANUSCRIPTS:
- LEIDEN, *Universiteitsbibliotheek*: Voss. Chym. Q.21, f. 193v–206v (171v–184v) [German]
- KASSEL, *Landesbibliothek*: 4° Ms. chem. 19, f. 82r–91r [German]

FIRST PRINTED:
- 1582 (Latin translation; in *Aureoli Theophrasti Paracelsi Archidoxorum seu de Secretis Naturae mysteriis libri decem* (Basel: Pietro Perna, 1582), VD16 P 399; Sudhoff, *Bibliographia Paracelsica*, 331–334 n° 191)
- 1605 (German; in Huser, *Chir.* 1605; VD17 3:313744Z; Sudhoff, *Bibliographia Paracelsica*, 463 n° 267)

HISTORICAL MANUSCRIPT CATALOGUES: Toxites, *Zugesagte Bücher* (Dresden), n° 12; *Libri Theophrasti* (Dresden), f. 6rb; Widemann, *Verzeichnisse* (Kassel), n° II, 83 / V, 139

ESSENTIAL BIBLIOGRAPHY: Sudhoff, *Bibliographia Paracelsica*, 333, 463; Sudhoff, *Paracelsus-Handschriften*, 777; CP 1: 282; CP 2: 247.

FURTHER BIBLIOGRAPHICAL REFERENCES:
Boeren, *Codices Vossiani Chymici* (1975), 152.
Udo Benzenhöfer, *Studien zum Frühwerk des Paracelsus im Bereich Medizin und Naturkunde* (Münster, 2005), 96, 106–109, 113, 133, 207.
Broszinski, *Manuscripta chemica in quarto* (2011), 60.

3.7 *Dritter Teil der Großen Wundartzney*

1. *Printing History, Manuscripts.* First printed in 1579, edited by Georg Forberger, in a volume entitled *Kleine Wundartzney Theophrasti von Hohenheim drey Bücher begreiffendt.* No manuscripts known.
2. *Editions.* Edited by Huser in *Chirurgische Bücher vnnd Schriften* (1591), 335–399. Edited by Sudhoff in Paracelsus, *Sämtliche Werke*, I/10: 423–479.
3. *Relationship between different versions.* Only one known version.
4. *Structure, genre/form, perspective, style.* The "third part of the Great Surgery" consists of (1) a dedication to Ferdinand (1503–1564), Archduke of Austria, King of Bohemia, later Holy Roman Emperor, dated 4th June 1537 in Mährisch Kromau (Moravský Krumlov); (2) a long preface; and (3) the text itself, composed of two sections: the first section, divided into 9 chapters, discusses the "French disease";

the second section ("Der Ander Tractat" in Forberger's edition, but "Der Dritte Tractat" in Huser's edition) is divided into 4 chapters and deals with the "French disease" in its relation to man and his conditions. Both sections are written in the first person and addressed to the reader.

5. *Relationship to other texts.* The first and second part of the *Große Wundarznei* were brought into print by Paracelsus himself in 1536. The first part was published by Hans Varnier in Ulm. Annoyed by many misprintings, Paracelsus turned to another printer, Heinrich Steiner in Augsburg, and published a revised version of the first part as well as the second part of the *Große Wundarznei* in the same year with the new printer.[155] A second printing, not supervised by Paracelsus,[156] appeared in 1537. The third, fourth and fifth part of the *Große Wundarznei* remained unpublished, and possibly in part unwritten. Georg Forberger published a "third part" of the *Große Wundarznei* in 1579, having been notified by Johannes Montanus that the hitherto unpublished three last parts of the "Große Wundartzney" survived in manuscript at Mährisch Kromau.[157] Huser obviously used the same manuscript as Forberger, but was able to read many words unreadable to Forberger.[158]
6. *Authenticity, authorship.* The authenticity of the *Dritte Teil der großen Wundartzney* was not questioned by Forberger or Huser, however seriously questioned by Sudhoff: the dedication to Ferdinand seems to him clearly spurious,[159] the text itself highly suspicious or even spurious.[160] Achelis regarded the whole work as a forgery.[161]
7. *Time of writing.* Allegedly written in 1537 (date of the dedication). Probably written or compiled in the 1570s to replace the missing books of the *Große Wundarznei.*

Word count: ca. 22.000 words

Incipit: "Es ist ein vnlustigs schreiben vnd gar ein vnflätigs ..."

---

155 Sudhoff, "Vorwort," in Paracelsus, *Sämtliche Werke*, I/10: VI–XV.
156 *Ibid.*, XXIV.
157 *Ibid.*, XXVIII; CP 3: 372, 378–379.
158 Sudhoff, *Bibliographia Paracelsica*, 317.
159 Sudhoff, "Vorwort," in Paracelsus, *Sämtliche Werke*, I/10: XXXII: "schwere Bedenken", "[untergeschoben], wie ich dringend vermute."
160 *Ibid.*, XXXIV: "ich halte dies dritte Buch [...] für sehr verdächtig, ja geradezu für untergeschoben."
161 Achelis, "Zur Frage der Echtheit des dritten Buches der *Großen Wundarznei*" (1939), 34, 40, 41: "Fälschung."

Manuscripts: no manuscripts known

First printed:
- 1579 (in: *Kleine Wundartzney Theophrasti von Hohenheim drey Bücher begreiffendt*, ed. Georg Forberger (Basel: Pietro Perna, 1579); VD16 P 480; Sudhoff, *Bibliographia Paracelsica*, 315–318 n° 182)

Essential bibliography: Sudhoff, *Bibliographia Paracelsica*, 317/318, 357, 413; Sudhoff, *Paracelsus-Handschriften*, 777; Sudhoff, "Vorwort," in Paracelsus, *Sämtliche Werke*, I/10: XXIV–XXXIV; CP 3: 370–382 n° 112.

Further bibliographical references:
Johann Daniel Achelis, "Zur Frage der Echtheit des dritten Buches der *Großen Wundarznei*," in Achelis, *Über die Syphilisschriften Theophrasts von Hohenheim* (Heidelberg, 1939), 34–42.

3.8 *Elixir und Tinctura* also: *Labor vegetabilis*

1. *Printing History, Manuscripts.* First printed in 1598. One manuscript.
2. *Editions.* Not edited by Huser or Sudhoff.
3. *Relationship between different versions.* The text in the printed version and in the manuscript is almost word by word the same, but the Kassel manuscript has extra paragraphs. Both texts have an illustration showing glass vessels. The manuscript does not explain how to use the remedies. The printed version has: "Curiert lepram, podagram, hydropisin, schlag vnd allerley zufallende kranckheyten" in the title (which would mean that it cures all severe diseases), but has no details about it in the text either.
4. *Structure, genre/form, perspective, style.* Detailed instructions for the preparation of several remedies.
5. *Relationship to other texts.*
6. *Authenticity, authorship.* According to Sudhoff, with no sign of authenticity.[162] The Kassel manuscript is not attributed to Paracelsus.
7. *Time of writing.* Further study required.

Word count: ca. 2.000 words (1598 edition); ca. 3.000 words (Kassel manuscript)

Incipit: "Nimb ein guten wein, den thu in ein cucurbit …" [ed. 1598] / "Du solt nehmen gutten weinn den thue in ein cucurbith …" [MS Kassel]

162 Sudhoff, *Bibliographia Paracelsica*, 426: "ohne Zeichen der Echtheit."

MANUSCRIPTS:
- KASSEL, *Landesbibliothek*: 4° Ms. chem. 32, f. 64v–72r [not attributed to Paracelsus]

FIRST PRINTED:
- 1598 (in: *Aurei Velleris Oder Der Guldin Schatz vnd Kunstkammer Tractatus II* (Rorschach, 1598); VD16 T 1961; Sudhoff, *Bibliographia Paracelsica*, 424–427 n° 244)

ESSENTIAL BIBLIOGRAPHY: Sudhoff, *Bibliographia Paracelsica*, 426, 695 n° 12.

FURTHER BIBLIOGRAPHICAL REFERENCES:
Broszinski, *Manuscripta chemica in quarto* (2011), 104.

3.9 *Für Pestilenz*

1. *Printing History, Manuscripts. Für Pestilenz* is so far the earliest known pseudo-Paracelsian medical writing ever printed.[163] It was printed in 1554 by Hans Baumann in Salzburg, edited by Egidius Karl of Salzburg. No later editions. No known manuscripts.
2. *Editions.* Not edited by Huser or Sudhoff.
3. *Relationship between different versions.* Only one known version.
4. *Structure, genre/form, perspective, style.* The text is divided into six parts, consisting entirely of recipes. Two of the recipes are explicitly ascribed to Paracelsus.[164]
5. *Relationship to other texts.*
6. *Authenticity, authorship.* According to the title page of the only printed edition of this text, this collection of recipes against pestilentia was "drawn from the book of Doctor Philippus Theophrastus Paracelsus" ("auß deß weitberüembten vnd hocherfarnen Doctoris Philippj Theophrastj Paracelsj Bůch gezogen"). On the backside of the title page is printed the earliest known transcript of Paracelsus's Salzburg epitaph together with a reproduction of his coat-of-arms. The rest

163 Sepp Domandl, "Paracelsus und Salzburg" (1982), 175: "wie angesehen er [Paracelsus] aber war, ergibt sich schon daraus, daß bereits 13 Jahre nach seinem Tode, 1554, ein Salzburger Arzt sein Pestbüchlein aus Reklamegründen unserem Paracelsus in die Schuhe schiebt."

164 Fol. XXIX *recto*: Another pill used and tried by Doctor Theophrastus and others during a plague in Bern in Switzerland ("Aber annder Pillel gleicherweiß wie die ersten zůgebrauchen/ seind vast köstlich vnd güt/ gebraucht vnd bewärt durch Doctor Theophrasten/ vnd annder in ainem seer grossen Sterben zů Bern in schweitz"); fol. XXIX *verso*: This pill was often used and tried by the late Doctor Theophrastus, like during a plague in Antwerp in Brabant ("Dise Pillel hat Doctor Theophrastus Pie Memorie (wie er dann von jm selb schreibt) vil vnnd offt bewärt/ vnd im brauch gehabt/ vnd vil gůts auch grossen nutz zů Antorff im Brabant/ in ainem grossen lauff damit geschafft").

of the text has, according to Sudhoff, "no Paracelsian spirit" whatsoever;[165] to Sudhoff the recipes do not "taste of Paracelsus" at all.[166]

7. *Time of writing.* The editor, Ägidius Karl, prior of the Salzburg Benedictine Abbey of St. Peter, had no known medical training or further medical interests at all. According to Telle it is possible that the present collection of recipes against pestilentia has its origin in the Salzburg library of Paracelsus himself, not as a writer but as a book owner.[167] Therefore probably written before 1541.

Word count: ca. 18.000 words

Incipit: "Alle die, so sich vor der pestilentz oder bösen lüfften wöllen bewaren …"

Manuscripts: no manuscripts known.

First printed:
- **1554** (Für Pestilentz. Ain seer nützlicher vnnd bewerter Tractat […] auß deß weitberüembten vnd hocherfarnen Doctoris Philippj Theophrastj Paracelsj Bůch gezogen. Welches Er/ von diser Khranckhait beschriben, ed. Egidius Karl (Salzburg: Hans Baumann, 1554); VD16 P 643; Sudhoff, *Bibliographia Paracelsica*, 50–53 n° 33)

Essential bibliography: Sudhoff, *Bibliographia Paracelsica*, 50–53.

Further bibliographical references:

Sepp Domandl, "Paracelsus und Salzburg," *Nova Acta Paracelsica*, 10 (1982), 173–180, on 175.

Jürgen Strein and Joachim Telle, "Deutsche Pseudoparacelsica über die Pest: Ein 'Begriff' zur Pestdiagnose (1553) und die 'Tabula de pestilitate' von Bartholomäus Scultetus (1578)," in Dominik Groß and Monika Reininger, eds., *Medizin in Geschichte, Philologie und Ethnologie: Festschrift für Gundolf Keil* (Würzburg, 2003), 349–370.

165 Sudhoff, *Bibliographia Paracelsica*, 52: "athmet nirgends Paracelsischen Geist."

166 Sudhoff, *Bibliographia Paracelsica*, 52: die "Recepte selbst schmecken durchaus nicht nach Paracelsus."

167 Strein and Telle, "Deutsche Pseudoparacelsica über die Pest" (2003), 351.

3.10 *Holzbüchlein*

1. *Printing History, Manuscripts.* First printed in 1564, edited by Michael Toxites as a separate publication. The Heidelberg manuscript is probably a copy from the 1564 printed edition with very few variant readings.[168]
2. *Editions.* Edited by Huser, *Chir.* 1605, 327–329. Not edited by Sudhoff.[169]
3. *Relationship between different versions.* In 1565, one year after his first edition, Toxites published a second edition which left the first three sections mostly untouched but had heavy changes in sections four and five.
4. *Structure, genre/form, perspective, style.* The *Holtzbüchlein* is a compilation from authentic texts of Paracelsus on the medical use of Guaiacum with a short introduction and some other small portions not taken from authentic texts. Sources are *Drei Bücher/ von den Frantzosen* and *De xylohebeno* [otherwise known in a Latin translation by Johannes Oporinus].
5. *Relationship to other texts.*
6. *Authenticity, authorship.* Although Huser included the text in his edition of the *Chirurgische Bücher vnd Schrifften*, he included a warning in the heading: "I do not consider this text to have been outlined in this manner by the author [i.e. Paracelsus], but rather to have been compiled by some other from his [i.e. Paracelsus's] writings."[170]
7. *Time of writing.* The *Holzbüchlein* was compiled by Hieronymus Wolf, probably in the early 1560s, though his name was suppressed by the editor, Michael Toxites.[171]

Word count: ca. 2.000 words

Incipit: "Es ist nit ohn das holtz welches ich nenne Xilohebenum ..."

Manuscripts:
- Heidelberg, *Universitätsbibliothek*: Cod. Pal. germ. 220, f. 3v–8r

---

168 On this manuscript see Wilhelm Ganzenmüller, "Glastechnisches" (1936).

169 Sudhoff, "Einleitung," in Paracelsus, *Sämtliche Werke*, 1/7: 10: "[es handelt sich] nur um eine Zurechtmachung von Paracelsus-Auszügen;" ibid.: "aus einschlägigen Schriften Hohenheims zusammengelesen und zwar aus den drei Büchern von der französischen Krankheit [...], aus dem Büchlein 'Vom holz Guaiaco' (nur einige Sätze) und dem Baseler Entwurf 'De xylohebeno libellus', dessen Huser nur in lateinischer Bearbeitung des Joh. Oporinus habhaft zu werden vermochte."

170 Huser, *Chir.* 1605, 327.

171 CP 1: 632.

First printed:

- 1564 (*Holtzbüchlein Des theuren/ Hocherfarnen/ von Gott hochgelehrten/ weisen Theophrasti Paracelsi*, ed. Michael Toxites (Strasburg: Christian Müller, 1564); VD16 P 471; Sudhoff, *Bibliographia Paracelsica*, 90–92 n° 61)
- 1565 (in: *Drey Bücher des Theuren/ Hocherfarnen/ von Gott Hochgelehrten/ weisen Theophrasti Paracelsi*, ed. Michael Toxites (Strasburg: Christian Müller, 1565); VD16 P 472; Sudhoff, *Bibliographia Paracelsica*, 106–109 n° 68)

Essential bibliography: Sudhoff, *Bibliographia Paracelsica*, 458; Sudhoff, "Vorwort," in Paracelsus, *Sämtliche Werke*, I/3: XLIV–XLVI; Sudhoff, "Einleitung," in Paracelsus, *Sämtliche Werke*, I/7: 10; CP 1: 632, 637; CP 2: 68, 81, 295.

Further bibliographical references:

Wilhelm Ganzenmüller, "Glastechnisches aus einem 'Kunstbuch' des 16. Jahrhunderts," *Glastechnische Berichte*, 14 (1936), 321–326; reprinted in Ganzenmüller, *Beiträge zur Geschichte der Technologie und der Alchemie* (Weinheim, 1956), 61–71.

Wolfram Brechtold, *Dr. Heinrich Wolff (1520–1581)*, MD thesis (Würzburg, 1959), 153–157.

Carlos Gilly, *Die Manuskripte in der Bibliothek des Johannes Oporinus* (Basel, 2001), 33 n° 5.

Labor vegetabilis: see *Elixir und Tinctura* (§ 3.8)

Laudanum metallicum für Königin Maria von Ungarn: see *Wahrhaftes Laudanum metallicum* (§ 3.14)

Theophrastia: see *De archidoxis* (§ 3.4)

Thesaurus chymiatricus: see *Clavis chymiae Philippi Theophrasti* (§ 3.2)

3.11 *Kurzer Begriff, ob ein Mensch an Pest erkrankt ist*

1. *Printing History, Manuscripts.* First printed in 1553 as a broadsheet by the printer Hans Baumann in Salzburg. Next printed as part of *Für Pestilenz* (§ 3.9) by the same Hans Baumann.
2. *Editions.* Not edited by Huser. Edited by Sudhoff in *Sämtliche Werke*, I/9: appendix, 699.
3. *Relationship between different versions.* The two printed texts of 1553 and 1554 are essentially the same with minor variations in the wording.
4. *Structure, genre/form, perspective, style.* A list of eleven signs that may indicate infection with the plague.

5. *Relationship to other texts.* The text is not found in *Regiment/ Gestelt allain für die/ so vnuermeydenlich in Pestilentzischen lüfften verharren vnd beleiben müssen*, another plague pamphlet issued by Hans Baumann in 1553.
6. *Authenticity, authorship.* Similar lists are found in many contemporary plague treatises.[172] The 1553 broadsheet is attributed to Paracelsus presumably as an advertising measure. Not authentic according to Sudhoff: "Someone with some medical knowledge had this diagnostic document printed […], which only remotely resembles the Sterzing Plague booklet and certainly does not originate from Paracelsus."[173]
7. *Time of writing.* Probably written or compiled in 1553 when the plague raged again, after a hundred years, in Salzburg.[174] Further information may or may not be gathered from a recently discovered and unpublished *Chronik des Erzstifts Salzburg*, written in 1551/1552 by Helias Brottbeyhel and continued for the years 1553–1561 by Hans Baumann, the printer of the broadsheet.[175]

Word count: ca. 380 words (1553 edition), ca. 500 words (1554 edition)

Incipit: Dise khranckhait der Pestílentz [1553 edition] / Es schreiben die altenn hocherfarnen Doctores vnd Maiser der Ertzney [1554 edition]

Manuscripts: no manuscripts known

---

172 For example: Hans Saltzmann, *Ein nutzliche ordnung vnd regiment wider die Pestilentz* (Vienna: Johannes Singriener, 1521), sig. e3r–e4v: "Das .vj. capitel zaigt an: die zaychen do mit man erkhennen mag/ ob der mensch mit der pestilentz beschwert sey oder nit/ vnd die zaichen des tods/ vnnd hoffnung des lebens"; Guglielmo Gratarolo, *Pestis descriptio, causae, signa omnigena, & praeseruatio* (Basel: Ludwig Lucius, 1554), 3–6: "Signa futurae pestis perquirere, ut ad praecaeuendum utilißimum, ita pro uirili conabor: qualem autem hic adhibuerim diligentiam, iudicet doctus & sincerus lector."

173 Sudhoff, "Vorwort," in Paracelsus, *Sämtliche Werke*, 1/9: 698.

174 Judas Thaddäus Zauner, *Chronik von Salzburg*, vol. 5 (Salzburg, 1803), 297–298; Maria Vinzenz Süss, *Die Bürgermeister in Salzburg von 1433 bis 1840* (Salzburg, 1840), 59; Franz Valentin Zillner, *Geschichte der Stadt Salzburg*, vol. 2 (Salzburg, 1890), 353; Karl Aberle, "Grabdenkmal, Schädel und Abbildungen des Theophrastus Paracelsus," *Mittheilungen der Gesellschaft für Salzburger Landeskunde*, 31 (1891), 1–224, on 171–172, quoting local manuscripts sources.

175 Karl-Heinz Ludwig, "Eine Salzburger Chronik in Bremen," Mitteilungen der Gesellschaft für Salzburger Landeskunde, 136 (1996), 19–26; Armin Hetzer, Die neuzeitlichen Handschriften der Ms.-Aufstellung (Die Handschriften der Staats- und Universitätsbibliothek Bremen, 2) (Wiesbaden: Harrassowitz, 2008), 31.

First printed:
- 1553 (Ain kurtzer begrif zůerkhennen/ Ob ain Mensch in disen geschwinden leuffen den gebrechen der Pestilentz hab oder nit. D. Philippum Theophrastum beschriben (Salzburg: Hans Baumann, 1553); not in Sudhoff, *Bibliographia Paracelsica*)
- 1554 ("Wie vnnd ihn was gestalt zůerkhennen sey/ ob der Mensch (so er sich besorgt) mit diser kranckhait der Pestilentz beladen sey oder nit", in *Für Pestilentz. Ain seer nützlicher vnnd bewerter Tractat* [...] *auß deß weitberüembten vnd hocherfarnen Doctoris Philippj Theophrastj Paracelsj Bůch gezogen. Welches Er/ von diser Khranckhait beschriben*, ed. Egidius Karl (Salzburg: Hans Baumann, 1554); VD16 P 643; Sudhoff, *Bibliographia Paracelsica*, 50–53 n° 33)

Essential bibliography: Sudhoff, "Vorwort," in Paracelsus, *Sämtliche Werke*, 1/9: 698–699.

Further bibliographical references:

Felix F. Strauss, "Herzog Ernst von Bayern und Hofbuchdrucker Hans Baumann", in *Salzburger Museum Carolino Augusteum, Jahresschrift* 5 (1959), 193–203, on 201.

Jürgen Strein and Joachim Telle, "Deutsche Pseudoparacelsica über die Pest: Ein 'Begriff' zur Pestdiagnose (1553) und die 'Tabula de pestilitate' von Bartholomäus Scultetus (1578)," in Dominik Groß and Monika Reininger, eds., *Medizin in Geschichte, Philologie und Ethnologie: Festschrift für Gundolf Keil* (Würzburg, 2003), 349–370.

Rudolf Werner Soukup, *Chemie in Österreich von den Anfängen bis zum Ende des 18. Jahrhunderts*, vol. 1: *Bergbau, Alchemie und frühe Chemie* (Vienna, 2007), 243.

3.12 *Von dem Vitriol*

1. *Printing History, Manuscripts.* First printed in 1565 in the second edition of the *Holzbüchlein* (§ 3.10), edited by Michael Toxites.
2. *Editions.* Not edited by Huser or Sudhoff.
3. *Relationship between different versions.* The first edition of the *Holzbüchlein* (§ 3.10), printed in 1564, announced on the title-page a "useful treatise on the vitriol and its virtues." Inside, appended to the *Holzbüchlein*, the reader found the "Proceß/ vnd ardt Olei Vitrioli" (process and usage of the oil of vitriol) as well as excerpts from two unpublished texts by Paracelsus concerning the vitriol. In the second edition this appendix was renamed "Von dem Vitriol/ vnd seinen kranckheyten/ Tractatus Theophrasti Paracelsi." The "Proceß Olei Vitrioli," incomplete in the first edition, was now printed in full, and the excerpts from other works were considerably expanded.
4. *Structure, genre/form, perspective, style.*

5. *Relationship to other texts.* Adam von Bodenstein published the "Process vnd art olei vitrioli" as well as some excerpts on vitriol from other texts in *Deß hocherfahrnesten Medici Aureoli Theophrasti Paracelsi schreyben, von den kranckheyten, so die vernunfft berauben* (Basel: Pietro Perna, 1567; VD16 P 634; Sudhoff, *Bibliographia Paracelsica*, 145–146 n° 91). This may have been an answer to Toxites's edition. Similarities and differences to Toxites's publication have not yet been investigated.
6. *Authenticity, authorship.* A compilation of authentic Paracelsian passages about vitriol.[176]
7. *Time of writing.* Probably compiled in the 1560s.

Word count:

Incipit: "Auß der minera vitrioli würdt der spiritus herausz gezogen ..."

Manuscripts: no manuscripts known

First printed:
- 1564 ("Proceß/ vnd ardt Olei Vitrioli, dardurch geheylet werden vier kranckheyten", in *Holtzbüchlein Des theuren/ Hocherfarnen/ von Gott hochgelehrten/ weisen Theophrasti Paracelsi*, ed. Michael Toxites (Strasburg: Christian Müller, 1564); VD16 P 471; Sudhoff, *Bibliographia Paracelsica*, 91–92 n° 61)
- 1565 ("Von dem Vitriol, vnd seinen kranckheyten, Tractatus Theophrasti Paracelsi", in *Drey Bücher des Theuren/ Hocherfarnen/ von Gott Hochgelehrten/ weisen Theophrasti Paracelsi*, ed. Michael Toxites (Strasburg: Christian Müller, 1565); VD16 P 472; Sudhoff, *Bibliographia Paracelsica*, 106–109 n° 68)

Historical Manuscript Catalogues: *Catalogus* Osek (Prague) 1, n° 17; cf. Widemann, *Verzeichnisse* (Kassel), n° II, 26; cf. Widemann, *Verzeichnisse* (Kassel), n° V, 134.

Essential bibliography: Sudhoff, *Bibliographia Paracelsica*, 91–92, 145, 392; Sudhoff, *Paracelsus-Handschriften*, 773.

---

176 Sudhoff, *Bibliographia Paracelsica*, 145: "Eine Zusammenstellung verschiedener Paracelsischer Anweisungen für Darstellung und Anwendung von Heilmitteln aus dem Vitriol." – The exact sources can be found in Sudhoff, *Bibliographia Paracelsica*, 108.

3.13 *Von der Scotomia oder Hauptsucht*

1. *Printing History, Manuscripts.* A short (one-page) text in a Kassel manuscript on eyes disease that purports to be taken from a work of Paracelsus.
2. *Editions.* Not edited by Huser. Tacitly edited by Sudhoff in Paracelsus, *Sämtliche Werke*, 1/14: 388.[177]
3. *Relationship between different versions.* Only one known version.
4. *Structure, genre/form, perspective, style.*
5. *Relationship to other texts.*
6. *Authenticity, authorship.* Sudhoff does not exactly deny its authenticity but says that he does not know of any such work.[178]
7. *Time of writing.* Further study required.

WORD COUNT: ca. 200 words

INCIPIT: "Theophrastus schreibt das ein sucht des haupts sei …"

MANUSCRIPTS:
- KASSEL, *Landesbibliothek*: 2° Ms. chem. 3, f. 167r

FIRST PRINTED: not printed before Sudhoff (1933)

ESSENTIAL BIBLIOGRAPHY: Sudhoff, *Paracelsus-Handschriften*, 704, 778 n° 1/6.

Von Ursprung und Herkommen der Pest: see *De pestilitate* (§ 3.5)

3.14 *Wahrhaftes Laudanum metallicum* also: *Laudanum metallicum für Königin Maria von Ungarn*

1. *Printing History, Manuscripts.* Printed in a heavily abridged version.[179] Four manuscripts. Another manuscript was for sale in 1793.[180]
2. *Editions.* Not edited by Huser or Sudhoff.

177 Not mentioned in Sudhoff's "Vorwort" or even in the table of contents.

178 Sudhoff, *Bibliographia Paracelsica*, 705: "Dies ist mir als Paracelsisch sonst nicht begegnet."

179 "Das Laudanum Metallicum," in *Medicina experimentalis Digbaeana, Das ist: Außerlesener und bewährter Artzney-Mittel Zweyter Theil* (Frankfurt a.M.: Johann Peter Zubrod, 1676). The *Laudanum metallicum* has been printed probably many more times and even earlier. See also Zedler, *Grosses vollständiges Universal-Lexicon* (1739), 20: col. 961: "Mercurius (roth sublimirter), Mercurius sublimatus ruber. Diesen Mercurium beschreibet Theophrastus, Crollius, und andere mehre, der eine nennet ihn Arcanum Corallinum, der andere Laudanum Metallicum."

180 "Berühmte, als in ihrer Art so seltene alchymistische uralte Manuscripten zu verkaufen," in *Der Anzeiger. Ein Tagblatt zum Behuf der Justiz, der Polizey und aller bürglichen Gewerbe*,

3. *Relationship between different versions.* Further study required.
4. *Structure, genre/form, perspective, style.* The full title in the Kassel manuscript is "Das warhafftige Laudanum Metallicum Theophrasti, Dardurch Die Königin auß Oësterreich Maria, Von Theophrasti Selbsten Curiert worden."
5. *Relationship to other texts.*
6. *Authenticity, authorship.* Allegedly written by Paracelsus for Queen Mary of Hungary (1505–1558). No direct judgement by Sudhoff, but he held all the recipes in this collection to be spurious. According to Szathmáry authentic "without doubt."[181]
7. *Time of writing.* Further study required.

Word count:

Incipit: "Recipe ein pfund wol gerainigten mercurij, thue jhn in ein wolbeschlagenen kolben ..."

Manuscripts:
- Kassel, *Landesbibliothek*: 4° Ms. chem. 58, f. 43v–44r
- Dresden, *Sächsische Landesbibliothek*: Mscr.Dresd.J.454, f. 22v–25r
- Cambridge/MA, *Harvard Library – Houghton Library*: MS Lat 221, circa f. 134r
- Budapest, *Országos Széchényi Könyvtár* [*National Library*]: shelfmark unknown[182]

First printed: Printed only in a heavily abridged version (1676) (see above, section 1)

Essential bibliography: Sudhoff, *Paracelsus-Handschriften*, 725, 778 n° 1/15; CP 2: 480.

Further bibliographical references:

Johann Heinrich Zedler, *Grosses vollständiges Universal-Lexicon aller Wissenschaften und Künste*, vol. 20 (Halle-Leipzig, 1739), col. 961.

1793, vol. 1, n° 50, col. 402–403, on col. 402 n° 5: "Das wahrhafte Laudanum Metallicum, dadurch die Königin Maria aus Oesterreich selbst von Theophrasto curirt worden."

181 Szathmáry, "Die Königin Maria und Paracelsus" (1932), 303: "Das Rezept ist eine späte Abschrift, an deren Glaubwürdigkeit man nicht zweifeln kann."

182 Quoted by Szathmáry, "Die Königin Maria und Paracelsus" (1932). It is *not* quoted in László Szathmáry, *Magyar alkémistak* (Budapest, 1928). – Bobory, *The Sword and the Crucible* (2009), 167 n. 369: "I have not managed to find this document which may have been transferred to the National Széchényi Library or the National Archives of Hungary at the beginning of the 20th century."

László Szathmáry, "Mária királyné és Paracelsus," *A Magyar Gyógyszerésztudományi Társaság Értesítője*, 8 (1932), n° 2, 274–282 (*non vidi*).

László Szathmáry, "Die Königin Maria und Paracelsus," *Berichte der Ungarischen Pharmazeutischen Gesellschaft*, 8 (1932), n° 4, 302–303 [German abstract].

William Jerome Wilson, "Catalogue of Latin and Vernacular Alchemical Manuscripts in the United States and Canada," *Osiris*, 6 (1939), 1–836, on 246.

Friedrich Back, "Zur Wirkungsgeschichte des Paracelsus und seiner Persönlichkeit in Ungarn," in *Paracelsus. Salzburger Vorträge 1997* (Vienna, 1998) (*Salzburger Beiträge zur Paracelsusforschung*, 31), 68–86, on 72.

Dóra Bobory, *The Sword and the Crucible: Count Boldizsár Batthyány and Natural Philosophy in Sixteenth-Century Hungary* (Newcastle upon Tyne, 2009), 55, 167.

Dóra Bobory, "Paracelsus in Pannonia: Alchemy and Medicine in Count Boldizsár Batthyány's Circle," in Miguel López Pérez, Didier Kahn and Mar Rey Bueno, eds., *Chymia: Science and Nature in Medieval and Early Modern Europe* (Newcastle upon Tyne, 2010), 115–137, on 116 and 131 n. 10.

Broszinski, *Manuscripta chemica in quarto* (2011), 192.

3.15 *Wunder Artzney*

1. *Printing History, Manuscripts.* First printed in 1573, not attributed to Paracelsus. Printed again in 1586, now attributed to Paracelsus.
2. *Editions.* Not edited by Huser or Sudhoff.
3. *Relationship between different versions.* Further study required.
4. *Structure, genre/form, perspective, style.*
5. *Relationship to other texts.*
6. *Authenticity, authorship.* The *Wunder Artzney*, when first published in 1573, did not claim to be a work of Paracelsus, but was, according to the title page, "auff die Theophrastisch Medicin gestelt," i.e. a kind of complement to the Paracelsian writings, and it was written "vor etlich hundert jaren," i.e. a few hundred years ago. It is actually nothing else than a German version of the *Liber de consideratione quintae essentiae* by Johannes de Rupescissa (d. after 1365). The 1573 edition was probably printed from the 16th century Basel manuscript cod. D II 24.[183] However, the second edition, printed in 1586, explicitly named Paracelsus as author of the text: *Theophrasti Paracelsi, Wunder Artzney Vnnd Verborgine Geheimnisse Allergeheimnissen* and even boasted about being "now, in this year 1586, edited for the first time from the autograph manuscripts of Theophrastus Paracelsus," which constitutes a double misdirection.
7. *Time of writing.* Text written in the 14th century by Johannes de Rupescissa. First attributed to Paracelsus in 1586.

183 Herkommer, "Johannes de Rupescissa", 1983, col. 726.

Word count:

Incipit: "Die zwey größte laster, so vnder der sonnen [...] / Salomon der weiß spricht am anderen capittel im bůch der weyßheit ..."

Manuscripts:
- Basel, *Universitätsbibliothek*: Cod. D II 24; 46 fols. [not attributed to Paracelsus]

First printed:
- 1573 (*Wunder Artzney/ Vonn allerley leibs gebrüchen/ vnnd zů fallenden kranckheiten* [...] *Alles mit fleiß beschrieben/ vnd auff die Theophrastisch Medicin gestelt: Vor etlich hundert jaren beschrieben/ vnd est diß M.D. Lxxiij. jar/ zů nutz vnd gutem allen liebhabern dieser kunst/ im Truck außgangen* (Basel: Sebastian Henricpetri, 1573); VD16 P 725; Sudhoff, *Bibliographia Paracelsica*, 250–252 n° 149) [not attributed to Paracelsus]
- 1586 (*Theophrasti Paracelsi, WunderArtzney Vnnd Verborgine Geheimnisse Allergeheimnissen/ etc.* [...] *Alles erst jetzt im jar 1586. auß den eygnen Handtgeschrifften des Hochermelten Herrn Theophrasti Paracelsi genommen/ vnd zuentledigung grosser Arbeit aller LIebhabern der Theophrastischen Medicin an Tag gegeben* (Basel: Sebastian Henricpetri, 1586); VD16 P 726; Sudhoff, *Bibliographia Paracelsica*, 360–362 n° 209) [attributed to Paracelsus]

Essential bibliography: Sudhoff, *Bibliographia Paracelsica*, 250–252, 360–362, 695 n° 8; CP 2: 247.

Further bibliographical references:

Hubert Herkommer, "Johannes de Rupescissa," in Kurt Ruh, ed., *Die deutsche Literatur des Mittelalters – Verfasserlexikon*, vol. 4 (Berlin, 1983), col. 724–729.

Udo Benzenhöfer, *Johannes' de Rupescissa 'Liber de consideratione quintae essentiae omnium rerum' deutsch: Studien zur Alchemia medica des 15. bis 17. Jahrhunderts* (Stuttgart, 1989), 47–48 n° 35.

4 Alchemy

Alchimia Theophrasti Paracelsi: see *De spiritibus metallorum* (§ 4.29)

Antwort Philippi Theophrasti: see *Brief an Wilhelm von Hohenheim* (§ 4.18)

4.1 *Apocalypsis Hermetis* also: *Apocalypsis spiritus secreti*

1. *Printing History, Manuscripts.* Two versions: one in Latin, the other in German. The Latin version is never attributed to Paracelsus; the German version most often is. First printed in Latin in 1566, not ascribed to Paracelsus, but explicitly "incerto Authore." The Latin text is accompanied by an Italian commentary by Giovan Battista Agnello (who was also the editor of the whole volume) and was published in London.[184] – German version A first printed in 1603; edited from other sources by I.P.S.M.S. in 1604, and Benedictus Figulus in 1608. – German version B first printed in 1606, edited by Franz Kieser. – Four Latin manuscripts (not ascribed to Paracelsus). Many German manuscripts of version A, most of them ascribed to Paracelsus. One German manuscript of version B.
2. *Editions.* Edited by Ps.-Huser, *Opera Bücher vnd Schrifften* (1603), 2: 668–671. Not edited by Sudhoff.
3. *Relationship between different versions.* First written in Latin, then translated into German.[185] The Stuttgart *Catalogus* lists two German manuscripts, but the second only is attributed to Paracelsus and is described as "mit zuhintrest angehenkten figuren" (with illustrations at the end).
4. *Structure, genre/form, perspective, style.*
5. *Relationship to other texts.* Cited in *De tinctura physicorum* (§ 4.30):[186] "Wiewol nuh die erst Experientz der Tinctur diesen Proceß vor der Fixation geben hat/ und unsere Voreltern auch vilmals seind ihres begerens vollmechtig worden: So hetten sie doch vil einen neheren weg zu des rhoten Löwens Schatzkammer gehabt/ wann sie der *Astronomiae* Concordantz mit der Alchimey gelehrnet hetten/ als ich sie in *Apocalypsi Hermetis* angezeigt hab." (Huser, 6: 368). – Paracelsus is named "Hermes Secundus" for the first time in the Ps.-Huser edition

184 On Agnello see James Stuart Campbell, *The Alchemical Patronage of Sir William Cecil, Lord Burghley*, Master's thesis (Wellington 2009), online (https://core.ac.uk/display/41336603, accessed 18 April 2021), 118–119, 121–124; Jennifer Rampling, *The Experimental Fire. Inventing English Alchemy, 1300–1700* (Chicago-London: Univ. of Chicago Press 2020), 174, 227.

185 German version A is a translation from the Latin; the position of version B is unclear.

186 For this reason Peuckert, *Pansophie* (1936), 317, attributed the *Apocalypsis* to the same author as *De tinctura physicorum*, ignoring the Latin version not attributed to Paracelsus. The same is true for Gilly, "Vom ägyptischen Hermes zum Trismegistus Germanus" (2010), 80. – Later Peuckert relativizes this direct attribution and speaks only of the origin of the two texts in the circles from which the early forgeries originated; Peuckert, *Pansophie* (1956), 285–286.

of the *Apocalypsis*;[187] he has the same title in one of the Kassel manuscripts of the *Apocalypsis*.[188]

A text by Johannes de Sole, *Die fünf Wesen des verborgenen Geists Theophrasti* (part of Johannes de Sole, *Arcana arcanorum*), adopts part of the *Apocalypsis Hermetis*.

6. *Authenticity, authorship*. This is a rewriting of the beginning of *De consideratione quintae essentiae* by Johannes de Rupescissa, followed by other alchemical considerations drawing from medieval sources and revolving around a medieval alchemical topic, that of the *spiritus occultus* ("hidden spirit"). Only the German version was attributed to Paracelsus.
7. *Time of writing*. Probably written in the first half of the 16th century (or earlier?). Later attributed to Paracelsus in the 1560s.

Word count: ***Latin:*** ca. 2.300 words. – ***German (version A):*** ca. 3.000 words. – ***German (version B):*** ca. 1.700 words.

Incipit: "Hermes, Plato, Aristoteles vnnd die anndern philosophi die verschiner zeitt geblüet haben ..." [version A] / "Sie haben verstanden, daß der himmel sey vber das wesen der vier elementen ..." [version B]

Manuscripts:

Latin
- Firenze, *Biblioteca Medicea Laurenziana*: Plut.89 sup.36, f. 230r–231v
- London, *Wellcome Library*: MS 1700, 217–227
- Oxford, *Bodleian Library*: MS. Ashmole 1467, f. 202v–208v
- Oxford, *Bodleian Library*: MS. Ashmole 1490, f. 15r–17v

German (version A)
- Dresden, *Sächsische Landesbibliothek*: Mscr.Dresd.N.54, f. 14r–18r
- Kassel, *Landesbibliothek*: 4° Ms. chem. 22, f. 270r–276v
- Kassel, *Landesbibliothek*: 4° Ms. chem. 60[7,1, f. 1r–6r
- Kassel, *Landesbibliothek*: 4° Ms. chem. 87, f. 129r–138v
- Lindau, *Ehemals Reichsstädtische Bibliothek*: Hs. P II 39; 27 pages
- Lindau, *Ehemals Reichsstädtische Bibliothek*: Hs. P IV 1, f. 1r–13v
- Lübeck, *Stadtbibliothek*: Ms. math. 4° 9, f. 37r–40r
- Solothurn, *Zentralbibliothek*: Cod. S I 185, f. 188r–191v
- Wolfenbüttel, *Herzog August Bibliothek*: Cod. Guelf. 74 Blank, f. 753r–759r

187 CP 3: 67; Ebeling, "Alchemical Hermetism" (2015), 80–81.
188 Gilly, "Vom ägyptischen Hermes zum Trismegistus Germanus" (2010), 80, 86–87.

German (version B)

- HAMBURG, *Staats- und Universitätsbibliothek*: Cod. alchim. 661, 341–351 [copied from the printed 1606 edition]
- KASSEL, *Landesbibliothek*: 4° Ms. chem. 22, f. 267r–269v

Johannes de Sole, Arcana arcanorum

- BAMBERG, *Staatsbibliothek*: Msc. Nat. 13
- PHILADELPHIA (PA), *Penn Libraries*: Ms. Codex 128
- PRAHA, *Strahovská knihovna*: Cod. DF IV 36, f. 52r–83r
- WOLFENBÜTTEL, *Herzog August Bibliothek*, Cod. 264.31 Extr., ca. 668–691

FIRST PRINTED:

- **1566** (Latin; in *Espositione di Giouanbatista Agnello Venetiano supra vn libro intitolato Apocalypsis spiritus secreti* (London, 1566))
- **1603** (German version A; in Ps.-Huser, *Opera Bücher vnd Schrifften* (1603); VD17 12:168390P; Sudhoff, *Bibliographia Paracelsica*, 443–446 n° 257)
- **1604** (German version A; in I.P.S.M.S., ed., *Alchimia vera, Das ist: Der wahren vnd von Gott hochbenedeyten/ Natur gemessen Edlen Kunst Alchimia wahre beschreibung* (no place, 1604); VD17 12:132982U; Sudhoff, *Bibliographia Paracelsica*, 453–454 n° 265)
- **1606** (German version B; in Franz Kieser, ed., *Cabala Chymica. Concordantia Chymica. Azot Philosoph. Solificatum. Drey vnnterschiedliche/ Nützliche/ vnd zuvor nie außgegangene Tractätlein* (Mühlhausen: Martin Spieß for Johann Spieß, 1606); VD17 12:650901T; Sudhoff, *Bibliographia Paracelsica*, 473–474 n° 277)
- **1608** (German version A; in Benedictus Figulus, ed., *Pandora Magnalium Naturalium Aurea Benedicta, De Benedicto Lapidis Philosoph. Mysterio* (Strasburg: Lazarus Zetzner, 1608); VD17 23:277709W; Sudhoff, *Bibliographia Paracelsica*, 480–482 n° 282)

HISTORICAL MANUSCRIPT CATALOGUES: *Catalogus* (Stuttgart), II, f. 1r, n° [5] and f. 1v, [12]; Widemann, *Verzeichnisse* (Kassel), n° V, 127; Uffenbach (1720), VIII, col. 655–658 n° 12/XIV [=Hamburg, alchim. 661]; von Linden (1788), part 2, 273 n° 357

ESSENTIAL BIBLIOGRAPHY: Sudhoff, *Bibliographia Paracelsica*, 445–446; Sudhoff, *Paracelsus-Handschriften*, 798 n° IV/18; CP 3: 67, 1109.

FURTHER BIBLIOGRAPHICAL REFERENCES:

Lynn Thorndike, *A History of magic and experimental Science*, vol. 5 (New York, 1951), 624.

Peuckert, *Pansophie* (1956), 285–286, 290, 414, 431, 465, 470, 471.

Mario Rosa, "Giovan Battista Agnello," in *Dizionario Biografico degli Italiani*, 1 (1960).

Gilly, *Paracelsus in der BPH* (Amsterdam, 1993), 49–50.

Gilly, "Das Bekenntnis zur Gnosis" (2000), 397.

Florian Ebeling, "'Geheimnis' und 'Geheimhaltung' in den Hermetica der Frühen Neuzeit," in Anne-Charlott Trepp and Hartmut Lehmann, eds., *Antike Weisheit und kulturelle Praxis. Hermetismus in der frühen Neuzeit* (Göttingen, 2001), 63–80, on 68–69.

James Stuart Campbell, *The Alchemical Patronage of Sir William Cecil, Lord Burghley*, Master's thesis (Wellington 2009), online (https://core.ac.uk/display/41336603, accessed 18 April 2021), 121–124.

Kahn, "Le retour de Gérard Dorn" (2010–2011), 149.

Gilly, "Vom ägyptischen Hermes zum Trismegistus Germanus" (2010), 80, 81, 86–87, 114.

Broszinski, *Manuscripta chemica in quarto* (2011), 75, 264, 365.

Volkhard Wels, *Manifestationen des Geistes. Frömmigkeit, Spiritualismus und Dichtung in der Frühen Neuzeit* (Göttingen, 2014), 137–138, 141, 149, 153, 176.

Florian Ebeling, "Alchemical Hermetism," in Christopher Partridge, ed., *The Occult World* (Abingdon-New York, 2015), 74–91.

Florian Ebeling, *Das Geheimnis des Hermes Trismegistos. Geschichte des Hermetismus von der Antike bis zur Gegenwart*, 3rd. rev. ed. (Munich, 2018), 107.

4.2 *Arcana arcanorum arcanissima*

1. *Printing History, Manuscripts.* First printed in 1780, not attributed to Paracelsus. Two manuscripts.
2. *Editions.* Not edited by Huser or Sudhoff.
3. *Relationship between different versions.* No information about different versions.
4. *Structure, genre/form, perspective, style.*
5. *Relationship to other texts.*
6. *Authenticity, authorship.* Allegedly written by "Theophrastus Junior" in 1546 and later found in the library of the Benedictine Monastery of Erfurt. The author recounts that he found his secrets in a manuscript of Johannes Trithemius called *Das allergrößte Geheimniß der gantzen Natur sive Materia Universalissima*. Accompanied by a portrait of Paracelsus "in the year of his decease."
7. *Time of writing.* Probably written in the 18th century.

WORD COUNT:

INCIPIT: "Es werden in den elementen zwey spermata generiret …"

MANUSCRIPTS:
- GLASGOW, *University Library*: MS Ferguson 219; 25 fols.
- ÜBERLINGEN, *Leopold Sophien Bibliothek*: Ms. 269; 7 fols.

First printed:

- 1780 ("Arcanum Arcanorum Arcanissimum", in *Geheimnisse einiger Philosophen und Adepten, aus der Verlassenschaft eines alten Mannes*, Erster Theil (Leipzig: Christian Gottlob Hilscher, 1780), 72–101

Historical Manuscript Catalogues: Dost (1791–1798): 1792, col. 595, n° 6; *idem* 1797, col. 395, n° 19 & 20; *idem* 1798, col. 307 n° 15

Essential bibliography: CP 2: 607.

Further bibliographical references:

Telle, *Alchemie und Poesie* (2013), 467 n. 25, 483.

Franziska Schaudeck, *Die alchemische Handschriftensammlung der Leopold-Sophien-Bibliothek in Überlingen am Bodensee* (Wiesbaden, 2020), 295–296.

4.3 *Arcana divina* also: *Zweites Silentium Dei/Geheimen Himmelsschlüssels zweiter Teil/Universal-Anleitung/Dei Magia oder Magia Divina*

1. *Printing History, Manuscripts.* Several manuscripts. – The writer of all four manuscripts of version B (see below) was Johann Salomon Haussen (1729–1802) of Glücksbrunn, who probably operated a mail-order supply of alchemical and magical texts, which would explain the extreme length of some of his manuscripts texts, filled with pages of empty words (see also *De septem stellis* (§ 5.9), where Haussen's version was blown up from 1.500 to 30.000 words without addition of any notable content).[189] First edited by Cölestin Fuchs in 1913.
2. *Editions.* Not edited by Huser or Sudhoff.
3. *Relationship between different versions.* There are two known versions: version A, not attributed to Paracelsus; version B, an enlarged version attributed to Paracelsus. The main title of this second version is "Das zweyte Silentium Dei;" section titles include "Arcana divina" and "Universal-Anleitung," while the whole version is styled as "Geheimen Himmelsschlüssels zweiter Teil" (the Second Part of the Secret Celestial Key).
4. *Structure, genre/form, perspective, style.* Description of the construction and usage of a "magical machine" (with a series of three lenses used as energy source) for alchemical purposes. Directly addresses the reader. The enlarged version contains drawings of the "machine" and is written partly in cipher (the key to the cipher being supplied at the end of the text).

189 For other bookselling distributors of magical and alchemical manuscripts see Schaudeck, *Die alchemische Handschriftensammlung* (2020), 117–123.

5. *Relationship to other texts.* Related to an 18th century German Rosicrucian text, *Thesaurus Thesaurorum a Fraternitate Rosae et aureae Crucis*.[190] The relationship to L.v.H., ed., *Magia Divina* (no place, 1745; reprinted in Scheible's *Das Kloster*, vol. 3),[191] remains to be studied.
6. *Authenticity, authorship.*
7. *Time of writing.* Allegedly written in Schaffhausen in 1535 or 1551 or 1555. With a preface allegedly written by Johann Arndt in 1599. Actually rather written in the 18th century.

WORD COUNT:

INCIPIT: "Alles und alles bestehet in Gott und durch das reine Gebet können wir Menschen ..." [Mellon MS 88] / "Der aus göttlicher vorsehung wahrhafte gundbrunnquell und weg der weisheit ..." [Kiesewetteriana 1e]

MANUSCRIPTS:

version A

- DARMSTADT, *Universitäts- und Landesbibliothek*: Hs. 1703
- DRESDEN, *Sächsische Landesbibliothek*: Mscr.Dresd.App.953
- LONDON, *Wellcome Library*: MS 971
- NEW HAVEN, *Beinecke Library*: Mellon MS 88
- TROGEN, *Kantonsbibliothek*: Sammlung Carl Meyer, CM Ms. 20
- WIEN, *Österreichische Nationalbibliothek*: Cod. Ser. n. 4135
- *olim* Nürnberg (Library of Adam Rudolph Solger, 18th century) Ms. Nr. 37;[192] 63 pages
- olim Ossegg/Osek (Stiftsbibliothek)[193]
- *olim* PRAHA, *Ústřední knihovna ČVUT*: Cod. C 1177[194]
- *olim* Antiquariat Karl und Faber, offered for auction in 1937[195]

---

190 Tilton, "The Urim and Thummim" (2016), 38 n. 3.

191 "Magia Divina", in J. Scheible, *Das Kloster. Weltlich und geistlich*, vol. 3 (Stuttgart: Scheible, 1846), n° 10, 524–563. See Peuckert, *Pansophie* (1956), 435, 471, 472, 481.

192 Solger, *Bibliotheca* (1760), 1: 250 n° 37; von Murr, *Memorabilia* (1786), 1: 410. – Solger's library was sold in 1766 to the city of Nuremberg. The manuscript probably still exists in the Stadtbibliothek Nürnberg.

193 Cölestin Fuchs prepared his edition from a "Manuscript des Stiftes Ossegg;" however, it is not listed in Bernard Wohlmann, "Verzeichniss der Handschriften in der Bibliothek des Stiftes Ossegg," in *Xenia Bernardina*, pars secunda: *Handschriftenverzeichnisse* (Vienna, 1891), 2: 115–164.

194 *Katalog Knihovny c.k. vysokých škol technických v Praze* (1898), 524.

195 Karl & Faber, *Auktion XV* (1937), 1, n° 3.

version B

- AMSTERDAM, *Bibliotheca Philosophica Hermetica*: M126
- MÜNCHEN, *Bayerische Staatsbibliothek*: Kiesewetteriana 1e[196]
- MÜNCHEN, *Bayerische Staatsbibliothek*: Kiesewetteriana 25d
- NEW HAVEN, *Beinecke Library*: Mellon MS 136[197]

FIRST PRINTED: Allegedly printed, according to an unreliable source, in Strasburg in 1618 as a work of Johann Arndt.[198] – Not edited before Fuchs (1913).

HISTORICAL MANUSCRIPT CATALOGUES: von Linden (1786), 19 n° 30, 44 n° 299; von Linden (1788), part 2, 13 n° 12 & 13, 273 n° 360

ESSENTIAL BIBLIOGRAPHY: Sudhoff, *Paracelsus-Handschriften*, 24–25, 684–686, 798 n° IV/13, IV/14 and IV/16.

FURTHER BIBLIOGRAPHICAL REFERENCES:

Adam Rudolf Solger, *Bibliotheca sive supellex librorum impressorum* [...] *et codicum manuscriptorum*, pars I (Nürnberg: Johann Andreas Endter, 1760), 250 n° 37.

Christoph Gottlieb von Murr, *Memorabilia Bibliothecarum publicarum Norimbergensium et Universitatis Altdorfinae*, Pars I (Nürnberg: Johann Hösch, 1786), 410.

Katalog Knihovny c.k. vysokých škol technických v Praze (Praha, 1898), 524.

Cölestin Fuchs, *Arcana divina: Manuscript des Stiftes Ossegg* (Komotau: Staats-Gymnasium, 1912) ("Schulprogramm"), 18 pages.

F. Leigh Gardner, *A Catalogue Raisonné of Works on the Occult Sciences*, vol. 1: *Rosicrucian Books*, 2nd. ed. (no place, 1923), 9 n° 60.

Karl & Faber, München, *Auktion XV: Bibliothek Dr. J. Häberlin (Frankfurt a.M.) und andere Beiträge* (November 1937), 1, n° 3.

Peuckert, *Pansophie* (1956), 476, 481.

Paul M. Allen, *A Christian Rosenkreutz Anthology* (Blauvelt (NY), 1968), 605.

Joachim Telle, "Manuscripta alchemica der Sammlung Mellon. Bemerkungen zum Katalog," *Sudhoffs Archiv*, 65 (1981), 79–96, on 83, 95.

Gilly, *Paracelsus in der BPH* (1993), 67–69 n° 60.

Franziska Schaudeck, *Die alchemische Handschriftensammlung der Leopold-Sophien-Bibliothek in Überlingen am Bodensee* (Wiesbaden, 2020).

196 Description in Sudhoff, *Paracelsus-Handschriften*, 684–686 n° 136.

197 Partial transcription and English translation in Ashrowan, *Alchemical Catoptrics* (2016), vol. 2, Appendix 4.

198 Gardner, *A Catalogue Raisonné* (1923), 9 n° 60; also Allen, *A Christian Rosenkreutz Anthology* (1968), 605.

Telle, *Alchemie und Poesie* (2013), 467.

Hereward Tilton, "Bells and Spells: Rosicrucianism and the Invocation of Planetary Spirits in Early Modern Germany", in *Celestial Magic*, special issue of *Culture and Cosmos*, 19 (2015), 167–188, on 181–182 n. 51.

Hereward Tilton, "The Urim and Thummim and the Origins of the Gold- und Rosenkreuz", in Hans Thomas Hakl, ed., *Octagon: Die Suche nach Vollkommenheit im Spiegel einer religionswissenschaftlichen, philosophischen und im besonderen Maße esoterischen Bibliothek*, 2 vols. (Gaggenau, 2016), 2: 35–70.

Richard Ashrowan, *Alchemical Catoptrics: Light, matter and methodologies of transformation in moving image practice*, PhD Thesis (Edinburgh College of Art, Univ. of Edinburgh, 2016).

4.4 *Augmentum Solis* (*Die roten Juden*) also: *Einen großen Schatz in geringer Zeit zu überkommen/ Vermehrung des Theophrasti Sonnen*

1. *Printing History, Manuscripts.* First printed in 1608, edited by Benedictus Figulus. Two manuscripts.
2. *Editions.* Not edited by Huser or Sudhoff.
3. *Relationship between different versions.* Further study required.
4. *Structure, genre/form, perspective, style.* A strange alchemical process using the term "die roten jüden" (the Red Jews) as a *Deckname* for a certain substance.[199]
5. *Relationship to other texts.*
6. *Authenticity, authorship.* The Munich manuscript is dated 1553 on the binding and is not attributed to Paracelsus.
7. *Time of writing.* Possibly written in the 1540s or earlier.

Word count: ca. 500 words

Incipit: "Zeuch hinweg in das gebürg, genannt Montes Caspii …"

Manuscripts:
- München, *Bayerische Staatsbibliothek*: cgm 3741
- Kassel, *Landesbibliothek*: 4° Ms. chem. 96, 255

199 *Für Pestilenz* (§ 3.9) has a recipe that uses amongst other ingredients "Römisch kerschen hütlein/ mit den roten Juden hütlein/ zů latein Alkekengj genant/ die sich puluern lassen" ("Roman cherry hats, with the red jews' hats, called Alkekengi in Latin, that can be pulverised"): *Für Pestilenz* (Salzburg, 1554), f. xxxvi. – The modern name of this plant is *Physalis alkekengi*.

First printed:

- 1608 (in Benedictus Figulus, ed., *Rosarium novum olympicum et benedictum. Das ist: Ein newer Gebenedeyter Philosophischer Rosengart* (Basel, 1608); VD17 12:133068G; Sudhoff, *Bibliographia Paracelsica*, 482–485 n° 283)

Historical Manuscript Catalogues: von Linden (1788), part 2, 52 n° 54

Essential bibliography: Sudhoff, *Bibliographia Paracelsica*, 696 n° 22.

Further bibliographical references:

Joachim Telle, "Was ist ein 'roter Jude'? Bemerkungen zu einem pseudoparacelsischen Arkantext bei Benedictus Figulus," in *Beiträge zur Geschichte der Pharmazie* (1976, n° 1), 5–8.

Andrew Colin Gow, *The Red Jews. Antisemitism in an Apocalyptic Age 1200–1600* (Leiden, 1995), 173–175.

Broszinski, *Manuscripta chemica in quarto* (2011), 383.

Rebekka Voß, "Die roten Juden: christliche und jüdische Färbung," in Rebekka Voß, *Umstrittene Erlöser. Politik, Ideologie und jüdisch-christlicher Messianismus in Deutschland, 1500–1600* (Göttingen, 2011), 92–122.

4.5 *Aurora philosophorum*

1. *Printing History, Manuscripts.* First printed in a shortened Latin translation in 1577, edited by Gerhard Dorn. Several manuscripts. – The complete German text as found in manuscripts was never printed, since Huser presented a German translation taken from Dorn's edition.
2. *Editions.* Edited by Huser, *Chir.* 1605, Appendix, 78–92 [shortened version; re-translated from Latin; with additions]. Not edited by Sudhoff.
3. *Relationship between different versions.* There are two German versions. Version A (the long version) is found in three manuscripts, one of which is lost. This version is not attributed to Paracelsus or any other author. It is generally assumed that Heidelberg cpg 303 is a direct copy of Heidelberg cpg 600.[200] However, there are minor differences between the two Heidelberg manuscripts, and there is textual evidence that contradicts this assumption: for example in cpg 303, f. 40r, there is omitted almost exactly one line of the text. This text is present in cpg. 600, f. 57v, but it is distributed over two lines.

   Version B (the shorter version) is found in three manuscripts, one of which (the Kassel manuscript) was obviously written in the same hand as a Leiden manuscript (Voss. Chym. F.24) that contains many shortened versions of

200 Redl, "Aurora Philosophorum" (2008), 698.

authentic Paracelsian theological works (see above, § 11). This shorter version is attributed to Paracelsus, and it is shortened in a way that reminds of the shortening of authentic theological works: while whole paragraphs were omitted and others were shortened, the order of paragraphs was always preserved; in many cases sentenced were simplified.

4. *Structure, genre/form, perspective, style.*
5. *Relationship to other texts.* The longer version is – at least in part – composed of chapters from different works. – Chapter 12 of the longer version was printed separately in a different context[201] and also appears in some manuscripts without the context of the *Aurora*.[202] – Chapter 14 refers to one "Isaac", so there may be some connection with the Isaac Hollandus alchemical corpus.[203] In addition, this chapter is connected with the pseudo-Paracelsian *De antimonio* (§ 4.27), a connection already noticed in the 17th century.[204] – According to Peuckert there is a citation from Alexander von Suchten in chapter 17.[205]
6. *Authenticity, authorship.* Rejected as spurious by Huser and Sudhoff.
7. *Time of writing.* Probably written in the 1560s, shortened and attributed to Paracelsus in the early 1570s.

Word count: Version A: ca. 20.000 words. – Version B: ca. 10.000 words.

Incipit: "Aller kunsten vnd naturlichen geheimnuß wissenschaft ist erstlich gewesen Adam ..." [version A] / "Adam ist der erste erfinder aller kunst ..." [version B]

Manuscripts:

version A

- Heidelberg, *Universitätsbibliothek*: Cod. Pal. germ. 303; 45 fols.
- Heidelberg, *Universitätsbibliothek*: Cod. Pal. germ. 600; 62 fols.
- Wrocław/Breslau, *Biblioteka Kapitulna/Dombibliothek*: Ms. 87 [lost]; 63 fols.

version B

- Dresden, *Sächsische Landesbibliothek*: Mscr.Dresd.e.7, 247–313
- Kassel, *Landesbibliothek*: 4° Ms. chem. 8, f. 1r–27v
- New Haven, *Beinecke Library*: Beinecke MS 309, f. 51r–71v

---

201 In Aurei Velleris Oder Der Guldin Schatz: un[d] Kunstkam[m]er Tractatus Quintus & Ultimus (Basel: Treuw, 1604), 73–75 (Oleum Vitrioli).

202 Kassel (LB) 4° Ms. chem. 31 (*De oleo Vitrioli*); Kassel (LB) 4° Ms. chem. 50 (*De oleo Victrioli*); Leiden, *Universiteitsbibliotheek*: Voss. chym. Q.17 (*Oleum de Vitriolo Rogerii Bachonis Angli*).

203 Van Gijsen, "Isaac Hollandus Revisited" (2010), 317.

204 Kassel (*LB*) 4° Ms. chem. 60[5,2, fol. 173r, headline: "Tractatus Deo [sic] oleo Stibij Rogerij Bachi De Antimonij Arcano Videatur Capite 14 in aurora philosophi Polacelsi [sic]: et Gerhardi Dornij <antimonium> plumbum philosophorum."

205 Peuckert, *Pansophie* (1956), 317.

FIRST PRINTED:

- 1577 (Latin; Version B; in *Aurora thesaurusque philosophorum Theophrasti Paracelsi*, ed. Gerhard Dorn (Basel: [Thomas Guarin], no date [1577]); VD16 P 405; Sudhoff, *Bibliographia Paracelsica*, 302–304 n° 177)

HISTORICAL MANUSCRIPT CATALOGUES: Widemann, *Verzeichnisse* (Kassel), n° v, 769; *Scripta Theophrasti* (Wien)

ESSENTIAL BIBLIOGRAPHY: Sudhoff, *Bibliographia Paracelsica*, 463; Sudhoff, *Paracelsus-Handschriften*, 777; CP 1: 131; CP 2: 30–31, 911; CP 3: 375–376, 607, 885, 1109.

FURTHER BIBLIOGRAPHICAL REFERENCES:

Peuckert, *Pansophie* (1956), 286–287, 289, 317, 386, 465, 487.

Peuckert, *Rosenkreutz* (1973), 65.

Philipp Redl, "Aurora Philosophorum. Zur Überlieferung eines pseudo-paracelsischen Textes aus dem 16. Jahrhundert," *Daphnis*, 37 (2008), 689–712.

Kahn, "Le retour de Gérard Dorn" (2010–2011), 129, 132–139, 146, 149–156.

Broszinski, *Manuscripta chemica in quarto* (2011), 21.

Katharina Dück, "Zum 'Limbus'-Begriff als 'Materia prima' im Corpus Paracelsicum," *Nova Acta Paracelsica*, N.F. 28 (2018), 83–100, on 84.

Florian Ebeling, *Das Geheimnis des Hermes Trismegistos. Geschichte des Hermetismus von der Antike bis zur Gegenwart*, 3rd. rev. ed. (Munich, 2018), 110–111.

Annelies van Gijsen, "Isaac Hollandus Revisited," in Miguel López Pérez, Didier Kahn and Mar Rey Bueno, *Chymia: Science and Nature in Medieval and Early Modern Europe* (Newcastle upon Tyne, 2010), 310–330.

Tobias Bulang, "Genealogy of Knowledge and Delegitimization of Universities: The Pseudo-Paracelsian *Aurora Philosophorum*," *Early Science and Medicine*, 24 (2019), 473–484; also published in this volume, 61–72.

Tobias Bulang, "Wissensgenealogien der frühen Neuzeit im Vergleich. Epistemische Entwürfe des Paracelsismus im wissensgeschichtlichen Kontext," *Daphnis*, 48 (2020), 38–64.

4.6 *Auslegung über Tabulam Smaragdinam Hermetis Trismegisti*

1. *Printing History, Manuscripts.* Not printed. Possibly one manuscript (current location unknown, formerly in the library of Erich Pfeffer, d. 1680).
2. *Editions.* Not edited by Huser or Sudhoff.
3. *Relationship between different versions.*
4. *Structure, genre/form, perspective, style.*
5. *Relationship to other texts.*
6. *Authenticity, authorship.* The *Auslegung* was mentioned in the *Tractätlein von der Reise Friedrich Galli nach der Einöde S. Michael* (first printed in *Traum-gesicht/ Welches Ben-Adam [...] gehabt*, Amsterdam: Hans Fabel, 1648) as part of a

manuscript with theological and alchemical texts by Paracelsus ("Theophrasti Paracelsi Schrifften, ohngefehr 6 Bogen"). – This *Auslegung* may be identified with (1) a manuscript with a similar title[206] – though not attributed to Paracelsus – that was in the library of the alchemist, quaker and spiritualist Erich (or Ulrich) Pfeffer, who lived in Amsterdam in the 1660s and 1670s.[207] Or it may (2) be a reference to a German version of the *Emerald Table* with a short commentary that forms part of the treatise *Von Bereitung des gebenedeiten philosophischen Steins* (§ 4.53), written by Caspar Hartung and later erroneously attributed to Paracelsus by Ps.-Huser, *Opera Bücher vnd Schrifften* (1603), 2: 688.

7. *Time of writing.*

Word count:

Incipit:

Manuscripts: no existing manuscripts known.

First printed: not printed.

Historical Manuscript Catalogues: *Catalogus* Pfeffer (1680), C4r n° 1; *Catalogus* Pfeffer (1683), 124 n° 15

Further bibliographical references:

Karl Sudhoff, "Hat wirklich Paracelsus die *Tabula smaragdina* ins Deutsche übersetzt?", *Mitteilungen zur Geschichte der Medizin und der Naturwissenschaften*, 26 (1927), 163.

Gilly, "Vom ägyptischen Hermes zum Trismegistus Germanus" (2010), 98.

Telle, *Alchemie und Poesie* (2013), 1: 267.

Hereward Tilton, "The Urim and Thummim and the Origins of the Gold- und Rosenkreuz", in Hans Thomas Hakl, ed., *Octagon: Die Suche nach Vollkommenheit im Spiegel einer religionswissenschaftlichen, philosophischen und im besonderen Maße esoterischen Bibliothek*, 2 vols. (Gaggenau, 2016), 2: 35–70, on 46–48.

206 "Auslegung über die Tabulam Hermetis Trismegisti, nach Magischer/ Cabalischer und Philosophischer Auslegung".

207 On Pfeffer se Tilton, "The Urim and Thummim" (2016), 46–48.

4.7 *Beschreibung der deutschen Alchimia* also: *Wahrhaftige Beschreibung der deutschen Alchimia durch Theophrastus Paracelsus erfunden / Deutsche Alchimie*

1. *Printing History, Manuscripts.* German version not printed. One manuscript. Translated from the Dutch version first printed in 1551, not attributed to Paracelsus.
2. *Editions.* Not edited by Huser or Sudhoff.
3. *Relationship between different versions.* This text was written by Jean Saulnier in French in 1432 (numerous manuscripts).[208] First edition in French by Gabriel de Castaigne: *Œuvre philosophique de J. Saunier*, title-page and 55 pp., appended to Gabriel de Castaigne, *Le Grand Miracle de nature metallique* (Paris: Charles Sevestre, 1615); re-edition in *Les Œuvres du R.P. Gabriel de Castaigne, tant medicinales que chymiques* (Paris: Jean d'Houry, 1660). Orléans, ms. 291 was edited by Sylvain Matton in 2015.[209] Latin translation by Guillaume Rascalon (Munich, BSB, clm 25110, fol. 80r–167r). Another Latin translation: *Magni lapidis naturalis philosophia et vera ars per Joannem Saignier* [*sic*] (Bremen: Jacob Koehler, 1664) (text erroneously dated 1632 instead of 1432). Dutch translation published in 1551: *Duytsche Alchimye. Daer in begrepen is/ die rechte ende warachtige compositie vanden steen der philosophen/ ende andere elexiren Wit ende Root/ so als die warachtelick binne Parijs ghewrocht/ ende ten eynde is ghebracht gewest/ in voorleden Jaren* (Amsterdam: Cornelis Karelsen, 1551; re-ed. Leiden: Gillis van Cranenbroeck, 1600). The German manuscript of the *Beschreibung der deutschen Alchimia* (Kassel, 8° Ms. chem. 15, written in 1613 by Raphael Egli) was probably translated from the Dutch.
4. *Structure, genre/form, perspective, style.* The treatise is addressed by the author to his son.
5. *Relationship to other texts.*
6. *Authenticity, authorship.* Attributed to Paracelsus only in the German Kassel manuscript, but not in the printed Dutch version of 1551 from which it was translated.[210]
7. *Time of writing.* 1432.

---

208 Pereira, *The Alchemical Corpus attributed to Raymond Lull* (1989), 44 n. 42.

209 Matton, "Une version inédite de *L'Œuvre de Jean Saulnier*" (2015).

210 Sudhoff, *Paracelsus-Handschriften*, 731: "Hohenheim's Name auf dem Titel soll nur besagen, dass diese Alchemie nach seinen Anregungen gearbeitet ist, verfasst hat er das Werkchen gewiss nicht."

Word count:

Incipit: "Lieber leser, jhr wolltt ewer hertze nichtt lassen abwenden von der edlen kunst der alchimien [...] / Nemet jn dem nahmen Gottes grob saltz also wie es aus der sehe kompt ..."

Manuscripts:
- Kassel, *Landesbibliothek*: 4° Ms. chem. 19, f. 1r–46v
- München, *Bayerische Staatsbibliothek*: Kiesewetteriana 29 (?)[211]

First printed: German version not printed
- 1551 [*Duytsche Alchimye. Daer in begrepen is/ die rechte ende warachtige compositie vanden steen der philosophen/ ende andere elexiren Wit ende Root/ so als die warachtelick binne Parijs ghewrocht/ ende ten eynde is ghebracht gewest/ in voorleden Jaren* (Amsterdam: Cornelis Karelsen, 1551)]

Essential bibliography: Sudhoff, *Paracelsus-Handschriften*, 730–731, 780 n° II/23; CP 3: 1183.

Further bibliographical references:

Broszinski, *Manuscripta chemica in quarto* (2011), 60.

Klaas Hoogendoorn, *Bibliography of the Exact Sciences in the Low Countries from ca. 1470 to the Golden Age (1700)* (Leiden, 2018), 777.

Thomas Denys, "Duytsche alchimye: Een kennismaking met het oudste gedrukte alchemistische traktaat in het Nederlands," *Queeste*, 24:1 (2017), 27–51 (*non vidi*).

Didier Kahn, "Recherches sur le *Livre* attribué au prétendu Bernard le Trévisan (fin du XVe siècle)," in Chiara Crisciani, ed., *Alchimia e medicina nel Medioevo* (Florence, 2003), 265–335, on 291 n. 78.

Michela Pereira, *The Alchemical Corpus attributed to Raymond Lull* (London, 1989), 44, n. 42.

Sylvain Matton, "Une version inédite de *L'Œuvre de Jean Saulnier*," in Sylvain Matton, ed., *Alchimie et philosophie mécaniste. Expérimentateurs et faussaires à l'âge classique* (Paris-Milano, 2015), 17–32 (edition by Matton of Orléans, ms. 291).

Borras Theophrasti: see *Figierung auf den Mercurium Saturni* (§ 4.31)

211 München, *Bayerische Staatsbibliothek*: Kiesewetteriana 29 contains according to a manuscript list: "Joh. Saignier. Alchimistische Tractate. 1502."

4.8 *Brief an Augustin Sattler* also: *Paracelsicae / Büchlein belangend lapidem/Von Bereitung des Steins der Weisen und von der Heilung des Steins und Podagras*

1. *Printing History, Manuscripts.* First printed in 1574, edited by Adam von Bodenstein. Several manuscripts.
2. *Editions.* Edited by Huser, *Chir.* 1605, Appendix, 100–101. Not edited by Sudhoff. Edited by Kühlmann and Telle in CP 3: 252–265 n° 105.
3. *Relationship between different versions.* Only one version known. The Karlsruhe manuscript is missing a few sentences. The first Kassel manuscript and the 1689 edition by Erbinäus von Brandau are both missing the first paragraphs containing the letter part.
4. *Structure, genre/form, perspective, style.* The first part is written in the first person in the form of a letter from Paracelsus to Augustin Sattler, directly addressing the recipient. The next part, divided into two "books," contains several alchemical processes. The text closes with greeting formula and date.
5. *Relationship to other texts.*
6. *Authenticity, authorship.* Rejected as spurious by Huser and Sudhoff.[212]
7. *Time of writing.* The text is allegedly dated from Krainburg (Carinthia) in 1540 in the manuscripts and 1551 in the printed text. Probably written in the 1560s.

WORD COUNT: ca. 1.200 words

INCIPIT: "Mein gemüt steht dermassen zu dir, lieber gefatter Augustin / Die alten haben mancherley art vnnd eigenschafften an jnen gehabt / Der sulphur hat die mutter zu gebären dise frucht ..."

MANUSCRIPTS:
- KARLSRUHE, *Badische Landesbibliothek*: Hs. Allerheiligen 3, 184–188
- KASSEL, *Landesbibliothek*: 4° Ms. chem. 60[2,6, f. 60r–63r
- KASSEL, *Landesbibliothek*: 4° Ms. med. 11, f. 176r–179v
- PRAHA, *Národní knihovna*: Osek 43, f. 1r–3v
- WIEN, *Österreichische Nationalbibliothek*: Cod. 11323b, f. 675r–678v
- WOLFENBÜTTEL, *Herzog August Bibliothek*: Cod. Guelf. 48.1 Aug. 4°, f. 62r–68r

FIRST PRINTED:
- 1574 (in: *Metamorphosis Theophrasti Paracelsi*, ed. Adam von Bodenstein (Basel: Samuel Apiarius for Pietro Perna, 1574); VD16 P 695; Sudhoff, *Bibliographia Paracelsica*, 267–268 n° 157; cf. ibid., 351–352 n° 204)

212 "contains nothing reminding of Paracelsus", Sudhoff, *Bibliographia Paracelsica*, 268.

HISTORICAL MANUSCRIPT CATALOGUES: von Linden (1788), part 2, 48 n° 54; Binz (1791), 139 [Metamorphosis]

ESSENTIAL BIBLIOGRAPHY: Sudhoff, *Bibliographia Paracelsica*, 463; Sudhoff, *Paracelsus-Handschriften*, 777; CP 3: 252–265 n° 105.

FURTHER BIBLIOGRAPHICAL REFERENCES:

Telle, "Paracelsus in pseudoparacelsischen Briefen" (2006–2007), 154 n. 26.

Broszinski, *Manuscripta chemica in quarto* (2011), 223.

Sven Limbeck, "Bild und Text in alchemischen Handschriften", in Petra Feuerstein-Herz and Stefan Laube, eds., *Goldes Wissen: Die Alchemie – Substanzen, Synthesen, Symbolik* (Wiesbaden, 2014) (*Ausstellungskataloge der Herzog August Bibliothek* Nr. 98), 239–276, on 271.

4.9 *Brief an Bartholomäus Korndörffer / Wie man die sieben Metalle clarificieren soll*

1. *Printing History, Manuscripts.* First printed in 1598 in *Aurei Velleris Oder Der Guldin Schatz vnd Kunstkammer Tractatus II*. Several manuscripts.
2. *Editions.* Not edited by Huser or Sudhoff. Edited by Kühlmann-Telle in CP 3: 279–299 n° 107.
3. *Relationship between different versions.*
4. *Structure, genre/form, perspective, style.* The first section is written in the form of a letter from Paracelsus to Bartholomäus Korndörffer, replying to a letter from Korndörffer, and dated 1537 (August 1537 in the Vienna manuscript). Only this first section is written in the first person, directly addressing the recipient. The second section has instructions to "precipitate" the seven metals. The third section is an "explanation of the aforesaid precipitations" in the form of an alchemical process. – According to the *Frühneuhochdeutsches Wörterbuch* (*FWB*) edited by Oskar Reichmann the term "clarificieren" has two meanings: (1) to perfect, or glorify, someone; (2) to purify something. The latter meaning is found only in Paracelsus according to the *FWB*.
5. *Relationship to other texts.* There are several pseudo-Paracelsian texts connected with the name of Bartholomäus Korndörffer: (1) *Brief an Bartholomäus Korndörffer / Wie man die sieben Metalle clarificieren soll* (§ 4.9); (2) *Aus einem Brief an Bartholomäus Korndörffer über den Sulphur* (§ 4.10); (3) *Aus einem Brief an Bartholomäus Korndörffer über die Metalle* (§ 4.11); (4) *Secretum magnum* (§ 7.19).
6. *Authenticity, authorship.*
7. *Time of writing.* Allegedly written in 1537. Possibly written in the 1590s.

Word count: ca. 1.000

Incipit: "Ihr begert an mich, lieber doctor Bartholomea, auß großem eifer …"

Manuscripts:
- Hamburg, *Staats- und Universitätsbibliothek*: Cod. alchim. 661, 355
- Hamburg, *Staats- und Universitätsbibliothek*: Cod. alchim. 715, f. 94r
- Heidelberg, *Deutsches Apotheken-Museum*: Ms. M 275, f. 132–136
- Kassel, *Landesbibliothek*: 4° Ms. chem. 2, f. 23r–25v
- Třeboň/Wittingau, *Státní oblastní archiv*: z Rožmberka 25, no shelfmark
- Wien, *Österreichische Nationalbibliothek*: Cod. 11259, f. 215r–220v, 233r–238v
- Wien, *Österreichische Nationalbibliothek*: Cod. 11347, f. 92r–99r
- Wolfenbüttel, *Herzog August Bibliothek*: Cod. Guelf. 264.31 Extrav., 944–945

First printed:
- 1598 (in: *Aurei Velleris Oder Der Guldin Schatz vnd Kunstkammer Tractatus II* (Rorschach, 1598); VD16 T 1961; Sudhoff, *Bibliographia Paracelsica*, 424–427 n° 244)

Historical Manuscript Catalogues: Widemann, *Verzeichnisse* (Kassel), n° II, 37 / V, 125; Uffenbach (1720), VIII, col. 655–658 n° 12/XIV [=Hamburg, alchim. 661]

Essential bibliography: Sudhoff, *Paracelsus-Handschriften*, 779 n° II/4, 802–803 (addendum to *Bibliographia Paracelsica*); CP 3: 279–299 n° 107, 318.

Further bibliographical references:

Peuckert, *Pansophie* (1956), 255, 259, 459.

Telle, "Paracelsus in pseudoparacelsischen Briefen" (2006–2007), 155 n. 32.

Ivo Purš, "Zbytečné cennosti a cenné zbytečnosti? K povaze a osudu alchymických rukopisů Rudolfa II. v majetku královny Kristýny a Isaaca Vossia" [Useless valuables and valuable uselessness? To nature and history of alchemic manuscripts of Rudolf II possessed by the Queen Christine and Isaac Vossius], *Miscellanea: Oddělení rukopisů a starých tisků*, 20 (2008), 33–62, on 53 (*non vidi*).

Broszinski, *Manuscripta chemica in quarto* (2011), 7.

4.10 *Aus einem Brief an Bartholomäus Korndörffer über den Sulphur* also: *Paracelsus schreibt an Bartholomaeum Korndörffer*

1. *Printing History, Manuscripts.* This text consists only of quotations from an alleged letter from Bartholomäus Korndörffer to Paracelsus. First printed in 1606. Full text unknown. Another edition in 1680; another one in 1706.

2. *Editions.* Not edited by Huser or Sudhoff.
3. *Relationship between different versions.* The 1680 edition of the text ends with a 16-line poem. The quotations printed in 1706 preserve basically the same text, but there are small differences in wording and some of the quotations of the 1680 edition are missing, especially the poem.
4. *Structure, genre/form, perspective, style.* Written in part in the first person and sometimes directly addressing the reader. The 1680 edition of the text ends with a short poem.
5. *Relationship to other texts.* The quotations are obviously connected with the *Letter to Bartholomäus Korndörffer* (§ 4.9), but the quotations are not from this text. Immediately following them there is a section with the heading "Folget von den praecipitaten Paracelsi" that is, however, an actual quotation from the *Letter to Korndörffer*. – For a list of texts associated with Korndörffer see *Brief an Bartholomäus Korndörffer* (§ 4.9).
6. *Authenticity, authorship.*
7. *Time of writing.* Possibly written in the 1590s or early 1600s.

Word count: ca. 300 words

Incipit: "Der metallische sulphur ist hundert mal besser in jhrer substantz / Wer das metall trifft, dessen sulphur sich mit dem golde und silber vergleicht …"

Manuscripts: no manuscripts known

First printed:

- **1606** (in: Franz Kieser, ed., *Cabala Chymica. Concordantia Chymica. Azot Philosoph. Solificatum. Drey vnnterschiedliche/ Nützliche/ vnd zuvor nie außgegangene Tractätlein* (Mühlhausen: Martin Spieß for Johann Spieß, 1606); VD17 12:650901T; Sudhoff, *Bibliographia Paracelsica*, 473–474 n° 277)
- **1680** (in: Ulr. C. von Dagitza, ed., *Alexandri von Suchten/ Eines wahren Philosophi und der Artzneyen Doctoris Chymische Schrifften Alle*, Johann Görlin (Hamburg) for Georg Wolff (Frankfurt a.M.), 1680; VD17 23:239279L; Sudhoff, *Bibliographia Paracelsica*, 618–619 n° 411; reprinted from the 1606 edition)
- **1706** (in: Christian de Medices, *Concursus Philosophorum, Das ist* […] *Beschreibung/* […] *welche gestalt die* […] *Alten das* […] *Geheimnüß vom Stein der Weisen zu wercke gerichtet*, ed. Johann Friedrich Brebiss (Jena: Johann Bielcke, 1706); not in Sudhoff, *Bibliographia Paracelsica*)

Essential bibliography: CP 3: 285.

4.11 *Aus einem Brief an Bartholomäus Korndörffer über die Metalle* also: *Theophrastus in tractatu an Korndörffer*

1. *Printing History, Manuscripts.* Not printed. Two manuscripts.
2. *Editions.* Not edited by Huser or Sudhoff.
3. *Relationship between different versions.* Only one version known.
4. *Structure, genre/form, perspective, style.* Allegedly taken from a letter from Paracelsus to Bartholomäus Korndörffer. One single use of the first person ("wir"), many references to Paracelsus in the third person ("er spricht," "er beschreibt" etc.).
5. *Relationship to other texts.* Quotes or refers to: *De putrefactione, De transmutatione metallorum, De coagulatione mercurii metallorum, De resuscitatione rerum naturalium, Manual* (§ 4.40), *Archidoxis magica* (§ 5.1), *Liber vexationum* (§ 4.39). – Authorities mentioned: Arnoldus, Raimundus, Bernhardus. – According to Sudhoff, inspired by *Wie man die sieben Metalle clarificieren soll* (*Brief an Bartholomäus Korndörffer*; § 4.9).[213] – For a list of texts associated with Korndörffer see *Brief an Bartholomäus Korndörffer* (§ 4.9).
6. *Authenticity, authorship.*
7. *Time of writing.* Further study required.

Word count: ca. 1.000 words

Incipit: "Also müssen die metal in jhr erstes vnd letztes wesen gebracht werden …"

Manuscripts:
- Kassel, *Landesbibliothek*: 4° Ms. chem. 22, f. 89v–92r
- Kassel, *Landesbibliothek*: 4° Ms. chem. 60[6,4, f. 75r–77r

First printed: not printed

Essential bibliography: Sudhoff, *Paracelsus-Handschriften*, 733.

Further bibliographical references:
Broszinski, *Manuscripta chemica in quarto* (2011), 70, 260.

213 Sudhoff, *Paracelsus-Handschriften*, 733: "Knüpft an den untergeschobenen Tractat 'Wie man die 7 Metalle clarificieren soll' an."

4.12 *Brief an Brundner* also: *Was Theophrastus seinem liebsten Freund Brunder geschrieben hat*

1. *Printing History, Manuscripts.* Not printed before Manacorda (1907, 1947). Three German manuscripts (the Wolfenbüttel manuscript has three slightly different copies). One manuscript with a Dutch translation.
2. *Editions.* Not edited by Huser or Sudhoff.
3. *Relationship between different versions.* All manuscripts have basically the same text. The second and third copy of the letter in the Wolfenbüttel manuscript do not have a proper caption, but preserve the same information in a postscriptum marked "NB." (*nota bene*).
4. *Structure, genre/form, perspective, style.* Written in the form of a letter addressed to "Lieber Brunder" (or "Bründer") and dated from Salzburg, but without a specific year. In the caption "Brundner" is identified as a nobleman from Carinthia. The letter is said to have been written by Paracelsus in cipher "before his death." Written in the first person.
5. *Relationship to other texts.*
6. *Authenticity, authorship.*
7. *Time of writing.* Further study required.

WORD COUNT: ca. 200 words

INCIPIT: "Lieber Bründner, stets bin ich ingedenk deiner gutthat …"

MANUSCRIPTS:
- CATANIA, *Biblioteca regionale universitaria, Biblioteca Strano*: U.Ms.BS.050 (*olim* Ms. Vent. 28), f. 11r
- HAMBURG, *Staats- und Universitätsbibliothek*: Cod. alchim. 673, VII, f. 10v–11r [Dutch]
- LONDON, *Wellcome Library*: MS 1156, 21
- WOLFENBÜTTEL, *Herzog August Bibliothek*: Cod. Guelf. 9 Noviss. 2°, f. 1r, 2r, 4r [3 copies]

FIRST PRINTED: not printed before Manacorda (1907)

ESSENTIAL BIBLIOGRAPHY: CP 3: 317.

FURTHER BIBLIOGRAPHICAL REFERENCES:
Guido Manacorda, "Di un codicetto tedesco posseduto della R. Biblioteca Ventimiliana in Catania," *Rivista di letteratura tedesca*, 1 (1907), 222–246.

Anonimo (scuola di Paracelso), *Canto della montagna*, ed. Guido Manacorda (Florence, 1947).
Telle, "Paracelsus in pseudoparacelsischen Briefen" (2006–2007), 155 n. 30.

4.13 *Brief an eine geistliche Person / Practica des verkehrten Mondes* also: *Practica des verkehrten Mondes / Theophrastus schreibt an einen seinen guten Freund/Brief an Georg Staier*

1. *Printing History, Manuscripts.* First printed in 1604, edited by I.P.S.M.S. Quoted by Johann Bringer in his 1613 printed collection (second half only); quotation reproduced by [Daniel Mögling] in his *Prodromus Rhodo-stauroticus Parergi Philosophici* (no place [Prague?], 1620), 6. Three manuscripts.
2. *Editions.* Not edited by Huser or Sudhoff.
3. *Relationship between different versions.* In the Wolfenbüttel manuscript the first half of the text is missing. The recipient of the letter is variously named "a good friend" (1604 ed.), "a cleric," "a cleric called Georgius Staierus" and often dated 1534. The alternative title used by Sudhoff and Kühlmann-Telle, "Practica of the reversed moon", is only found at the end of the second half.
4. *Structure, genre/form, perspective, style.* Written in the first person.
5. *Relationship to other texts.*
6. *Authenticity, authorship.*
7. *Time of writing.* Allegedly written in 1534. Probably written in the 1570s, because some manuscripts (Wolfenbüttel, Kassel) have the notice "copied from the autograph of Theophrastus in 1579."

WORD COUNT: ca. 400 words

INCIPIT: "Ich Philippus Theophrastus sage noch, das viel künstler zu mir kommen sein / Derhalben sagen wir nicht mehr, denn das du nemest den mon vom firmament / Wir sagen nicht mehr als daß du nehmest den mond vom firmament ..."

MANUSCRIPTS:
- DRESDEN, *Sächsische Landesbibliothek*: Mscr.Dresd.N.54, f. 18r–18v
- DRESDEN, *Sächsische Landesbibliothek*: Mscr.Dresd.N.159, f. 29v
- KASSEL, *Landesbibliothek*: 4° Ms. chem. 72, f. 36r–36v
- WOLFENBÜTTEL, *Herzog August Bibliothek*: Cod. Guelf. 9 Noviss. 2°, f. 5r

FIRST PRINTED:
- **1604** (in: I.P.S.M.S., ed., *Alchimia vera, Das ist: Der wahren vnd von Gott hochbenedeyten/ Natur gemessen Edlen Kunst Alchimia wahre beschreibung* (no place, 1604), 58–60; VD17 12:132982U; Sudhoff, *Bibliographia Paracelsica*, 453–454 n° 265)

– 1613 (in: *Occulta philosophia Von den verborgenen Philosophischen Geheimnussen der heimlichen Goldblumen* (Frankfurt a.M.: Johann Bringer, 1613), 51–52; VD17 39:122887Y; cf. Sudhoff, *Bibliographia Paracelsica*, 501)

ESSENTIAL BIBLIOGRAPHY: Sudhoff, *Bibliographia Paracelsica*, 501, 504, 696 n° 15; CP 3: 231.

FURTHER BIBLIOGRAPHICAL REFERENCES:
Joachim Telle, "Paracelsus in pseudoparacelsischen Briefen", in *Berichte zur Wissenschaftsgeschichte* 27 (2004), 149–160; reprinted in *Nova Acta Paracelsica* N.F. 20 (2006/07), pp. 147–164, esp. 161 n. 53.
Broszinski, *Manuscripta chemica in quarto* (2011), 321.

4.14 *Brief an einen Gönner und Freund* (*Vom Wunderstein*) also: *Von dem Wunderstein*

1. *Printing History, Manuscripts.* Not printed before Telle (1993). Two manuscripts.
2. *Editions.* Not edited by Huser or Sudhoff. Critical edition of both versions (but not the *practica*) by Joachim Telle in 1993.
3. *Relationship between different versions.* The text exists in two versions, in prose and in verse, each found in one single manuscript. The Weimar manuscript has the prose version, a text called *Von dem Wunderstein*. The Kassel manuscript has a rhymed version. Telle seems to conclude that the rhymed version is an adaptation of the prose text.
4. *Structure, genre/form, perspective, style.* The prose version is written in the form of a letter to "his good patron and friend." The recipient is directly addressed in the first and in the last paragraph. In the Weimar manuscript the text is followed by a *practica*. In the Kassel manuscript the verses are followed by two paragraphs of a practical text that is, however, incomplete (missing folios in the manuscript).
5. *Relationship to other texts.*
6. *Authenticity, authorship.*
7. *Time of writing.* Further study required.

WORD COUNT: ca. 1.000 words (letter), 1.300 words (*practica*); ca. 900 words (rhymed version)

INCIPIT: "Lapis philosophorum vndt tincturen / Besonderß lieber herr vnnd freundt, daß jr denckt, ich sol euch neue zeitung zu schreiben …"

Manuscripts:
- Weimar, *Herzogin Anna Amalia Bibliothek*: Q 456, 361–367 [letter and *practica*]
- Kassel, *Landesbibliothek*: 4° Ms. chem. 86, f. 77v–80v [verses]

First printed: not printed before Telle (1993)

Essential bibliography: Sudhoff, *Paracelsus-Handschriften*, 226–228, 780 n° II/21; CP 3: 231, 319, 718.

Further bibliographical references:

Joachim Telle, "Das pseudoparacelsische Adler/Löwe-Sinnbild unter deutschen Lehrdichtern," in Joachim Telle, *Buchsignete und Alchemie im XVI. und XVII. Jahrhundert. Studien zur frühneuzeitlichen Sinnbildkunst* (Hürtgenwald, 2004), 95–119, on 96. – Reprinted with additions in Telle, *Alchemie und Poesie* (2013), 897–922.

Joachim Telle, "'Von der Wahrheit der alchemischen Kunst'. Der pseudoparacelsische Brieftraktat 'Vom Wunderstein' in einer frühneuzeitlichen Versfassung," in Peter Dilg and Hartmut Rudolph, eds., *Resultate und Desiderate der Paracelsus-Forschung* (Stuttgart, 1993), 57–78. – Reprinted in Telle, *Alchemie und Poesie* (2013), 865–888.

Telle, "Paracelsus in pseudoparacelsischen Briefen" (2006–2007), 159 n. 48.

Broszinski, *Manuscripta chemica in quarto* (2011), 362.

4.15 *Brief an einen guten Freund / Brief an Meister Wilhelm*

1. *Printing History, Manuscripts.* Not printed. One manuscript.
2. *Editions.* Not edited by Huser or Sudhoff.
3. *Relationship between different versions.* Only one known version.
4. *Structure, genre/form, perspective, style.* Written in the form of a letter from Paracelsus to a "master Wilhelm", in reply to a previous letter and referring to at least one more letter allegedly written by Paracelsus. Written in the first person, directly addressing "Meister Wilhelm."
5. *Relationship to other texts.*
6. *Authenticity, authorship.* A rather crude forgery, inspired by other apocryphal letters and probably by *De tinctura physicorum* (§ 4.30).
7. *Time of writing.* Probably written in the 1610s.

Word count: ca. 900 words

Incipit: "Ehrnhaffter lieber meister Wilhelm, auß eurem briefe vorneme jch."

MANUSCRIPTS:
- HALDENSLEBEN, *Alvenslebensche Bibliothek*: Alv. Mg 163, f. 217v–222v

FIRST PRINTED: not printed

Brief an einen Jünger: see *Summum naturae mysterium* (§ 4.49)

Brief an Georg Steier: see *Brief an eine geistliche Person* (§ 4.13)

4.16 *Brief an Poldörfer* also: *De reductione metallorum in primam materiam / Arcana de lapide philosophorum/Aqua Sallabri Paracelsi*

1. *Printing History, Manuscripts.* First edited in 1608 by Benedictus Figulus in his *Rosarium novum olympicum*. No manuscripts known. Translated into Latin from Figulus's edition and edited in 1684 by Johann Seger Weidenfeld in his *De secretis adeptorum.*
2. *Editions.* Not edited by Huser or Sudhoff.
3. *Relationship between different versions.* Only one known version.
4. *Structure, genre/form, perspective, style.* Written in form of a letter from Paracelsus allegedly written in Pressburg (Bratislava) to a "Doctor Poldörffer" in Lereckaw, a place not identified. Poldörffer is said to have been a former "Wandergesell" (fellow traveller) of Paracelsus. The text is written in the first person and directly addresses Poldörffer ("Du wilt von mir wissen …").
5. *Relationship to other texts.* Accompanied in Figulus's edition by other short texts: *Figierung auf Mercurium Saturni Theophrasti*, *Borras Theophrasti* and *Aurum diaphoreticum Theophrasti* (see § 4.31). These texts seem to have been originally independent and gathered as a group only by Figulus.
6. *Authenticity, authorship.*
7. *Time of writing.* Further study required.

WORD COUNT: ca. 800 words

INCIPIT: "Du wilt von mir wissen, das du selbs wol außnemmen möchtest / Scies nullam nobis esse cognitam methodum reducendi metalla in mercurium …"

MANUSCRIPTS: no manuscripts known

FIRST PRINTED:
- 1608 (in: Benedictus Figulus, ed., *Rosarium novum olympicum et benedictum. Das ist: Ein newer Gebenedeyter Philosophischer Rosengart* (Basel, 1608); VD17 12:133068G; Sudhoff, *Bibliographia Paracelsica*, 482–485 n° 283)

– 1684 [Latin; in Johann Seger Weidenfeld, *De secretis adeptorum sive de usu spiritus vini Lulliani libri IV* (London: H. Hills jun., 1684)]

Essential bibliography: Sudhoff, *Bibliographia Paracelsica*, 696 n° 20; CP 3: 317.

Further bibliographical references:
Telle, "Paracelsus in pseudoparacelsischen Briefen" (2006–2007), 155 n. 29.

Brief an seinen guten Freund: see *Brief an eine geistliche Person* (§ 4.13)

4.17 *Brief an Theodor*
also: *Processus tincturae antimonii / Tinctura antimonii*

1. *Printing History, Manuscripts.* First edited in 1608 by Benedictus Figulus. Five manuscripts.
2. *Editions.* Not edited by Huser or Sudhoff.
3. *Relationship between different versions.* The text consists of two parts: the letter addressed to Theodor, and a practical part ("processus tincturae antimonii"). The Hamburg manuscript has the practical part only, while the Berlin manuscript and Figulus's edition have the letter part only and lack the practical part. The practical part has many variations in the different copies.
4. *Structure, genre/form, perspective, style.* Written in the first person in the form of a letter. In some manuscripts the recipient is named only in the last paragraph of the letter. The text is said to have been copied by the recipient himself from the autograph letter of Paracelsus.
5. *Relationship to other texts.*
6. *Authenticity, authorship.*
7. *Time of writing.* The letter is dated 17 May 1535 in the first Kassel manuscript and 17 May 1536 in all other copies. Probably written in the late 16th century.

Word count: ca. 1.200 words [ca. 400 words without the *practica*]

Incipit: "Es haben wol recht vndt billich die altten spagyri gesagt / Nimm zwei pfundt asinat, der schön langspißig ist / Recipe antimonii der schön lang spießlecht ist zwei pfund …"

Manuscripts:
– Berlin, *Staatsbibliothek*: Ms. germ. oct. 927, 1–3
– Hamburg, *Staats- und Universitätsbibliothek*: Cod. alchim. 668, f. 199v–200v
– Kassel, *Landesbibliothek*: 4° Ms. chem. 8, f. 297r–298v
– Kassel, *Landesbibliothek*: 4° Ms. chem. 35[4, f. 41r–43v
– Kassel, *Landesbibliothek*: 4° Ms. chem. 72, f. 243r–245r

FIRST PRINTED:
- 1608 (in: Heinrich Khunrath, *De igne magorum philosophorumque secreto externo & visibili*, ed. Benedictus Figulus (Strasburg: Lazarus Zetzner, 1608); VD17 23:716355P; Sudhoff, *Bibliographia Paracelsica*, 489–490 n° 286)

ESSENTIAL BIBLIOGRAPHY: Sudhoff, *Bibliographia Paracelsica*, 489–490;[214] Sudhoff, *Paracelsus-Handschriften*, 721, 778 n° I/14; CP 3: 1071.

FURTHER BIBLIOGRAPHICAL REFERENCES:
Telle, "Paracelsus in pseudoparacelsischen Briefen" (2006–2007), 154 n. 25.
Broszinski, *Manuscripta chemica in quarto* (2011), 26, 129, 323.

4.18 *Brief an Wilhelm von Hohenheim* also: *Antwort Philippi Theophrasti / Brief an Wilhelm von Hohenheim*

1. *Printing History, Manuscripts.* First edited in 1608 by Benedictus Figulus. Two manuscripts. Critical edition in CP 3: 248–251 n° 104.
2. *Editions.* Not edited by Huser or Sudhoff.
3. *Relationship between different versions.* Only one version known.
4. *Structure, genre/form, perspective, style.*
5. *Relationship to other texts.* In Figulus's edition and in both manuscripts the letter is presented immediately after *Particular Wilhelm von Hohenheims* allegedly sent to his son. – For a list of more texts connected with Wilhelm von Hohenheim see *Brief von Wilhelm von Hohenheim an seinen Sohn* (§ 4.20).
6. *Authenticity, authorship.* Not authentic according to Sudhoff.[215]
7. *Time of writing.* Further study required.

WORD COUNT: ca. 200 words

INCIPIT: "Lieber herr vatter, euwer schreiben sampt dem vbersandten proceß hab ich empfangen / Recipe aquae fortis communis, solvir darinnen silber so viel du wilt …"

MANUSCRIPTS:
- LEIDEN, *Universiteitsbibliotheek*: Voss. Chym. Q.1, f. 43r–43v
- WOLFENBÜTTEL, *Herzog August Bibliothek*: Cod. Guelf. 49 Noviss. 8°, f. 42v

214 Sudhoff ignored the *Letter to Theodorus* and did not even mention it: a highly unusual oversight.

215 Sudhoff, *Bibliographia Paracelsica*, 484: "An der Unechtheit dieser Briefe ist wohl kein Zweifel, wenn auch der Ausdruck 'mein Pflug' für seine Thätigkeit bei Hohenheim des öfteren so gebraucht ist, wie in diesem Briefe."

First printed:
- 1608 (in: *Rosarium novum olympicum et benedictum. Das ist: Ein newer Gebenedeyter Philosophischer Rosengart*, ed. Benedictus Figulus (Basel, 1608); VD17 12:133068G; Sudhoff, *Bibliographia Paracelsica*, 482–485 n° 283)

Historical Manuscript Catalogues: von Linden (1788), part 2, 52 n° 54

Essential bibliography: Sudhoff, *Bibliographia Paracelsica*, 696 n° 21.

Further bibliographical references:
Boeren, *Codices Vossiani Chymici* (1975), 114.
Telle, "Paracelsus in pseudoparacelsischen Briefen" (2006–2007), 155 n. 27.

4.19 *Brief von Salomon Trismosin an Paracelsus* also: *Ein Schreiben an Theophrastus von seinem Lehrmeister zugeschrieben / Epistola Trismosini ad suum discipulum Theophrastum*

1. *Printing History, Manuscripts.* First printed in 1612. Five manuscripts.
2. *Editions.* Not edited by Huser or Sudhoff. Critical edition in CP 3: 219–238, n° 102.
3. *Relationship between different versions.* Only one version known.
4. *Structure, genre/form, perspective, style.* Written in the first person in the form of a letter from Salomon Trismosin to Paracelsus, dated 15 or 17 or 18 April 1515 from a place called variously "Marck Laußnitz," "Marc Lausen" or "Lusin" (not identified). According to the Kassel and the Leiden manuscript, Paracelsus stayed at Chur in Switzerland at the time of writing. Paracelsus is directly addressed as "Philippus" by the writer.
5. *Relationship to other texts.*
6. *Authenticity, authorship.*
7. *Time of writing.* Allegedly written in 1515. Possibly written in the 1590s or later.

Word count: ca. 400 words

Incipit: "Salomon Trismosin grüsset dich, jüngling Hohenheim [...] / Ich höre vnd laß mir sagen, daß du, Philippe, zu werck gehest ..."

Manuscripts:
- Dresden, *Sächsische Landesbibliothek*: Mscr.Dresd.J.409, f. 1r–2r
- Gotha, *Forschungsbibliothek*: Chart. B 373, f. 350v–352v
- Kassel, *Landesbibliothek*: 4° Ms. chem. 60[5,2, f. 123r–123v
- Kopenhagen, *Det Kongelige Bibliotek*: GKS 1722 kvart, f. 32r–33r
- Leiden, *Universiteitsbibliotheek*: Voss. Chym. Q.24, f. 177v–178v

First printed:
- **1612** ("Ein Schreiben/ so Theophrasto sein Lehrmeister zugeschrieben", in *Vier nützliche Chymische Tractat Vom Stein der Weisen* (Halle: Christoph Bismarck for Joachim Krusicke, 1612); VD17 23:242126X; not in Sudhoff, *Bibliographia Paracelsica*)

Essential bibliography: No mention of the text either in Sudhoff's *Bibliographia Paracelsica* or *Paracelsus-Handschriften*; CP 3: 219–238, n° 102.

Further bibliographical references:
Peuckert, *Pansophie* (1956), 315–316, 471.
Boeren, *Codices Vossiani Chymici* (1975), 159.
Telle, "Paracelsus in pseudoparacelsischen Briefen" (2006–2007), 155 n. 31.
Broszinski, *Manuscripta chemica in quarto* (2011), 248.

4.20 *Brief von Wilhelm von Hohenheim an seinen Sohn* also: *Litterae Wilhelmi Bombasts von Hohenheim / Particular Wilhelm von Hohenheims/Wilhelm Bombast von Hohenheim schreibt an seinen Sohn*

1. *Printing History, Manuscripts.* First edited in 1608 by Benedictus Figulus. Three manuscripts. Critical edition in CP 3: 239–247, n° 103.
2. *Editions.* Not edited by Huser or Sudhoff.
3. *Relationship between different versions.* Only one version known.
4. *Structure, genre/form, perspective, style.* Written in the first person in the form of a letter from Wilhelm von Hohenheim, directly addressing the son Paracelsus ("lieber Sohn Philippe").
5. *Relationship to other texts.* Paracelsus's father, Wilhelm von Hohenheim, seems to have attracted much attention from the authors of spurious Paracelsian texts: see (1) *Brief von Wilhelm von Hohenheim an seinen Sohn* (§ 4.20), (2) *Brief an Wilhelm von Hohenheim* (§ 4.18), (3) *Particularia des Wilhelm von Hohenheim für seinen Sohn* (§ 7.10), (4) *Eine gemeinsame Arbeit von Theophrastus mit seinem Vater* (§ 7.6), (5) *Die Metalle in die primam materiam zu bringen / Ein Process des Theophrastus für seinen Vater* (§ 7.8).
6. *Authenticity, authorship.* Not authentic according to Sudhoff.[216]
7. *Time of writing.* Further study required.

Word count: ca. 300 words

216 Sudhoff, *Bibliographia Paracelsica*, 484: "An der Unechtheit dieser Briefe ist wohl kein Zweifel."

Incipit: "Lieber sohn Philippe, wisse, das wir mittler göttlicher gnaden ein gerechte particular tinctur getroffen / Recipe solis ein lot, mercurii zwei lott / Recipe aquam fortem optime zwei theil …"

Manuscripts:
- Kremsmünster, *Stiftsbibliothek*: CC Cim. 8 f. 218r–218v
- Leiden, *Universiteitsbibliotheek*: Voss. Chym. Q.1, f. 43r
- Wolfenbüttel, *Herzog August Bibliothek*: Cod. Guelf. 49 Noviss. 8°, f. 42r

First printed:
- 1608 (in: *Rosarium novum olympicum et benedictum. Das ist: Ein newer Gebenedeyter Philosophischer Rosengart*, ed. Benedictus Figulus (Basel, 1608); VD17 12:133068G; Sudhoff, *Bibliographia Paracelsica*, 482–485 n° 283)

Historical Manuscript Catalogues: von Linden (1788), part 2, 52 n° 54

Essential bibliography: Sudhoff, *Bibliographia Paracelsica*, 484.

Further bibliographical references:
Peuckert, *Pansophie* (1956), 316.
Boeren, *Codices Vossiani Chymici* (1975), 114.
Telle, "Paracelsus in pseudoparacelsischen Briefen" (2006–2007), 155 n. 27.
Broszinski, *Manuscripta chemica in quarto* (2011), 170.
CP 3 (2013).

Brief von Wilwold an Paracelsus: see *Heimlichkeit des Herrn Wilwold* (§ 4.35)

Büchlein belangend lapidem: see *Brief an Augustin Sattler* (§ 4.8)

4.21 *Buch mit der Sackpfeife* also: *Büchlein mit der himmlischen Sackpfeife*

1. *Printing History, Manuscripts.* First edited in 1608 by Benedictus Figulus. Four manuscripts of the whole text. – The text consists of two parts that are probably originally independent. The first part, *Vom gelben und roten Mann* ("Of the yellow and the red man") was first printed in 1598 in a collection called *Aurum Vellus* and in 1610, edited by Joachim Tancke. Four manuscripts in Kassel. The second part, *Das wahre Aurum potabile* ("The true aurum potabile"), was first printed in 1604 in a continuation of the *Aureum Vellus* with the title *Tractatus de vitriolo philosophorum* or *Vom philosophischen Vitriol.*
2. *Editions.* Not edited by Huser or Sudhoff.

3. *Relationship between different versions.* In some copies the first part is addressed to a prince (Werl manuscript and Figulus: "eur fürstlich gnaden," also one manuscript of *Vom gelben und roten Mann*: "eur churfürstlichen genaden") while others directly address the reader (Heidelberg manuscript: "du," Bamberg manuscript: "ir"). The latter two are both written – at different times and with a different context – by Martin Sturtz.[217]
4. *Structure, genre/form, perspective, style.*
5. *Relationship to other texts. Vom gelben und roten Mann* has basically the same text as the first of part of the *Buch mit der Sackpfeife*, but the order of paragraphs is different in some cases, and there are also a few extra sentences. One manuscript of Vom *gelben und roten Mann* is rewritten and considerably enlarged. The same is true for *Vom philosophischen Vitriol*: comparing with the second part of the *Buch mit der Sackpfeife*, some sentences are missing, and there are some extra sentences as well.
6. *Authenticity, authorship.* The *Buch mit der Sackpfeife* is anonymous in the Werl manuscript, attributed to a "Comes a marca" (Bernardus Trevisanus?) in the Bamberg and Heidelberg manuscripts, and attributed to Paracelsus in the Kopenhagen manuscript and in Figulus's 1608 edition. The first part, *Vom gelben und roten Mann*, is – in the 1598 edition only – attributed to "Melchior cardinalis et episcopus Brixiensis," who can be identified as Melchior von Meckau (or Meggau), provost of Meißen, prince-bishop of Brixen (ca. 1440–1509). Melchior, however, would rather have been the prince addressed by the author, if this ascription is reliable at all. The title of the text, *Buch mit Sackpfeife*, is a complete mystery. Since there is no reference to a "bagpipe" in the text or a reference to any musical instrument at all, the only explanation that we can think of is that there was at some time a manuscript with an illustration containing a bagpipe.[218]
7. *Time of writing.* The oldest manuscripts are from the 1570s. Attributed to Paracelsus in the 1590s or 1600s. The first part may have been written ca. 1490–1510.

Word count: ca. 2.000 words (*Buch mit der Sackpfeife*), ca. 500 to 1.000 words (*Vom gelben und roten Mann*), ca. 2.000 words (*Vom philosophischen Vitriol*).

217 On Martin Sturtz see Joachim Telle, "Zu einer Lehrdichtung von Martin Sturtz," in Telle, *Alchemie und Poesie* (2013), 549–556. Telle, however, did not identify Sturtz as the writer of the Bamberg manuscript.

218 There exists an alchemical manuscript (not related to this text) with such an illustration in Florence, Biblioteca Medicea Laurenziana, Ms. Ashb. 1166: see Jacques van Lennep, *Alchimie. Contribution à l'histoire de l'art alchimique*, 2nd ed. (Paris, 1985), 91, ill. 96. – The same illustration in a 17th century Italian manuscript, Los Angeles (Getty Museum), Manly Palmer Hall collection, box 22 = MS 123, on 19.

INCIPIT: "Im namen Gottes wiel ich anheben die ware heymlickeit der philosophen anzuzeigen / Nehmet im namen des herren den stein, der allen weissen bekanth ist ..."

MANUSCRIPTS:

Buch mit der Sackpfeife

- BAMBERG, *Staatsbibliothek*: Msc. nat. 7, 66r–70v
- HEIDELBERG, *Universitätsbibliothek*: Cod. Sal. VIII, 96 Bd. 4, 36v–39v
- KOPENHAGEN, *Det Kongelige Bibliotek*: GKS 1720 kvart, f. 78r–85v
- WERL, *Stadtarchiv*: Familienarchiv v. Papen-Lohe, Akte A III g 1, Kasten 215, f. 85r–89r

Vom gelben und roten Mann[219]

- KASSEL, *Landesbibliothek*: 4° Ms. chem. 60[2,7, 23–26
- KASSEL, *Landesbibliothek*: 4° Ms. chem. 60[8,2, f. 191v–194v
- KASSEL, *Landesbibliothek*: 4° Ms. chem. 60[9,2, f. 230r–231r
- KASSEL, *Landesbibliothek*: 4° Ms. chem. 72, f. 275va–276ra

FIRST PRINTED:

- 1598 (in: *Aurei velleris Oder Der Guldin Schatz vnd Kunstkammer Tractatus III* (Rorschach, 1598); VD16 T 1961; Sudhoff, *Bibliographia Paracelsica*, 424–427 n° 244); reprinted 1610 (ed. Joachim Tancke) (*Vom gelben und roten Mann*)
- 1604 (in: *Aurei Velleris Oder Der Guldin Schatz: und Kunstkammer Tractatus Quintus & Vltimus* (Basel: Jacob Treuw, 1604); VD17 14:646499H; not in Sudhoff, *Bibliographia Paracelsica*) (*Vom philosophischen Vitriol*)
- 1608 (in: *Thesaurinella Olympica aurea tripartita. Das ist: Ein himmlisch güldenes Schatzkämmerlein*, ed. Benedictus Figulus (Frankfurt a.M.: Wolfgang Richter for Nicolaus Stein, 1608); VD17 3:302600E; Sudhoff, *Bibliographia Paracelsica*, 477–479 n° 280) (*Buch mit der Sackpfeife*)

ESSENTIAL BIBLIOGRAPHY: Sudhoff, *Bibliographia Paracelsica*, 696 n° 17; Sudhoff, *Paracelsus-Handschriften*, 225–226, 780 n° II/20.

FURTHER BIBLIOGRAPHICAL REFERENCES:

Gerhard Eis and Gundolf Keil, "Nachträge zum *Verfasserlexikon*," in *Studia neophilologica* 30 (1958), 232–250, on 236: "Bernhard von der Mark" (by Gerhard Eis).

Joachim Telle, "Ein altdeutsches Spruchgedicht nach der *Turba Philosophorum*," in *Zeitschrift für deutsche Philologie*, 95 (1976), 416–443, on 422.

219 Alternative titles: Vom gelben und roten Mann / Ein parabolisches Vergleichnis dieses würdigen Steins / Ein philosophisches Werk und Gespräch vom gelben und roten Mann / Processus aenigmaticus / Beschreibung lapidis aurei.

Rainer Rudolf, "Bernhard von der Mark," in *Verfasserlexikon. Die deutsche Literatur des Mittelalters*, vol. 1 (Berlin, 1978), col. 771–772.

Joachim Telle, Review of Rudolph Zaunick, *Der sächsische Paracelsist Georg Forberger* (Wiesbaden, 1977), *Sudhoffs Archiv*, 63 (1979), 102–104, on 103.

Broszinski, *Manuscripta chemica in quarto* (2011), 226, 287, 294, 325.

Volker Zapf, "Bernardus Trevisanus," in *Deutsches Literatur-Lexikon. Das Mittelalter*, ed. Wolfgang Achnitz, vol. 6 (Berlin-Boston, 2014), col. 1055–1057.

4.22 *Carboantes*
also: *Theophrastia*, book three

1. *Printing History, Manuscripts.* Not printed, no known manuscripts.[220] *Carboantes* is the last of three works composing the book *Theophrastia*, according to Valentius de Retiis (see § 1.24). Like *Parasarchum* (1.11), it is apparently spurious, although nowhere to be found. There are many lengthy quotations from *Carboantes* in Leonhart Thurneisser's מְלִיצָה καὶ Ἑρμενεία *Das ist ein Onomasticon* published in 1583. Sudhoff holds these quotations to be fictitious.[221] Was it also the opinion of Huser, who did not include any of it in his edition?
2. *Editions.* Not edited by Huser or Sudhoff.
3. *Relationship between different versions.* It is not certain that the text existed at all.
4. *Structure, genre/form, perspective, style.* Valentius de Retiis described it thus: "The third [is] called *Carboantes*, in which he [Paracelsus] explains the transmutations in form and in being (*in forma et esse*)."[222] Karl Widemann characterized the text as follows: "The 72 books of *Carboantes* deal with the transmutation of metals."[223] The Dresden *Catalogus* summarizes the text: "Carboantes de Occultis lib 66. Separatio Vtilis ab inutili. Transmutatio in forma et esse. <Aurum> potabile. Quinta essentia. Lapis Philosophorum et Margaritae."
5. *Relationship to other texts.*
6. *Authenticity, authorship.* Both *Parasarchus* and *Carboantes* are mentioned in different 16th and 17th century lists of works written by Paracelsus.[224] Gerhard Dorn[225] and Michael Toxites both believed in the existence of *Theophrastia*,

220 Adam Haslmayr wrote in 1611 in a letter to Karl Widemann: "Das Buech Carboantes und Parasarchum hab ich nicht gesehen. Ich hör, es habs der Thurneisser gehabtt." (Gilly, *Adam Haslmayr*, 1994, 97).

221 Sudhoff, *Bibliographia Paracelsica*, 388.

222 "Tertium Carboantes dictum, in quo transmutationes declarat in forma et esse," edited in CP 1: 590.

223 "Carboantes libri 72, tractant transmutationem metallorum," quoted in Gilly, *Adam Haslmayr* (1994), 103 n. 20.

224 CP 1: 596.

225 Gilly, *Paracelsus in der BPH* (1993), 37.

Sudhoff denotes it as "more than legendary,"[226] Gilly doubts its existence[227] while Kühlmann and Telle reject any such doubts as "unjustified."[228]

7. *Time of writing.* Further study required.

Word count:

Incipit:

Manuscripts: no manuscripts known

First printed: not printed

Historical Manuscript Catalogues: Toxites, *Zugesagte Bücher* (Dresden), n° 6; *Libri Theophrasti* (Dresden), fol. 5vb; *Scripta Theophrasti* (Wien); Widemann, *Unausgangene Bücher* (Hannover), n° 1

Essential bibliography: Sudhoff, *Bibliographia Paracelsica*, 142, 338; Sudhoff, *Paracelsus-Handschriften*, 188, 546; CP 1: 585, 590, 593, 595–596, 617; CP 2: 26, 165, 234, 245, 324, 326, 330; CP 3: 388.

Further bibliographical references:

Peuckert, *Pansophie* (1956), 325, 460.

Joachim Telle, "Johann Huser in seinen Briefen. Zum schlesischen Paracelsismus im 16. Jahrhundert," in Joachim Telle, ed., *Parerga Paracelsica* (Stuttgart, 1991), 159–248, on 196, 220, 230 n. 135.

Gilly, *Paracelsus in der BPH* (1993), 37, 38.

Gilly, *Adam Haslmayr* (1994), 97, 103.

4.23 *Chymischer Psalter*

1. *Printing History, Manuscripts.* First printed as a work of Paracelsus in 1771 with a preface dated 1522. Originally published in Latin by Bernard Gilles Penot in *Philippi Aureoli Theophrasti Paracelsi centum quindecim curationes experimentaque* ([Lyon]: Jean Lertout, 1582), 50–67 under the title "Regulae seu Canones aliquot Philosophici de Lapide Philosophico". Next printed in Latin in 1608 in Benedictus Figulus's *Pandora*, in a section with works by Alexander von Suchten, but not expressly ascribed to von Suchten. Figulus obviously edited it from a

226 Sudhoff, *Bibliographia Paracelsica*, 338.
227 Gilly, *Adam Haslmayr* (1994), 103 n. 20.
228 CP 1: 595.

manuscript with many misreadings. Reprinted from Figulus's edition in 1680 in *Alexandri von Suchten Chymische Schrifften Alle* with corrections of the misreadings. The first German translation (unnoticed by Sudhoff), rather crude and not attributed to Paracelsus (nor to Alexander von Suchten), appeared in 1760 in the second edition of *R. Abrahami Eleazaris Uraltes Chymisches Werk.* The 1771 German translation under the title "Chymischer Psalter" is a new and better translation. – Isaac Newton copied the "Regulae seu Canones" from Penot's 1582 edition.

2. *Editions.* Not edited by Huser or Sudhoff.
3. *Relationship between different versions.* Further study required.
4. *Structure, genre/form, perspective, style.* 157 canons ("Grundsätze"), originally numbered 1 through 159 omitting numbers 93 and 129; now numbered 1 through 157.
5. *Relationship to other texts.* The German *Psalterium Chymicum, seu Manuale Paracelsi* published in the same year 1771 in *Der hermetische Nord-Stern* (Sudhoff, *Bibliographia Paracelsica*, 661–662 n° 462) is a different work, namely the *Manuale de lapide philosophico medicinali* (§ 4.40) first published in 1572.
6. *Authenticity, authorship.* Written by Bernard Gilles Penot according to Sudhoff. Attributed to Alexander von Suchten as an authentic work of his by Carlos Gilly.[229]
7. *Time of writing.* Allegedly written in 1522. Probably written in the 1570s.

Word count:

Incipit: "Id quod perfectioni proximum est, facilius ad perfectionem deducitur / Das was der vollkommenheit am nächsten ist, wird am leichtsten zur Vollkommenheit gebracht …"

Manuscripts:
- Cambridge, *King's College Library*: Keynes ms. 43, f. 1r–4r [Latin; autograph by Isaac Newton; copy from the 1582 printed edition]
- Kassel, *Landesbibliothek*: 4° Ms. chem. 34[4, f. 47r–55r [Latin; not attributed to Paracelsus]
- Kassel, *Landesbibliothek*: 4° Ms. chem. 60[1,2, f. 123r–130r [Latin; not attributed to Paracelsus]
- Leiden, *Universiteitsbibliotheek*: Voss. Chym. F.23, f. 242r–252v [German; not attributed to Paracelsus]

229 Gilly, "Un bel trattato ermetico del paracelsismo" (2002), 186, 194: "After having analyzed its contents, I consider the *156 Regulae seu Canones de lapide philosophico* (first published anonymously by Penot in 1582, subsequently printed under Suchten's name in Figulus' *Pandora* in 1608), to be an authentic work by Suchten."

FIRST PRINTED:
- 1582 (Latin; "Regulae seu Canones aliquot Philosophici de Lapide Philosophico", in *Philippi Aureoli Theophrasti Paracelsi centum quindecim curationes experimentaque*, ed. B.G. Penot ([Lyon]: Jean Lertout, 1582)
- 1608 (Latin; in *Pandora Magnalium Naturalium Aurea Benedicta, De Benedicto Lapidis Philosoph. Mysterio*, ed. Benedictus Figulus (Strasburg: Lazarus Zetzner, 1608); VD17 23:277709W; Sudhoff, *Bibliographia Paracelsica*, 480–482 n° 282)
- 1680 (Latin; in *Alexandri von Suchten/ Eines wahren Philosophi und der Artzneyen Doctoris Chymische Schrifften Alle*, ed. Ulr. C. von Dagitza, Johann Görlin (Hamburg) für Georg Wolff (Frankfurt a.M.), 1680; VD17 23:239279L; Sudhoff, *Bibliographia Paracelsica*, 618–619 n° 411)
- 1760 (German; "Hier folgen einige philosophische Regeln oder Canones von dem Stein der Weisen", in *R. Abrahami Eleazaris Uraltes Chymisches Werk*, ed. Julius Gervasius, 2nd ed. (Leipzig: Lankisch, 1760); VD18, 10213538; not in the first edition; not attributed to Paracelsus)
- 1771 (German; *Philippi Aureoli Theophrasti Paracelsi Chymischer Psalter, oder Philosophische Grundsätze vom Stein derer Weisen Anno 1522* (Berlin: Johann Friedrich Vieweg, 1771); Sudhoff, *Bibliographia Paracelsica*, 662–663 n° 463; reprinted in 1791)

ESSENTIAL BIBLIOGRAPHY: Sudhoff, *Bibliographia Paracelsica*, 330, 662, 697 n° 56; CP 3: 580.

FURTHER BIBLIOGRAPHICAL REFERENCES:
Peuckert, *Pansophie* (1956), 484.
Boeren, *Codices Vossiani Chymici* (1975), 73.
Gilly, "Un bel trattato ermetico del paracelsismo" (2002), 186, 194.
Broszinski, *Manuscripta chemica in quarto* (2011), 116, 207.

4.24 *Chymisches medizinisches Vademecum*

1. *Printing History, Manuscripts.* Not printed. One manuscript known.
2. *Editions.* Not edited by Huser or Sudhoff.
3. *Relationship between different versions.* Only one known version.
4. *Structure, genre/form, perspective, style.* Written in the first person, directly addressing the reader.
5. *Relationship to other texts.*
6. *Authenticity, authorship.* The *Vademecum* is a conventional practical alchemical text with no connection to Paracelsus.
7. *Time of writing.* Further study required.

WORD COUNT: ca. 4.500 words

INCIPIT: "Allhier hebet sich an das allerhöchste geheimnüß der gantzen natur …"

MANUSCRIPTS:
– LEIPZIG, *Universtitätsbibliothek*: Cod. mag. 82; 11 fols.

FIRST PRINTED: not printed

HISTORICAL MANUSCRIPT CATALOGUES: Schröer (1710), 4 n° 135; Schröer/Roth-Scholtz (1732), 16 n° 137.

ESSENTIAL BIBLIOGRAPHY:

FURTHER BIBLIOGRAPHICAL REFERENCES:
Bellingradt and Otto, *Magical Manuscripts* (2017), 150 n° 135.

4.25 *Clavis oder Das zehnte Buch der Archidoxen* (*Lux lucens in tenebris*) also: *Das zehnte Buch der Archidoxen/Liber X. Archidoxorum sive Lux Lucens in Tenebris et Clavis Librorum/Hauptschlüssel der paracelsischen Arcanen oder Das zehnte Buch der Archidoxen/Lux in Tenebris Lucens Archidoxorum praecipuorumque medicinalium spagyricorum clavis/Liber Lux lucens in tenebris/Decimus liber Archidoxorum Theophrasti secretus*

1. *Printing History, Manuscripts.* First edited by Johannes Staricius in 1624. Several manuscripts.
2. *Editions.* Not edited by Huser or Sudhoff.
3. *Relationship between different versions.* Only one version known.
4. *Structure, genre/form, perspective, style.*
5. *Relationship to other texts.* Not connected in any way to *Das zehnte Buch der Archidoxen vom Stein der Weisen* (§ 4.56) first edited by Friedrich Moller in 1618.
6. *Authenticity, authorship.* According to a legend, the manuscript of this text was written in cipher[230] and in Latin by Paracelsus himself, and was later found hidden in a wall in Tyrol. On behalf of archduke Maximilian of Austria the text would have been translated into German by Adam Haslmayr and the translation later revised by Andreas Orthelius of Rudolstadt.[231]

230 For this cipher alphabet ("Theophrasti Alphabethum") used by Haslmayr, Paul Nagel and others, see Gilly, *Adam Haslmayr* (1994), 101 n. 11.

231 Legend found in ÜBERLINGEN, *Leopold Sophien Bibliothek*: Ms. 177, fol. 91–93, cited *in extenso* in CP 3: 1172–1173 and also by Gilly, *Adam Haslmayr* (1994), 94–96 and 101 n. 10.

7. *Time of writing*. Brought into circulation by Adam Haslmayr[232] ca. 1610 and then by Benedictus Figulus in 1611.[233] Written possibly immediately before.

Word count: ca. 5.000 words (ed. 1624); ca. 6.000 words (manuscripts)

Incipit: "Ob wir wol vnsere Archidoxa vnnd andere bücher die medicinam betreffendt …"

Manuscripts:

- Amsterdam, *Bibliotheca Philosophica Hermetica*: PH559 [manuscript in a copy of vol. 6 of Huser's edition]
- Göttingen, *Niedersächsische Staats- und Universitätsbibliothek*: 8 Cod. Ms. hist. nat. 14, f. 1–15 [copied from the 1684 printed edition]
- Gotha, *Staatsarchiv*: Geheimes Archiv E XI n° 83, f. 1–24
- Hamburg, *Staats- und Universitätsbibliothek*: Cod. alchim. 555; 30 pages
- Karlsruhe, *Badische Landesbibliothek*: Hs. Bernus 149, f. 2r–34v
- Kassel, *Landesbibliothek*: 4° Ms. chem. 39, f. 10r–11r [only summaries of the ten chapters; written by Benedictus Figulus]
- Kassel, *Landesbibliothek*: 4° Ms. chem. 58, f. 125r–134v
- London, *Royal Society*: Boyle Papers, RB/1/14, f. 166–177 [English?]
- London, *Wellcome Library*: MS 3563, 530–538 [partly in shorthand]
- Strasbourg, *Bibliothèque nationale et universitaire*: ms. 199, f. 277–298
- Weimar, *Herzogin Anna Amalia Bibliothek*: Q 462, f. 63r–81r
- Wolfenbüttel, *Herzog August Bibliothek*: Cod. Guelf. 94 Blank., f. 1r–9r

First printed:

- **1624** (*Clavis, Oder/ Das Zehende Buch der Archidoxen. D: Philippi Theophrati Paracelsi, Bombast von Hohenheim/ etc.*, ed. Johannes Staricius (Magdeburg: Johann Francke, 1624); VD17 14:628169Z; Sudhoff, *Bibliographia Paracelsica*, 537–538 n° 328)
- **1681** (in: *Lapis vegetabilis, Oder Die höchste Artzney* (Strasburg: Georg Andreas Dolhopff, 1681); VD17 1:000141S; Sudhoff, *Bibliographia Paracelsica*, 620–622 n° 414)
- **1684** (in: *Aureoli Theophrasti Paracelsi kleine Hand- und Denck-Bibel* (Frankfurt a.M.-Leipzig: Andreas Luppius, 1684); VD17 12:103261E; Sudhoff, *Bibliographia Paracelsica*, 623–626 n° 418)

232 Gilly, *Paracelsus in der BPH* (1993), 32: "das nicht von Paracelsus, sondern wohl von einem Bekannten des Adam Haslmayr stammt."

233 See the autograph letter from Benedictus Figulus to Count Moritz of Hessen-Kassel, dated 6 July 1611, edited in CP 3: 1153–1194 n° 166, esp. 1172–1174.

Historical Manuscript Catalogues: Widemann, *Unausgangene Bücher* (Hannover), n° 41; *Catalogus* Pfeffer (1683), 136

Essential bibliography: Sudhoff, *Bibliographia Paracelsica*, 537, 696 n° 48; Sudhoff, *Paracelsus-Handschriften*, 779 n° II/1b; CP 3: 1153–1194 n° 166, esp. 1172–1174.

Further bibliographical references:
Peuckert, *Pansophie* (1956), 432, 465.
Peuckert, *Gabalia* (1967), 322.
Peuckert, *Rosenkreutz* (1973), 243, 375.
Helmut Möller, *Staricius und sein 'HeldenSchatz'. Episoden eines Akademikerlebens* (Göttingen, 2003), 145–146.
Gilly, *Paracelsus in der BPH*, 32, 56, 68.
Gilly, *Adam Haslmayr* (1994), 94, 96, 97, 101–103 n. 7–16.
Broszinski, *Manuscripta chemica in quarto* (2011), 193.

4.26 *Clavis vel praxis manualis* also: *Clavis aut manuductio in suos libros*

1. *Printing History, Manuscripts.* First edited in Latin in 1613 by Johannes Rhenanus in his *Solis e puteo emergentis* [...] *libri tres*. No manuscripts known.
2. *Editions.* Not edited by Huser or Sudhoff.
3. *Relationship between different versions.* Only one known version.
4. *Structure, genre/form, perspective, style.*
5. *Relationship to other texts.*
6. *Authenticity, authorship.* According to Rhenanus, Paracelsus wrote this treatise, which he entitled *Clavis aut manuductio in suos libros*, in favor of a friend who, instead of publishing it, hid it until now. In Sudhoff's words: "No authenticity can be claimed here whatsoever."[234]
7. *Time of writing.* Possibly written in the 1600s or early 1610s. Further study required.

Word count: ca. 2.500 words

Incipit: "Quamvis in nostris libri veram auri potabilis ..."

Manuscripts: no manuscripts known

234 Sudhoff, *Bibliographia Paracelsica*, 499: "von 'Echtheit' kann hier keine Rede sein."

First printed:

- **1613** (in: Johannes Rhenanus, *Solis e puteo emergentis: sive dissertationis chymiotechnicae libri tres* (Frankfurt a.M.: Humm, 1613); VD17 12:160113P; Sudhoff, *Bibliographia Paracelsica*, 497–499 n° 294)

Essential bibliography: Sudhoff, *Bibliographia Paracelsica*, 498–499, 696 n° 25.

4.27 *De antimonio*
also: *De virtutibus antimonii/Tractatus de oleo stibii seu antimonii / Von der Tinctur oder Oleo Stibii/Von der Tinctur Antimonii*

1. *Printing History, Manuscripts.* First edited by Jonas Kitzkatz in 1583 (this text is different from all other manuscript and printed versions). Several manuscripts.
2. *Editions.* Not edited by Huser or Sudhoff.
3. *Relationship between different versions.* Attributed to Roger Bacon in some manuscripts and in a 1604 edition by Johann Thoelde.
4. *Structure, genre/form, perspective, style.*
5. *Relationship to other texts.* According to van Gijsen *De antimonio* is connected to chapter 14 of the *Aurora philosophorum* (§ 4.5).[235] There is, however, no direct consistency with the latter whatsoever. The relationship between *De antimonio* and an assumed treatise of Isaac Hollandus on antimony[236] is equally unclear. – Whether there is any connection with authentic Paracelsian texts on antimony[237] is yet to be established.

---

235 A. van Gijsen, "Isaac Hollandus revisited" (2010), 317. – Already noted by an owner or reader of 4° Ms. chem. 72 (Kassel, LB), fol. 173r: "videatur capite 14 in aurora philosophi Polacelsi et Gerhardi Dornii <antimonium> plumbum philosophorum."

236 A. van Gijsen, *ibid.*, 317.

237 E.g. *De vita longa*, book 3, chap. 6: "De antimonio", in Huser, 6: 177–178 (Latin) and 209 (German); *Große Wundarznei*, part 2, treatise 3, chap. 5: "Von der Tinctur Antimonij" (ed. 1536, fol. 42r–43r). – On antimony and Paracelsian medicine see John Brian Christopherson, "Discussion on the Special Use of Antimony," *Proceedings of the Royal Society of Medicine*, 22 (1929), 559–568; Hans Schmidt, "Antimon in der Arzneimittelsynthese," *Zeitschrift für angewandte Chemie*, 43 (1930), 963–970; Paul Walden, "Zur Geschichte der Antimonpräparate und des Basilius Valentinus," *Scientia Pharmaceutica*, 13 (1942), 5–8; Wilhelm Ganzenmüller, "Geschichtliches," in *Gmelins Handbuch der anorganischen Chemie*, vol. 18: *Antimon*, part A (Weinheim, 1950), 1–48; Friedrich Dobler, "Die chemische Arzneibereitung bei Theophrastus Paracelsus am Beispiel seiner Antimonpräparate," *Pharmaceutica Acta Helvetiae*, 32 (1957), 181–193, 226–252; Friedrich Dobler, "Die chemische Fundierung der Heilkunde durch Theophrastus Paracelsus. Experimentelle Überprüfung seiner Antimonpräparate," in *Die Vorträge der Hauptversammlung der Internationalen Gesellschaft für Geschichte der Pharmazie (Luzern 1956)* (Vienna, 1957), 76–86; Wolfgang Schneider, "Paracelsus und das Antimon," in *Die Vorträge der Hauptversammlung der Internationalen Gesellschaft für Geschichte der*

6. *Authenticity, authorship. De antimonio* is an anonymous text, variously attributed to Paracelsus or Roger Bacon. None of them is the author.
7. *Time of writing.* Possibly written in the 1560s.

Word count: ca. 3.000 words (1583 edition); ca. 1000–2.000 words (Kassel manuscripts); ca. 4000 words (1604 and 1608 editions)

Incipit: "Labahot ist ein schwefel damit zu calcioniren die metallen …" [ed. 1583] / "Stibium alß die philosophi sagen ist componirt aus dem edlen mineralischen sulphur …" [ed. 1604]

Manuscripts:

anonymous
- Kassel, *Landesbibliothek*: 4° Ms. chem. 19, f. 92r–100v
- Kassel, *Landesbibliothek*: 4° Ms. chem. 72, f. 245v–252v

attributed to Roger Bacon
- Kassel, *Landesbibliothek*: 2° Ms. chem. 3, f. 192v–198r
- Kassel, *Landesbibliothek*: 4° Ms. chem. 60[5,2, f. 173r–181v
- Kopenhagen, *Det Kongelige Bibliotek*: GKS 1724 kvart
- Leiden, *Universiteitsbibliotheek*: Voss. Chym. Q.17, 188v–196v

First printed:
- **1583** (in: *De antimonio tractatus Philippi Theophrati Paracelsi ab Hohenheim*, ed. Jonas Kitzkatz (Hof: Matthäus Pfeilschmidt, 1583); not in VD16; Sudhoff, *Bibliographia Paracelsica*, 340–341 n° 195) [attributed to Paracelsus]
- **1604** (in: *TriumphWagen Antimonii, Fratris Basilii Valentini*, ed. Johann Thoelde (Leipzig: Jacob Popporeich for Jacob Apel, 1604); VD17 12:111147B; not in Sudhoff, *Bibliographia Paracelsica*) [attributed to Roger Bacon]
- **1608** (in: Heinrich Khunrath, *De igne magorum philosophorumque secreto externo & visibili*, ed. Benedictus Figulus (Strasburg: Lazarus Zetzner, 1608); VD17 23:716355P; Sudhoff, *Bibliographia Paracelsica*, 489–490 n° 286)

Historical Manuscript Catalogues: *Libri Theophrasti* (Dresden), f. 6rb; Thomasius (1699), 140; Binz (1791), 153

Pharmazie (Dubrovnik 1959) (Stuttgart, 1960), 157–166; Gottfried Roth, "Der 'Antimonstreit' und die Wiener medizinische Fakultät," *Gesnerus*, 20 (1963), 165–169; Norbert Marxer, "Heilen mit Antimon: Von der Chemiatrie zur Chemotherapie," *Pharmazeutische Zeitung*, 145 (2000), 731–736.

Essential bibliography: Sudhoff, *Bibliographia Paracelsica*, 340–341, 695 n° 10; Sudhoff, *Paracelsus-Handschriften*, 186; CP 3: 985, 1005, 1066.

Further bibliographical references:
Peuckert, *Pansophie* (1956), 427–428.
Boeren, *Codices Vossiani Chymici* (1975), 145.
Annelies van Gijsen, "Isaac Hollandus Revisited," in Miguel López Pérez, Didier Kahn and Mar Rey Bueno, eds., *Chymia: Science and Nature in Medieval and Early Modern Europe* (Newcastle upon Tyne, 2010), 310–330, on 317.
Broszinski, *Manuscripta chemica in quarto* (2011), 61, 250, 323.

4.28 *De lumine naturae* also: *Vom Licht der Natur/De mercurio alchimistarum*

1. *Printing History, Manuscripts.* First published by Ps.-Huser in 1603. No manuscripts known.
2. *Editions.* Edited by Ps.-Huser in *Opera Bücher vnd Schrifften*, II, 1603, 682–686. Not edited by Sudhoff.
3. *Relationship between different versions.* The texts of Ps.-Huser's 1603 edition and the 1612 edition in *Zwey schöne Chymische Tractetlein* are essentially the same, although there are several misreadings in the 1603 edition; the latter also tends to add German translations to Latin expressions ("effectum oder wirckung" for "effectum"). Ps.-Huser's edition really consists of two separate treatises (*De lumine naturae* and *De mercurio alchimistarum*) by the same author printed one after the other without any separating heading.
4. *Structure, genre/form, perspective, style.* Written entirely in the third person, using no figurative, allegoric or idiomatic expressions.
5. *Relationship to other texts.*
6. *Authenticity, authorship.* The anonymous editor of *Zwey schöne Chymische Tractetlein* (1612) reported in his preface that he received *De lumine naturae* in 1588 in Prague from the hand of the author, Martin Faber from Königsberg in Prussia, doctor of medicine and lately deceased. The editor mentions the printing of *De lumine naturae* in Paracelsus's *Opera* of 1603, but adds that (1) *De lumine naturae* does not match the style of Paracelsus and (2) its author "was not in rerum natura" (i.e. did not live) during the lifetime of Paracelsus.[238]
7. *Time of writing.* Probably written in the 1580s by Martin Faber.

238 *Zwey schöne Chymische Tractetlein* (Jena-Leipzig, 1612), Vorrede, sig. A3r: "Vnd wiewol das Tractetlein de Lumine Naturae in die opera Paracelsi gesetzt/ vnd jme zugeeignet/ mag ich mit Warheit schreiben/ daß es Theophrasti Stylo nicht gemeß/ vnd dieser Doctor/ der es gemacht/ wie Paracelsus noch gelebt habe/ in rerum Natura nicht gewesen."

Word count: ca. 1.100 words (*De lumine naturae*, 1612); ca. 3.500 words (*De lumine naturae* and *De mercurio alchimistarum*, 1603)

Incipit: "So wir mit nutz die studia secretae philosophiae tractiren wollen …"

Manuscripts: no manuscripts known

First printed: not printed before Ps.-Huser (1603)
- **1603** (in: Ps.-Huser, *Opera Bücher vnd Schrifften* (1603); VD17 12:168390P)
- **1612** (in: *Zwey schöne Chymische Tractetlein: I. De Mercurio Alchimistarum, II. De Lumine Naturae* (Jena-Leipzig, 1612); VD17 23:296106T; reprinted in *Heptas Alchymica*, ed. Theophil Neander (Leipzig, 1621); VD17 23:240185V).

Historical Manuscript Catalogues: cf. *Catalogus* (Stuttgart), II, f. 1v, n° [12]

Essential bibliography: Sudhoff, *Bibliographia Paracelsica*, 445; CP 3: 788–789.

Further bibliographical references:
Florian Ebeling, "'Geheimnis' und 'Geheimhaltung' in den Hermetica der Frühen Neuzeit," in Anne-Charlott Trepp and Hartmut Lehmann, eds., *Antike Weisheit und kulturelle Praxis. Hermetismus in der frühen Neuzeit* (Göttingen, 2001), 63–80, on 74–75.

De reductione metallorum in primam materiam: see *Aus einem Brief an Poldörfer* (§ 4.16)

4.29 *De spiritibus metallorum* also: *Alchimia Theophrasti Paracelsi vom einfachen Feuer/Alchimia oder De spiritibus metallorum/Liber de generatione metallorum*

1. *Printing History, Manuscripts.* First printed in Latin in 1570, edited by Gerhard Dorn. First printed in German from two different but closely related sources in 1571 and 1572. Four manuscripts.
2. *Editions.* Edited by Huser, *Chir.* 1605, Appendix, 71–77. Not edited by Sudhoff.
3. *Relationship between different versions.* The 1571 version is almost word-by-word parallel to the Latin version printed in 1570, but some minor differences suggest that the text was translated from German to Latin and not vice versa.
4. *Structure, genre/form, perspective, style.*
5. *Relationship to other texts.*

6. *Authenticity, authorship.* Huser (1605) seems to be the first to reject this treatise as spurious. Sudhoff confirms this statement: "this text is without doubt not written by Hohenheim."[239]
7. *Time of writing.* Probably written in the 1560s.

Word count: ca. 4.500 [Latin text]; ca. 6.000 words [German text]

Incipit: "Invocato prius nomine domini Iesu Christi nostri servatoris opusculum hoc auspicabimur / In nomine Domini nostri Iesu Christi wöllen wir hie anfangen zů schreiben diß büchlein / Nun zubeschreiben was die kunst alchimia in sich begreifft ..."

Manuscripts:
- Budapest, *Országos Széchényi Könyvtár* [*National Library*]: Quart. Germ. 246 [copied from the 1584 printed edition]
- Dresden, *Sächsische Landesbibliothek*: Msrc.Dresd.J.346, f. 1r–28r
- Hamburg, *Staats- und Universitätsbibliothek*: Cod. alchim. 192, f. 274r–279r
- Karlsruhe, *Badische Landesbibliothek*: Hs. Bernus 148, f. 2r–5r
- Kassel, *Landesbibliothek*: 4° Ms. chem. 16, f. 37r–51v [copied from the 1572 printed edition]
- Kassel, *Landesbibliothek*: 4° Ms. chem. 25, f. 221r–240r
- Wien, *Österreichische Nationalbibliothek*: Cod. 11466, f. 93r–103r [copied from the 1618 Huser edition]
- Wolfenbüttel, *Herzog August Bibliothek*: Cod. Guelf. 30.4 Aug. 4°, 316, 318–348
- Wrocław/Breslau, *Biblioteka Uniwersytecka*: R 2317, 1–44 [probably copied from the 1571 printed edition]

First printed:
- 1570 (Latin; in *Aureoli Philippi Theophrasti Paracelsi De summis Naturae Mysterijs Libri tres*, ed. Gerhard Dorn (Basel, 1570); VD16 P 565; Sudhoff, *Bibliographia Paracelsica*, 209–211 n° 125)
- 1571 (German; in *De spiritibus planetarum sive metallorum Doctoris Theophrasti Paracelsi ab Hohenheim libri III* (Basel: Pietro Perna, 1571); VD16 P 567; Sudhoff, *Bibliographia Paracelsica*, 224–226 n° 134)
- 1572 (German; in *Metamorphosis. Doctoris Theophrasti von Hohenheim/ der zerstörten gůten küsnten vnnd artzney/ restauratoris/ gewaltigs vnnd nutzlichs schreiben*, ed. Adam von Bodenstein (no place [Basel: Samuel Apiarius], 1572); VD16

239 Sudhoff, *Bibliographia Paracelsica*, 209.

P 694; Sudhoff, *Bibliographia Paracelsica*, 229–231 n° 137; cf. 231 n° 138, and 351–352 n° 204)

HISTORICAL MANUSCRIPT CATALOGUES: Binz (1791), 139 [*Metamorphosis*]

ESSENTIAL BIBLIOGRAPHY: Sudhoff, *Bibliographia Paracelsica*, 209–210, 225, 463; Sudhoff, *Paracelsus-Handschriften*, 777.

FURTHER BIBLIOGRAPHICAL REFERENCES:

Peuckert, *Pansophie* (1956), 452.

Broszinski, *Manuscripta chemica in quarto* (2011), 86.

Urs Leo Gantenbein, "Die 'Alchimia Theophrasti Paracelsi': Eine pseudoparacelsische Rezeption der 'Visio Arislei' und des 'Donum Dei'," in *Nachlese zum 50. Paracelsustag 2001* (Vienna, 2002) (*Salzburger Beiträge zur Paracelsusforschung*, 35), 36–60.

Kathrin Pfister, "Kenelm Digby (1603–1665) und das 'sympathetische' Pulver – ein Beitrag zur Geschichte der pseudo-paracelsischen Waffensalbe," *Nova Acta Paracelsica*, N.F. 27 (2016), 47–66.

4.30 *De tinctura physicorum*
also: *Liber de tinctura physicorum contra sophistas natos post diluvium/Büchlin von der tinctura physica/Wider die sophisten der mittel alter*

1. *Printing History, Manuscripts.* First edited, all from different manuscripts, by Michael Toxites, Johann Albert Wimpinaeus, Pietro Perna (all three in 1570), Georg Phaedro (1571), Adam von Bodenstein (1574). Huser (1590) took his text from yet another manuscript. Gerhard Dorn published a Latin translation as early as 1570. At least seven further manuscripts.
2. *Editions.* Edited by Huser, 6 (1590): 363–374. Edited by Sudhoff in Paracelsus, *Sämtliche Werke*, I/14: 391–399.
3. *Relationship between different versions.* Only one version known.
4. *Structure, genre/form, perspective, style.*
5. *Relationship to other texts. De tinctura physicorum* cites two other pseudo-paracelsian alchemical texts: *Thesaurus thesaurorum* (§ 4.52) and *Liber vexationum* (§ 4.39). Sudhoff was inclined to ascribe all three texts to one single, unknown author;[240] however, this view seems to have no real foundation other than plagiarism of Paracelsus's wrath in *Paragranum* in the first two treatises. However, the motif of the 'red lion' connects *De tinctura physicorum* with *Thesaurus thesaurorum* (§ 4.52), *Brief an eine geistliche Person / Practica des*

240 Sudhoff, "Vorwort," in Paracelsus, *Sämtliche Werke*, I/14: XII.

verkehrten Mondes (§ 4.13) and especially *Brief an einen Gönner und Freund* (*Vom Wunderstein*) (§ 4.14), that also has the motif of the 'white eagle.'[241]

6. *Authenticity, authorship. De tinctura physicorum* is one of the most influential pseudo-Paracelsian alchemical writings. According to Kühlmann-Telle it "contributed significantly to the early modern legend of the successful alchemist Paracelsus. Its author presented it expressly as a response to the 'defamation' of Paracelsus as walking 'beggarly' through the country, as a 'vagrant beggar'."[242]
7. *Time of writing.* Probably written in the 1560s.

Word count: ca. 3.500 words

Incipit: "Dieweil du sophist mich mit so vil verleimbten worten hin vnd her tregst ..."

Manuscripts:

- Wrocław/Breslau, *Biblioteka Uniwersytecka*: R 2317, 45–72
- Dresden, *Sächsische Landesbibliothek*: Mscr.Dresd.N.107, f. 147r–156v
- Herdringen, *Bibliotheca Fürstenbergiana*: Ms. Fü 3550, 121–133
- Kassel, *Landesbibliothek*: 4° Ms. chem. 16, f. 65r–80v
- Kassel, *Landesbibliothek*: 4° Ms. chem. 17, f. 159v–168v
- Kassel, *Landesbibliothek*: 4° Ms. chem. 25, f. 210r–220v
- Wien, *Österreichische Nationalbibliothek*: Cod. 11343 [copy from a printed Huser edition]

First printed:
- 1570 (in: *Archidoxa Philippi Theophrasti Bombast*, ed. Michael Toxites (Strasburg: Theodosius Rihel, 1570); VD16 P 395; Sudhoff, *Bibliographia Paracelsica*, 187–190 n° 118; cf. Sudhoff, *Bibliographia Paracelsica*, 235–237 n° 140, 268–272 n° 158)
- 1570 (in: *D. Theophrasti Paracelsi von Hohenheim Archidoxa ex Theophrastia*, ed. Johann Albert Wimpinaeus (Munich: Adam Berg, 1570); VD16 P 392; Sudhoff, *Bibliographia Paracelsica*, 190–196 n° 119; cf. Sudhoff, *Bibliographia Paracelsica*, 215–217 n° 129)
- 1570 (Latin, in *Archidoxorum Aureolo Ph. Theophrasti Paracelsi de secretis naturae*, ed. Gerhard Dorn (Basel: Pietro Perna, 1570); VD16 P 398; Sudhoff, *Bibliographia Paracelsica*, 205–207 n° 123)
- 1570 (*D. Theophrasti Paracelsi Büchlin von der Tinctura Physica* (Basel: Pietro Perna, 1570); VD16 P 649; Sudhoff, *Bibliographia Paracelsica*, 212–213 n° 127)

241 CP 3: 231.
242 CP 3: 319.

- 1571 (in: *De spiritibus planetarum sive metallorum Doctoris Theophrati Paracelsi ab Hohenheim libri III*, ed. Georg Phaedro (Basel: [Pietro Perna], 1571); VD16 P 567; Sudhoff, *Bibliographia Paracelsica*, 224–226 n° 134)
- 1574 (in: *Metamorphosis Theophrasti Paracelsi*, ed. Adam von Bodenstein (Basel: Samuel Apiarius for Pietro Perna, 1574); VD16 P 695; Sudhoff, *Bibliographia Paracelsica*, 267–268 n° 157; cf. Sudhoff, *Bibliographia Paracelsica*, 351–352 n° 204)

HISTORICAL MANUSCRIPT CATALOGUES: *Catalogus* Osek (Prague) 1, n° 38; *Scripta Theophrasti* (Wien)

ESSENTIAL BIBLIOGRAPHY: Sudhoff, *Bibliographia Paracelsica*, 392–393; Sudhoff, *Paracelsus-Handschriften*, 773; Sudhoff, "Vorwort," in Paracelsus, *Sämtliche Werke*, I/14: XII–XVI; CP 1: 172, 465; CP 2: 30, 318, 319–320, 703, 1009–1010; CP 3: 292, 296–297, 425, 925.

FURTHER BIBLIOGRAPHICAL REFERENCES:

Peuckert, *Pansophie* (1956), 210, 252–254, 258–259, 281, 285–287, 458, 459, 493.

Peuckert, *Gabalia* (1967), 161.

Peuckert, *Rosenkreutz* (1973), 358.

Pagel, "The Paracelsian Elias Artista," 8, 12, 16.

Wolf-Dieter Müller-Jahncke and Joachim Telle, "Numismatik und Alchemie. Mitteilungen zu Münzen und Medaillen des 17. und 18. Jahrhunderts," in Christoph Meinel, ed., *Die Alchemie in der europäischen Kultur- und Wissenschaftsgeschichte* (Wiesbaden, 1986), 229–275, on 248 (on the term "lili").

Urs Leo Gantenbein, "*Separatio puri ab impuro*: Die Alchemie des Paracelsus," *Nova Acta Paracelsica*, N.F. 11 (1997), 3–59, on 14.

Carlos Gilly, "Johann Arndt und die 'dritte Reformation' im Zeichen des Paracelsus," *Nova Acta Paracelsica*, N.F. 11 (1997), 60–77, on 63.

Goldammer, *Der göttliche Magier* (1991), 105.

Gilly, "Vom ägyptischen Hermes zum Trismegistus Germanus" (2010), 80–81, 87.

Kahn, "Le retour de Gérard Dorn" (2010–2011), 149.

Broszinski, *Manuscripta chemica in quarto* (2011), 51, 54, 86.

Didier Kahn, "Le contexte alchimique des *Noces Chymiques*," *Recherches germaniques*, hors-série n° 13 (2018), 79–89.

4.31 *Figierung auf den Mercurium Saturni/Borras Theophrasti*

1. *Printing History, Manuscripts.* First printed in 1608, edited by Benedictus Figulus. Three manuscripts. The Kassel copy was not identified by Sudhoff or Broszinski.
2. *Editions.* Not edited by Huser or Sudhoff.

3. *Relationship between different versions.* The text printed by Figulus has a "verständlichere Erklärung" at the end not present in any manuscript. The Kassel copy is missing a few paragraphs and has some extra paragraphs as well. The section on "Borras Theophrasti" is incomplete in Hamburg alchim. 661 and missing both Hamburg alchim. 664 and Kassel 4° chem. 20.
4. *Structure, genre/form, perspective, style.* Written in the first person, directly addressing the reader. According to Karl Widemann addressed to a "D. Polmen" (reading unsure).
5. *Relationship to other texts.* Related to the Czech *Secretum Theophrasti* (§ 4.46). Possibly related to a text "De fixationibus mercurii saturnini" allegedly taken from an autograph of [Bartholomäus] Korndörffer in LEIDEN, *Universiteitsbibliotheek*: Voss. Chym. Q.38, f. 63r–64r.[243] Further study required.
6. *Authenticity, authorship.*
7. *Time of writing.* Further study required.

WORD COUNT: ca. 450 words

INCIPIT: "Obwol diß geheimnus bey sich selbst was hochschetziges ist / Nimpt drei teil mercurius saturni, ein teil gold …"

MANUSCRIPTS:
- HAMBURG, *Staats- und Universitätsbibliothek*: Cod. alchim. 661, on 353
- HAMBURG, *Staats- und Universitätsbibliothek*: Cod. alchim. 664, on 107–108
- KASSEL, *Landesbibliothek*: 4° Ms. chem. 20, f. 100v–101v

FIRST PRINTED:
- 1608 (in: *Rosarium novum olympicum et benedictum. Das ist: Ein newer Gebenedeyter Philosophischer Rosengart*, ed. Benedictus Figulus (Basel, 1608); VD17 12:133068G; Sudhoff, *Bibliographia Paracelsica*, 482–485 n° 283)

HISTORICAL MANUSCRIPT CATALOGUES: Widemann, *Verzeichnisse* (Kassel), n° V, 156; Uffenbach (1720), VIII, col. 655–658 n° 12/XIV [= HAMBURG, alchim. 661]

ESSENTIAL BIBLIOGRAPHY: Sudhoff, *Bibliographia Paracelsica*, 483; Sudhoff, *Paracelsus-Handschriften*, 716–717 [not identified], 779 n° I/22.

243 Lederer, *Der Kölner Kurfürst Herzog Ernst von Bayern* (1992), 110.

Further bibliographical references:

Boeren, *Codices Vossiani Chymici* (1975), 195.

Thomas Lederer, *Der Kölner Kurfürst Herzog Ernst von Bayern (1554–1612) und sein Rat Johann Grasse (um 1560–1618) als Alchemiker der frühen Neuzeit. Ein Beitrag zur Geschichte des Paracelsismus*, Diss. phil. (Heidelberg, 1992), 110.

Broszinski, *Manuscripta chemica in quarto* (2011), 66 [not identified].

4.32 *Geheimes und vollständiges Wunsch-Hütlein*

1. *Printing History, Manuscripts.* First printed in 1738, edited by Sincerus Aletophilus (pseudonym).
2. *Editions.* Not edited by Huser or Sudhoff.
3. *Relationship between different versions.* Only one known version.
4. *Structure, genre/form, perspective, style.* Written in the first person.
5. *Relationship to other texts.*
6. *Authenticity, authorship.* Sincerus Aletophilus was the pseudonym of an alchemist or physician who also edited (or rather wrote) in 1738 a *Tractatus chymicus de quinta essentia Das ist: Chymisches Werck von dem Fünfften Wesen* under the name of Basilius Valentinus.[244] Both texts were on the list of eleven manuscripts (1735) that the publisher, Augustinus Crusius in Erfurt, was about to edit.[245] The title may have been inspired by a sentence in the *Clavis oder Das zehnte Buch der Archidoxen* (§ 4.25) mentioning "des Fortunati Hütlein."[246]
7. *Time of writing.* The long preface was allegedly dated from Basel, 4 April 1523. Probably not written before the late 17th century.

Word count:

Incipit: "Ich habe zwar in unsern Archidoxis und andern Schriften [...] / Ich habe in meinem büchlein Archidoxorum von scheidung derer elementen ..."

Manuscripts:

– Praha, *Národní muzeum*: Cod. XI C 15 [probably copied from the printed edition]

First printed:

– **1738** (*Phil. Aureoli Theophrasti Paracelsi Bombast von Hohenheim Geheimes und vollständiges Wunsch-Hütlein*, ed. by Sincerus Aletophilus (Erfurt: August Crusius, 1738); VD18, 11358149; Sudhoff, *Bibliographia Paracelsica*, 648–649 n° 443)

---

244 Loth, "Das Medizinalwesen" (1904), 411.

245 Crusius (1735), two unnumbered pages (fol.)()(6rv), n° 10.

246 Gilly, *Paracelsus in der BPH* (1993), 59.

Historical Manuscript Catalogues: Crusius (1735), n° 10

Essential bibliography: Sudhoff, *Bibliographia Paracelsica*, 648–649, 697 n° 52; CP 3: 222, 229, 235, 297.

Further bibliographical references:

Richard Loth, "Das Medizinalwesen, der ärztliche Stand und die medizinische Fakultät bis zum Anfang des 17. Jahrhunderts in Erfurt," *Jahrbücher der Königlichen Akademie gemeinnütziger Wissenschaften zu Erfurt*, N.F. 30 (1904), 383–466.

Peuckert, *Pansophie* (1956), 457.

Gilly, *Paracelsus in der BPH*, 59.

4.33 *Glücksrute zu Paracelsi chymischem Schatz*

1. *Printing History, Manuscripts.* First edited in 1679 by A.A. in *Mercurii Zweyfacher Schlangen-stab.*
2. *Editions.* Not edited by Huser or Sudhoff.
3. *Relationship between different versions.* Only one known version.
4. *Structure, genre/form, perspective, style.* Written in the form of a dream or vision.
5. *Relationship to other texts.* A "key" to the treasure prophecy in the *Prophezeiung von den drei Schätzen und dem Löwen von Mitternacht* (§ 6.3). Opens with a short quotation from the *Prophezeiung* concerning the three treasures.
6. *Authenticity, authorship.* Not attributed to Paracelsus.
7. *Time of writing.* Probably written in the mid-17th century. Further study required.

Word count:

Incipit: "Es hat dem könig Alexandro Magno, der kunstreiche baumeister auß Macedonien, Dinocrates, vorgebracht [...] / Vber diesem herrlichen schatz hatte ich anfangs mancherley gedancken ..."

Manuscripts:

- Washington, *Smithsonian Libraries*: MSS 000939 B

First printed:

- 1679 (in: *Mercurii Zweyfacher Schlangen-stab*, ed. by "A.A. Naturalium Cultor" (Ulm: Balthasar Kühns Witwe, 1679); VD17 39:117815F; Sudhoff, *Bibliographia Paracelsica*, 617–618 n° 410)

Essential bibliography: Sudhoff, *Bibliographia Paracelsica*, 617–618, 697 n° 50.

4.34 *Goldener Schlüssel*

1. *Printing History, Manuscripts.* Not printed. One manuscript (current location unknown, formerly in the library of Max Joseph Freiherr von Linden). Full title: "Der goldne Schlüssel Philippi Theophrasti aller seiner gehabten Arcanen, aus einem uralten Manuscript aufs fleißigste abkopiert, und dem Kunstbegierigen treuherzig entdeckt. 1583."
2. *Editions.* Not edited by Huser or Sudhoff.
3. *Relationship between different versions.* Only one version known.
4. *Structure, genre/form, perspective, style.* A collection of 55 alchemical processes, listed in the 1788 *Catalogus.*
5. *Relationship to other texts.*
6. *Authenticity, authorship.*
7. *Time of writing.* Von Linden's copy was dated 1583.

Word count:

Incipit:

Manuscripts: no manuscripts available

First printed: not printed

Historical Manuscript Catalogues: von Linden (1786), 19 n° 35; von Linden (1788), part 2, 125–129 n° 109

Essential bibliography: Sudhoff, *Paracelsus-Handschriften*, 24, 25.

4.35 *Heimlichkeit des Herrn Wilwold*
also: *Brief von Wilwold an Paracelsus/Tinctura Doctoris Wilibaldi Parisiensis*

1. *Printing History, Manuscripts.* First printed in 1756, edited by "B.A." in *Chymischer Versuche zweyte Sammlung.* One manuscript, dating from ca. 1570/1575.
2. *Editions.* Not edited by Huser or Sudhoff.
3. *Relationship between different versions.* The 16th century manuscript and the 18th century printed text have the same text, only the language of the printed version is modernized.
4. *Structure, genre/form, perspective, style.* Written in the form of an alchemical process allegedly sent by Wilwoldt (1756 edition: Weilwalden), doctor in Paris, to Paracelsus and dated 1 January 1529.

5. *Relationship to other texts.* Basically the same text is found in two manuscripts in Vienna, not connected with Paracelsus, but entitled "Opus magistri Willibaldini doctoris Parisicus"[247] and "Tinctura Doct. Willibaldi Parisiensis. Fixatio Solis,"[248] respectively. The text is Latin with a few German words. The first of these was written in the late 15th century.
6. *Authenticity, authorship.* Karl Widemann, in one of his catalogues of manuscripts, associated the *Heimlichkeit des Herrn Wilwold* with yet another Willibald: "Practica Wilibaldi Storch Itali, Doctoris & Rectoris vniuersitatis Parisiensis pro veritate laboris tam Vniuersalis quam Particularis proprio tractatu."[249] This leads to a text found in a Kassel manuscript by "Willibaldus Storch Italus Doctor ac Rector Alme Vniuersitatis Parisiensium […] Anno 1480"[250] which is especially interesting since it mentions "hoheim", identified by Hartmut Broszinski as Paracelsus.[251] It is not known if there is any connection with "Wilbold (alias Willibald) Storch," who was in imperial service in 1504 and was granted by Emperor Maximilian I the castle of "Clauß am Piern" (today probably: Klause am Pyhrn) in 1512, where he died in 1523.[252]
7. *Time of writing.* Allegedly written in 1529. – Only the manuscript has a note at the end, dated 4 May 1547, stating that the text was copied by "Christoff Esch" (?) from a manuscript owned by "Marttin Mulpacher." It is not clear, however, if the text was attributed to Paracelsus at this time already. The oldest manuscript of the original text is from the late 15th century.

Word count: ca. 300 words

247 Wien, ÖNB, Cod. 5230, f. 39v. From the possession of Nicolaus Pol (1494). For Pol see Anke Timmermann, "Prescriptions of Alchemy. Two Austrian Medical Doctors and their Alchemical Manuscripts," in Karen Hunger Parshall, Michael T. Walton and Bruce T. Moran, eds., *Bridging Traditions. Alchemy, Chemistry, and Paracelsian Practices in the Early Modern Era* (Kirksville, MO, 2015), 159–185, on 169, 173, 180. The text ascribed to Willibaldus is not mentioned.

248 Wien, ÖNB, Cod. 11355, f. 342r–v (17th century).

249 Karl Widemann, "Chymici manu scripti libri omnis generis secretiores de summa Veterum Medicina, et alij Philosophici" (Kassel, LB, 2° Ms. chem. 7[3, f. 55r–58r), f. 58r n° 109.

250 Kassel, LB, 4° Ms. chem. 61, f. 44v–56v.

251 Broszinski, *Manuscripta chemica in quarto* (2011), 296.

252 Johann Georg Adam von Hoheneck, *Die Löbliche Herren Herren Stände, von Herren- und Ritterstand, in dem Ertz-Hertzogthum Oesterreich ob der Ennß* (Passau, 1747), 720: "Herr Wilbold (alias Willibald) Storch […] ward Anno 1504. Kayserl. Verweser zu Ausse/ und überkame seiner treuen Dienst willen von Kayser Maximiliano I. das Schloß Clauß am Piern geschencket […] Er Herr Willibald Storch zu Clauß stirbt hierauf Anno 1523 […]."

INCIPIT: "Recipe unam partem ducatorum tenuissime laminati …" [Wien 11355] / "Recipe ain tail dün geplecht gold, das rein ist …" [Wolfenbüttel] / "Nimm von einem sehr feinen Blatgold ein Theil …" [ed. 1756]

MANUSCRIPTS:
- WOLFENBÜTTEL, *Herzog August Bibliothek*: Cod. Guelf. 18.13 Aug. 4°, f. 210v–211r
- WIEN, *Österreichische Nationalbibliothek*: Cod. 5230, f. 39v [15th century, not attributed to Paracelsus]
- WIEN, *Österreichische Nationalbibliothek*: Cod. 11355, f. 342r–? [not attributed to Paracelsus]

FIRST PRINTED:
- 1756 (*Chymischer Versuche zweyte Sammlung*, ed. by "B.A." (Frankfurt a.M.-Leipzig, 1756); not in Sudhoff, *Bibliographia Paracelsica*)

ESSENTIAL BIBLIOGRAPHY: missing in both Sudhoff's *Bibliographia Paracelsica* and *Paracelsus-Handschriften.*

FURTHER BIBLIOGRAPHICAL REFERENCES:
Telle, "Paracelsus in pseudoparacelsischen Briefen" (2006–2007), 155 n. 28.
Broszinski, *Manuscripta chemica in quarto* (2011), 296.

4.36 *Kurzer Weg den Lapidem Philosophorum zu bereiten*

1. *Printing History, Manuscripts.* First edited as a work of Paracelsus in 1749 by Johann Georg Wagentruz in *Lumen Artis Prudentiae Intelligentiae Sapientiae.* One manuscript. Another manuscript was offered in 1798.
2. *Editions.* Not edited by Huser or Sudhoff.
3. *Relationship between different versions.* Only one known version.
4. *Structure, genre/form, perspective, style.*
5. *Relationship to other texts.* The beginning is the same text as chapter 21 plus the "clavis" to chapter 21 of J.G. Toeltius, *Coelum reseratum chymicum* (Frankfurt a.M.-Leipzig, 1737). Toeltius's *Coelum* was one of the works announced for publication by the Erfurt publisher Augustinus Crusius in 1735.[253] There are, however, many differences in respect to single words. It is not known if the *Kurze Weg* is compiled from a manuscript of the *Coelum reseratum chymicum* or if both texts have the same sources. On the other hand, Wagentruz's publication is known for being compiled from many different sources.[254]

253 Crusius (1735), two unnumbered pages (fol.)()(6rv).
254 Telle, *Alchemie und Poesie* (2013), 521, 608, 756, 977, 981.

6. *Authenticity, authorship.*
7. *Time of writing.* Further study required.

Word count:

Incipit: "Du sollst wissen, daß fast in keinem ding in der welt mehr feuer verborgen ..."

Manuscripts:
- Praha, *Národní muzeum*: Cod. XI C 15 [probably copied from the printed edition]

First printed:
- 1749 (in: *Lumen Artis Prudentiae Intelligentiae Sapientiae Das ist: Wahre, klare und aus dem rechten Philosophischen Grund fliessende Unterweisung* [...], ed. Johann Georg Wagentruz (Frankfurt a.M.-Leipzig, 1749)

Historical Manuscript Catalogues: Dost (1791–1798): 1798, col. 309 n° 56

Essential bibliography: Sudhoff, *Bibliographia Paracelsica*, 655–656, 697 n° 55.

Further bibliographical references:

4.37 *Lapis philosophorum*

1. *Printing History, Manuscripts.* Not printed. Two manuscripts.
2. *Editions.* Not edited by Huser or Sudhoff.
3. *Relationship between different versions.* Only one version known.
4. *Structure, genre/form, perspective, style.* Seven chapters ("septem gradus"). Written in the third person.
5. *Relationship to other texts.* The first sentence only is a quotation from book 4 of the *Archidoxis* (Huser, 6: 33).
6. *Authenticity, authorship.* Allegedly acquired in 1573 by "J. Erasmus Rulandt" from Emperor Maximilian II's physician "Doctor Suevus" in Vienna. None of these persons is listed in a 1566 "Hofstaatsverzeichnis" by Nicolaus Mameranus.[255] It is not known if there is any connection with Bernhard Suevus, doctor of medicine, 1623 physicus in Rotenburg ob der Tauber, deceased in 1636.[256]
7. *Time of writing.* Possibly written in the 1570s.

---

255 Nicolaus Mameranus, *Kurtze vnd eigentliche verzeychnus der Römischen Kayserlichen Mayestat/vnnd jhrer Mayestat Gemahels Hofstats/vnnd aller anwesenden Churfürsten* [...] *vnd der abwesenden Räthe* (Augsburg: Matthaeus Franck, 1566); VD16 M 443.

256 Wilhelm Ludwell, *Sambt und Sondere Rechts-Aussprüche und Belehrungen über Zerschiedene und rare Rechts-Fälle* (Nuremberg-Brünn: Georg Lehmann, 1712), 230.

Word count: ca. 2.000 words

Incipit: "Vnndt wie wol das ist, das solches arcanum mehr englisch dann menschlich jst zu melden …"

Manuscripts:
- Kassel, *Landesbibliothek*: 4° Ms. chem. 58, f. 47r–49v
- Kassel, *Landesbibliothek*: 8° Ms. chem. 28[4, f. 3v–10v

First printed: not printed

Essential bibliography: Sudhoff, *Paracelsus-Handschriften*, 725, 729–730, 780 n° II/14.
Broszinski, *Manuscripta chemica in quarto* (2011), 192.

4.38 *Liber mercuriorum* also: *De mercuriis metallorum*

1. *Printing History, Manuscripts.* First printed in Latin in 1582, along with the *Manualia* (§ 8.9). A German translation was printed in 1605, edited by Johann Huser.[257] Two German, one Latin manuscript.
2. *Editions.* Edited by Huser, *Chir.* 1605, Appendix, 98–99. Not edited by Sudhoff.
3. *Relationship between different versions.* Probably originally written in German: the Latin text printed in 1582 preserves some German words. The Latin text of the Kassel manuscript is different from the Latin text printed in 1582, according to Sudhoff.
4. *Structure, genre/form, perspective, style.*
5. *Relationship to other texts.* A short process in Kassel, *Landesbibliothek*: 4° Ms. chem. 8, f. 59v may be related to the "Mercurium solis" process of the *Liber mercuriorum*. Further study required.
6. *Authenticity, authorship.* Rejected as spurious by the writer of the Lübeck manuscript,[258] by Huser[259] and Sudhoff.
7. *Time of writing.* Probably written in the 1570s.

Word count: ca. 1.700 words [Huser's German text and Heidelberg manuscript]

---

257 Sudhoff, *Bibliographia Paracelsica*, 333: "Sie erscheinen hier lateinisch, Huser hat sie ins Deutsche hiernach übersetzen lassen."

258 "Non credo Theophrasti esse hunc librum, sed alterius cuiusdam docti Viri" (quoted by Sudhoff, *Paracelsus-Handschriften*, 118).

259 "für Theophrasti Werck vnd Arbeit nicht zu halten" (Huser, quoted by Sudhoff, *Bibliographia Paracelsica*, 333).

INCIPIT: "Mach dir ein aquafort von vier pfunden salpeter / Extrahe aquam fortem ex libris quatuor salis nitri …" [ed. 1582]

MANUSCRIPTS:

German

- HEIDELBERG, *Universitätsbibliothek*: Cod. Pal. germ. 598, f. 151r–157r [German]
- LÜBECK, *Stadtbibliothek*: Ms. med. 4° 12 [German; lost in war]

Latin

- KASSEL, *Landesbibliothek*: 4° Ms. chem. 18, f. 37v–38r [Latin]

FIRST PRINTED:

- 1582 (Latin; in *Aureoli Theophrasti Paracelsi Archidoxorum seu de Secretis Naturae mysteriis libri decem* (Basel: Pietro Perna, 1582), 173–180; VD16 P 399; Sudhoff, *Bibliographia Paracelsica*, 331–334 n° 191)

HISTORICAL MANUSCRIPT CATALOGUES: cf. *Libri Theophrasti* (Dresden), f. 6rb; cf. Widemann, *Unausgangene Bücher* (Hannover), n° 35, 127; Widemann, *Verzeichnisse* (Kassel), n° v, 156; Thomasius (1699), 140; Binz (1791), 153

ESSENTIAL BIBLIOGRAPHY: Sudhoff, *Bibliographia Paracelsica*, 463; Sudhoff, *Paracelsus-Handschriften*, 118, 156, 735, 777.

FURTHER BIBLIOGRAPHICAL REFERENCES:

Broszinski, *Manuscripta chemica in quarto* (2011), 57.

Pamela Kalnig, Matthias Miller and Karin Zimmermann, *Die Codices Palatini germanici in der Universitätsbibliothek Heidelberg (Cod. Pal. germ. 496–670)* (Wiesbaden, 2014), 263–264.

4.39 *Liber vexationum* also: *Coelum philosophorum/Kunst und Natur der Alchemie/De vexationibus alchimistarum/Liber fixationum*

1. *Printing History, Manuscripts.* First edited in 1567 by Adam von Bodenstein as a separate work. In 1574 Bodenstein published a revised edition that corrected some misreadings. The first Latin translation was published in 1568 by Gerhard Dorn. A partial Latin translation was prepared by Georg Joachim Rheticus (autograph manuscript in Florence).[260] Three manuscripts.

260 Sudhoff, "Rheticus und Paracelsus" (1903), 1850: "eine flotte lateinische Eingewandung dieses alchemistischen Traktates, die sehr zu ihrem Vorteil absticht von der ungelenken Latinisierung Gerhard Dorns." – Burmeister, *Georg Joachim Rhetikus* (1967), 2: 23 n° 8;

2. *Editions.* Edited by Huser, 6 (1590): 375–395. Edited by Sudhoff in Paracelsus, *Sämtliche Werke*, I/14: 405–420.
3. *Relationship between different versions.* Further study required. – A fragment (230 words) is found in KASSEL, *Landesbibliothek*: 4° Ms. chem. 2, f. 20r–20v.
4. *Structure, genre/form, perspective, style.*
5. *Relationship to other texts.*
6. *Authenticity, authorship.* In Sudhoff's eyes, the author was "clearly not the brightest of sparks,"[261] which clearly differentiates this author from that of *De tinctura physicorum* (§ 4.30), described by Sudhoff as much at ease and familiar with the works of Paracelsus.[262] Finally, Sudhoff tends to consider the *Liber vexationum* as intrinsically close to (and potentially to be ascribed to the same author as) the *Manuale de lapide philosophico* (§ 4.40).[263]
7. *Time of writing.* Probably written in the 1560s. – A manuscript (now lost and not seen by Sudhoff himself) contained the text under the title *Kunst vnd natur der Alchymia vnd was darauff zuhalten sy* and ascribed it to Paracelsus, which is remarkable since the binding of the manuscript is said to be dated 1542. However it is questionable how far this second-hand and somewhat imprecise information is reliable.

WORD COUNT: ca. 6.500 words

INCIPIT: "Ir lieben vnd erfarnen meister diser kunst alchimiae …"

MANUSCRIPTS:

German
- HERZOGENBURG, *Stiftsbibliothek*: Ms 363 [lost]
- KOPENHAGEN, *Det Kongelige Bibliotek*: Thott 737 kvart [probably copied from a printed edition]
- WOLFENBÜTTEL, *Herzog August Bibliothek*: Cod. Guelf. 60.6 Aug. 8°, f. 43r–67r

Latin
- FIRENZE, *Biblioteca Nazionale Centrale*: Magl. XVI 113 (Latin translation by Georg Joachim Rheticus, autograph)

Figala, "Die sogenannten Sieben Bücher" (1971); Burmeister, "Die chemischen Schriften des G.J. Rhetikus" (1974), 180–182; Burmeister, "Philippus Theophrastus Paracelsus" (2015), 440, 441 n. 19.

261 Sudhoff, "Vorwort," in Paracelsus, *Sämtliche Werke*, I/14: XXIV: "eine Leuchte war er nicht."

262 *Ibid.*, XVI and XVII.

263 *Ibid.*, XXIV.

First printed:
- 1567 (in: *Liber Vexationum D. Phil. Theophrasti Paracelsi*, ed. Adam von Bodenstein (no place and date [preface: October 1567; printed by Perna in Basel]); VD16 P 595; Sudhoff, *Bibliographia Paracelsica*, 143–144 n° 90; cf. 257–260 n° 153)
- 1568 (Latin; *Pyrophilia vexationumque liber. D. Phil. Theophrasti Paracelsi*, ed. Gerhard Dorn (Basel: Pietro Perna, 1568); VD16 P 597; Sudhoff, *Bibliographia Paracelsica*, 160–162 n° 101)

Historical Manuscript Catalogues: *Scripta Theophrasti* (Wien); Widemann, *Verzeichnisse* (Kassel), n° II, 122; von Linden (1786), 31 n° 174

Essential bibliography: Sudhoff, *Bibliographia Paracelsica*, 393; Sudhoff, *Paracelsus-Handschriften*, 773–774; Sudhoff, "Vorwort," in Paracelsus, *Sämtliche Werke*, I/14: XX, XXIV; CP 1: 98, 496; CP 3: 708.

Further bibliographical references:

Karl Sudhoff, "Rheticus und Paracelsus, in neuer Beitrag zu ihren Beziehungen," *Münchener Medizinische Wochenschrift*, 50 (1903), vol. 2, n° 42, on 1850 (short notice).

Karl Sudhoff, "Noch einmal Rheticus und Paracelsus," *Verhandlungen der Naturforschenden Gesellschaft in Basel*, 15 (1904), 329–333.

Peuckert, *Pansophie* (1956), 252–253.

Peuckert, *Gabalia* (1967), 96.

Karl Heinz Burmeister, *Georg Joachim Rhetikus 1514–1574. Eine Bio-Bibliographie*, 3 vols. (Wiesbaden, 1967), 1: 23, 156; 2: 23 n° 8.

Karin Figala, "Die sogenannten Sieben Bücher über die Fundamente der chemischen Kunst von Joachim Rhetikus (1514–1576)," *Sudhoffs Archiv*, 55 (1971), 247–256, on 255–256.

Karl Heinz Burmeister, "Georg Joachim Rhetikus als Paracelsist," *Montfort*, 24 (1972), 619–629.

Karl Heinz Burmeister, "Die chemischen Schriften des Georg Joachim Rhetikus," *Organon*, 10 (1974), 177–185.

Gilly, *Adam Haslmayr* (1994), 164, 206.

Kahn, "Le retour de Gérard Dorn" (2010–2011), 147.

Karl Heinz Burmeister, "Philippus Theophrastus Paracelsus," in Burmeister, *Magister Rheticus und seine Schulgesellen. Das Ringen um Kenntnis und Durchsetzung des heliozentrischen Weltsystems des Kopernikus um 1540/50* (Konstanz-München, 2015), 437–442.

Lux lucens in tenebris: see *Clavis oder Das zehnte Buch der Archidoxen* (§ 4.25)

4.40 *Manuale de lapide philosophico medicinali* also: *De lapide philosophico medicinali/Psalterium chymicum*

1. *Printing History, Manuscripts.* First edited from different sources in 1572 by Adam von Bodenstein and Michael Toxites. Printed in 1771 under the title *Psalterium chymicum.* One manuscript.
2. *Editions.* Edited by Huser, 6 (1590): 421–436. Edited by Sudhoff in Paracelsus, *Sämtliche Werke*, I/14: 421–432.
3. *Relationship between different versions.* No information about different versions.
4. *Structure, genre/form, perspective, style.*
5. *Relationship to other texts.*
6. *Authenticity, authorship.* In a note to the reader preceding the text, Huser points out that "many people" consider the *Manuale* spurious; his own view remains unclear.
7. *Time of writing.* Probably written in the late 1560s.

Word count: ca. 5.000 words

Incipit: "Dieweil Gott den geist der artzney durch Machaonem Podalyrium Apollinem Hyppocratem etc. grundtlich hatt angefangen zů würcken …"

Manuscripts:
- Berlin, *Staatsbibliothek*: Ms. lat. oct. 182
- Budapest, *Országos Széchényi Könyvtár* [*National Library*]: Quart. Germ. 246 [copied from the 1572 printed edition]
- Heidelberg, *Universitätsbibliothek*: Cod. Pal. germ. 295, f. 66r–71v, 73r–79r [copied from the 1572 printed edition]
- Kassel, *Landesbibliothek*: 4° Ms. chem. 78, f. 48r–65v
- Weimar, *Herzogin Anna Amalia Bibliothek*: Q 462, f. 173r–174v [fragment, with additional text]

First printed:
- **1572** (in: *Metamorphosis. Doctoris Theophrasti von Hohenheim/ der zerstörten gůten küsnten vnnd artzney/ restauratoris/ gewaltigs vnnd nutzlichs schreiben*, ed. Adam von Bodenstein (no place [Basel: Samuel Apiarius], 1572); VD16 P 694; Sudhoff, *Bibliographia Paracelsica*, 229–231 n° 137)
- **1572** (in: *De Lapide Philosophorum, Drey Tractat/ Philippi Theophrasti Paracelsi*, ed. Michael Toxites (Strasburg: Nicolaus Wiriot, 1572); VD16 P 488; Sudhoff, *Bibliographia Paracelsica*, 235–237 n° 140; cf. 268–272 n° 158)

Historical Manuscript Catalogues: *Scripta Theophrasti* (Wien); Binz (1791), on 134

Essential bibliography: Sudhoff, *Bibliographia Paracelsica*, 394; Sudhoff, *Paracelsus-Handschriften*, 774; Sudhoff, "Vorwort," in Paracelsus, *Sämtliche Werke*, I/14: XXIV–XXV; CP 1: 599; CP 2: 285–286; CP 3: 50, 708, 788.

Further bibliographical references:

Peuckert, *Pansophie* (1956), 210–211, 252–253, 256, 259, 465.

Peuckert, *Gabalia* (1967), 322.

Pagel, "The Paracelsian Elias Artista," 13.

Achim Aurnhammer, "Zum Hermaphroditen in der Sinnbildkunst der Alchemisten", in Christoph Meinel, ed., *Die Alchemie in der europäischen Kultur- und Wissenschaftsgeschichte* (Wiesbaden, 1986), 179–200, on 194.

Broszinski, *Manuscripta chemica in quarto* (2011), 342.

Carlos Gilly, *500 лет гнозиса в Европе. гностическая традиция в печатных и рукописных книгах* (Amsterdam, 1993), 118 n° 21a.

Manuel de Paracelse: see *Short Catechism of Alchemy* (§ 4.47)

4.41 *Methodus chymica*

1. *Printing History, Manuscripts.* One manuscript.
2. *Editions.* Not edited by Huser or Sudhoff.
3. *Relationship between different versions.* Only one known version.
4. *Structure, genre/form, perspective, style.*
5. *Relationship to other texts.*
6. *Authenticity, authorship.* Alchemical treatise with instructions for the preparation of a transmuting substance. Attributed to Paracelsus both in the heading and closing. Has "nothing to do with Paracelsus," according to Sudhoff.[264]
7. *Time of writing.* Further study required.

Word count:

Incipit: "Regula seu methodus, qua ratione puteis iste conficiendus […] / Daß die augmentation oder tinctur schwerlich oder gar nit kann geschehen …"

Manuscripts:

– Kassel, *Landesbibliothek*: 4° Ms. chem. 58, f. 118v–123r

First printed: not printed

264 Sudhoff, *Paracelsus-Handschriften*, 728, 727: "Es gilt dasselbe, wie bei den vorhergehenden Abschnitten," i.e. "Hat mit Hohenheim nichts zu thun."

Essential bibliography: Sudhoff, *Paracelsus-Handschriften*, 727–728, 780 n° II/17.

Further bibliographical references:
Broszinski, *Manuscripta chemica in quarto* (2011), 193.

4.42 *Monarchia Tincturae Physicorum*

1. *Printing History, Manuscripts.* Not printed. One manuscript
2. *Editions.* Not edited by Huser or Sudhoff.
3. *Relationship between different versions.* Only one version known.
4. *Structure, genre/form, perspective, style.* Divided into 21 chapters according to von Linden's *Catalogus*.
5. *Relationship to other texts.*
6. *Authenticity, authorship.* According to von Linden's *Catalogus* the author was Dr. med. Thomas Grünberger; however, Grünberger was probably only the compiler or writer of the manuscript.
7. *Time of writing.*

Word count:

Incipit:

Manuscripts:
- Bratislava, *Univerzitná knižnica*: Ms 1170 (E 1213), f. 6r–39v
- *olim* Tovačov, *Zámecká knihovna*: R 331[265]

First printed: not printed

Historical Manuscript Catalogues: von Linden (1788), part 2, 278 n° 365

Further bibliographical references:
Imrich Kotvan, *Codices manuscripti Bibliothecae Universitatis Bratislavensis/Rukopisy Univerzitnej knižnice v Bratislave* (Bratislava, 1970), 403.
Stanislav Petr, "Rukopisné fondy zámeckých a hradních knihoven na Moravě a ve Slezsku," *Problematika historických a vzácných knižních fondů října*, 4 (1994), 6–52, on 35.
Eva Beňová, *Theatrum chemicum, praecipuos selectorum avctorum tractatus de chemiae et lapidis philosophici antiqvitate, veritate, jure, praestantia, & operationibus,*

265 The Tovačov Castle Library is currently stored in the Olomouc branch of the Provincial Archive in Opava, see Beňová, *Theatrum chemicum* (2009), 13.

continens. Küenburská knižní sbírka 1532–1815, master thesis (Olomouc, 2009), online (https://theses.cz/id/na62ww, accessed 18 January 2021), 13.

4.43 *Mysterium lapidis philosophorum*

1. *Printing History, Manuscripts.* First edited in 1625 by Joachim Morsius (using the pseudonym Anastasius Philaretus Cosmopolita) in his *Magische Propheceyung Aureoli Philippi Theophrasti Paracelsi, Von Entdeckung seiner 3. Schätzen.* The manuscript in Lübeck was written ca. 1589/1590 by Balthasar Walther,[266] who provided Morsius with this text.
2. *Editions.* Not edited by Huser or Sudhoff.
3. *Relationship between different versions.* Only one version known.
4. *Structure, genre/form, perspective, style.*
5. *Relationship to other texts.* The last part of the text is written in cipher. This cipher does not seem to be related to the cipher used by Haslmayr, Paul Nagel and others; see *Clavis oder Das zehnte Buch der Archidoxen* (§ 4.25).
6. *Authenticity, authorship.* "Spurious without doubt" according to Sudhoff.[267]
7. *Time of writing.* Allegedly written in 1527. Probably written in the 1580s.

WORD COUNT: ca. 300 words

INCIPIT: "Diß jahr hab ich Theophrastus von Hohenheim aus vielerlittener muhe vnd arbeit vnd nachreden …" [MS Lübeck]

MANUSCRIPTS:
- KARLSRUHE, *Badische Landesbibliothek*: Hs. Allerheiligen 3, on 382–384
- LÜBECK, *Stadtbibliothek*: Ms. math. 4° 9, f. 35r
- NEW HAVEN, *Beinecke Library*: Beinecke MS 309, f. 75r–?

FIRST PRINTED:
- **1625** (in: *Magische Propheceyung Aureoli Philippi Theophrasti Paracelsi, Von Entdeckung seiner 3. Schätzen*, ed. Anastasius Philaretus Cosmopolita [=Joachim Morsius] ("Philadelphiae", 1625); not in VD17; Sudhoff, *Bibliographia Paracelsica*, 539–541 n° 330)

ESSENTIAL BIBLIOGRAPHY: Sudhoff, *Bibliographia Paracelsica*, 540–541, 697 n° 49.

266 Penman, "'Ein Liebhaber des Mysterii'" (2010), esp. 79, 94. – Penman quotes the Karlsruhe MS with wrong pages, it should read "pp. 382–384" (not "pp. 348–349").

267 Sudhoff, *Bibliographia Paracelsica*, 541: "zweifellos unächt."

FURTHER BIBLIOGRAPHICAL REFERENCES:

Sten Lindroth, *Paracelsismen i Sverige till 1600-talets mitt* (Uppsala, 1943), 175.

Peuckert, *Pansophie* (1956), 493.

Leigh T.I. Penman, "'Ein Liebhaber des Mysterii, und ein großer Verwandter desselben.' Toward the Life of Balthasar Walther: Kabbalist, Alchemist and Wandering Paracelsian Physician," *Sudhoffs Archiv*, 94 (2010), 73–99, on 94.

Leigh T.I. Penman, "Böhme's Student and Mentor: the Liegnitz Physician Balthasar Walther (c. 1558–c. 1630)," in Wilhelm Kühlmann and Friedrich Vollhardt, eds., *Offenbarung und Episteme. Zur europäischen Wirkung Jakob Böhmes im 17. und 18. Jahrhundert* (Berlin-Boston, 2012), 47–65, on 61.

Paracelsicae Büchlein belangend lapidem: see *Brief an Augustin Sattler* (§ 4.8)

Practica des verkehrten Mondes: see *Brief an eine geistliche Person* (§ 4.13)

4.44 *Praefiguratio in lapidem philosophorum*

1. *Printing History, Manuscripts.* Not printed. One manuscript, attributed to "Theophrastus von Hohenhaimb."
2. *Editions.* Not edited by Huser or Sudhoff.
3. *Relationship between different versions.* Only one version known.
4. *Structure, genre/form, perspective, style.* A series of illustrations accompanied by short texts in German.
5. *Relationship to other texts.* This series of illustrations was first printed in the *Pretiosa margarita*, ed. by Giano Lacinio (Venice, 1546),[268] sig. **8v–***4v and is also found (with Latin texts) in LEIDEN, *Universiteitsbibliotheek*: Voss. Chym. F.29, f. 97v–99v [untitled, anonymous].
6. *Authenticity, authorship.*
7. *Time of writing.*

WORD COUNT:

INCIPIT: "Zuwissen ist, dz zu anfang diser figur sitzt ein konig auff einem koniglichen thron …"

MANUSCRIPTS:

– NÜRNBERG, *Germanisches Nationalmuseum*: Hs. 16752, f. 43r–50v

---

268 On Lacinio see Francesco Vizza, "Giano Lacinio Alchimista Francescano del Cinquecento," in *Rendiconti della Accademia Nazionale delle Scienze detta dei XL*, Memorie di Scienze Fisiche e Naturali, serie V, vol. XXXIX, parte II, tomo II (2015), 37–46.

First printed: not printed

Further bibliographical references:

Norbert H. Ott, "Alchemie," in Hella Frühmorgen-Voss and Norbert H. Ott, eds., *Katalog der deutschsprachigen illustrierten Handschriften des Mittelalters*, vol. 1 (Munich, 1991).

Processus tincturae antimonii: see *Brief an Theodorus* (§ 4.17)

4.45 *Ratio extrahendi ex omnibus metallis mercurium*

1. *Printing History, Manuscripts.* First published by Huser in 1590 in Latin. Two German manuscripts, one Latin manuscript.
2. *Editions.* Edited by Huser, 6 (1590): 437–440 [Latin]. Edited by Sudhoff in Paracelsus, *Sämtliche Werke*, 1/14: 433–435 [Latin].
3. *Relationship between different versions.* The Kassel manuscript (Latin) is very close to Huser's Latin text, while the German manuscripts seem to be translated from Latin.
4. *Structure, genre/form, perspective, style.*
5. *Relationship to other texts.*
6. *Authenticity, authorship.* Text considered by Huser as clearly spurious.[269]
7. *Time of writing.* Probably written in the 1560s or earlier.

Word count: ca. 900 words

Incipit: "Extrahere corporibus metallicis mercurium nihil aliud est quam ipsa resolvere / Außziehen des mercurii aus den metallen ist nichts anders dann derselben resolvierung …"

Manuscripts:

German

- Praha, *Národní knihovna*: Osek 43, f. 156r–158r
- Wolfenbüttel, *Herzog August Bibliothek*: Cod. Guelf. 264.31 Extrav., on 284–285 [fragment]

Latin

- Kassel, *Landesbibliothek*: 4° Ms. chem. 54, f. 148r–151r

First printed: not printed before Huser (1590)

269 Huser, 6: 436.

ESSENTIAL BIBLIOGRAPHY: Sudhoff, *Bibliographia Paracelsica*, 394–395; Sudhoff, *Paracelsus-Handschriften*, 774; Sudhoff, "Vorwort," in Paracelsus, *Sämtliche Werke*, I/14: XXV; CP 2: 247, 703.

FURTHER BIBLIOGRAPHICAL REFERENCES:
Peuckert, *Pansophie* (1956), 252, 257.
Broszinski, *Manuscripta chemica in quarto* (2011), 184–185 [this text not mentioned].

Rote Juden: see *Augmentum Solis* (§ 4.4)

4.46 *Secretum Theophrasti*

1. *Printing History, Manuscripts.* In Czech. One manuscript. The full title is "Kniha dokonaleho Teynost Theoffrasta Paraczelsa." Above this title some 16th century owner has written "Secretum Theophrasti." Sudhoff could not obtain a translation of the text. Boeren does not give the Czech title but rather: "Secretum Theophrasti Paracelsi." According to Boeren it was edited by Otakar Zachar, a Czech historian of chemistry and alchemy, in a book or chapter called *Paracelsismus in Böhmen im XVI. Jahrhundert*. This, alas, was never published as far as we know. The text is, however, mentioned by Zachar in other publications.[270]
2. *Editions.* Not edited by Huser or Sudhoff.
3. *Relationship between different versions.* Only one known version.
4. *Structure, genre/form, perspective, style.*
5. *Relationship to other texts.* The Czech text contains a section on the "Borras Theophrasti" ("Burras Theoffrastuw") made of safran that is also part of *Figierung auf Mercurium Saturni Theophrasti* (§ 4.46).
6. *Authenticity, authorship.*
7. *Time of writing.* Further study required.

WORD COUNT:

INCIPIT: "Acz toto teynost welika gt adraha sama wsobie …"

MANUSCRIPTS:
– LEIDEN, *Universiteitsbibliotheek*: Voss. Chym. F.3, f. 360v–361r

FIRST PRINTED: not printed

270 Zachar, *O alchymii a českých alchymistech* (1911), 176–177.

Essential bibliography: Sudhoff, *Paracelsus-Handschriften*, 693, 770, 779 n° 1/24.

Further bibliographical references:
Otakar Zachar, *O alchymii a českých alchymistech. Dle staročeských pramenů a rukopisů alchymických* (Prague, 1911), 176–177.
Boeren, *Codices Vossiani Chymici* (1975), 11.

4.47 *Short Catechism of Alchemy* also: *Manuel de Paracelse*

1. *Printing History, Manuscripts.* First printed anonymously in 1766 [by Théodore Henri de Tschudy (1727–1769)] in *L'Étoile Flamboyante, ou la Société des Francs-Maçons*, vol. 2 (Frankfurt-Paris: Antoine Boudet, 1766) as *Catéchisme ou instruction pour le grade d'adepte ou apprentif Philosophe sublime et inconnu* (not attributed to Paracelsus). A German translation of the French original appeared in *Der flammende Stern. Oder die Gesellschaft der Freymäurer*, vol. 2 (no place, 1779) as *Catechismus, oder Unterricht für den Adeptengrad, oder den Lehrling der erhabenen und unbekannten Philosophen.* – First printed as a work of Paracelsus by Arthur Edward Waite in 1894.
2. *Editions.* Not edited by Huser or Sudhoff.
3. *Relationship between different versions.* Only one known version.
4. *Structure, genre/form, perspective, style.*
5. *Relationship to other texts.* Tschudy actually plagiarised the *Statuts des Philosophes inconnus* loosely attributed to Sendivogius and published in pseudo-Michael Sendivogius, *Traitez du Cosmopolite Nouvellement découverts* (Paris: Laurent d'Houry, 1691), 10–40.[271]
6. *Authenticity, authorship.* The Paracelsification of the text was initiated by Éliphas Lévi (1810–1875). In the second volume of his *Dogme et rituel de la haute magie* (1861) he praised the *Manuel de Paracelse*, a work that according to him contained "all the mysteries" of alchemy and cabala and was only to be found in the Vatican library where Sendivogius prepared a copy that in due course served baron Tschoudy for his publication of the *Catéchisme*.[272] Sudhoff searched in

271 See Kahn, "Alchimie et franc-maçonnerie au XVIIIe siècle" (2012), on [4].

272 Éliphas Lévi, *Dogme et rituel de la haute magie* (1861), 2: 296: "Parmi les livres rares et précieux qui contiennent les mystères du grand arcane, il faut compter au premier rang le *Sentier chimique* ou *Manuel de Paracelse*, qui contient tous les mystères de la physique démonstrative et de la plus secrète cabale. Ce livre manuscrit, précieux et original, ne se trouve que dans la bibliothèque du Vatican. Sendivogius en a tiré une copie dont le baron de Tschoudy s'est servi pour composer le catéchisme hermétique contenu dans son ouvrage intitulé: *L'Étoile flamboyante*."

vain for this manuscript.[273] Waite included an English translation in his collection of mostly pseudo-Paracelsian texts.

7. *Time of writing.* Probably written in the late 17th century.

Word count: ca. 8.500 words [1766 French edition]; ca. 8.000 words [1894 English translation ed. Waite]

Incipit: "What is the chief study of a philosopher?..."

Manuscripts: no manuscripts known

First printed:
- **1766** (in: [Théodore Henri de Tschudy], *L'Étoile Flamboyante, ou la Société des Francs-Maçons* (Frankfurt-Paris: Antoine Boudet, 1766), 2: 234–302) [not attributed to Paracelsus]
- **1779** (German translation; in *Der flammende Stern. Oder die Gesellschaft der Freymäurer* (no place, 1779, 2: 162–203) [not attributed to Paracelsus]
- **1894** (in: *The Hermetic and Alchemical writings of Aureolus Philippus Theophrastus Bombast, of Hohenheim, called Paracelsus the Great*, ed. Arthur Edward Waite (London: James Elliot, 1894), 1: 288–305)

Essential bibliography: Sudhoff, *Paracelsus-Handschriften*, 808.

Further bibliographical references:

Éliphas Lévi, *Dogme et rituel de la haute magie*, (Paris: Germer Ballière, 1861), 2: 296.

Harold Fisch, "Bacon and Paracelsus," *The Cambridge Journal*, 5, n° 12 (1952), 752–758, on 754.

Didier Kahn, "Alchimie et franc-maçonnerie au XVIIIe siècle," revised version (February 2012), online (http://halshs.archives-ouvertes.fr/halshs-00674236, accessed 20 November 2020), on [4].

4.48 *Speculum sapientiae* also: *Güldenes Testament/Testament d'or*

1. *Printing History, Manuscripts.* First printed in 1705. Several manuscripts.
2. *Editions.* Not edited by Huser or Sudhoff.
3. *Relationship between different versions.* Only one known version.
4. *Structure, genre/form, perspective, style.*

273 Sudhoff, *Paracelsus-Handschriften*, 808: "ein heute im Vatican unbekanntes, sagenhaftes [Manuscript]."

5. *Relationship to other texts.*
6. *Authenticity, authorship.* Written by "J.G.M." to his "dear friend and son" ("an meinen lieben Vetter und Sohn") Johann Heinrich Vierordt. The author was identified by Mike Zuber as Johann Gottmann.[274] In at least one manuscript (London, Society of Antiquaries), not known to Zuber, there is a "statement, dated 27 March 1672, […] (fol. 242v) at the end of a final paragraph describing the transmission of the text from Paracelsus via a German count, and a MS found in a chapel in the Thuringian Forest by two young German doctors who took it to Leiden."[275] This is an obvious allusion to the anonymous *Von der Reise Friederich Galli Nach der Einöde S. MichaEl* [*sic*], in: Floretus a Bethabor, *Traum-gesicht/ Welches Ben-Adam/ zur Zeit der Regierung Rucharetz/ des Königes von Adama / gehabt* (Amsterdam: Hans Fabel, 1648), 11–24.[276]
7. *Time of writing.* Dated 1672

WORD COUNT:

INCIPIT: "Im Anfang da der Allerhöchste Gott […] / Wenn wir Menschen nach angebohrner Pflicht […] / Ob zwar nie beschlossen gehabt"

MANUSCRIPTS:
- LONDON, *Society of Antiquaries of London*: MS 225, f. 235r–242v

  further manuscripts
- GOTHA, *Forschungsbibliothek*: Membr. II 175
- HAMBURG, *Staats- und Universitätsbibliothek*: Cod. alchim. 649, 644–85
- HAMBURG, *Staats- und Universitätsbibliothek*: Cod. alchim. 671, 484–509
- HANNOVER, *Gottfried Wilhelm Leibniz Bibliothek/Niedersächsische Landesbibliothek*: LH 37,6, f. 32r–45v
- LONDON, *British Library*: Sloane MS 3797, f. 53r–95v
- STRASBOURG, *Bibliothèque nationale et universitaire*: ms. 2.034, 1–44. Several other French manuscripts are found in a number of public libraries in France and abroad.

FIRST PRINTED:
- 1705 (in: *Quadratum Alchymisticum: Das ist: Vier auserlesene rare Tractätgen Vom Stein der Weisen* (Hamburg: Christian Liebezeit, 1705), 3–54; VD18 10197648) [not attributed to Paracelsus]

274 Zuber, "The Duke" (2018), 123.

275 "Catalogue description", online (https://discovery.nationalarchives.gov.uk/details/r/9b7b9fee-dd73-40d6-b5bc-9177917eecc9, accessed 10 May 2021).

276 Kahn, "Paracelsisme, alchimie et diplomatie" (2008), 107 n. 20.

Further bibliographical references:

Didier Kahn, "Paracelsisme, alchimie et diplomatie dans le contexte de la Paix de Westphalie", in Christian Mouchel and Colette Nativel, eds., *République des lettres, République des arts. Mélanges en l'honneur de Marc Fumaroli* (Geneva: Droz, 2008), 103–121.

Mike A. Zuber, "The Duke, the Soldier of Fortune, and a Rosicrucian Legacy: Exploring the Roles of Manuscripts in Early-Modern Alchemy", in *Ambix* 65 (2018), 122–142.

4.49 *Summum naturae mysterium* also: *Brief an einen Jünger*

1. *Printing History, Manuscripts.* Not printed. One manuscript.
2. *Editions.* Not edited by Huser or Sudhoff.
3. *Relationship between different versions.* Only one known version.
4. *Structure, genre/form, perspective, style.* Written in the form of a letter, directly addressing the "beloved follower and god-fearing disciple" (no name given).
5. *Relationship to other texts.* Refers to the *Turba Philosophorum* and *Rosarium* and cites *De tinctura physicorum* (§ 4.30) and *Apocalypsis* (§ 4.1). The title may refer to a volume edited by Gerhard Dorn, *De summis Naturae Mysterijs Libri tres* (Basel: Pietro Perna, 1570; Sudhoff, *Bibliographia Paracelsica*, 209–211 n° 125), containing Latin translations of *De spiritibus planetarum* (§ 4.29), *De occulta philosophia* (§ 5.7) and *Archidoxis magica* (§ 5.1). – A text in Wien, *Schottenstift*: Cod. 722 (*olim* Hübl 539) has a similar title: *Summum mysterium naturae philosophiae.* Further study required. – Karl Widemann lists the following text in one of his catalogues: "Secreta operationes duodecim signorum caelestium quoad medicinam, das ist Haimlichkaitten vnd natürliche Würckungen der Sigillen der zwölf Zaichen Zodiaci, das ist die Himmlische Artzenei und Wunder-Würckung der zwölf Zaichen des Thierkraises, dis Buech ist ain Auslegung des Bueches Theophrasti de summis naturae mysteriis von ime intituliertt, seu Commentariorum trium seu tractatuum medicinae caelestis, seu mysteriorum signorum zodiaci" (Widemann, *Verzeichnisse* (Kassel), n° IV, 28).
6. *Authenticity, authorship.*
7. *Time of writing.* Further study required.

Word count: ca. 3.500 words

Incipit: "Dier alß meinem geliebten jünger vnd gottfürchtigen discipulo […] / Ich Theophrastus Paracellsus ein fürst aller philosophen bekenne vor mich …"

Manuscripts:

– Kassel, *Landesbibliothek*: 4° Ms. chem. 31, f. 75r–80v

First printed: not printed

Further bibliographical references:
Broszinski, *Manuscripta chemica in quarto* (2011), 100.

4.50 *Testamentum, darin die Transmutatio der sieben Metulle angezeigt wird*

1. *Printing History, Manuscripts.* First printed in 1598 in the collection *Aureum vellus.*
2. *Editions.* Not edited by Huser or Sudhoff
3. *Relationship between different versions.* Only one known version.
4. *Structure, genre/form, perspective, style.*
5. *Relationship to other texts.*
6. *Authenticity, authorship.* According to Sudhoff, without any trace of Paracelsian authenticity.[277]
7. *Time of writing.* Further study required.

Word count: ca. 800 words

Incipit: "Nimb des liquoris ein theyl …"

Manuscripts: no manuscripts known

First printed:
– 1598 (in: *Aurei Velleris Oder Der Guldin Schatz vnd Kunstkammer Tractatus II* (Rorschach, 1598); VD16 T 1961; Sudhoff, *Bibliographia Paracelsica*, 424–427 n° 244)

Essential bibliography: Sudhoff, *Bibliographia Paracelsica*, 696 n° 14.

Theophrastia: see *Carboantes* (§ 4.22)

4.51 *Theoria aus den cabalischen Gezeugnissen* also: *Eine wahrhafte gegründete Theoria aus den cabalischen Gezeugnissen Theophrasti/Von dem Ursprung und Höffel der Quinta Essentia*

1. *Printing History, Manuscripts.* First edited in 1608 by Benedictus Figulus. No manuscripts known.

277 Sudhoff, *Bibliographia Paracelsica*, 426: "Auch dies ohne Spuren Paracelsischer Autorschaft."

2. *Editions.* Not edited by Huser or Sudhoff.
3. *Relationship between different versions.* Only one known version.
4. *Structure, genre/form, perspective, style.* Written in the third person.
5. *Relationship to other texts.* Connected by the usage of terms like "Höffel" and "Paradeiß Wasser" to *Wahres Zeugnis von der Materia Lapidis Philosophorum* (§ 4.55), also edited by Figulus.
6. *Authenticity, authorship.* Not directly attributed to Paracelsus, but allegedly an extract from "cabalistic testimonies of Theophrastus." Probably written or compiled by Figulus himself; the text is signed on the last page: "Dixi ego B. F. P. L. C.," i.e. "Stated by me, Benedictus Figulus, Poeta Laureatus Coronatus."[278]
7. *Time of writing.* Probably written in the 1600s.

Word count: ca. 6.400 words

Incipit: "Im nammen der heiligen vnzertheilten dreyfaltigkeit wöllen wir anfahen …"

Manuscripts: no manuscripts known

First printed:
- 1608 (in: *Rosarium novum olympicum et benedictum. Das ist: Ein newer Gebenedeyter Philosophischer Rosengart*, ed. Benedictus Figulus (Basel, 1608); VD17 12:133068G; Sudhoff, *Bibliographia Paracelsica*, 482–485 n° 283)

Essential bibliography: Sudhoff, *Bibliographia Paracelsica*, 696 n° 23; CP 1: 451; CP 3: 1118.

Further bibliographical references:
Peuckert, *Pansophie* (1956), 314, 360.
Peuckert, *Gabalia* (1967), 20.
Joachim Telle, "Benedictus Figulus. Zu Leben und Werk eines deutschen Paracelsisten," *Medizinhistorisches Journal*, 22 (1987), 303–326, on 16.

4.52 *Thesaurus thesaurorum* also: *Thesaurus alchimistarum*

1. *Printing History, Manuscripts.* First printed from different manuscripts in 1574, edited by Adam von Bodenstein and shortly thereafter by Michael Toxites. Four manuscripts.

278 Sudhoff, *Bibliographia Paracelsica*, 484: "von Figulus selbst verfasst;" CP 1: 451: von "Figulus […] kompiliert worden;" CP 3: 1118: "von Figulus verfaßt[e] Kompilation."

2. *Editions.* Edited by Huser, 6 (1590): 396–401. Edited by Sudhoff in Paracelsus, *Sämtliche Werke*, I/14: 401–404.
3. *Relationship between different versions.* The motif of the 'red lion' connects the *Thesaurus thesaurorum* with *De tinctura physicorum* (§ 4.30), *Brief an eine geistliche Person / Practica des verkehrten Mondes* (§ 4.13) and *Brief an einen Gönner und Freund (Vom Wunderstein)* (§ 4.14).[279]
4. *Structure, genre/form, perspective, style.*
5. *Relationship to other texts.* Toxites insisted on the close relationship between the *Thesaurus*, *De tinctura physicorum* (§ 4.30) and *Manuale* (§ 4.40), "for these three have in common one scopus by which Theophrastus created much good as it will show in due course."[280] For Toxites the *Thesaurus* is "nearly an explanation of *De tinctura physicorum*, and has the same focus,"[281] which does not mean that the *Thesaurus* is a commentary on the *De tinctura physicorum*, but that they share the same ideas and arguments. – Sudhoff was inclined to ascribe *De tinctura physicorum*, the *Thesaurus thesaurorum* and the *Liber vexationum* (§ 4.39) to one single, unknown author.[282] – Now it happens that the *Thesaurus* mentions *De natura rerum* (§ 2.2) along with the *Archidoxis* and refers to its own alchemical process as that of the *Tinctura Alchimistarum*.[283] Whether this can lead to anything conclusive is still to be determined. – In any case, the *Thesaurus* and *De tinctura physicorum* share some topics, especially the doctrine of "red lion" and "white eagle" that can also be found in the *Brief an eine geistliche Person / Practica des verkehrten Mondes* (§ 4.13) and in the *Brief an einen Gönner und Freund* (*Vom Wunderstein*) (§ 4.14).[284]
6. *Authenticity, authorship.*
7. *Time of writing.* Probably written in the 1560s.

Word count: ca. 1.500 words

Incipit: "Die natur gebüret ein mineral in visceribus terre …"

279 CP 3: 231.
280 Toxites to the reader, in Sudhoff, *Bibliographia Paracelsica*, n° 158, on 366.
281 Second preface, in Sudhoff, *Bibliographia Paracelsica*, n° 158, fol. B2r–v: "vast ein außlegung tincturae Physicorum vnnd eines arguments."
282 Sudhoff, "Vorwort," in Paracelsus, *Sämtliche Werke*, I/14: XII. – However, on XVI, Sudhoff does not mention the *Liber vexationum* any longer and even contradicts himself: he tends rather to ascribe both the *Thesaurus thesaurorum* and *De tinctura physicorum* to one single author, and suggests (on XX) to date the *Thesaurus* from the years 1560, slightly before *De tinctura physicorum*.
283 Huser, 6: 400.
284 CP 3: 231.

Manuscripts:
- Dresden, *Sächsische Landesbibliothek*: Mscr.Dresd.N.107, f. 157r–161v
- Kassel, *Landesbibliothek*: 2° Ms. chem. 3, f. 184r
- Kopenhagen, *Det Kongelige Bibliotek*: Thott 737 kvart [probably copied from the 1574 printed edition]
- Leiden, *Universiteitsbibliotheek*: Voss. Chym. Q.17, f. 70v–74v
- Leiden, *Universiteitsbibliotheek*: Voss. Chym. Q.21, f. 182v–183r [fragment]
- München, *Bayerische Staatsbibliothek*: cgm 4228
- Wolfenbüttel, *Herzog August Bibliothek*: Cod. Guelf. 11 Aug. 2°, f. 16–17

First printed:
- **1574** (in: *Theophrasti Paracelsi schreiben von tribus Principiis aller Generaten*, ed. Adam von Bodenstein (Basel: Samuel Apiarius for Pietro Perna, 1574); VD16 P 672; Sudhoff, *Bibliographia Paracelsica*, 257–260 n° 153)
- **1574** (in: *Archidoxa Philippi Theophrasti Bombast Paracelsi Magni Zehen Bücher*, ed. Michael Toxites (Strasburg: Christian Müller, 1574); VD16 P 397; Sudhoff, *Bibliographia Paracelsica*, 268–272 n° 158)

Historical Manuscript Catalogues: *Scripta Theophrasti* (Wien); Thomasius (1699), on 141

Essential bibliography: Sudhoff, *Bibliographia Paracelsica*, 393; Sudhoff, *Paracelsus-Handschriften*, 774; Sudhoff, "Vorwort," in Paracelsus, *Sämtliche Werke*, I/14: XVI; CP 1: 496; CP 2: 703; CP 3: 231.

Further bibliographical references:

Richard Weber, *Philosophia Adepta. Eine Untersuchung der frühen Schriften Theophrasts von Hohenheim gen. Paracelsus*, PhD thesis, Göttingen 1953, 172–179.

Richard Weber, "Der grüne Löwe," *Abhandlungen der Braunschweigischen Wissenschaftlichen Gesellschaft*, 6 (1954), 288–302.

Peuckert, *Pansophie* (1956), 252, 253, 258, 259, 286, 287.

Boeren, *Codices Vossiani Chymici* (1975), 144, 152.

Tinctura antimonii: see *Brief an Theodor* (§ 4.17)

Vom Stein der Weisen der zehnte Tractat Archidoxorum: see *Zehntes Buch der Archidoxen* (§ 4.56)

Vom Wunderstein: see *Brief an einen Gönner und Freund* (§ 4.14)

4.53 *Von Bereitung des gebenedeiten philosophischen Steins*

1. *Printing History, Manuscripts.* First published by Ps.-Huser in 1603. Four manuscripts (not attributed to Paracelsus, but to Caspar Hartung vom Hoff). One manuscript with a related text.
2. *Editions.* Edited by Ps.-Huser in *Opera Bücher vnd Schrifften* (1603), 2: 686–691. Not edited by Sudhoff.
3. *Relationship between different versions.* The Kassel manuscript preserves (under the title "Nun facht an das recht puch jn disser kunst") a different but connected text illustrated with seven carefully executed pen drawings referred to as "figur" in the text. This last fact is especially interesting since according to the title in Figulus' 1608 edition *Von Bereitung* is equipped "mit schönen Figuren" and "Figuren" are also mentioned in the preface of *Von Bereitung* although there are no illustrations present. Further study required.
4. *Structure, genre/form, perspective, style.*
5. *Relationship to other texts.* Part of *Von Bereitung* are two texts that have been of interest to historians: (1) the so-called *Tägliches Gebet* (§ 10.15) of Paracelsus, a short 11-line prayer;[285] (2) the "oldest German translation of the *Tabula Smaragdina*," discovered by Julius Ruska in the 1616 Ps.-Huser edition of Paracelsus's *Opera Bücher vnd Schrifften*,[286] and therefore attributed by Ruska to Paracelsus. It is, however, part of *Von Bereitung*.
6. *Authenticity, authorship. Von Bereitung* is connected with the name of Caspar Hartung vom Hoff auß der Gastein (i.e. Hofgastein/Austria).[287] Ps.-Huser's edition of 1603 has "ex Manuscripto exemplari D. Caspari vom Hoff" in the heading (table of contents, fol. (:) 4r), while the editions of 1604 and 1608 name "Caspar Hartung vom Hoff" as author. The Leiden and Wien manuscripts originally did not name any author: in both cases "Caspar Hartung vom Hoff" was added at a later time. The London manuscript has "Casparus Hartung von Hoff," but whether this is a later addition or not has yet to be determined. The Kassel manuscript is anonymous.

   Caspar Hartung is known to historians of alchemy as the owner or writer (which of these we do not know) of two compilations of alchemical texts and

285 CP 1: 450.

286 Ruska, *Tabula Smaragdina* (1926), 207–208. See Sudhoff, "Hat wirklich Paracelsus die Tabula Smaragdina ins Deutsche übersetzt?" (1927).

287 Most modern authors believe "vom Hoff" to denote from "Hof im Vogtland" (today Bavaria). The "Vademecum" in New Haven, however, clearly says: "Hoff auß der Gastein" – today Hofgastein (Austria).

recipes: the so-called *Kunstbüchlein*, a manuscript now in Kassel,[288] and the so-called *Vademecum*, a manuscript now in New Haven.[289]

The *Kunstbüchlein* has a dedication from Caspar Hartung to Wilhelm Gunzhofer/Guntzkofler, Probst (provost) in Kraig,[290] dated 1549. The *Vademecum* is dated 1557. Both the *Kunstbüchlein* and the *Vademecum* seem to contain no original texts written by Caspar Hartung. Therefore it is best to follow Huser who believed that the text was taken "from a manuscript from the possession of Caspar [Hartung] vom Hoff."

7. *Time of writing.* Possibly written in the 1550s. Attributed to Paracelsus only in Ps.-Huser's edition (1603).

Word count: ca. 4.500 words

Incipit: "Es ist zu wissen daß alle ding von Gott kommen [...] / Nun fahet an in Gottes nammen das buch der weißheit ..."

Manuscripts:
- Leiden, *Universiteitsbibliotheek*: Voss. Chym. Q.17, f. 53r–65v
- London, *Wellcome Library*: MS 524, f. 98–122
- Wien, *Österreichische Nationalbibliothek*: Cod. 11428, f. 157–163
- *olim* Johann Georg Binz, bookseller in Vienna (1791); current location unknown

related text
- Kassel, *Landesbibliothek*: 4° Ms. chem. 72, f. 276ra–278rb

First printed: not printed before Ps.-Huser (1603)
- **1603** (in: Paracelsus, *Opera Bücher vnd Schrifften*, ed. Ps.-Huser (1603); VD17 12:168390P)
- **1604** (in: *Avrei Velleris, Oder Der Guldin Schatz: vnd Kunstkammern Tomi Secvndi Tractatvs I. Philosophischer vnd Spagirischer Geschrifften/ Alter vnnd Newer Philosophorvm, Medicorvm vnd Spagyrorvm* (Basel: J. Exertier, 1604); VD17 14:646496L)
- **1608** (in: *Thesaurinella Olympica aurea tripartita. Das ist: Ein himmlisch güldenes Schatzkämmerlein*, ed. Benedictus Figulus (Frankfurt a.M.: Wolfgang Richter for Nicolaus Stein, 1608); VD17 3:302600E; Sudhoff, *Bibliographia Paracelsica*, 477–479 n° 280)

---

288 Kassel, *Landesbibliothek*, 8° Ms. chem. 32; see *Das 'Kunstbüchlein' des Alchemisten Caspar Hartung vom Hoff*, ed. Haage (1975).

289 New Haven, *Beinecke Library*, Mellon MS 38.

290 See Friedrich Wilhelm Leitner, "Zur Geschichte der Propstei in Kraig," *Rudolfinum*, 3 (2002), 227–246; idem, *Die Inschriften des Bundeslandes Kärnten*, vol. 2: *Die Inschriften des politischen Bezirks St. Veit an der Glan* (Vienna, 2008), XXXIII.

HISTORICAL MANUSCRIPT CATALOGUES: Binz (1791), on 92

ESSENTIAL BIBLIOGRAPHY: Sudhoff, *Bibliographia Paracelsica*, 445; Sudhoff, *Paracelsus-Handschriften*, 776; CP 1: 450.

FURTHER BIBLIOGRAPHICAL REFERENCES:

Julius Ruska, *Tabula Smaragdina. Ein Beitrag zur Geschichte der hermetischen Literatur* (Heidelberg, 1926), 207–208.

Karl Sudhoff, "Hat wirklich Paracelsus die Tabula Smaragdina ins Deutsche übersetzt?," *Mitteilungen zur Geschichte der Medizin und der Naturwissenschaften*, 26 (1927), 163.

Peuckert, *Pansophie* (1956), 319–320, 472.

Peuckert, *Rosenkreutz* (1973), 84.

Das 'Kunstbüchlein' des Alchemisten Caspar Hartung vom Hoff. In Abbildung und Transkription herausgegeben und erläutert von Bernhard Haage (Göppingen, 1975).

Boeren, *Codices Vossiani Chymici* (1975), 143.

Joachim Telle, "Das 'Kunstbüchlein' des Alchemisten Caspar Hartung vom Hoff," Review of Bernhard Haage, ed., *Das 'Kunstbüchlein' des Alchemisten Caspar Hartung vom Hoff*, in *Beiträge zur Geschichte der Pharmazie*, 28/4 (1976), 29–30.

Herwig Buntz, Review of Bernhard Haage, ed., *Das 'Kunstbüchlein' des Alchemisten Caspar Hartung vom Hoff*, in *Zeitschrift für deutsche Philologie*, 96 (1977), 138–140.

Wolfgang Hirth, Review of Bernhard Haage, ed., *Das 'Kunstbüchlein' des Alchemisten Caspar Hartung vom Hoff*, in *Beiträge zur Geschichte der deutschen Sprache und Literatur*, 99 (1977), 158–159.

Joachim Telle, *Sol und Luna. Literar- und alchemiegeschichtliche Studien zu einem altdeutschen Bildgedicht* (Hürtgenwald, 1980), 109, 126–127.

4.54 *Von den vier Arcana*
also: *Die vier arcana Theophrasti sambt derselben explication*

1. *Printing History, Manuscripts.* Not printed. Several manuscripts.
2. *Editions.* Not edited by Huser or Sudhoff.
3. *Relationship between different versions.* Only one known version.
4. *Structure, genre/form, perspective, style.* Explains four Paracelsian arcana: *Prima materia, lapis philosophorum, mercurius vitae,* and *tinctura.*
5. *Relationship to other texts.* The text is also part of Johannes de Sole's *Arcana arcanorum.* Further study required.
6. *Authenticity, authorship.* Obviously not attributed to Paracelsus, but rather an explanation or interpretation of some spurious works (not named).
7. *Time of writing.* Further study required.

WORD COUNT: ca. 800 words

INCIPIT: "Recipe der primae materiae inflavum …"

MANUSCRIPTS:

Von den vier Arcana

- HAMBURG, *Staats- und Universitätsbibliothek*: Cod. alchim. 664, 153–156

Johannes de Sole, Arcana arcanorum

- BAMBERG, *Staatsbibliothek*: Msc. Nat. 13
- PHILADELPHIA (PA), *Penn Libraries*: Ms. Codex 128
- PRAHA, *Strahovská knihovna*: Cod. DF IV 36, f. 52r–83r
- WOLFENBÜTTEL, *Herzog August Bibliothek*, Cod. 264.31 Extr., ca. 668–691

FIRST PRINTED: not printed

4.55 *Von der Materia lapidis philosophorum* also: *Wahres Zeugnis von der Materia lapidis philosophorum*

1. *Printing History, Manuscripts.* First edited in 1608 by Benedictus Figulus. Reprinted twice from Figulus's edition, but no other editions or manuscripts known.
2. *Editions.* Not edited by Huser or Sudhoff.
3. *Relationship between different versions.* Only one known version.
4. *Structure, genre/form, perspective, style.*
5. *Relationship to other texts.* Not attributed to Paracelsus, but allegedly excerpts and quotations from Paracelsus; these excerpts are otherwise unknown. There are, however, ideas and concepts that are present also in other pseudo-Paracelsian texts. – *Von der Materia lapidis philosophorum* inspired the German writer Joachim Veil (born 1950) who used part of the text in his *Die Wiederkehr des Bumerangs*.[291]
6. *Authenticity, authorship.*
7. *Time of writing.* Probably written in the early 1600s.

WORD COUNT: ca. 6.500 words

INCIPIT: "Daß es sey die höchste artzney von wegen seiner innwendigen fewrigen krafft …"

291 Joachim Veil, "Die Wiederkehr des Bumerangs. Vom Spurenlesen oder Wo man den Ochsen suchen soll, auf dem man sitzt," in *Literaturmagazin*, vol. 14, ed. Günter Kunert (Reinbek, 1981), 245–251, on 251; Joachim Veil, *Die Wiederkehr des Bumerangs* (Frankfurt a.M., 1982), 33.

Manuscripts: no manuscripts known

First printed:
- 1608 (in: *Thesaurinella Olympica aurea tripartita. Das ist: Ein himmlisch güldenes Schatzkämmerlein*, ed. Benedictus Figulus (Frankfurt a.M.: Wolfgang Richter for Nicolaus Stein, 1608); VD17 3:302600E; Sudhoff, *Bibliographia Paracelsica*, 477–479 n° 280)

Essential bibliography: Sudhoff, *Bibliographia Paracelsica*, 696 n° 18.

Further bibliographical references:
Peuckert, *Pansophie* (1956), 319.

Wie man die sieben Metalle clarificieren soll: see *Brief an Bartholomäus Korndörffer* (§ 4.9)

4.56 *Zehntes Buch der Archidoxen vom Stein der Weisen* also: *Vom Stein der Weisen der zehnte Tractat Archidoxorum*

1. *Printing History, Manuscripts.* First edited from different sources in 1618 by Friedrich Moller, and in 1621 by Theophil Neander. One manuscript.
2. *Editions.* Not edited by Huser or Sudhoff.
3. *Relationship between different versions.*
4. *Structure, genre/form, perspective, style.*
5. *Relationship to other texts.* Not connected in any way to *Clavis oder Das zehnte Buch der Archidoxen* (§ 4.25) first edited by Johannes Staricius in 1624.
6. *Authenticity, authorship.* Written to supply the supposedly missing tenth book of the *Archidoxis*. The fact that Paracelsus called his text "Ten Books of Archidoxes" while there are only nine books present puzzled the early editors, namely Michael Toxites and Gerhard Dorn. Later, Johannes Huser explained in the preface to his edition of the *Archidoxis* that Paracelsus himself wrote "Decem Libri Archidoxis" on the vellum cover of his autograph manuscript; in the table of contents, however, he listed only nine books, while explaining in the first book that he divided his work into ten books, but would not put the tenth on paper; finally, Paracelsus wrote his colophon as follows: "Archidoxis ex Theophrastia pars prima finit […] in nouem libros diuisa," i.e. "End of the First Part of the Archidoxes divided into nine books."[292]
7. *Time of writing.* Possibly written in the 1610s or earlier. Further study required.

292 Johann Huser, Preface to the Reader ("Lectori S[alutem]"), in Huser, 6: [VI]–[VII], quoted by Sudhoff, *Bibliographia Paracelsica*, 389–390.

WORD COUNT: 3.000 words

INCIPIT: "Nun wollen wir beschreiben des zehende buch vnser Archidoxen [...] / Es ist weiter kundt vnd offenbar das viel vnbeständige gemühter ..."

MANUSCRIPTS:
- LEIDEN, *Universiteitsbibliotheek*: Voss. Chym. F.6, f. 90r–99v

FIRST PRINTED:
- 1618 (in: *Ternio reliquiarum alchymiae Oder Drey Schöne Tractätlein/ Vom Stein der Weisen*, ed. Friedrich Moller (Berlin, 1618); VD17 39:116580T; Sudhoff, *Bibliographia Paracelsica*, 511–512 n° 304)
- 1621 (in: *Heptas Alchymica Das ist: Auszerlesene Philosophische Tractätlein*, ed. Theophil Neander (Leipzig: Henning Groß, 1621); VD17 23:240185V; Sudhoff, *Bibliographia Paracelsica*, 527–528 n° 319)

ESSENTIAL BIBLIOGRAPHY: Sudhoff, *Bibliographia Paracelsica*, 512, 527. 696 n° 32; Sudhoff, *Paracelsus-Handschriften*, 779 n° II/1a; CP 3: 1173.

FURTHER BIBLIOGRAPHICAL REFERENCES:
Boeren, *Codices Vossiani Chymici* (1975), 23.

5 Magic

Arbatel: see *De magia veterum* (§ 5.6)

Arcanum arcanorum seu magisterium philosophorum de sacra campanula angelorum: see *Vom Modus des Gebrauchs mit der Tinctur* (§ 5.25)

5.1 *Archidoxis magica* also: *Archidoxis magice/Archidoxia magica/Medicina coelestis/ Magische Archidoxen/Magia Theophrasti*

1. *Printing History, Manuscripts.* A work composed of seven books. The first four books were first printed in a Latin translation by Gerhrard Dorn in 1570 under the title *Medicina cœlestis sive de signis Zodiaci et mysteriis eorum* in *Aureoli Philippi Theophrasti Paracelsi De summis Naturae Mysterijs Libri tres*. They were next published in German in 1571 by Pietro Perna under the title *De signis Zodiaci et ejus mysteriis* in *De spiritibus planetarum sive metallorum Doctoris*

Theophrasti Paracelsi ab Hohenheim: Libri III. – Book VI was first edited in 1572 by Michael Toxites under the title *Liber sextus Archidoxiae Magicae de Compositione Metallorum* in *Zwen Tractatus Philippi Theophrasti Paracelsi.* – Books V and VII were first printed (together with Book VI) in 1572 by Pietro Perna under the title *Liber V De Speculi Constellatione* and *Liber VII Archidoxis Magice De Sigillis Planetarum* in *Archidoxorum* [...] *Theophrasti Paracelsi X. Bücher.* Only Huser printed all seven books together as a whole. – Many manuscripts. According to Sudhoff the manuscripts in Nürnberg and Wolfenbüttel have a more complete text than the printed versions. The Praha (*olim* Osseg) manuscript is closely related to the sources used by the early editors in a way that has yet to be determined. The Braunschweig manuscript (owned by historian of pharmacy Wolfgang Schneider during the 20th century) has a Hutterite background not known to Schneider; it is a copy from a manuscript written by Leonhard Gagasser.[293] Many manuscripts have only single books of the *Archidoxis magica.*

2. *Editions.* Edited by Huser, 10 (1590): Appendix, 66–138.[294] Edited by Sudhoff in Paracelsus, *Sämtliche Werke*, I/14: 437–498.
3. *Relationship between different versions.* Further study required.
4. *Structure, genre/form, perspective, style.*
5. *Relationship to other texts.* Further study required.
6. *Authenticity, authorship.* Huser expressed doubts on the authenticity of this treatise, echoing the opinion of others,[295] but decided to include it amongst the other texts of the Appendix of vol. 10, at least "until we have more certainty about who the author is," "as it is not alien to Paracelsus's material and many have accepted and acknowledged it as his work."[296] Sudhoff had "not the least doubt" that the *Archidoxis magica* were spurious.[297] Wolfgang Schneider held at least parts of the *Archidoxis magica* to be authentic,[298] while Goldammer con-

293 Rauert and Rothkegel, *Katalog der hutterischen Handschriften* (2011), 654–660.

294 Johann Huser, *Ad Lectorem*, in Huser, 10 (1590): Appendix, 66: "Freundtlicher Leser/ wisse/ das die Autographa dieser Magischen Bücher/ jetzt folgend/ mir noch niemahls zu sehen worden/ auch von keinem andern gehöret/ der sie gesehen hette. Seindt derwegen die Ersten 4. aüß dem zuvor außgangenem Exemplar/ die andern 3. mit etlichen Manuscriptis Conferiert/ widerumb in Truck geben."

295 *Ibid.*: "Es soll aber auch nicht vngemeldet bleiben/ das etliche an diesen Büchern Archidoxis Magicae dubitieren/ ob sie Theophrasti seien."

296 *Ibid.*: "Jedoch/ weil sie Theophrasti Sachen nicht vngemeß/ vnd von vielen für seine Bücher angenommen vnd erkennt werden/ mögen sie auff dieß mahl neben den andern vnterlauffen/ biß man deß Authoris gewisser werde."

297 Sudhoff, "Vorwort," in Paracelsus, *Sämtliche Werke*, I/14: XXVIII: "an ihrer Unechtheit [besteht] nicht der mindeste Zweifel."

298 Schneider, *Paracelsus – Autor der Archidoxis Magica?* (1982).

sidered the *Archidoxis magica* a "compilation from authentic Paracelsian works" and at the same time "a creative achievement on its own".[299]

7. *Time of writing.* Probably written in the 1560s.

Word count: ca. 18.000 words [Huser edition]

Incipit: "Es wirt nicht on sein, so diese vnser schrifften in tag kommen werden"

Manuscripts:

German

- Braunschweig, *Universitätsbibliothek*: Nat. A 139, f. 145r–170r [written by Adam Bien after a copy from Leonhard Gagasser; in the 20th century in the library of Wolfgang Schneider]
- Dessau, *Landesarchiv Sachsen-Anhalt*: Z 70, A 17a Nr. 105c
- Gotha, *Forschungsbibliothek*: Chart. B 1211, f. 10r–12v
- Hamburg, *Staats- und Universitätsbibliothek*: Cod. alchim. 192, f. 57r–88r
- Heiligenkreuz, *Stiftsbibliothek*: Ms. 334, f. 63r–83v
- Kassel, *Landesbibliothek*: 4° Ms. chem. 16, f. 87v–88r [fragment]
- Kassel, *Landesbibliothek*: 4° Ms. chem. 24, f. 144r–145r
- Kassel, *Landesbibliothek*: 4° Ms. chem. 35[4, f. 29v [fragment]
- Kassel, *Landesbibliothek*: 4° Ms. chem. 60[8,2, f. 35r [fragment]
- Kassel, *Landesbibliothek*: 4° Ms. chem. 72, f. 237v–243r
- Kassel, *Landesbibliothek*: 4° Ms. chem. 87, f. 61r–62r [fragment]
- Kassel, *Landesbibliothek*: 4° Ms. chem. 96, 169–187
- Kopenhagen, *Det Kongelige Bibliotek*: Thott 737 kvart
- München, *Bayerische Staatsbibliothek*: cgm 6145; 52 fols.
- München, *Bayerische Staatsbibliothek*: cgm 9544, f. 4r–77r
- Nürnberg, *Germanisches Nationalmuseum*: Hs. 9829; 150 fols.
- Paris, *Bibliothèque de l'Arsenal*: Ms. 8007; 78 fols.
- Paris, *Bibliothèque de l'Arsenal*: Ms. 8008

299 Goldammer, "Probleme der deutero-paracelsischen Schriften am Beispiel der Magie" (1993), 353: "Es ist wohl eine Kompilation aus den echten Paracelsus-Werken, Illustrierung und Begründung andersartiger magischmedizinischer Anweisungen im Namen des Paracelsus, weniger die Verwendung verlorener paracelsischer Originale – eine eigenschöpferische Leistung des Kompilators beim Zusammenbau älterer Quellen, teils aus den echten paracelsischen Schriften, teils und vorwiegend anderer Herkunft, die Paracelsus als Autor und Autorität beansprucht." See also Goldammer, *Der göttliche Magier und die Magierin Natur*, 104: "es handelt sich wohl nur – wie in den meisten anderen pseudoparacelsischen Spuria – um eine Verarbeitung Paracelsischer Begriffe und Gedankengänge."

- PRAHA, *Národní knihovna*: Osek 43, f. 17r–19v, 44r–49r, 66r–70v, 73r–78v
- SEITENSTETTEN, *Stiftsbibliothek*: Cod. 69 [lost]
- WOLFENBÜTTEL, *Herzog August Bibliothek*: Cod. Guelf. 115 Blank.; 78 fols.
- WROCŁAW/BRESLAU, *Biblioteka Uniwersytecka*: R 2317, on 108–184 [probably copied from a printed edition]

English

- LONDON, *British Library*: Sloane MS 3822
- LONDON, *British Library*: Sloane MS 3648

FIRST PRINTED:

- 1570 (Latin; "Medicina coelestis sive de signis Zodiaci et mysteriis eorum", in *Aureoli Philippi Theophrasti Paracelsi De summis Naturae Mysterijs Libri tres*, ed. Gerhard Dorn (Basel: Pietro Perna, 1570); VD16 P 565; Sudhoff, *Bibliographia Paracelsica*, 209–211 n° 125) [books 1–4]
- 1571 (German; "*De signis Zodiaci et ejus mysteriis*", in *De spiritibus planetarum sive metallorum Doctoris Theophrasti Paracelsi ab Hohenheim: Libri III* (Basel: Pietro Perna, 1571); VD16 P 567; Sudhoff, *Bibliographia Paracelsica*, 224–226 n° 134) [books 1–4]
- 1572 (German; "Liber sextus Archidoxiae Magicae de Compositine Metallorum", in *Zwen Tractatus Philippi Theophrasti Paracelsi*, ed. Michael Toxites (Strasburg: Bernhard Jobin, 1572); VD16 P 611; Sudhoff, *Bibliographia Paracelsica*, 237–239 n° 141) [book 6]
- 1572 (German; "Liber VI. Archidoxis Magice" and "Liber VII Archidoxis Magice De Sigillis Planetarum", in *Archidoxorum* [...] *Theophrasti Paracelsi X. Bücher* (Basel: Pietro Perna, 1572); VD16 P 396; Sudhoff, *Bibliographia Paracelsica*, 239–241 n° 142) [books 5–7]

HISTORICAL MANUSCRIPT CATALOGUES: *Catalogus* Osek (Prague) 2, n° 5; Widemann, *Verzeichnisse* (Kassel), n° II, 24 / V, 6, 206; Thomasius (1699), on 137; Arpe (1726), on 121; Jahn (1754–1758), 3: 152 n° 206

ESSENTIAL BIBLIOGRAPHY: Sudhoff, *Bibliographia Paracelsica*, 407; Sudhoff, *Paracelsus-Handschriften*, 776; Sudhoff, "Vorwort," in Paracelsus, *Sämtliche Werke*, I/14: XXV–XXVIII; CP 3: 503.

FURTHER BIBLIOGRAPHICAL REFERENCES:

Franz Spunda, *Das Weltbild des Paracelsus* (Vienna, 1941), 218–244.

Karl Anton Nowotny, "Zur Geschichte der astrologischen Medaillen," *Numismatische Zeitschrift*, 74 (1951), 100–104 + Tafel III.

Peuckert, *Pansophie* (1956), 476, 500.

Peuckert, *Gabalia* (1967), 162, 197–205, 212, 216, 219, 224–232, 280, 328, 429, 547, 566, 572, 573.

Will-Erich Peuckert, "Paracelsische Zauberei," *Nova Acta Paracelsica*, 8 (1957), 71–94, on 86.

Hartmut Rudolph, "Bemerkungen zum sogenannten Jenichen-Flugblatt," *Nova Acta Paracelsica*, 9 (1977), 205–211, on 207.

Wolfgang Schneider, *Paracelsus – Autor der Archidoxis Magica?* (Braunschweig, 1982).

Kurt Goldammer, "Neues zur Frage der Paracelsischen Spuria," Review of Wolfgang Schneider, *Paracelsus – Autor der Archidoxis Magica?*, *Sudhoffs Archiv*, 68 (1984), 235–239.

Willem Frans Daems, Review of Wolfgang Schneider, *Paracelsus – Autor der Archidoxis Magica?*, *Gesnerus*, 41 (1984), 179.

Carlos Gilly, *Spanien und der Basler Buchdruck bis 1600. Ein Querschnitt durch die spanische Geistesgeschichte aus der Sicht einer europäischen Buchdruckerstadt* (Basel, 1985), 34.

Wolf-Dieter Müller-Jahncke, *Astrologisch-magische Theorie und Praxis in der Heilkunde der frühen Neuzeit* (Stuttgart, 1985), 84–85.

Louise Gnädinger, "Paracelsus, der Magier vom Etzel. Deutungsversuch eines Flugblattes (Anfang 17. Jahrhundert)," *Nova Acta Paracelsica*, N.F. 6 (1991–1992), 2–20, on 9–11.

Goldammer, *Der göttliche Magier* (1991), 65, 104–105.

Wolf-Dieter Müller-Jahncke, "Magische Medizin bei Paracelsus und den Paracelsisten," in Peter Dilg and Hartmut Rudolph, eds., *Resultate und Desiderate der Paracelsus-Forschung* (Stuttgart, 1993), 43–55.

Kurt Goldammer, "Probleme der deutero-paracelsischen Schriften am Beispiel der Magie," in Heinz Dopsch, Kurt Goldammer and Peter F. Kramml, eds., *Paracelsus (1493–1541)* (Salzburg, 1993), 353–355.

Wolf-Dieter Müller-Jahncke, "Die Magie des 16. und 17. Jahrhunderts im Spannungsfeld von Naturwissenschaft und Dämonologie," in Herbert Jaumann, ed., *Domänen der Literaturwissenschaft* (Tübingen, 2001), 269–280, on 271–272.

Bruce T. Moran, *Andreas Libavius and the Transformation of Alchemy. Separating Chemical Cultures with Polemical Fire* (Sagamore Beach, MA, 2007), 271–273 (on weapon salve).

Ute Frietsch, "Zwischen Transmutation und Transsubstantiation. Zum theologischen Subtext der *Archidoxis*-Schrift des Paracelsus," *Nova Acta Paracelsica*, N.F. 19 (2005), 29–51, on 39–40.

Hereward Tilton, "Of Electrum and the Armour of Achilles: Myth and Magic in a Manuscript of Heinrich Khunrath (1560–1605)," *Aries*, 6 (2006), 117–157, on 128–129.

Karl Möseneder, *Paracelsus und die Bilder. Über Glauben, Magie und Astrologie im Reformationszeitalter* (Tübingen, 2009), 142–143.

Broszinski, *Manuscripta chemica in quarto* (2011), 52, 82, 129, 285, 297, 323, 363, 382.
Matthias H. Rauert and Martin Rothkegel, *Katalog der hutterischen Handschriften und der Drucke aus hutterischem Besitz in Europa*, vol. 1 (Gütersloh, 2011), 654–660.
Kathrin Pfister, "Kenelm Digby (1603–1665) und das 'sympathetische' Pulver – ein Beitrag zur Geschichte der pseudo-paracelsischen Waffensalbe," *Nova Acta Paracelsica*, N.F. 27 (2016), 47–66.

5.2 *Ars Gaballiae* also: *Buch der großen Gabalae / Ars Cabala*

1. *Printing History, Manuscripts.* Lost. Not printed. No manuscripts known. – Mentioned in *Libri Theophrasti*, a catalogue of manuscripts written ca. 1560: "Ars Gaballiae. das buch der grossen Gabalae" and in Karl Widemann's list of "Unausgangene Bücher" [unpublished manuscripts]: "Ars cabala, dz Buech der grossen Cabal."[300]
2. *Editions.* Not edited by Huser or Sudhoff.
3. *Relationship between different versions.* Text not preserved.
4. *Structure, genre/form, perspective, style.*
5. *Relationship to other texts.* 'Gabalia' was a term used by Paracelsus in different contexts and without a clear definition.[301]
6. *Authenticity, authorship.*
7. *Time of writing.* Probably written in the 1550s if it ever existed. Further study required.

Word count:

Incipit:

Manuscripts: no manuscripts known

First printed: not printed

Historical Manuscript Catalogues: *Libri Theophrasti* (Dresden), fol. 4rb; Widemann, *Unausgangene Bücher* (Hannover), n° 114

300 CP 1: 451. See also Gilly, "Il primo prontuario di magia bianca" (2002), 202, 211.

301 The most important quotations can be found in Urs Leo Gantenbein, ed., Paracelsus, *Theologische Werke 1* (Berlin-New York, 2008), 311 n. 158–159; see also introduction, on 95; Paracelsus, *Liber de religione perpetua, ibid.*, 307, 311–315; and Martin Žemla, "The *Astronomia Olympi novi* and the *Theologia Cabalistica*" (2021). – See also CP 1: 336–337; Goldammer, *Der göttliche Magier* (1991), 50.

ESSENTIAL BIBLIOGRAPHY: CP 1: 451.

FURTHER BIBLIOGRAPHICAL REFERENCES:
Gilly, "Il primo prontuario di magia bianca" (2002), 199–217.
Martin Žemla, "The *Astronomia Olympi novi* and the *Theologia Cabalistica*: Two Pseudo-Paracelsian Works of the *Philosophia Mystica* (1618)," *Early Science and Medicine*, 24 (2019), 527–548 (in the present volume, 115–136).

5.3 *Beschreibung des wunderbaren Bechers*

1. *Printing History, Manuscripts.* Not printed. One manuscript, written ca. 1635. The full title is, according to Sudhoff: "Beschreibung des wunderbahren bächers, dessen im Ersten buch Moyse am 44. […] gedacht würde. Samb einem gantzen Cabalistischen Ordinadt, darzugehörig, vnd wie man denselbigen brauchen soll Auß Herrn Theoph: Paracelsi eigenen Hand Schrifft."
2. *Editions.* Not edited by Huser or Sudhoff.
3. *Relationship between different versions.* Only one known version.
4. *Structure, genre/form, perspective, style.*
5. *Relationship to other texts.*
6. *Authenticity, authorship.* Allegedly from an autograph of Paracelsus. On a magical cup that is said to be in the possession of Johannes Montanus the Younger at Strigau. With a drawing of the cup.
7. *Time of writing.* Probably written in the beginning of the 17th century. Further study required.

WORD COUNT:

INCIPIT: "Anfenglichen soll man im aequinoctio verne, eben in der stund vnd minut wan die sonne …"

MANUSCRIPTS:
– KOPENHAGEN, *Det Kongelige Bibliotek*: GKS 259 folio, 229–234

FIRST PRINTED: not printed

ESSENTIAL BIBLIOGRAPHY: Sudhoff, *Paracelsus-Handschriften*, 668–669.

FURTHER BIBLIOGRAPHICAL REFERENCES:
Peuckert, *Pansophie* (1956), 476.

Brief an einen Schüler: see *Vom Modus des Gebrauchs mit der Tinctur* (§ 5.25)

Brief an Ulrich Beck: see *Von der Heilung zauberischer Schäden* (§ 5.29)

Büchlein von olympischer Geister Zitierung: see *Vom Modus des Gebrauchs mit der Tinctur* (§ 5.25)

5.4 *De fundamento scientiarum sapientiaeque* also: *Liber de fundamento scientiarum sapientiaeque/Vom Fundament der Weißheit und Künsten/Fundamenta scientiarum et sapientiae/Buch beider menschlicher Weisheit/Von Weisheit beider Seelen und Leibs*

1. *Printing History, Manuscripts.* First edited by Adam von Bodenstein in 1565.
2. *Editions.* Not edited by Huser or Sudhoff. Edited by Huser, 9 (1590): 414–457. Edited by Sudhoff in Paracelsus, *Sämtliche Werke*, I/13: 287–334.
3. *Relationship between different versions.*
4. *Structure, genre/form, perspective, style.* The first book is missing; the second (and only extant) book consists of three "tractatus."[302]
5. *Relationship to other texts.* According to Peuckert, there are quotations from the (now missing) first book of *De fundamento scientiarum sapientiaequae* in the *Schatzkammer der Natur* (§ 5.23).[303] – *De fundamento* is one of the few Paracelsian texts that mentions the magician Vergilius, the others being *De natura rerum* (§ 2.2), *Liber de imaginibus* (§ 5.17), *Archidoxis magica* (§ 5.1) and the *Liber principiorum* (§ 5.18).[304]
6. *Authenticity, authorship.* Edited by Huser from a manuscript provided by Johannes Montanus. "Uncertified and probably spurious," according to Rudolf Eucken.[305] To this Sudhoff commented: "If they are necessarily 'spurious' [only because they were edited from Montanus's manuscripts] seems to be doubtful".[306] According to Neumann, the authenticity of *De fundamento scientiarum*

302 See Huser, 9: 413, "To the Reader": "Dieses folgende Buch de Fundamento Scientiarum, wirdt vom Authore gennenet das ANDER Buch: welches aber das Erste sey (darinnen er des Leibs Notturfft vnd Gebresten sonderlich beschrieben) vnd zu diesem gehörig ist/ weiß man nicht. Soll derwegen der Leser mit diesem für gut haben/ biß das Erste auch gefunden werde/ etc."

303 Peuckert, *Gabalia* (1967), 466.

304 Peuckert, *Gabalia* (1967), 162.

305 Eucken (1880), 334.

306 Sudhoff, "Vorwort," in Paracelsus, *Sämtliche Werke*, I/13: IX.

sapientiaeque "is still not certain. In terms of content and style, however, the references to other writings by Hohenheim cannot be overlooked. In the middle of the 16th century, no one doubted its authenticity."[307]

7. *Time of writing.* Uncertain ("unbestimmt"), according to Sudhoff.

Word count: ca. 18.000 words

Incipit: "Einer der da will vonn künsten der weißheit schreiben ..."

Manuscripts: no manuscripts known

First printed:

- *1565* (*in: Das Bůch Paramirum Aureoli Theophrasti Paracelsi* [...] *Item, Vom Fundament vnd weißheit der künsten, der seelen vnd leibs kranckheiten* (Frankfurt am Main: Heirs of Christian Egenolf, 1565), f. 85v–124r; VD16 P 519; Sudhoff, *Bibliographia Paracelsica*, 106 n° 67)

Historical Manuscript Catalogues:

Essential bibliography: Sudhoff, *Bibliographia Paracelsica*, 106, 405, 510; Sudhoff, "Vorwort," in Paracelsus, *Sämtliche Werke*, I/13: IX; Goldammer, "Einleitung", in Paracelsus, Sämtliche Werke, II/2: XXXVIII; CP 1: 24, 364–366, 370, 379; CP 3: 834, 1021/1022.

Further bibliographical references:

Rudolf Eucken, "Des Paracelsus Lehren von der Entwicklung", in *Philosophische Monatshefte*, 16 (1880), 321–338, on 334; reprinted in Eucken, *Beiträge zur Einführung in die Geschichte der Philosophie*, 2nd ed. (Leipzig, 1906), 22–38, on 34.

Josef Strebel, "Zur Echtheitsfrage des *Liber de fundamento scientiarum sapientaeque*", in *Nova Acta Paracelsica*, 2 (1945), 187–199.

Peuckert, *Pansophie* (1956), 190.

Peuckert, *Gabalia* (1967), 162, 466.

Hanns-Peter Neumann, *Natura sagax – Die geistige Natur. Zum Zusammenhang von Naturphilosophie und Mystik in der frühen Neuzeit am Beispiel Johann Arndts* (Tübingen: Niemeyer, 2004), 33.

5.5 *De lapide philosophorum, qui lapis amoris appellatur*

1. *Printing History, Manuscripts.* Not printed. Two manuscripts.
2. *Editions.* Not edited by Huser or Sudhoff.

307 Neumann, *Natura sagax* (2004), 33.

3. *Relationship between different versions.*
4. *Structure, genre/form, perspective, style.*
5. *Relationship to other texts.*
6. *Authenticity, authorship.*
7. *Time of writing.* Karl Widemann received the text in 1588 from Gabriel Örtl (Leiden manuscript). According to Peuckert written in 1576 (no sources given).[308]

WORD COUNT:

INCIPIT: "Lapis philosophicus wirtt also genanndt von etlichen ain stain der lib ..."

MANUSCRIPTS:
- LEIDEN, *Universiteitsbibliotheek*: Voss. Chym. Q.17, f. 65v–66v
- WIEN, *Österreichische Nationalbibliothek*: Cod. 11259, f. 209r–214r

FIRST PRINTED: not printed

ESSENTIAL BIBLIOGRAPHY: Sudhoff, *Paracelsus-Handschriften*, 797 n° IV/5.

FURTHER BIBLIOGRAPHICAL REFERENCES:
Peuckert, *Pansophie* (1956), 321, 472.
Julius Ruska, "Pseudepigraphe Rasis-Schriften," *Osiris*, 7 (1939), 31–94, on 66.
Gerhard Eis, "Eine medizinisch-alchemistische Sammelhandschrift des Augsburger Paracelsisten Gabriel Örtel," *Archiv für Kulturgeschichte*, 40 (1958), 251–262, on 251; reprinted in Gerhard Eis, *Vor und nach Paracelsus. Untersuchungen über Hohenheims Traditionsverbundenheit und Nachrichten über seine Anhänger* (Stuttgart, 1965), 113–127, on 114.
Boeren, *Codices Vossiani Chymici* (1975), 143.

5.6 *De magia veterum* also: *Magia veterum/Magiae Theophrasti neun Bücher/Arbatel/Von der Magia*

1. *Printing History, Manuscripts.* The *Arbatel* was first printed in Latin in 1575 by Pietro Perna,[309] the most important printer-publisher of Paracelsus in Basel at that time,[310] but it was not attributed to Paracelsus. It was reprinted in 1579 in

308 Will-Erich Peuckert, *Pansophie* (Stuttgart, 1936), 356.

309 The most important study on the *Arbatel* and its relationship to paracelsism is Gilly, "Il primo prontuario di magia bianca" (2002). Most of what is said here derives from this publication.

310 Gilly, "Zwischen Erfahrung und Spekulation" (1977), 65.

the undated edition of Agrippa von Nettesheim's *Opera* that appeared with the false imprint "Lugduni, per Beringos fratres" (actually Basel, Thomas Guarin). This false imprint is the reason why the *Arbatel* is often said to have been first printed as early as 1510 or 1531. The Latin manuscripts are copied from the first edition. Several German manuscripts are known with at least four different translations.[311] Version C, which was also partially printed by Gottfried Arnold (see below), is attributed to Paracelsus.

2. *Editions.* Not edited by Huser or Sudhoff.
3. *Relationship between different versions.* The German version C, which is probably the oldest according to Gilly, has the 49 aphorisms in a different order, enriched with many citations from Agrippa von Nettesheim, according to Gilly.[312]
4. *Structure, genre/form, perspective, style.* Originally written in Latin in 49 (7 × 7) aphorisms forming an *Isagoge* which was only the first of nine books planned. No further book was ever published (or written[313]).
5. *Relationship to other texts.* The *Arbatel*'s doctrine of Olympic spirits was influential among pseudo-Paracelsian writers. Even dates were given, for example by Benedictus Figulus, for the reigns of spirits as outlined in the *Arbatel*.[314]
6. *Authenticity, authorship.* The author is an anonymous Paracelsian who styled Paracelsus along with Hermes Trismegistus as "secretorum pater" in aphorism 26.[315] Attributed to Paracelsus in version C of the German manuscripts and in the summary edited by Gottfried Arnold in his *Kirchen- and Ketzerhistorie*.[316] It was "one of the most influential and most contested manuals of Renaissance

311 Gilly, "Il primo prontuario di magia bianca" (2002), 209.

312 Gilly, "Vom ägyptischen Hermes zum Trismegistus Germanus" (2010), 84 n. 42.

313 Gilly, "Il primo prontuario di magia bianca" (2002), 210.

314 Gilly, "Vom ägyptischen Hermes zum Trismegistus Germanus" (2010), 72.

315 Gilly, "Zwischen Erfahrung und Spekulation" (1977), 66; Gilly, "Paracelsus, Hermes, Ficino" (1999), 302–306 n° 63, 306; Gilly, "Vom ägyptischen Hermes zum Trismegistus Germanus" (2010), 83.

316 Arnold, *Unpartheyische Kirchen- und Ketzer-Historie*, vol. 2 (1700), 159 (part IV, section II, num. XXII): "so mögen auch hier einige beschreibungen und anmerckungen hievon füglich statt haben/ und zwar aus einem solchen buche/ welches um den anfang des 17. seculi herausgekommen unter dem titul: Magia Veterum oder ein buch von den allerhöchsten secreten und heimlichkeiten der magiae, welche dem menschen gebühren zu wissen/ und sich ohne sünde deren gottselig gebrauchen mag/ ohne erzürnung Gottes durch Christliche und gebührliche mittel und Göttlichen beystand in kurtzer zeit zu lernen und zu begreiffen. Durch D.T.P.E.M.T.P. 1612." The initials can be read as: "Durch D[en] T[euren] P[hilosophum] E[t] M[edicum] T[heophrastum] P[aracelsum]:" Gilly, *Paracelsus in der BPH* (1993), 70. Sudhoff, *Paracelsus-Handschriften*, 682, reads: "Doctor Theophrastus Paracelsus Eremita Magus etc. (oder Magiae T . . Professor)." – In the table of contents, however, Arnold writes: "aus dem Anonymo Auctore de Magia Veterum." See Gilly, "Paracelsus, Hermes, Ficino" (1999), 306.

magic."[317] It also was the first text to revive the long-forgotten terms *theosophia* and *anthroposophia*.[318]

7. *Time of writing.* The *Arbatel* was probably written after 1571, the year of the first publication of Paracelsus's *Astronomia magna*.[319] The German version attributed to Paracelsus may have been prepared in the late 16th century. Further study required.

WORD COUNT: ca. 7.500 words (Latin text), ca. 20.000 words (German version C, Erlangen manuscript)

INCIPIT: "Hie fähet an das buch von der magia oder von der weißheit der alten [...] / Nachdem diß erste und vornembste buch des gantzen voluminis ..."

MANUSCRIPTS:

version C (attributed to Paracelsus)

- DARMSTADT, *Universitäts- und Landesbibliothek*: Hs. 1720/1, 1–61
- DESSAU, *Landesarchiv Sachsen-Anhalt*: Z 70, A 17a Nr. 105c, 1–119
- ERLANGEN, *Universitätsbibliothek*: MS B 244 (*olim* Irm. 1508), 1–60
- HAMBURG, *Staats- und Universitätsbibliothek*: Cod. alchim. 728
- KOPENHAGEN, *Det Kongelige Bibliotek*: Thott 630 kvart
- LONDON, *British Library*: Harley MS 514; 122 fols. [Latin; translated from German; attributed to Paracelsus]

FIRST PRINTED:

- 1700 ("Von der Magia", in *Gottfried Arnolds Fortsetzung und Erläuterung Oder Dritter und Vierdter Theil der unpartheyischen Kirchen- und Ketzer-Historie* (Frankfurt a.M.: Thomas Fritsch, 1700), 158–168)

HISTORICAL MANUSCRIPT CATALOGUES: Widemann, *Verzeichnisse* (Kassel), n° I, 5 / IV, 1; Arpe (1726), 121; Schröer/Roth-Scholtz (1732), 16 add. n° 4; von Linden (1786), 40 n° 265; Linck (1787), 243 n° 5; von Linden (1788), 4 n° 8 & 9, 29 n° 146, 43 n° 293 & 294; Ponickau (1791), col. 1232 n° Quarto 22

ESSENTIAL BIBLIOGRAPHY: Sudhoff, *Paracelsus-Handschriften*, 681–682, 798 n° IV/12; CP 2: 625; CP 3: 1087–1088.

---

317 Gilly, "Zwischen Erfahrung und Spekulation" (1977), 65.

318 *Ibid.*, 88–89; Gilly, "Il primo prontuario di magia bianca" (2002), 213; CP 2: 625.

319 Gilly, "Il primo prontuario di magia bianca" (2002), 211; CP 3: 1087.

Further bibliographical references:

Gottfried Arnold, *Unpartheyische Kirchen- und Ketzer-Historie*, vol. 2 (part 3/4) (Frankfurt a.M., 1700), 159 (part IV, section II, num. XXII).

Peuckert, *Pansophie* (1956), 333–337, 338, 479–481.

Peuckert, *Gabalia* (1967), 397.

Peuckert, *Rosenkreutz* (1973), 161.

Antonio Rotondò, "Pietro Perna e la vita culturale e religiosa di Basilea fra il 1570 e il 1580," in Rotondò, *Studi e ricerche di storia ereticale italiana del cinquecento* (Torino, 1974), 273–392, on 337–391; reprinted in Rotondò, *Studi di storia ereticale del Cinquecento* (Florence, 2008), 1: 479–576.

Gilly, "Zwischen Erfahrung und Spekulation" (1977), 65–66, 68–69, 73, 88–89; (1979), 185–187.

Hans Schneider, "Johann Arndts Studienzeit," *Jahrbuch der Gesellschaft für niedersächsische Kirchengeschichte*, 89 (1991), 133–175, on 150–154; reprinted in Hans Schneider, *Der fremde Arndt. Studien zu Leben, Werk und Wirkung Johann Arndts (1555–1621)* (Göttingen, 2006), 83–129, on 102–106.

Gilly, *Paracelsus in der BPH* (1993), 68, 70.

Gilly, "Theophrastia Sancta" (1994), 433 n. 20, 436 n. 26.

Carlos Gilly, "Paracelsus, Hermes, Ficino, and the 'Picatrix deutsch'", in Sebastiano Gentile and Carlos Gilly, eds., *Marsilio Ficino e il ritorno di Ermete Trismegisto* (Florence, 1999), 302–306 n° 63, 306.

Gilly, "Das Bekenntnis zur Gnosis" (2000), 397–399.

Gilly, "Il primo prontuario di magia bianca" (2002).

Gilly, "Vom ägyptischen Hermes zum Trismegistus Germanus" (2010), 72, 82–84.

Beket Bukovinská and Ivo Purš, "Die Tischglocke Rudolfs II.: über ihren Urheber und ihre Bedeutung," *Studia Rudolphina*, 10 (2010), 89–104, on 96.

Hereward Tilton, "Bells and Spells: Rosicrucianism and the Invocation of Planetary Spirits in Early Modern Germany," in *Celestial Magic*, special issue of *Culture and Cosmos*, 19 (2015), 167–188, on 178, 179.

Bellingradt and Otto, *Magical Manuscripts* (2017), 21–22 n. 71 and 72.

5.7 *De occulta philosophia* also: *Philosophia occulta/Liber occultae philosophiae*

1. *Printing History, Manuscripts.* First edited in 1570 by Michael Toxites in *Archidoxa Philippi Theophrasti Bombast*, and several months later by Johann Albert Wimpinaeus in *Etliche Tractetlein zur Archidoxa gehörig*, clearly from another manuscript source. In Toxites's edition some paragraphs were missing in two places (an error of the printer?); these were added in the 1574 re-edition.[320] No

320 CP 2: 415.

full manuscripts; excerpts in two manuscripts. Manuscripts (now lost) were circulating already in 1569.[321] – Because of his edition of *De occulta philosophia* Toxites was accused of being a "nigromanticus" and practicing "forbidden arts."[322]

2. *Editions.* Edited by Huser, 9 (1590): 329–368. Edited by Sudhoff in Paracelsus, *Sämtliche Werke*, I/14: 513–542.
3. *Relationship between different versions.*
4. *Structure, genre/form, perspective, style.*
5. *Relationship to other texts.* For related texts on "pygmaei" see *Revelatio in somnio* (§ 5.22) and *Von den Erdgeistern* (§ 5.26).
6. *Authenticity, authorship.* Considered as spurious by Sudhoff, who included this text among the spuria of vol. 14 of his edition.[323] One may notice that Paracelsus himself referred once to a work of his entitled *De occulta philosophia* in his treatise *Von dem Schlaff und Wachen der Leiber und Geister*.[324] – Andrew Weeks has recently come to the conclusion that there is "no reason to dispute the authenticity [...] since its themes and thoughts are nearly all found in the undisputed [*De causis morborum invisibilium*]".[325] However, at least the mentioning by Paracelsus of "Henricus Cornelius Agrippa [...] Petrus de Abano [...] Trithemius" at the beginning of the treatise is of highly questionable authenticity, if not blatantly spurious.
7. *Time of writing.* Probably written in the late 1560s.

Word count: ca. 14.000 words

Incipit: "Von den hohen verborgnen dingen ist hie mein fürnemmen [...] / Von heyligung der örter, der instrumenten vnnd aller geschöpff auff erden ..."

Manuscripts:
- Amsterdam, *Bibliotheca Philosophica Hermetica*: M 127
- Dresden, *Sächsische Landesbibliothek*: Mscr.Dresd.N.43, f. 34r–64r [German]
- Kassel, *Landesbibliothek*: 4° Ms. chem. 17, f. 126v–129r, 169r–170v [excerpts]

321 A manuscript of the *Liber Occultae Philosophiae Theophrasti* was mentioned and owned in 1569 by Johann Hiller; see CP 3: 55–56, 61, 62.

322 CP 2: 489–490.

323 Sudhoff, "Vorwort," in Paracelsus, *Sämtliche Werke*, I/14: XII: "Ich selbst halte alles hier Zusammengestellte für sicher unecht."

324 Huser, 9: 412: "Darumb hab fleissig acht darauff/ was du die Nacht siehest/ oder waß dir von diesem oder jenem *Spiritu humano* gesagt oder angezeigt wirdt/ daß du es auch Morgens wissest: (Wie wir auch *in libro de Occulta Philosophia* beschrieben haben)."

325 Weeks, in Jacob Boehme, *Aurora* (*Morgen Röte im auffgang, 1612*), ed. Andrew Weeks (Leiden, 2013), 485. The complete discussion in Weeks, "Paracelsus and the Idea of the Renaissance" (2013).

- KASSEL, *Landesbibliothek*: 4° Ms. chem. 25, f. 102v
- KASSEL, *Landesbibliothek*: 4° Ms. chem. 72, f. 227r, 236v
- MÜNCHEN, *Bayerische Staatsbibliothek*: cgm 3737, f. 29v–33r [probably copied from the printed 1570 edition]
- WROCŁAW/BRESLAU, *Biblioteka Uniwersytecka*: Ms. 6610 (*olim* Görlitz), f. 9v–11r [excerpts]

FIRST PRINTED:
- 1570 (in: *Archidoxa Philippi Theophrasti Bombast*, ed. Michael Toxites (Strasburg: Theodosius Rihel, 1570), sig. a4r–g1r; VD16 P 395; Sudhoff, *Bibliographia Paracelsica*, 187–190 n° 118; Sudhoff, *Bibliographia Paracelsica*, 268–272 n° 158)
- 1570 (in: *Philippi Theophrasti Paracelsi von Hohenhaim etliche Tractetlein zur Archidoxa gehörig*, ed. Johann Albert Wimpinaeus (Munich: Adam Berg, 1570), sig. B2r–G1r; VD16 P 629; Sudhoff, *Bibliographia Paracelsica*, 213–215 n° 128)

HISTORICAL MANUSCRIPT CATALOGUES: *Catalogus* Osek (Prague) 1, n° 29; *Scripta Theophrasti* (Wien); Widemann, *Verzeichnisse* (Kassel), n° V, 242

ESSENTIAL BIBLIOGRAPHY: Sudhoff, *Bibliographia Paracelsica*, 404; Sudhoff, *Paracelsus-Handschriften*, 150, 215, 711, 714, 663, 776; Sudhoff, "Vorwort," in Paracelsus, *Sämtliche Werke*, I/14: XXIX; CP 1: 421; CP 2: 201, 415, 489–492, 1008.

FURTHER BIBLIOGRAPHICAL REFERENCES:

Josef Strebel, "Zur Echtheitsfrage der paracelsischen Geheimphilosophie (*Ph. Occulta*) und die *Philosophia ad Athenienses*," *Nova Acta Paracelsica*, 4 (1947), 48–54 [questionable].

Peuckert, *Pansophie* (1956), 339, 454, 474, 478.

Peuckert, "Paracelsische Zauberei," *Nova Acta Paracelsica*, 8 (1957), 71–94, on 88, 90.

Peuckert, *Gabalia* (1967), 447.

Wolf-Dieter Müller-Jahncke, *Astrologisch-magische Theorie und Praxis in der Heilkunde der frühen Neuzeit* (Stuttgart: Franz Steiner, 1985), 85.

Goldammer, *Der göttliche Magier* (1991), 105–106.

Thomas Hofmeier, "Paracelsus und Ägypten," *Nova Acta Paracelsica*, N.F. 15 (2001), 41–54, on 52.

Tried and tested: The appreciation of Hermetic and alchemical manuscripts from the 15th–20th centuries (Amsterdam: Bibliotheca Philosophica Hermetica, 2004), 41

Broszinski, *Manuscripta chemica in quarto* (2011), 53, 54, 322.

Andrew Weeks, ed., Jacob Boehme, *Aurora* (*Morgen Röte im auffgang, 1612*) (Leiden, 2013).

Andrew Weeks, "Paracelsus and the Idea of the Renaissance", in *Die Wiederkehr der Renaissance im 19. und 20. Jahrhundert*, ed. Helmut Koopmann and Frank Baron (Münster, 2013), 89–113.

5.8 *De philomela*

1. *Printing History, Manuscripts.* Not printed. Three manuscripts.
2. *Editions.* Not edited by Huser or Sudhoff.
3. *Relationship between different versions.* There are two versions of the text with almost no differences; version 2, however, directly addresses "e. khw. mt." ("ewer khuniglich mayestat", your royal majesty) and is not directly associated with Paracelsus. The Vienna manuscript has both versions, the Leiden manuscript only version 2.
4. *Structure, genre/form, perspective, style.* An instruction to transport a nightingale for one mile on a twig so that it sings all the way before it falls down dead.
5. *Relationship to other texts.*
6. *Authenticity, authorship.* The Vienna manuscript is associated with a "D[octor] Koplerus" who may or may not be identical with Dr. Wolfgang Kappler (see § 10.3). The text of the Wolfenbüttel manuscript is associated with the name of Georg Phaedro, according to CP 2: 544. – A "foolish stunt" according to Sudhoff.[326] – Mentioned in the catalogues of Jonas Adelwert's library.[327]
7. *Time of writing.* The Wolfenbüttel manuscript is dated 1567.

Word count: ca. 130 words [version 1], ca. 240 words [version 2].

Incipit: "Recipe zibetam non sophisticatam grana sex ..."

Manuscripts:
- Leiden, *Universiteitsbibliotheek*: Voss. Chym. F.16, f. 10v [anonymous]
- Wien, *Österreichische Nationalbibliothek*: Cod. 11259, f. 238v–240r
- Wolfenbüttel, *Herzog August Bibliothek*: Cod. Guelf. 33.6 Aug. 4°, f. 444–446

First printed: not printed

326 Sudhoff, *Paracelsus-Handschriften*, 198: "Solchen läppischen Kunststücklein musste eben der Name des Wundermannes erst das richtige Ansehen verleihen!"

327 Sudhoff, *Paracelsus-Handschriften*, 28: "Philomela;" also in the Adelwert catalogue in Stuttgart, Hauptstaatsarchiv, A 47, Bü. 3: "Eiusdem Philomela."

HISTORICAL MANUSCRIPT CATALOGUES: *Catalogus* (Stuttgart), I, f. 4v n° [151]; *Catalogus* Adelwert (Leiden), n° [108]

ESSENTIAL BIBLIOGRAPHY: Sudhoff, *Paracelsus-Handschriften*, 111, 197, *797* n° IV/4; CP 2: 544.

FURTHER BIBLIOGRAPHICAL REFERENCES:
Boeren, *Codices Vossiani Chymici* (1975), 52.

5.9 *De septem stellis*
also: *Liber de septem stellis/Septem stellae*

1. *Printing History, Manuscripts.* Not printed. Seven manuscripts.
2. *Editions.* Not edited by Huser or Sudhoff.
3. *Relationship between different versions.* The *Liber de septem stellis* preserved among the Kiesewetter manuscripts in Munich is an extremely enlarged version. The text itself requires further study.
4. *Structure, genre/form, perspective, style.*
5. *Relationship to other texts.*
6. *Authenticity, authorship.*
7. *Time of writing.* Probably written in the 1570s. Mentioned in 1582 in a letter by Johann Franke to Leonhart Thurneisser.[328] Next mentioned in a list of manuscripts by Karl Widemann.[329] Mentioned again in 1726 in a list of manuscripts by Peter Friedrich Arpe.[330]

WORD COUNT: ca. 1.500 words (Lübeck manuscript), ca. 30.000 words (Munich manuscript)

INCIPIT: "Wir Theophrastus sagen bey unserer wahrheit …"

MANUSCRIPTS:
- DRESDEN, *Sächsische Landesbibliothek*: Mscr.Dresd.N.119, f. 8r–11v
- LEIPZIG, *Universtitätsbibliothek*: Cod. mag. 39; 11 fols.
- LEIPZIG, *Universtitätsbibliothek*: Ms. 0392; 41 fols.

---

328 CP 3: 508.
329 Sudhoff, *Paracelsus-Handschriften*, 749, 751; CP 3: 508.
330 Sudhoff, *Paracelsus-Handschriften*, 22; Julian Paulus, "Eine unbeachtete Paracelsus-Handschrift aus Franken" (1991), 144.

- LÜBECK, *Stadtbibliothek*: Ms. math. 4° 9, f. 143ra–144rb
- MÜNCHEN, *Bayerische Staatsbibliothek*: Kiesewetteriana 1e, 180–343 [enlarged version][331]
- PRAHA, *Strahovská knihovna*: Cod. DF IV 59, f. 137v–154r
- *olim* Ichtershausen, library of Duke Johann Friedrich VI. von Sachsen; confiscated in 1627; current location unknown

FIRST PRINTED: not printed

HISTORICAL MANUSCRIPT CATALOGUES: Widemann, *Verzeichnisse* (Kassel), n° IV, 6 / V, 124; Herzog Johann Friedrich von Sachsen-Weimar (1627), 217 n° 8; Schröer (1710), 3 n° 69; Arpe (1726), 123; Schröer/Roth-Scholtz (1732), 10 n° 71; Jahn (1754–1758), 3: 152 n° 204; von Linden (1788), 45 n° 304

ESSENTIAL BIBLIOGRAPHY: Sudhoff, *Paracelsus-Handschriften*, 22, 685, 798 n° IV/16; CP 3: 494, 508.

FURTHER BIBLIOGRAPHICAL REFERENCES:

Peuckert, *Pansophie* (1956), 481, 490.

Julian Paulus, "Eine unbeachtete Paracelsus-Handschrift aus Franken", in Joachim Telle, ed., *Parerga Paracelsica* (Stuttgart, 1991), 141–147, on 144.

Carlos Gilly, "Il ritrovamento dell'originale del *Calendarium Naturale Magicum Perpetuum* di Großschedel" / "The rediscovery of the original of Großschedel's *Calendarium Naturale Magicum Perpetuum*," in Carlos Gilly and Cis van Heertum, eds., *Magia, alchimia, scienza dal '400 al '700. L'influsso di Ermete Trismegisto* (Venice, 2002), 295–315, on 311 and 315 n. 15.

Gerhard Powitz, "Mittelalterliche Handschriften aus dem Besitz der Senckenbergischen Bibliothek in Frankfurt am Main," *Codices Manuscripti*, 60/61 (2007), 1–22, on 4 n° X39.

Leigh T.I. Penman, "A Second Christian Rosencreuz? Jakob Böhme's Disciple Balthasar Walther (1558–c. 1630) and the Kabbalah. With a Bibliography of Walther's Printed Works," *Scripta Instituti Donneriani Aboensis*, 20 (2008), 154–172, on 161.

Leigh T.I. Penman, "'Ein Liebhaber des Mysterii, und ein großer Verwandter desselben.' Toward the Life of Balthasar Walther: Kabbalist, Alchemist and Wandering Paracelsian Physician," *Sudhoffs Archiv*, 94 (2010), 73–99, on 79.

331 Description in Sudhoff, *Paracelsus-Handschriften*, 684–686 n° 136.

Hereward Tilton, "Bells and Spells: Rosicrucianism and the Invocation of Planetary Spirits in Early Modern Germany," in *Celestial Magic*, special issue of *Culture and Cosmos*, 19 (2015), 167–188, on 183 n. 54.
Bellingradt and Otto, *Magical Manuscripts* (2017), 117 n° 69.

De signis zodiaci et ejus mysteriis: see *Archidoxis magica* (§ 5.1)

5.10 *Experimentum Theophrasti Paracelsi*

1. *Printing History, Manuscripts.* Not printed. One manuscript.
2. *Editions.* Not edited by Huser or Sudhoff.
3. *Relationship between different versions.* Only one known version.
4. *Structure, genre/form, perspective, style.* Written in the first person, directly addressing the reader.
5. *Relationship to other texts.* An instruction to conjure spirits. Relationship to other such texts unclear.
6. *Authenticity, authorship.*
7. *Time of writing.* Possibly written in the 1600s or 1610s. Further study required.

WORD COUNT: ca. 2.500 words

INCIPIT: "Neyge deine ohren und werkhe auf mein wort …"

MANUSCRIPTS:
– WIEN, *Österreichische Nationalbibliothek*: Cod. 11266, f. 91v–105r

FIRST PRINTED: not printed before Spunda (1924)

HISTORICAL MANUSCRIPT CATALOGUES: von Linden (1788), part 2, 38 n° 32

MODERN EDITIONS:
– *Magische Unterweisungen des edlen und hochgelehrten Philosophi und Medici Philippi Theophrasti Bombasti von Hohenheim Paracelsus genannt*, ed. Franz Spunda (Leipzig, 1924), 22–33 [from Wien, ÖNB, Cod. 11266]

ESSENTIAL BIBLIOGRAPHY: Sudhoff, *Paracelsus-Handschriften*, 675–677, 797 n° IV/11.

FURTHER BIBLIOGRAPHICAL REFERENCES:
Peuckert, *Pansophie* (1956), 474–475.

5.11 *Fragment über Geisterbann*

1. *Printing History, Manuscripts.* One manuscript, with some lost folios; only the final sentences survive.
2. *Editions.* Not edited by Huser or Sudhoff.
3. *Relationship between different versions.* Only one known version.
4. *Structure, genre/form, perspective, style.*
5. *Relationship to other texts.*
6. *Authenticity, authorship.* Original title unknown. May or may not be attributed to Paracelsus. The last sentence runs: "Further notice may be asked from the aforementioned franciscan brother Georg Dreßler in Neuhaus since he has tried it. And this was confided by Theophrastus to Jacob Beyer."[332]
7. *Time of writing.* Further study required.

WORD COUNT:

INCIPIT:

MANUSCRIPTS:
- WIEN, *Österreichische Nationalbibliothek*: Cod. 11259, f. 241r

ESSENTIAL BIBLIOGRAPHY: Sudhoff, *Paracelsus-Handschriften*, 198–199, 797 n° IV/6.

FURTHER BIBLIOGRAPHICAL REFERENCES:

Gerhard Eis, "Zur Paracelsusnachfolge im Sudetenraum," *Südostforschungen*, 6 (1941), 440–462, on 443; reprinted in Eis, *Vor und nach Paracelsus. Untersuchungen über Hohenheims Traditionsverbundenheit und Nachrichten über seine Anhänger* (Stuttgart, 1965), 136–161, on 139.

5.12 *Geheime astrologische Wirkungen Regis Salomonis*

1. *Printing History, Manuscripts.* Not printed. One manuscript.
2. *Editions.* Not edited by Huser or Sudhoff.
3. *Relationship between different versions.* Only one known version.

332 Sudhoff, *Paracelsus-Handschriften*, 198: "Mehrer Bericht ist von obgedachtem franciscaner Münch oder Brueder zu Newhauß Georgen Dreßler zuerfragen, dann er es probiert, vnnd hatt solches Theophrastus Herrn Jacob Beyer, vertraut." Neither Dressler nor Beyer are mentioned by Broszinski, *Manuscripta chemica in quarto* (2011) or Boeren, *Codices Vossiani Chymici* (1975). – An illustrated alchemical manuscript (Vatican, Cod. pal. lat. 1336) was written in 1560/1561 in Neuhausen near Worms and dedicated to Kurfürst Friedrich III. von der Pfalz, see CP 2: 128. Digital facsimile available via University Library Heidelberg.

4. *Structure, genre/form, perspective, style.*
5. *Relationship to other texts.* Contains an excerpt of the Paracelsian *Von Schlafen und Wachen* (Huser, 9: 407 ff.) and excerpts of some cabalistic-magical texts according to Sudhoff.
6. *Authenticity, authorship.*
7. *Time of writing.* Further study required.

Word count:

Incipit: "Die geheime göttliche geistliche sternkunst giebt sichtiglich und augenscheinlich die verborgenen rathschläge der heiligen dreyfaltigkeit an den tag …"

Manuscripts:
- München, *Bayerische Staatsbibliothek*: Kiesewetteriana 1e, 156–179[333]

First printed: not printed

Essential bibliography: Sudhoff, *Paracelsus-Handschriften*, 685, 798 n° iv/15.

Geheimnis der Campana philosophica: see *Vom Modus des Gebrauchs mit der Tinctur* (§ 5.25)

Heiliges Gebetbuch: see *Psalterium magicum* (§ 5.21)

Geheimnis aller seiner Geheimnisse: see *Vom Modus des Gebrauchs mit der Tinctur* (§ 5.25)

5.13 *Heiliges Sales-Büchlein oder die Glücks-Ruthe*

1. *Printing History, Manuscripts.* Probably first printed as a work of Paracelsus ca. 1900. The full title of the ca. 1920 edition is: "Erstlich von der Ruthe in's Gemeine, wie solches der hochgelehrte Herr Dr. Johannes Theophrastus Paracelsus beschreibt."
2. *Editions.* Not edited by Huser or Sudhoff.
3. *Relationship between different versions.* Only one known version.
4. *Structure, genre/form, perspective, style.*
5. *Relationship to other texts.*
6. *Authenticity, authorship.*
7. *Time of writing.* It is not known if the text goes back to an older edition under a different name.

333 Description in Sudhoff, *Paracelsus-Handschriften*, 684–686 n° 136.

Word count:

Incipit: "Es sind zwar viele Ruthengeher, die sagen, daß jede Ruthe in ihrem Planetenstand geschnitten werden sollte …"

Manuscripts: no manuscripts known

First printed:
- ca. 1900 (in: *Die große Moses-Bibel das ist das 6 und 7. Buch Mosis* (Philadelphia [Berlin], no date)[334]
- ca. 1900 (in: *Das 6. und 7. Buch Mosis* (no place, no date [Berlin ca. 1900])[335]
- ca. 1920 ("Erstlich von der Ruthe in's Gemeine, wie solches der hochgelehrte Herr Dr. Johannes Theophrastus Paracelsus beschreibt," in *Das heilige Sales-Büchlein oder die Glücks-Ruthe* (no place, no date [Berlin, ca. 1920]), 26–31)

Further bibliographical references:
Mitteilungen der Gesellschaft für jüdische Volkskunde, 1900, n° 1, 19.
Richard Andree, *Braunschweiger Volkskunde*, 2nd ed. (Braunschweig, 1901), 399.
Karl Brunner, "Aus den Sitzungsprotokollen des Vereins für Volkskunde. Freitag, den 22. Januar 1909," *Zeitschrift des Vereins für Volkskunde*, 19 (1909), 245–248, on 245–246.
Wolfgang Bauer, ed., *Das sechste und siebente Buch Mosis. Sein wahrer Wert und was das Volk darin sucht* (Berlin, 1996).

5.14 *Hundertachtunddreißig ganz neu entdeckte Geheimnisse*

1. *Printing History, Manuscripts.* A collection of "secrets," first printed in 1717, reprinted at least in 1720, 1726, 1729, 1732, 1737, 1756, in some cases with newly appended secrets.
2. *Editions.* Not edited by Huser or Sudhoff.
3. *Relationship between different versions.*
4. *Structure, genre/form, perspective, style.*
5. *Relationship to other texts.* Parts of the *138 Geheimnisse* are parallel to *Experimenta* attributed to Andreas Tentzel.[336] These *Experimenta* are dated 1615, but were probably published for the first time in 1725.
6. *Authenticity, authorship.* Not a pseudo-Paracelsian work, but a collection of short texts including these texts that are attributed to or mention Paracelsus: "Daß

334 Mentioned in *Mitteilungen der Gesellschaft für jüdische Volkskunde* (1900), n° 1, 19.

335 Mentioned by Karl Brunner, "Aus den Sitzungsprotokollen," (1909), 245–246. This edition or an earlier one is also mentioned by Richard Andree, *Braunschweiger Volkskunde* (1901), 399.

336 Peuckert, *Gabalia* (1967), 354–356.

Granat-Aepffel von allerhand Arten zusammen gesetzt/ aneinander wachsen" (23, n° 20), "Das Senechthon Theophrasti" (46, n° 41), "Von denen Olympischen Geistern des Theophrasti" (113–114, n° 122), "Von denen magischen Spiegeln" (121–122, n° 130), "Theophrasti magisches Glöcklein zu machen" (n° 131), "De Secretis Secretorum Thesauro Philosophor[um]" (127–129, n° 136).

7. *Time of writing.* According to the preface, the collector ("Der Herr Collector") was already deceased at the time of publication. The secret n° 45 mentions university studies in 1664 in Bern (Switzerland).[337]

Word count:

Incipit: "Man klopffet die Zweige (Virga) mit einem Hammer ..." [n° 20] / "Man siehet sich um, daß man eine sehr grosse Krote bekomme ..." [n° 41] / "Des Theophrasti Buch von denen Olympischen Geistern ist ein pur Nicomantisch Werck ..." [n° 122] / "Derer habe ich dreyerley Gattungen gesehen ..." [n° 130] / "Der Mensch hat in sich einen unaussprechlichen Magischen Magneten ..."

Manuscripts: no manuscripts known

First printed:
- 1717 (*Hundert Acht und Dreyßig Gantz Neu-entdeckte Geheimnüsse Oder Sonderbahre bißher unbekannte und noch nie an das Tages-Licht gegebene Secrete* (Frankfurt a.M.-Leipzig: Johann Martin Hagen, 1717); VD18, 12226580)

Essential bibliography:

Further bibliographical references:

Will-Erich Peuckert, "Paracelsische Zauberei," *Nova Acta Paracelsica*, 8 (1957), 71–94, on 92–93.

Peuckert, *Gabalia* (1967), 212–213, 259, 312, 337–357, 556–557.

5.15 *Judicium von den Pelagianischen, Libanischen, Trithemischen auch anderen Schriften*

1. *Printing History, Manuscripts.* Lost. Not printed. No manuscript known. This text circulated in the early 17th century.
2. *Editions.* Not edited by Huser or Sudhoff.
3. *Relationship between different versions.* Text not preserved.

337 All biographical or pseudo-biographical hints collected by Peuckert, *op. cit.*, 339 ff.

4. *Structure, genre/form, perspective, style.*
5. *Relationship to other texts.*
6. *Authenticity, authorship.*
7. *Time of writing.* Probably written in the beginning of the 17th century or earlier. Further study required.

Word count:

Incipit:

Manuscripts: no manuscripts known

First printed: not printed

Essential bibliography: Sudhoff, *Paracelsus-Handschriften*, 762; CP 2: 607.

Further bibliographical references:
Gilly, "Il primo prontuario di magia bianca" (2002), 210.

5.16 *Kleines Commando-Büchlein*

1. *Printing History, Manuscripts.* Not printed. Two manuscripts known, both written in the 18th century.
2. *Editions.* Not edited by Huser or Sudhoff.
3. *Relationship between different versions.* Only one version known.
4. *Structure, genre/form, perspective, style.* Instructions for the preparation of magical parchments, written in the perspective of a father or teacher directly addressing the reader in the second person ("du," "du mein Sohn").
5. *Relationship to other texts.* The "Geister-Commando" published by Johann Scheible in *Die Sage vom Faust* (*Das Kloster*, vol. 5, Stuttgart 1847) is a different text not related to Paracelsus.
6. *Authenticity, authorship.* In a trial before the court of the Saxonian town of Pirna (1817–1820) over the accusation of treasure hunting, a manuscript called *Klein Commando-Büchlein oder Deutlicher Unterricht Magiam zu erlangen und die Edle Kunst Alchÿmiam sammt der Kunst* was confiscated. On the top of the title page is written "Philippi Theophrasti," probably by a later hand.

   It is the same text as the *Klain Commando Büchlein oder deutlich Anzaig und Erklerung der unschatzbaren großen und vollkommen Kunst aller Künste und Magnalien, das größte Geheimnüß der Natur und verborgen Magia* […] *gleichsam zu commandiren* […] *als ein heilig Testament hinterlassen an seinen Sun von Johann Fust, berühmten Artist zu Maintz* MCCCCLIV, found among the Leipzig

University Library magical manuscripts, where it is attributed to Johannes Faust rather than Paracelsus.

7. *Time of writing.* Further study required.

Word count:

Incipit: "Die ganze natur des himmels bestehet in sieben haupt geistern …"

Manuscripts:
- Dresden, *Sächsisches Staatsarchiv*: 10062 Amt Pirna, Nr. 2688; 66 pages
- Leipzig, *Universtitätsbibliothek*: Cod. mag. 137; 15 fols. [not attributed to Paracelsus]

First printed: not printed

Further bibliographical references:
Bellingradt and Otto, *Magical Manuscripts* (2017), 60.

Kräftiges Gebetbuch: see *Psalterium magicum* (§ 5.21)

Liber de vermibus: see *Liber principiorum* (§ 5.18)

5.17 *Liber de imaginibus* auch: *De imaginibus*

1. *Printing History, Manuscripts.* First printed in 1572 by Pietro Perna; not reprinted before Huser. Three manuscripts.
2. *Editions.* Edited by Huser, 9 (1591): 369–393. Edited by Sudhoff in Paracelsus, *Sämtliche Werke*, 1/13: 359–386.
3. *Relationship between different versions.* Only one known version.
4. *Structure, genre/form, perspective, style.* Divided into 13 chapters.
5. *Relationship to other texts.* The *Liber de imaginibus* mentions a book "written by myself" on the 'Nuremberg figures', a pictorial prophecy about the papal succession.[338] Another text on images written by Paracelsus is the *Liber de imaginibus idolatriae* (ca. 1525/28). – Both in the 1572 printed edition and in all three manuscripts the *Liber de imaginibus* is found in close vicinity to chapters of *Archidoxis magica* (§ 5.1). In Huser's edition the *Liber de imaginibus* is preceded by *De occulta philosophia* (§ 5.7), both printed from a manuscript provided by Johannes Montanus. – It has been shown that chap. 12 of the *Liber de imaginibus* shares the same topics and specific terminology as the authentic *Vom langen*

338 Holenstein, "Sterne, Zeichen, Zukunft" (1994), 51. – See Huser, 9 (1591): 373.

Leben (which expressly refers to a treatise called *De imaginibus*) – not to mention another topic shared with *De vita longa*, IV, 3.[339]

6. *Authenticity, authorship.* The authenticity of the *Liber de imaginibus* has been cast in doubt by Sudhoff (I, 13, XI). Sudhoff was followed by Telle who attributed the *Liber de imaginibus* to "(Ps.-) Paracelsus".[340] Other authors like Strebel, Peuckert or Pagel disagreed with Sudhoff and explicitly regarded the text as genuine.[341] Gause regarded the *Liber de imaginibus* as authentic without giving arguments, while Möseneder speaks rather tentatively of "the presumed author Paracelsus" or a text "connected with his [i.e. Paracelsus's] name".[342] The most recent study on the *Liber de imaginibus* by Amadeo Murase regards the Liber as authentic.[343]
7. *Time of writing.* According to Gause, the *Liber de imaginibus* was written after Paracelsus's visit to Nuremberg in 1529,[344] while Rudolph considers the text to be not reliably datable[345] and Holenstein dates it rather generously to the years 1520 to 1535.[346] – If, on the other hand, the text is spurious it may have been written in the 1560s like the Archidoxis magica or De occulta philosophia.

WORD COUNT: ca. 9.200 words

INCIPIT: "Zubeschreiben die krafft vnd tugendt vnd wunderbarlichen wirckung der bilder ..."

MANUSCRIPTS:
- HALLE, *Marienbibliothek*: Hs. Nr. 70 (*olim* Nr. 34), no foliation
- HAMBURG, *Staats- und Universitätsbibliothek*: Cod. alchim. 192, f. 23r–37v
- PRAHA, *Národní knihovna*: Osek 43, f. 49v–65v

FIRST PRINTED:
- 1572 (in *Archidoxorum* [...] *Theophrasti Paracelsi X. Bücher* (Basel: Pietro Perna, 1572); VD16 P 396; Sudhoff, *Bibliographia Paracelsica*, 239–241 n° 142)

---

339 Kahn, "La Création *ex nihilo*" (forthcoming).
340 Telle, *Alchemie und Poesie*, 702; also CP 1, 274.
341 Strebel, "Über Wesen und Zweck" (1946), 45; Peuckert, "Paracelsische Zauberei" (1957), 85; Pagel and Winder, "Gnostisches bei Paracelsus" (1968), 361; Pagel, *Paracelsus* (1982), 369: "a genuine treatise".
342 Möseneder, Paracelsus und die Bilder (2009), 4.
343 Murase, "The Homunculus" (2020), 49.
344 Gause, *Paracelsus* (1993), 145.
345 Hartmut Rudolph, "Einige Gesichtspunkte" (1981), 12.
346 Holenstein, "Sprechende Steine" (2012–2013), 77.

Historical Manuscript Catalogues: *Catalogus* Osek (Prague) 2, n° 2; *Catalogus* (Stuttgart), I, f. 4r n° [146]; *Catalogus* Adelwert (Leiden), n° [112]

Essential bibliography: Sudhoff, *Bibliographia Paracelsica*, 239–241 n° 142; Sudhoff, *Paracelsus-Handschriften*, 28, 31, 142, 776; Sudhoff, "Vorwort," in Paracelsus, *Sämtliche Werke*, I/13, X–XI; CP 1 (2001): 274; CP 3 (2013): 712, 1139.

Further bibliographical references:

Josef Strebel, "Über Wesen und Zweck der sogenannten niederen Magie im Sinne von Paracelsus", *Nova Acta Paracelsica*, 3 (1946), 43–48.

Gustav Friedrich Hartlaub, "'Paracelsisches' in der Kunst der Paracelsuszeit", *Nova Acta Paracelsica*, 7 (1954), 132–163, on 144–145, 153.

Peuckert, *Pansophie* (1956), 212–213, 236, 291, 355, 468, 485.

Will-Erich Peuckert, "Paracelsische Zauberei," *Nova Acta Paracelsica*, 8 (1957), 71–94, on 76, 85.

Walter Pagel, "Paracelsus and Techellus the Jew", *Bulletin of the History of Medicine*, 34 (1960), 274–277.

Peuckert, *Gabalia* (1967), 81, 161–162, 330.

Walter Pagel and Marianne Winder, "Gnostisches bei Paracelsus und Konrad von Megenberg", in Gundolf Keil et al., eds., *Fachliteratur des Mittelalters. Festschrift für Gerhard Eis* (Stuttgart, 1968), 359–371.

Hartmut Rudolph, "Kosmosspekulation und Trinitätslehre. Ein Beitrag zur Beziehung zwischen Weltbild und Theologie bei Paracelsus", in *Paracelsus in der Tradition. Vorträge Paracelsustag 1978* (Vienna, 1980) (*Salzburger Beiträge zur Paracelsusforschung*, 21), 32–47, on 34.

Arlene Miller Guinsburg, "Die Ideenwelt des Paracelsus und seiner Anhänger in Hinsicht auf das Thema des christlichen Magus und dessen Wirken", in *Von Paracelsus zu Goethe und Wilhelm von Humboldt* (Vienna, 1981) (*Salzburger Beiträge zur Paracelsusforschung*, 22), 27–54, on 40–41.

Hartmut Rudolph, "Einige Gesichtspunkte zum Thema 'Paracelsus und Luther'", in *Von Paracelsus zu Goethe und Wilhelm von Humboldt* (Vienna, 1981) (*Salzburger Beiträge zur Paracelsusforschung*, 22), 9–26, on 12.

Walter Pagel, *Paracelsus: An Introduction to Philosophical Medicine in the Era of the Renaissance*, 2nd rev. ed. (Basel, 1982), 149, 361, 367, 369.

Walter Pagel, *The Smiling Spleen. Paracelsianism in Storm and Stress* (Basel, 1984), 186.

Ute Gause, *Paracelsus (1493–1541). Genese und Entfaltung seiner frühen Theologie* (Tübingen, 1993), 145, 146, 156–158.

Gilly, *Paracelsus in der BPH* (1993), 39.

Gilly, "*Theophrastia Sancta*" (1994), 428–429.

Pia Holenstein Weidemann, "Sterne, Zeichen, Zukunft. Paracelsus' astrologische Theorien im 'Liber Philosophiae de arte praesaga' und im 'Liber de imaginibus'", *Nova Acta Paracelsica*, N.F. 8 (1994), 37–55.

Thomas Hofmeier, "Paracelsus und Ägypten," *Nova Acta Paracelsica*, N.F. 15 (2001), 41–54, on 49–50.

Pia Holenstein Weidemann, "Endzeitliche Vorstellungen bei Paracelsus und seinen Zeitgenossen", *Nova Acta Paracelsica*, N.F. 18 (2004), 33–60, on 50–55, 59.

Karl Möseneder, *Paracelsus und die Bilder. Über Glauben, Magie und Astrologie im Reformationszeitalter* (Tübingen, 2009), 4–6, 71–162, 196–207.

Karl Möseneder, "Paracelsus: Über die 'kraft und tugent' von Bildern, schattenhafte Kunstwerke und den vollkommenen Künstler als 'signator perfectus'", in *Paracelsus gestern und heute. Forschung – Interpretation – Vermarktung. 59. Paracelsustag 2010* (Salzburg, 2012) (*Salzburger Beiträge zur Paracelsusforschung*, 44), 33–58, on 40, 44, 46.

Pia Holenstein Weidemann, "Sprechende Steine und streitende Fische: Göttliche Botschaften als Endzeitwarnungen im 16. Jahrhundert", *Nova Acta Paracelsica*, N.F. 26 (2012–2013), 69–101, on 71–72, 77, 84–85.

Telle, *Alchemie und Poesie* (2013), 702 n. 70.

Amadeo Murase, "The Homunculus and the Paracelsian *Liber de imaginibus*," *Ambix*, 67 (2020), 47–61.

Didier Kahn, "La Création *ex nihilo* et la notion d'*increatum* chez Paracelse", in Edouard Mehl and Isabelle Pantin, eds., *"De mundi recentioribus phœnomenis." Cosmologie et science dans l'Europe des Temps modernes, XVe–XVIIe siècles. Essais en l'honneur de Miguel Ángel Granada* (Turnhout: Brepols, forthcoming).

5.18 *Liber principiorum* also: *Vom Ursprung der Künste/Büchlein vom Ursprung der Künste/ Liber de vermibus*

1. *Printing History, Manuscripts.* First printed in 1570 in two different editions from different sources. Michael Toxites included the text in his *Trei Tractat Philippi Theophrasti Paracelsi Bombast*, Johann Albert Wimpinaeus included it in his *Philippi Theophrasti Paracelsi von Hohenhaim etliche Tractetlein zur Archidoxa gehörig*. There is a manuscript in Salzburg and manuscripts with excerpts in Bern and Kassel.
2. *Editions.* Edited by Huser, 7 (1590): 278–295. Edited by Sudhoff in Paracelsus, *Sämtliche Werke*, I/14: 499–512.
3. *Relationship between different versions.*
4. *Structure, genre/form, perspective, style.* The text deals with the healing powers of snakes, spiders, worms and toads. It is divided in 12 chapters and written in the first person.

5. *Relationship to other texts.* The *Schatzkammer der Natur* (§ 5.23) names the *Büchlein vom Ursprung der Künste* as one its two main sources.[347]
6. *Authenticity, authorship.* "Certainly spurious" according to Sudhoff.[348]
7. *Time of writing.* Probably written in the 1560s.

Word count: ca. 6.000 words

Incipit: "Dieweil der allmächtig Gott nit alleyn alle notturft zur speiß vnd tranck …"

Manuscripts:
- Salzburg, *Salzburg Museum*: Hs 2136 (*olim* Paracelsiana 155), f. 199r–237r [copied from printed edition?]
- Bern, *Burgerbibliothek*: Cod. 664, f. 39v–40v [excerpts]
- Giessen, *Universitätsbibliothek*: S. 64,656, f. 18r–80r [lost; copied from printed edition]
- Kassel, *Landesbibliothek*: 4° Ms. chem. 17, f. 170v [excerpts]

First printed:
- 1570 (in: *Trei Tractat Philippi Theophrasti Paracelsi Bombast*, ed. Michael Toxites (Strasburg: Theodosius Rihel, 1570); VD16 P 702; Sudhoff, *Bibliographia Paracelsica*, 180–181 n° 113, 222–224 n° 133)
- 1570 (in: *Philippi Theophrasti Paracelsi von Hohenhaim etliche Tractetlein zur Archidoxa gehörig*, ed. Johann Albert Wimpinaeus (Munich: Adam Berg, 1570); VD16 P 629; Sudhoff, *Bibliographia Paracelsica*, 213–215 n° 128, 215–217 n° 129)

Essential bibliography: Sudhoff, *Paracelsus-Forschungen*, II (1889), 137–138; Sudhoff, *Bibliographia Paracelsica*, 181, 397; Sudhoff, *Paracelsus-Handschriften*, 775; Sudhoff, "Vorwort," in Paracelsus, *Sämtliche Werke*, I/14: XXVIII–XXIX.

Further bibliographical references:

Peuckert, *Pansophie* (1956), 449.

Will-Erich Peuckert, "Paracelsische Zauberei," *Nova Acta Paracelsica*, 8 (1957), 71–94, on 87–88.

Peuckert, *Gabalia* (1967), 162, 463, 466–467.

Wolf-Dieter Müller-Jahncke, *Astrologisch-magische Theorie und Praxis in der Heilkunde der frühen Neuzeit* (Stuttgart, 1985), 85–86.

See Gilly, *Paracelsus in der BPH* (1993), 46.

---

347 See Gilly, *Paracelsus in der BPH* (1993), 46.

348 Sudhoff, "Vorwort," in Paracelsus, *Sämtliche Werke*, I/14: XXVIII.

Goldammer, *Der göttliche Magier* (1991), 87, 105–106.
Urs A.F. Heim, "Die *Große Wundartzney* (1536) des Paracelsus und die Chirurgie seiner Zeit," *Nova Acta Paracelsica*, 20 (2006–2007), 7–53, on 24 (Regenwurmsalbe).
Broszinski, *Manuscripta chemica in quarto* (2011), 55.

Magia veterum: see *De magia veterum* (§ 5.6)

Magisches Arcanum mit der Malerei: see *Oraculum Oneiromanticum* (§ 5.19)

Medicina coelestis: see *Archidoxis magica* (§ 5.1)

Modus der Tinctur: see *Vom Modus des Gebrauchs mit der Tinctur* (§ 5.25)

5.19 *Oraculum Oneiromanticum* also: *Magisches Arcanum mit der Malerei zu erfahren was du willst durch den Schlaf/Das magialische arcanum mit der Materie zu erfahren was du willst durch den Schlaf*

1. *Printing History, Manuscripts.* Two manuscripts. Full title: "Oraculum oneiromanticum, daß ist, 2. Curieuse Experimenta, in den Schlaff zu erfahren waß du wilst." Two short texts. The first has the heading: "Theophrasti Paracelsi Magisches Arcanum mit der Mahlerey zu erfahren waß du wilst durch den Schlaf," the second, "Trithemij Arcanum eines jeglichen Dinges Warheit zu erfahren in Traume." Alternative title: "Das magialische arcanum, mit der Materie zu erfahren, was du willst, durch den Schlaf."
2. *Editions.* Not edited by Huser or Sudhoff.
3. *Relationship between different versions.* The Leipzig manuscript has both parts, the Wien manuscript only the first part.
4. *Structure, genre/form, perspective, style.* Instructions for acquiring knowledge of secrets while asleep. The first part gives instructions for the conjuration of a spirit, the second part instructs to have one's questions answered in a dream. Directly addresses the reader.
5. *Relationship to other texts.* Not related to the *Revelatio in somnio* (§ 5.22). – A manuscript in the possession of Max Joseph Freiherr von Linden (see Catalogus 1788) had the title: "Gebeth des unvergleichlichen Theophrasti Paracelsi, welcher die Geister gezwungen, ihm im Schlaf zu erscheinen, und seine verlangten Geheimnisse zu offenbaren." A manusript offered by Johann Georg Binz in 1791 had the title: "Vision, oder Erscheinung alles dasjenige im Schlaf zu erfahren, was man zu wissen und zu erfahren verlangt." These are probably the same text.
6. *Authenticity, authorship.*
7. *Time of writing.* Probably written in the early 17th century. Further study required.

WORD COUNT: ca. 600 words (Leipzig manuscript), ca. 350 words (Wien manuscript)

INCIPIT: "Mache dir eine dinte von saffran darunter mische ambra zibeth und einhorn …" [Cod. mag. 41] / "Nachdem du ohne zweifel die materiam zur erkenntnis haben wirst …" [Wien 11266]

MANUSCRIPTS:
- LEIPZIG, *Universtitätsbibliothek*: Cod. mag. 41, f. 1r–3r
- WIEN, *Österreichische Nationalbibliothek*: Cod. 11266, f. 81r–83r

FIRST PRINTED: not printed before Spunda (1924)

HISTORICAL MANUSCRIPT CATALOGUES: Schröer (1710), 4 n° 119; Schröer/Roth-Scholtz (1732), 14 n° 121; von Linden (1788), 24 n° 110, part 2: 38 n° 32; Binz (1791), 226

MODERN EDITIONS:
- *Magische Unterweisungen des edlen und hochgelehrten Philosophi und Medici Philippi Theophrasti Bombasti von Hohenheim Paracelsus genannt*, ed. Franz Spunda (Leipzig, 1924), 12–13 [from Wien, ÖNB, Cod. 11266]

ESSENTIAL BIBLIOGRAPHY: Sudhoff, *Paracelsus-Handschriften*, 671, 797 n° IV/10.

FURTHER BIBLIOGRAPHICAL REFERENCES:
Peuckert, *Pansophie* (1956), 474.
Bellingradt and Otto, *Magical Manuscripts* (2017), 142–143 n° 119.

5.20 *Oratio generalis magica Philippi Theophrasti Paracelsi* also: *Oratio magico cabalistica/Oratio magica*

1. *Printing History, Manuscripts.* Three manuscripts.
2. *Editions.* Not edited by Huser or Sudhoff.
3. *Relationship between different versions.* The relationship between the Latin and the German *Oratio* is not clear.
4. *Structure, genre/form, perspective, style.*
5. *Relationship to other texts.* Not related to *Theophrasti Paracelsi tägliches Gebet* (§ 10.15). – There are three magical prayers in Widemann's *Verzeichnisse* that are, however, not attributed to Paracelsus: *Oratio cabalistica vel psalmus cabalisticus constans 72 versibus* (IV, 18); *Gebet auf jeden Tag in der Woche zu sprechen aus der Cabala* (IV, 42); *Magisches Gebet* (V, 474). Further study required.
6. *Authenticity, authorship.*
7. *Time of writing.* Possibly written in the early 17th century. Further study required.

Word count: ca. 700 words [Latin text]

Incipit: "Domine Deus El Sadai qui diem hunc septimum ..." [Bamberg] / "O herr, du starcker, allmächtiger, wahrhaftiger, heiliger und gerechter Gott ..." [Amsterdam, Dresden]

Manuscripts:

German

- Amsterdam, *Bibliotheca Philosophica Hermetica*: M126, 131
- Dresden, *Sächsische Landesbibliothek*: Mscr.Dresd.N.43, f. 31r–33v

Latin

- Bamberg, *Staatsbibliothek*: Msc. theol. 191, f. 27v–33r

First printed: not printed

Historical Manuscript Catalogues: Widemann, *Verzeichnisse* (Kassel), n° iv, 18, 42 / v, 474

Essential bibliography: Not mentioned by Sudhoff.
Gilly, *Paracelsus in der BPH* (1993), 68.

Picatrix: see *Von der ewigen Heimlichkeit* (§ 5.27)

5.21 *Psalterium magicum* also: *Heiliges Gebetbuch oder Psalterium magicum/Kräftiges Gebetbuch/Psalterium Davidis magico-cabalisticum/Usus practicus Psalmorum Regis Davidis*

1. *Printing History, Manuscripts.* Not printed. Several manuscripts.
2. *Editions.* Not edited by Huser or Sudhoff.
3. *Relationship between different versions.* Not all manuscripts have the complete text: Dresden N.119 has a selection of 50 psalms only.
4. *Structure, genre/form, perspective, style.* According to Gilly, "magical formula and prayers similar to the *Picatrix* [cf. § 5.27], Pietro d'Abano or the *Clavicula Salomonis*, each appended to a verse from the Psalms and used to attain some practical aim."[349]
5. *Relationship to other texts.*
6. *Authenticity, authorship.*
7. *Time of writing.* The oldest traces of the text are excerpts in the late 16th century manuscript Breslau R 2316, already noted but not identified by Sudhoff. The

349 Gilly, *Paracelsus in der BPH* (1993), 68.

title was "Auß der Magena [?] Theophrastina. Vber etliche psalm." A manuscript (now lost) in the library of Johann Christian Gottfried Jahn was dated 1576.

Word count: ca. 8.000 words (Leipzig manuscript), ca. 500 words (Breslau manuscript; excerpts), ca. 1.600 words (Dresden manuscript N.119).

Incipit: "Nach dem unsere vorfahren genau erforschet und geglaubet […] / Wie viell heilige väter haben den psalter alß das hertz und kern […] / Daß eine frau die schwanger ist nicht umb die frucht komme …"

Manuscripts:
- Amsterdam, *Bibliotheca Philosophica Hermetica*: M126
- Dessau, *Anhaltische Landesbücherei*: BB 2101
- Dresden, *Sächsische Landesbibliothek*: Mscr.Dresd.N.43, f. 3r–26v
- Dresden, *Sächsische Landesbibliothek*: Mscr.Dresd.N.119, f. 1r–7r
- Dresden, *Sächsische Landesbibliothek*: Mscr.Dresd.N.166,2; 9 fols.
- Dresden, *Sächsische Landesbibliothek*: Mscr.Dresd.App.738; 59 fols.
- Leipzig, *Universtitätsbibliothek*: Cod. mag. 9; 91 pages
- *olim* Dresden, Verein für sächsische Volkskunde (1900); current location unknown[350]
- *olim* Ichtershausen, library of Duke Johann Friedrich of Sachsen-Weimar; confiscated in 1627; current location unknown

First printed: not printed

Historical Manuscript Catalogues: Widemann, *Verzeichnisse* (Kassel), n° IV, 25; Herzog Johann Friedrich von Sachsen-Weimar (1627), 217 n° 20; Schröer (1710), 2 n° 57; Arpe (1726), 122; Schröer/Roth-Scholtz (1732), 9 n° 59, 16 add. n° 5; Jahn (1754–1758), 148 n° 166; Linck (1787), 243 n° 7; von Linden (1788), part 2, 27 n° 22; cf. Widemann, *Unausgangene Bücher* (Hannover), n° 113; Ponickau (1791), col. 1232 n° Quarto 12; Dost (1791–1798): 1797, col. 396, n° 59

Essential bibliography: Sudhoff, *Paracelsus-Handschriften*, 26, 679; Goldammer, "Einleitendes zum Psalmenkommentar," in Paracelsus, *Sämtliche Werke*, ed. Kurt Goldammer, vol. 4/1 (Wiesbaden, 1955): XXVIII–IL, on XLV.

350 *Mittheilungen des Vereins für Sächsische Volkskunde*, 2 (1900), n° 1, on 6: "Für das Archiv wurden folgende Beiträge eingesandt: Von Herrn Geh. Baurat Wanckel-Dresden: […] 110. Psalterium magicum. Vol. 1 (aus d. J. 1773 Hdsch.)"

Further bibliographical references:

W.R.B. Prideaux, "Hebrew Incantations," *Notes and Queries*, 9th series, 10 (1902), 355.

Gilly, *Paracelsus in der BPH* (1993), 67–68.

Ulrich Hecht, ed., *Ex Libris Christian I. und Christian II. von Anhalt-Bernburg. Aus den Beständen der Anhaltischen Landesbücherei Dessau* (Bernburg, 1993), 23.

Bellingradt and Otto, *Magical Manuscripts* (2017), 110 n° 57.

5.22 *Revelatio in somnio* also: *Revelatio per somnium*

1. *Printing History, Manuscripts.* Not printed. One manuscript. Both titles, "Revelatio in somnio" (f. II^r, 8v) and "Revelatio per somnium" (f. III^r) were added by a later hand in the Breslau manuscript.
2. *Editions.* Not edited by Huser or Sudhoff.
3. *Relationship between different versions.* Only one known version.
4. *Structure, genre/form, perspective, style.* In some parts the text directly addresses the reader. The order of the different sections is sometimes disturbed; the writer used symbols to indicate a different order.
5. *Relationship to other texts.* A text on earth spirits (*pygmaei, sylphae*) and treasures. The relationship to chapter 5 of the pseudo-Paracelsian *De occulta philosophia* (§ 5.7) and to the authentic *De nymphis, sylphis, pygmaeis* is unclear.[351] – Not related to the *Oraculum Oneiromanticum / Magisches Arcanum mit der Malerei zu erfahren was du willst durch den Schlaf* (§ 5.19). – See also *Von den Erdgeistern* (§ 5.26).
6. *Authenticity, authorship.*
7. *Time of writing.* Further study required.

Word count: ca. 3.800 words

Incipit: "Wiltu jnen werden bei der nacht waß zu wißen …"

Manuscripts:

- Wrocław/Breslau, *Biblioteka Uniwersytecka*: R 2316, f. 8v–17r, p. 35–72

First printed: not printed

351 Compiled from the *Occulta Philosophia* according to Peuckert, *Pansophie* (1936), 531: "eine Kompilation aus der Occulta Philosophia."

Essential bibliography: Sudhoff, *Paracelsus-Handschriften*, 663–665, 797 n° IV/7.

Further bibliographical references:
Peuckert, *Pansophie* (1936), 531.
Peuckert, *Pansophie* (1956), 474.

5.23 *Schatzkammer der Natur* also: *Secreta secretissimorum nach der Schatzkammer der Natur/Von der Extraction der Mumia spiritualis/Thesaurinella naturae/Libellus secretorum secretissimorum/Secreta secretorum Theophrasti/Erklärung der zwei großen Geheimnisse/De mumiis microcosmi*

1. *Printing History, Manuscripts.* First printed in Latin in 1629, edited by Andreas Tentzel as *Medicina diastatica*. First printed in German in 1635, edited by Jacob Lupius as *Schatzkammer der Natur*. Several manuscripts, one of them a presentation copy from Joachim Tancke to Landgraf Moritz von Hessen-Kassel in 1609. One lost manuscript associated with Karl Widemann.[352]
2. *Editions.* Not edited by Huser or Sudhoff.
3. *Relationship between different versions.* Written in German. Translated into Latin and modified in some places by Andreas Tentzel.[353] The printed German version of 1635 is mostly true to the original version.
4. *Structure, genre/form, perspective, style.* Divided into two parts, the first "on the extraction of the spiritual mumia," the second a 'magical' interpretation of Adam's fall. Two illustrations.[354]
5. *Relationship to other texts.* The *Schatzkammer der Natur* names *Libellus de tempore*[355] and *Büchlein vom Ursprung der Künste* as its two main sources.[356] The latter may refer to the *Liber principiorum* (§ 5.18) with the alternative title *Büchlein vom Ursprung der Künste* or to the *Liber de inventione artium*, edited by Huser, 9 (1591): 162–186 ("Auß Theophrasti eigner Handschrifft"). Further study required.

352 Sudhoff, *Paracelsus-Handschriften*, 661–662.

353 Benzenhöfer, "Joachim Tancke" (1987), 22–23.

354 Reproduced from the Kopenhagen manuscript by Scialdone, "Aller heilsamste Mum Iah" (2001), 24–25.

355 There are two texts with the title *De tempore*: (1) A Paracelsian work first printed in 1572, part of "Fünf Tractat Philosophiae" edited by Huser in vol. IX. See Sudhoff, *Bibliographia Paracelsica*, 240, 404. (2) A Weigelian or Pseudo-Weigelian work reported by Karl Widemann in 1612. See Carlos Gilly, "Iter Rosicrucianum. Auf der Suche nach unbekannten Quellen der frühen Rosenkreuzer," in *Das Erbe des Christian Rosenkreuz* (Amsterdam, 1988), 63–89, on 87 n. 29; Gilly, *Adam Haslmayr* (1994), 132 n. 21.

356 See Gilly, *Paracelsus in der BPH* (1993), 46.

6. *Authenticity, authorship.* Not directly attributed to Paracelsus both in the Latin and in the German printed edition, but rather quoting Paracelsus as an authority on the title-page. The Kiesewetter manuscript associates the text with Karl Widemann who is styled "vicarius perpetuus" and is said to have sent the text to Sigismund III, King of Poland. According to Gilly, Widemann was in the possession of the *Schatzkammer der Natur* as early as 1612 and may have acquired it, together with the *Clavis oder Das zehnte Buch der Archidoxen* (§ 4.25), from Adam Haslmayr.[357] The Kopenhagen manuscript was copied by Johann Friedrich Jung according to Gilly, but the handwriting was also attributed to Abraham von Franckenberg, according to an 18th century note quoted by Scialdone.[358] – The real author (i.e. allegedly Joachim Tancke) is revealed, according to Benzenhöfer and Broszinski, in the Kassel manuscript, which is a presentation copy for Landgraf Moritz von Hessen-Kassel with an autograph dedication of Tancke.[359] On the other hand, there is a second volume equally dedicated by Tancke to Landgraf Moritz and written by the same scribe with a text also based on Paracelsus and Pseudo-Paracelsus[360] where the text cannot be attributed to Tancke.
7. *Time of writing.* Probably written in the 1600s.

Word count:

Incipit: "Also nach anleitung gegebener exempell kan den menschen oder viehe jhre gesunde oder vngesunde natur [...] / Es ist ein großes, lieben filij doctrinae, wan der almechtige Gott vatter im himmell [...] / Gleich wie die vier elementa jn jhrer wirckung vngleich sein [...] / So ich von der natur der beume erkendtnüs gutes vnd böses ..." [Kassel]

Manuscripts:
- Amsterdam, *Bibliotheca Philosophica Hermetica*: M126
- Dresden, *Sächsische Landesbibliothek*: Mscr.Dresd.N.80c
- Hamburg, *Staats- und Universitätsbibliothek*: Cod. alchim. 596; 30 pages [probably copied from a printed edition]
- Hamburg, *Staats- und Universitätsbibliothek*: Cod. alchim. 722, f. 1r–23r
- Kassel, *Landesbibliothek*: 4° Ms. chem. 99; 59 fols.

---

357 *Ibid.*, 68.
358 Hamburg, *Staats- und Universitätsbibliothek*: Cod. alchim. 722, f. 1v; quoted by Scialdone, "Aller heilsamste Mum Iah" (2001), 15.
359 Broszinski, *Manuscripta chemica in quarto* (2011), 388.
360 *Ibid.*, 389.

- KOPENHAGEN, *Det Kongelige Bibliotek*: GKS 1765 kvart
- LONDON, *Wellcome Library*: MS 3563, 539–544 [extracts, partly in shorthand]
- LONDON, *Wellcome Library*: MS 3757; 298 pages
- MÜNCHEN, *Bayerische Staatsbibliothek*: Kiesewetteriana 31
- PARIS, *Bibliothèque Sainte-Geneviève*: Ms 2199, f. 57r–66r
- PRAHA, *Strahovská knihovna*: DF IV 59, f. 247r–283v
- STRASBOURG, *Bibliothèque nationale et universitaire*: ms. 199, f. 299–328
- WIEN, *Österreichische Nationalbibliothek*: Cod. 11330, f. 210r–247v
- WOLFENBÜTTEL, *Herzog August Bibliothek*: Cod. Guelf. 51.3 Aug. 4°, f. 1r–65v

FIRST PRINTED:
- **1629** (in: *Medicina Diastatica*, ed. Andreas Tentzelius (Jena: Johann Birckner, 1629); VD17 39:115825D; Sudhoff, *Bibliographia Paracelsica*, 541–542 n° 331)
- **1635** (*Schatzkammer der Natur: Gründliche Erkärung zweyer grossen Geheimnüsse, Vnd erstlichen die Extractio der spiritualischen Mumiae des Menschen, vnd anderer Thier etc.*, ed. Jacob Lupius (1635); not in VD17; Sudhoff, *Bibliographia Paracelsica*, 563–564 n° 358; – *id.* (no place, 1645); VD17 29:733616E)[361]

HISTORICAL MANUSCRIPT CATALOGUES: Widemann, *Unausgangene Bücher* (Hannover), n° 19, 41; Widemann, *Verzeichnisse* (Kassel), n° III, 36 / IV, 32 / V, 21, 33; Uffenbach (1747), 317 n° 98.

ESSENTIAL BIBLIOGRAPHY: Sudhoff, *Bibliographia Paracelsica*, 541–542 n° 331, 563 n° 358; Sudhoff, *Paracelsus-Handschriften*, 26, 658–661, 679, 797 n° IV/3.

FURTHER BIBLIOGRAPHICAL REFERENCES:

Peuckert, *Gabalia* (1967), 249, 257–259, 259–268, 465–470, 552–553, 562.

Gilly, *Paracelsus in der BPH* (1993), 46, 67–68.

Udo Benzenhöfer, "Joachim Tancke (1557–1609). Leben und Werk eines Leipziger Paracelsisten," in *Paracelsus und Paracelsisten. Vorträge 1984/85* (Vienna, 1987) (*Salzburger Beiträge zur Paracelsusforschung*, 25), 9–81, on 18–23, 29, 42–44.

Heinrich Schipperges, *Die Entienlehre des Paracelsus. Aufbau und Umriß seiner Theoretischen Pathologie* (Berlin-Heidelberg, 1988), 34.

Bruce T. Moran, *The Alchemical World of the German Court. Occult Philosophy and Chemical Medicine in the Circle of Moritz of Hessen (1572–1632)* (Stuttgart, 1991), 138–141.

361 No copy known of this first edition. Sudhoff, *Bibliographia Paracelsica*, does not name any library nor does he mark the entry with an asterisk for a book in his own possession. See Benzenhöfer, "Joachim Tancke" (1987), 70 n. 130.

Heinrich Schipperges, "Medicina occulta. Oberfläche und Hintergrund", in August Buck, ed., *Die okkulten Wissenschaften in der Renaissance* (Wiesbaden, 1992), 125–138, on 129, 131.

Maria Paola Scialdone, "'Aller heilsamste Mum Iah.' 'Lebensbalsam' (pseudo-) Paracelsiano e 'Selbsterlösung' nel 'Raphael' di Abraham von Franckenberg," *Studi germanici*, N.S. 39 (2001), 7–35, on 14–15.

Hania Siebenpfeiffer, "'Malgré la mort, je vis encore' – Mumien und Gespenster als Manifestationen des Unheimlichen im 17. Jahrhundert," in Moritz Baßler et al., *Gespenster. Erscheinungen – Medien – Theorien* (Würzburg, 2005), 105–125, on 117–119.

Broszinski, *Manuscripta chemica in quarto* (2011), 387–388.

Christoph Sander, *Magnes. Der Magnetstein und der Magnetismus in den Wissenschaften der Frühen Neuzeit* (Leiden, 2020), 58–61, 936–938.

Secreta secretissimorum nach der Schatzkammer der Natur: see *Schatzkammer der Natur* (§ 5.23)

Secretum magicum: see § 2.7

5.24 *Speculum sive lumen naturale*
also: *Speculum luminis naturae secundum lumen naturale*

1. *Printing History, Manuscripts.* Not printed. Two manuscripts.
2. *Editions.* Not edited by Huser or Sudhoff.
3. *Relationship between different versions.* Only one known version.
4. *Structure, genre/form, perspective, style.*
5. *Relationship to other texts.*
6. *Authenticity, authorship.*
7. *Time of writing.* Further study required. The Leiden copy is dated 1593. Probably written in the 1580s or early 1590s.

Word count:

Incipit: "Wir wollen vns deß behelffen, des vns in der natur jst ein gegeben …"

Manuscripts:
- Erlangen, *Universitätsbibliothek*: MS B 41 (*olim* Irm. 1439), f. 109v–118v
- Leiden, *Universiteitsbibliotheek*: Voss. Chym. Q.38, f. 138v–151v

First printed: not printed

HISTORICAL MANUSCRIPT CATALOGUES: cf. *Catalogus* (Stuttgart), I, f. 9v, n° [347]

ESSENTIAL BIBLIOGRAPHY: Sudhoff, *Paracelsus-Handschriften*, 546, 617, 796 n° III/129.

FURTHER BIBLIOGRAPHICAL REFERENCES:
Boeren, *Codices Vossiani Chymici* (1975), 196.

Thesaurinella naturae: see *Schatzkammer der Natur* (§ 5.23)

5.25 *Vom Modus des Gebrauchs mit der Tinctur* (*Über eine magische Glocke*) also: *Ein Modus der Tinctur/Ein Modus des Gebrauchs/Ein Modus so ich selbst gearbeitet/Ein Büchlein von olympischer Geister Zitierung/Geheimnis der Campana philosophica/Arcanum arcanorum seu magisterium philosophorum de sacra campanula angelorum/Von dem philosophischen Glöcklein/Brief an einen Schüler/Geheimnis aller seiner Geheimnisse*

1. *Printing History, Manuscripts.* First printed in 1686 in *Philippi Theophrasti Paracelsi Bombast Gröstes und höchstes Geheimnüsz aller seiner Geheimnüsse.* Another edition appeared in 1745 in *Magia divina Oder Gründ- und deutlicher Unterricht, Von denen fürnehmsten Caballistischen Kunst-Stücken.* Three accessible manuscripts, whereabouts of further manuscripts unknown.
2. *Editions.* Not edited by Huser or Sudhoff.
3. *Relationship between different versions.* The relationship between the different versions remains unclear. It seems that the text was partly rewritten, shortened, enlarged more than once. – The *Büchlein von olympischer Geister Zitierung* is a compilation from the *Gröstes und höchstes Geheimnüsz* printed in 1686 and the *Arbatel.*[362]
4. *Structure, genre/form, perspective, style.* Originally written in the form of a letter from Paracelsus to his disciple. Written in the first person, directly addressing the disciple. Later versions seem to have stripped the letter parts.
5. *Relationship to other texts.* Related to a passage of the *Archidoxis magica*, Book 6, about bells made for the same purpose.
6. *Authenticity, authorship.*
7. *Time of writing.* Probably written in the early 17th century. Further study required.

362 Gilly, "Il primo prontuario di magia bianca" (2002), 204, 214.

Word count: ca. 2.500 words

Incipit: "Erstlich mercke den tag darinnen du gebohren bist ..."

Manuscripts:

German

- Amsterdam, *Bibliotheca Philosophica Hermetica*: M126[363]
- Bamberg, *Staatsarchiv*: Markgraftum Brandenburg-Bayreuth, Geheimes Archiv Bayreuth n° 929; 3 fols.
- Wien, *Österreichische Nationalbibliothek*: Cod. 11266, f. 70r–81r
- *olim* library of Karl Sudhoff (1899, 1924);[364] current location unknown
- *olim* library of Friedrich Schwickert (1918);[365] current location unknown
- *olim* private collection (sold in 2013)[366]

French

- Paris, *Bibliothèque nationale de France*: Français 24243, f. 75r–80r

First printed:

- **1686** ("Folget ein Modus des Gebrauchs, so ich Theophrastus Bombast selber gearbeitet", in *Philippi Theophrasti Paracelsi Bombast Gröstes und höchstes Geheimnüsz aller seiner Geheimnüsse* (no place, 1686); VD17 14:628669B; Sudhoff, *Bibliographia Paracelsica*, 626 n° 419)
- **1745** (in: *Magia divina Oder Gründ- und deutlicher Unterricht, Von denen fürnehmsten Caballistischen Kunst-Stücken*, ed. L.v.H. (no place, 1745); not in Sudhoff, *Bibliographia Paracelsica*)
- **ca. 1870/90** (in: *Die egyptischen großen Offenbarungen, in sich begreifend die aufgefundenen Geheimnißbücher Mosis* (Köln: Peter Hammer, 1725 [i.e. Stuttgart: Scheible?, ca. 1870/1890?]); mentioned in Sudhoff, *Bibliographia Paracelsica*, 646; copy in Berlin, Humbold Univ.)

363 From the library of Karl Kiesewetter. See Sudhoff, *Paracelsus-Handschriften*, 26, 679. See Gilly, *Paracelsus in der BPH* (1993), 67–68.

364 See Sudhoff, *Paracelsus-Handschriften*, 678–679 n° 131.

365 Quoted in Sindbad, *Das Lebenselixier in Bulwers Romanen und in den Schriften wirklicher Adepten* (1918), 17–20.

366 See Zisska & Schauer Buch- und Kunstauktionshaus, München, Katalog Auktion 61 (2013), 9–10 n° 20a.

Modern editions:

- *Magische Unterweisungen des edlen und hochgelehrten Philosophi und Medici Philippi Theophrasti Bombasti von Hohenheim Paracelsus genannt*, ed. Franz Spunda (Leipzig, 1924), 1–11 [from Wien, ÖNB, Cod. 11266].[367]

Historical Manuscript Catalogues: Arpe (1726), 122; Jahn (1754–1758), 3: 148 n° 169, 172; von Linden (1786), 41 n° 266; Gräffer (1786), 11 n° 78; von Linden (1788), 4 n° 5, 29 n° 148, 32 n° 166, 42 n° 289

Essential bibliography: Sudhoff, *Bibliographia Paracelsica*, 626, 697 n° 51; Sudhoff, *Paracelsus-Handschriften*, 22–27, 671, 678, 797 n° IV/9; Sudhoff, "Vorwort," in Paracelsus, *Sämtliche Werke*, 1/14: XXXV.

Further bibliographical references:

Sindbad [i.e. Friedrich Schwickert], *Das Lebenselixier in Bulwers Romanen und in den Schriften wirklicher Adepten* (Leipzig, 1918).

Richard Weber, "Der Zauberer Paracelsus," in Helmut Dölker, ed., *Festschrift für Will-Erich Peuckert* (Berlin, 1955), 128–136.

Peuckert, *Pansophie* (1956), 475–476, 480–481.

Peuckert, *Gabalia* (1967), 216–224, 328, 395–400.

Gilly, *Paracelsus in der BPH* (1993), 67–68, 70.

André Lehr, "De necromantische klokjes van Paracelsus en Girardius," *Berichten uit het Nationaal Beiaardmuseum*, n° 5 (November 1992), 19–20 (*non vidi*); Danish translation: "Paracelsus' og Girardius' nekromantiske klokke," *Acta Campanologica*, 5 (1994), n° 6, 165–168 (*non vidi*).

Gilly, "Il primo prontuario di magia bianca" (2002), 204, 214.

Bellingradt and Otto, *Magical Manuscripts* (2017), 21–22 n. 72.

Vom Ursprung der Künste: see *Liber principiorum* (§ 5.18)

5.26 *Von den Erdgeistern*

1. *Printing History, Manuscripts.* Not printed. One manuscript.

2. *Editions.* Not edited by Huser or Sudhoff.

3. *Relationship between different versions.* Only one version known.

4. *Structure, genre/form, perspective, style.*

367 Also printed separately: "Magische Unterweisungen des Theophrastus Paracelsus. Zum erstenmal aus der Handschrift veröffentlicht," ed. Franz Spunda, in *Magische Blätter. Mitteilungen über praktische Geheimwissenschaften*, 4 (1923), n° 10/11, 293–296.

5. *Relationship to other texts.* The short text about "*pygmei*" and "*sylphae*" is related to part 5 of Pseudo-Paracelsus' *De occulta philosophia* and to some sections of Pseudo-Paracelsus' *Revelatio in somnio* (§ 5.22; immediately following *Von der Erdgeistern* in the Breslau manuscript). Further study required. – A manuscript in the library of Johann Christian Gottfried Jahn (present location unknown) had the title *VII. Stellae, oder ein Glöcklein zu machen, daß man die Schätze von denen Pygmaeis bekommen könne.* Another related text is found in KASSEL, *Landesbibliothek*, 4° Ms. chem. 25, f. 101r–110r.[368]
6. *Authenticity, authorship.* Without title and not attributed to Paracelsus in the Breslau manuscript.
7. *Time of writing.* Further study required.

WORD COUNT: ca. 650 words

INCIPIT: "In dem element der erdten sein die edelsten vnd menschlicher natur nechst zue gethane geister pygmei oder sylphas genant …"

MANUSCRIPTS:
- WROCŁAW/BRESLAU, *Biblioteka Uniwersytecka*: R 2316, f. 4r–8r

FIRST PRINTED: not printed

HISTORICAL MANUSCRIPT CATALOGUES: Jahn (1754–1758), 148 n° 169.

ESSENTIAL BIBLIOGRAPHY: Sudhoff, *Paracelsus-Handschriften*, 663.

FURTHER BIBLIOGRAPHICAL REFERENCES:
Broszinski, *Manuscripta chemica in quarto* (2011), 85.

5.27 *Von der ewigen Heimlichkeit und der heimlichen Ewigkeit* also: *Von der ewigen Heimlichkeit / Picatrix*

1. *Printing History, Manuscripts.* Not printed. At least the Leiden manuscript is attributed to Paracelsus.
2. *Editions.* Not edited by Huser or Sudhoff.
3. *Relationship between different versions.*
4. *Structure, genre/form, perspective, style.*

368 All these additional texts are not mentioned by Michael Toggweiler, *Die Odyssee der Pygmäen. Eine andere Geschichte der neuzeitlichen Anthropologie*, PhD thesis (Bern, 2012), online (http://permalink.snl.ch/bib/sz001923724, accessed 31 December 2020).

5. *Relationship to other texts.*
6. *Authenticity, authorship.*
7. *Time of writing.* One manuscript is dated 1597 (Erlangen), another one 1567 (Leiden). According to Gilly it is a German version of the entire first book and the first three chapters of the second book of the *Picatrix*.[369]

Word count:

Incipit: "In allen dingen so wir furnehmen zuthun …"

Manuscripts:
- Erlangen, *Universitätsbibliothek*: ms B 242 (*olim* Irm. 1707); 59 fols. [not attributed to Paracelsus]
- Halle, *Universitäts- und Landesbibliothek*: Ms. 21 A 10, f. 149–223
- Leiden, *Universiteitsbibliotheek*: Voss. Chym. F.14, f. 410r–458r

First printed: not printed

Essential bibliography: Sudhoff, *Paracelsus-Handschriften*, 686–688, 798 n° iv/17.

Further bibliographical references:

Boeren, *Codices Vossiani Chymici* (1975), 48.

Carlos Gilly, "Paracelsus, Hermes, Ficino, and the 'Picatrix deutsch'," in Sebastiano Gentile and Carlos Gilly, eds., *Marsilio Ficino e il ritorno di Ermete Trismegisto* (Florence, 1999), 302–306 n° 63.

Gilly, "Vom ägyptischen Hermes zum Trismegistus Germanus" (2010), 98–99.

Bellingradt and Otto, *Magical Manuscripts* (2017), 21 n. 71.

Von der Extraction der Mumia spiritualis: see *Schatzkammer der Natur* (§ 5.23)

5.28 *Von der göttlichen cabalistischen Astronomia* also: *Büchlein von der göttlichen geistlichen innerlichen und äußerlichen cabalistischen Astronomiae der großen sieben Thronengel oder Kräfte Gottes*

1. *Printing History, Manuscripts.* Not printed. One manuscript.
2. *Editions.* Not edited by Huser or Sudhoff.
3. *Relationship between different versions.* Only one known version.

---

369 Gilly, "Paracelsus, Hermes, Ficino" (1999), 302–306 n° 63, 304; Gilly, "Vom ägyptischen Hermes zum Trismegistus Germanus" (2010), 99 n. 97.

4. *Structure, genre/form, perspective, style.*
5. *Relationship to other texts.*
6. *Authenticity, authorship.*
7. *Time of writing.* Further study required.

Word count:

Incipit:

Manuscripts:
– München, *Bayerische Staatsbibliothek*: Kiesewetteriana 25a

First printed: not printed

Essential bibliography: Not mentioned by Sudhoff.

5.29 *Von der Heilung zauberischer Schäden* also: *Brief an Ulrich Beck*

1. *Printing History, Manuscripts.* First edited in 1608 by Benedictus Figulus as an appendix to *Philippi Theophrasti Paracelsi Kleine Wund-Artzney*. Although announced as a work of Bartholomaeus Carrichter on the main title-page and on the respective "Zwischentitelblatt" and although the dedicatory letter to Ulrich Beck, dated 16 January 1551, is signed by Carrichter, the text itself is dated 3 January 1551 and signed "Philippus Theophrastus Paracelsus, von Hohenheim" at the end. There are six manuscripts where the whole text (including the dedicatory letter) is expressly attributed to Paracelsus. In the Leiden manuscript the attribution to Paracelsus was added at a later time by Karl Widemann. – The first printed edition not associated with Paracelsus appeared one year later than that of Figulus, in 1609, as an appendix to a reprint of Michael Toxites's edition of Carrichter's *Kräutterbuch*. The text is a reprint of the Figulus edition with the same "Zwischentitelblatt." The text has the same date, 3 January 1551, but is not signed, neither by Carrichter nor by Paracelsus.
2. *Editions.* Not edited by Huser or Sudhoff.
3. *Relationship between different versions.*
4. *Structure, genre/form, perspective, style.*
5. *Relationship to other texts.*
6. *Authenticity, authorship.* According to Kühlmann/Telle, the whole text shows no influence of Paracelsus at all.[370]
7. *Time of writing.* Written in 1551. Attributed to Paracelsus probably in the 1580s.

---

370 CP 2: 340.

Word count:

Incipit: "Erbarer und kunstreicher meister Vlrich, lieber getrewer freund […] / Allhie will ich anheben zu erzehlen alle die stuck …"

Manuscripts:
- Gotha, *Forschungsbibliothek*: Chart. B 362, f. 203–214
- Kassel, *Landesbibliothek*: 4° Ms. chem. 74, f. 87r–98r
- Leiden, *Universiteitsbibliotheek*: Voss. Chym. Q.56, i, f. 101–120
- München, *Bayerische Staatsbibliothek*: Kiesewetteriana 25c
- Tübingen, *Universitätsbibliothek*: Jf ix 133 H
- Wien, *Österreichische Nationalbibliothek*: Cod. 11266, f. 13r–69r
- Wien, *Österreichische Nationalbibliothek*: Cod. 11330, f. 248r–257v

First printed:
- **1608** ("Von Heylung der Zauberischen Schäden," in *Philippi Theophrasti Paracelsi Kleine Wund-Artzney* […] *Sampt zweyen angehenckten außbündigen Tractätlein H. Bartholomaei Karrichters*, ed. Benedictus Figulus (Strasburg: Paul Ledertz, 1608); VD17 23:294271X; Sudhoff, *Bibliographia Paracelsica*, 485–487 n° 284)
- **1609** ("Von Heylung der Zauberischen Schäden", in *Kräutterbuch Des Edelen vnd Hochgelehrten Herren/ Doctoris Bartholomei Carrichters* (Strasburg: Anton Betram, 1609); VD17 12:657929X; reprinted 1610, 1615, 1617, 1618).

Historical Manuscript Catalogues: cf. Widemann, *Verzeichnisse* (Kassel), n° v, 484

Essential bibliography: Sudhoff, *Bibliographia Paracelsica*, 696 n° 24; Sudhoff, *Paracelsus-Handschriften*, 797 n° IV/1; CP 2: 335, 340.

Further bibliographical references:

Peuckert, *Pansophie* (1956), 487.

Peuckert, *Gabalia* (1967), 426–427.

Boeren, *Codices Vossiani Chymici* (1975), 227.

Joachim Telle and Julian Paulus, "Bartholomäus Carrichter. Zu Leben und Werk eines deutschen Fachschriftstellers des 16. Jahrhunderts. Mit einem Werkverzeichnis," in Heinz Schott and Ilana Zinguer, eds., *Paracelsus und seine internationale Rezeption in der frühen Neuzeit* (Leiden, 1998), 58–95.

Broszinski, *Manuscripta chemica in quarto* (2011), 335.

Von der Magia: see *De magia veterum* (§ 5.6)

5.30 *Wie man die sieben Sigilla der Planeten machen und bereiten soll*

1. *Printing History, Manuscripts.* Not printed. One manuscript, written ca. 1635.
2. *Editions.* Not edited by Huser or Sudhoff.
3. *Relationship between different versions.* Only one known version.
4. *Structure, genre/form, perspective, style.*
5. *Relationship to other texts.* According to Sudhoff, inspired by book 7 of the *Archidoxis magica* (§ 5.1).[371]
6. *Authenticity, authorship.* Allegedly from an autograph of Paracelsus.
7. *Time of writing.* Possibly written in the 1610s or 1620s. Further study required.

Word count:

Incipit: "Nim saturnum so auß Tyroll oder von Villach gebracht wird ..."

Manuscripts:
- Kopenhagen, *Det Kongelige Bibliotek*: GKS 259 folio, 171–201

First printed: not printed

Essential bibliography: Sudhoff, *Paracelsus-Handschriften*, 666–668, 797 n° IV/8.

5.31 *Zenechdon Paracelsi*
also: *Zenexton Paracelsi/Zenexdon Paracelsi/Zenechdon Theophrasti/Senechdon*

1. *Printing History, Manuscripts.* Description of a plague amulet that is often attributed to Paracelsus.[372] Printed in 1582 by Hiob Fincel in his *Von der Pestilentz*. Printed in 1597 by Andreas Starck in his *Von der Pestilentz Kurtzer vnd einfeltiger Bericht*. Printed in 1609 (with a woodcut showing the amulet) by Oswald Crollius in his *Basilica Chymica*. Printed in 1620 by Raymund Minderer, a Paracelsian physician (but not attributed to Paracelsus) in his *Medicina Militaris* [...] *Id est: Gemaine Handstücklein zur Kriegs-Artzney gehörig*.
2. *Editions.* Not edited by Huser or Sudhoff.
3. *Relationship between different versions.*
4. *Structure, genre/form, perspective, style.*

---

371 Sudhoff, *Paracelsus-Handschriften*, 667, 668.

372 Monika Klutz, *Die Rezepte in Oswald Crolls Basilica Chymica* (1974), 80; Bachour, *Oswaldus Crollius und Daniel Sennert im frühneuzeitlichen Istanbul* (2012), 170–171: 'Zenexton Paracelsi'; CP 3: 969.

5. *Relationship to other texts.* The "zenexton" is mentioned in the authentic Nördlingen treatise *Zwey Bücher von der Pestilentz vnnd jhren Zufellen* (Huser, 3: 147).
6. *Authenticity, authorship.* Associated with Paracelsus because the "Zenexton" is mentioned in his *Zwey Bücher von der Pestilentz vnnd jhren Zufellen.*[373]
7. *Time of writing.* Possibly written in the 1570s or earlier. Further study required.

Word count:

Incipit: "Recipe bufonum torrefactum et pulverizatum uncias duo ..." [ed. 1582]

Manuscripts:
- Wien, *Österreichische Nationalbibliothek*: Cod. 11114, f. 90v–91r

First printed:
- 1582 ("Zenechdon Paracelsi," in Hiob Fincel, *Von der Pestilentz Notwendiger Vnterricht der Praeseruation vnnd Cur* (Leipzig: Johann Beyer, 1582), f. D3r; VD16 F 1102)
- 1597 ("Bapst Adrianus VI. hat zur zeit der Pestilentz an seinem eigen Leib getragen dis Amuletum, vnd stets gelobet/ das man Zenechdon Paracelsi nennet/ vnd ist dieses die Description," in Andreas Starck, *Von der Pestilentz Kurtzer vnd einfeltiger Bericht/ wie man sich* [...] *dafur bewaren/ vnd die auch heilen möge* (Erfurt: Esaias Mechler, 1597), f. D3v–D4r; VD16 S 8627)
- 1609 ("Zenexton Paracelsi," in Oswald Crollius, *Basilica Chymica* (Frankfurt a.M.: Claude de Marne, 1609), 237–240; VD17 23:290172G; reprinted in Johann Hiskia Cardilucius, *Königlicher Chymischer und Artzneyischer Palast* (Nuremberg: Christoph Riegel, 1684), 679–681; VD17 23:303530Y)
- 1620 (in: Raymund Minderer, *Medicina Militaris* [...] *Id est: Gemaine Handstücklein zur Kriegs Artzney gehörig* (Augsburg: Andreas Aperger, 1620), 67–68; VD17 23:241205C)

Essential bibliography: Sudhoff, *Paracelsus-Handschriften*, 199, 797 n° IV/2; CP 3: 969.

Th. Engelmann, "Ethnographische Pharmazeutik," *Verhandlungen der Naturforschenden Gesellschaft in Basel*, 26 (1915), 273–279, on 277–278.

Peuckert, *Gabalia* (1967), 102, 237, 240, 279, 553.

Monika Klutz, *Die Rezepte in Oswald Crolls Basilica Chymica (1609) und ihre Beziehungen zu Paracelsus* (Braunschweig, 1974), 80.

---

373 Huser, 3: 147: "Also muß auch dz Praeseruatiff in solcher krafft sein/ dz Zenexton an Halß gehenckt werde/ diser Leib wirt von anderen nicht geletzt."

Udo Friedrich, "'[…] das wir selber künste könen erdencken […].' Magiediskussion und paracelsisches Wissen im Wagnerbuch," in Peter Dilg and Hartmut Rudolph, eds., *Neue Beiträge zur Paracelsus-Forschung* (Stuttgart, 1995), 169–194, on 184 n. 56.

Natalia Bachour, *Oswaldus Crollius und Daniel Sennert im frühneuzeitlichen Istanbul. Studien zur Rezeption des Paracelsismus im Werk des osmanischen Arztes Ṣāliḥ b. Naṣrullāh Ibn Sallūm al-Ḥalabī* (Freiburg i.Br., 2012), 170–171.

6 Prophecies and Treasures

Bergwerks- und Schatzbücher: see *Verzeichnis etlicher Schätze* (§ 6.4)

6.1 *Eine andere Prognostication* also: *Prognosticon*

1. *Printing History, Manuscripts.* First edited in 1590 by Johann Huser. Reprinted several times in collections of prophecies. One manuscript (and other manuscripts copied from printed editions).
2. *Editions.* Edited by Huser, 10 (1590): Appendix, 228–230. Edited by Sudhoff in Paracelsus, *Sämtliche Werke*, 1/14: 543–544.
3. *Relationship between different versions.* The 1613 edition has some variants to Huser's edition and was printed from a different source. Huser has one paragraph not present in any other edition. The original title seems to be "Prognosticon," while "Ein andere Prognostication" (another prognostication) seems to be a descriptive title used by Huser.
4. *Structure, genre/form, perspective, style.* Written in the first person, directly addressing the reader.
5. *Relationship to other texts. Die andere Prognostication* is often accompanied by the preface to the *Prognostication auf XXIIII Jahr zukünftig*. It was quoted by Paul Linck in his *Von den Dreyen Seculis* (1599/1602).[374]
6. *Authenticity, authorship.* According to Sudhoff, "obviously coming from another pen."[375] – *Die andere Prognostication* was edited by Huser from a manuscript of Johannes Montanus and printed between two authentic prophecies of Paracelsus: the *Prognostication auf XXIIII. Jahr zukünftig* and the interpretation of the Lichtenberger images.
7. *Time of writing.* Probably written in the 1590s. Further study required.

---

374 Murase, *Paracelsismus und Chiliasmus* (2013), 162 n. 576.

375 Sudhoff, "Vorwort," in Paracelsus, *Sämtliche Werke*, 1/11: XXVII: "offensichtlich fremder Feder entstammend."

Word count: ca. 800 words

Incipit: "Ich mag gedencken, das du mir nit glauben werdest …"

Manuscripts:
- Dresden, *Sächsische Landesbibliothek*: Mscr.Dresd.N.118, f. 196–200
- Karlsruhe, *Badische Landesbibliothek*: Hs. Durlach 146 [copied from the 1613 printed edition]

First printed:
- **1600** (in: Johann Wolf, *Lectionum memorabilium et reconditarum tomus secundus* (Lauingen: Leonhard Rheinmichel for the author, 1600) [Latin]; VD16 W 4211; Sudhoff, *Bibliographia Paracelsica*, 431 n° 250)
- **1613** (in: *VI. Prognostica Von Verenderung vnd zufälligem Glück vnd Vnglück der höchsten Potentaten im Römischen Reich, ed. Henricus* <*Neotechnos*> (Halle: Christoph Bißmarck for Thomas Schürer, 1613); VD17 32:660420Z; Sudhoff, *Bibliographia Paracelsica*, 499–501 n° 296; three reprints, 1620–1621)
- **1643** (in: *Neun vhralte Propheceyungen/ Prognostica, Visiones vnd Weissagungen* (Schmalkalden: Valentin Schmalhertz, 1643); VD17 1:077771B; not in Sudhoff, *Bibliographia Paracelsica*)
- **1664** (in: Johannes Praetorius, *Turci-Cida, Oder Der vielfach-vorgeschlagene Türcken-Schläger* (Zwickau: Melchior Göpner, 1664); VD17 23:235879A; not in Sudhoff, *Bibliographia Paracelsica*)
- **1697** (in: Wenceslaus Greger, *Praemissae Prognosticorum nov-antiqvorum! Oder Erstes Drittheil/ Etlicher uhralten* […] *Propheceyungen*, "Nachgedruckt" [i.e. new edition] (no place, 1697); VD17 1:712254H; not in Sudhoff, *Bibliographia Paracelsica*)

Essential bibliography: Sudhoff, *Paracelsus-Handschriften*, 149–150, 776; Sudhoff, *Bibliographia Paracelsica*, 408; Sudhoff, "Vorwort," in Paracelsus, *Sämtliche Werke*, I/11: XXVII; CP 2: 318; CP 3: 1185.

Further bibliographical references:
Amadeo Murase, *Paracelsismus und Chiliasmus im deutschsprachigen Raum um 1600*, PhD thesis (Heidelberg, 2013), 162.

Geheimes Inventarium: see *Prophezeiung von den drei Schätzen* (§ 6.3)

6.2 *Krankes Römisches Reich*

1. *Printing History, Manuscripts.* First printed in 1620, attributed to Paracelsus; reprinted in 1621 without attribution to Paracelsus. No manuscripts known.

2. *Editions.* Not edited by Huser or Sudhoff.
3. *Relationship between different versions.*
4. *Structure, genre/form, perspective, style.*
5. *Relationship to other texts.* The text was a response to Andrew Willet's *Papa Purus putus Antichristus sive Decima Quaestio de Antichristo* (Frankfurt: Hulsius, 1619). According to one of Karl Widemann's *Verzeichnisse*, Adam Haslmayr wrote a "Responsio" to the *Krankes römisches Reich.*
6. *Authenticity, authorship.* In the first edition the work was attributed to Paracelsus. According to Sudhoff, Paracelsus is not mentioned anywhere in the text, which has no connection to Paracelsian thought. Chapter 13 mentions the Rosicrucians.
7. *Time of writing.* Probably written ca. 1620.

Word count:

Incipit: "Das die oberst weltlich macht, monarchia vnd standt …"

Manuscripts: no manuscripts known

First printed:
- 1620 (*Das krancke Römische Reich Dessen Vrsach, vnd glaubwürdige Artzney dargegen, Authore Paracelso &c.* […] *Sampt anhengenden Schluß, wie deß Reichs Schwachheit vnd Fall, sey zuuerstehen. Gestellet durch M. Angelum de Grauen, Palatinum* (no place, 1620); Sudhoff, *Bibliographia Paracelsica*, 523–524 n° 315; VD17 14:003029B; reprinted as *Vier Bey diesem jetzigen Auffstande deß gantzen heiligen Römischen Reichs hoch nothwendige Fragen und discurs*, no place 1621)

Historical Manuscript Catalogues: Widemann, *Verzeichnisse* (Kassel), n° x, 10

Essential bibliography: Sudhoff, *Bibliographia Paracelsica*, 523–524 n° 315, 697 n° 46.

Further bibliographical references:

Carlos Gilly, "The 'Midnight Lion', the 'Eagle' and the 'Antichrist': Political, religious and chiliastic propaganda in the pamphlets, illustrated broadsheets and ballads of the Thirty Years War," *Nederlands Archief voor Kerkgeschiedenis*, 80 (2000), 46–77, on 51.

Carlos Gilly, "Der 'Löwe von Mitternacht', der 'Adler' und der 'Endchrist': Die politische, religiöse und chiliastische Publizistik in den Flugschriften, illustrierten Flugblättern und Volksliedern des Dreissigjährigen Krieges", in *Rosenkreuz als europäisches Phänomen im 17. Jahrhundert* (Amsterdam, 2002), 249–268 (with plates, 234–248), on 255–256.

Thomas Fuchs, *Der Dreißigjährige Krieg und seine Drucksachen*, Katalog zur gleichnamigen Ausstellung der Universitätsbibliothek Leipzig (Leipzig, 2018), 57.

Löwe aus Mitternacht: see *Prophezeiung von den drei Schätzen* (§ 6.3)

Magische Prophezeiung: see *Prophezeiung von den drei Schätzen* (§ 6.3)

Prognosticon: see *Eine andere Prognostication* (§ 6.1)

6.3 *Prophezeiung von den drei Schätzen und dem Löwen aus Mitternacht* also: *Extract einer Prophecei / Magische Prophezeiung* [...] *von Entdeckung seiner drei Schätze / Prophezeiung von der Entdeckung dreier Schätze Hohenheims / Der Löwe aus Mitternacht / Geheimes Inventarium*

1. *Printing History, Manuscripts.* First edited in 1622 by Georg Andreas Stein (version 1). Next edited in 1625 by Joachim Morsius (version 2), and anonymously in 1631 (version 3a/b). – Version 1 was again printed in 1632 and 1708. Version 2 was never reprinted. Versions 3a and 3b have small differences (plus version 3a is missing the last paragraphs) and were reprinted many times. – Several manuscripts.
2. *Editions.* Not edited by Huser. Edited by Sudhoff in Paracelsus, *Sämtliche Werke*, I/14: 545–546 [mixed text from all versions].
3. *Relationship between different versions.* The *Geheimes Inventarium* is an enlarged version of the *Prophezeiung* with many repetitions.
4. *Structure, genre/form, perspective, style.* Version 1 written in the third person, version 2 and 3 written in the first person.
5. *Relationship to other texts.* The *Prophezeiung* is obviously dependent on *De tinctura physicorum* (§ 4.30). It is quoted in the Rosicrucian *Fama Fraternitatis*. – There are authentic prognostications of Paracelsus, that have been printed at the same time as the spurious *Prophezeiung*.[376]
6. *Authenticity, authorship.* The Hannover manuscript was written by Karl Widemann who received the text from Adam Haslmayr.[377] The Nürnberg manuscript is connected with Haslmayr as well. Haslmayr also sent the *Prophezeiung* to August von Anhalt in 1611 or 1612.[378] The text of the first printed edition of 1622 was allegedly provided by Balthasar Brendel of Altensalz. If this Brendel

376 Gilly, *Paracelsus in der BPH* (1993), 41.
377 Gilly, *Adam Haslmayr* (1994), 207, 214.
378 CP 3: 930.

can be identified with a "Notarius Publicus Caesareus" Balthasar Brendel who was active at the same time and in the same region (Voigtland)[379] remains to be studied.

7. *Time of writing.* Allegedly written in 1549 (or 1546 in version 3b), but in reality probably in the 1600s.

WORD COUNT: ca. 800 words (version 1), ca. 950 words (version 2), ca. 750 words (version 3a), ca. 800 words (version 3b), ca. 1.600 words (*Geheimes Inventarium*)

INCIPIT: "Es propheceyet vnd schreibet Philippus Theophrastus ..." [version 1] / "Ich werde in meinem grabe nicht gelassen vnd verbleiben ..." [version 2] / "Das ich in meinen grabe nicht gelassen werde ..." [version 3] / "Es möchte mich vielleicht einer in meinen Schriften zu schulden anweisen ..." [*Geheimes Inventarium*]

MANUSCRIPTS:

version 1

- BASEL, *Universitätsbibliothek*: Frey-Gryn Mscr I 2:Nr.15; 3 pages [version 1]
- GÖTTINGEN, *Niedersächsische Staats- und Universitätsbibliothek*: 4 Cod. Ms. hist. 846, 5 pages
- HANNOVER, *Gottfried Wilhelm Leibniz Bibliothek/Niedersächsische Landesbibliothek*: Ms. IV 341, 689–689b
- STUTTGART, *Hauptstaatsarchiv*: A47 Bü 12/1; 9 pages
- WOLFENBÜTTEL, *Herzog August Bibliothek*: Cod. Guelf. 264.32 Extrav., f. 81–84

version 3

- BERN, *Burgerbibliothek*: Mss.h.h.x.95, fol. 80–83
- DRESDEN, *Sächsisches Staatsarchiv*: 10026 Geheimes Kabinett, Nr. Loc. 00799/06
- NÜRNBERG, *Stadtbibliothek*: Nor.H.266(2); 4 fols. [version 3; attributed to Adam Haslmayr]

***enlarged version* (*Geheimes Inventarium*)**

- WIEN, *Österreichische Nationalbibliothek*: Cod. 11266

not classified

- BUDAPEST, *Egyetemi Könyvtár* [*University Library*]: Cod. A 219
- DARMSTADT, *Hessisches Staatsarchiv*: Bestand D 4 Nr. 239/1
- HAMBURG, *Staats- und Universitätsbibliothek*: Cod. jur. 2594, 409–417
- LEIPZIG, *Universtitätsbibliothek*: Ms. Rep. IV n. 106, vol. IV, f. 92r–93r
- MÜNCHEN, *Bayerische Staatsbibliothek*: cgm 1172, f. 146r, 156r–158r
- UPPSALA, *Universitetsbiblioteket*: Cod. Ups. Palmsk. 172, 617–618

---

379 Johann Seifert, *Stamm-Taffeln Gelehrter Leute Zweyter Theil* (Regensburg: Johann Georg Hofmann for the author, 1723), Stammtafel "Brendel".

- WOLFENBÜTTEL, *Herzog August Bibliothek*: Cod. Guelf. 9 Noviss. 2°, f. 6r–7v, 9r–13v, 14r–17r [three copies]

French

- PARIS, *Bibliothèque de l'Arsenal*: Ms. 3012, 146–150
- PARIS, *Bibliothèque de l'Arsenal*: Ms. 3027, f. 20r–30r
- *olim* Librairie Dorbon-aîné (1938);[380] current location unknown

FIRST PRINTED:

- **1622** ("Extract einer Prophecey Doctoris Theophrasti, etc.," in *Extract Einer Prophecey Doctor Theophrasti Paracelsi*, ed. Georg Andreas Stein (no place, 1622); VD17 39:132166T; Sudhoff, *Bibliographia Paracelsica*, 530–532 n° 322) [**version 1**]
- **1625** ("Extract Magischer Propheceyung vnnd Beschreibung/ von Entdeckung der 3. Schätzen Theophrasti Paracelsi," in *Magische Propheceyung Aureoli Philippi Theophrasti Paracelsi, Von Entdeckung seiner 3. Schätzen*, ed. Anastasius Philaretus Cosmopolita [= Joachim Morsius] ("Philadelphiae," 1625); not in VD17; Sudhoff, *Bibliographia Paracelsica*, 539–541 n° 330) [**version 2**]
- **1631** ("Extract vnd gründlicher Bericht/ oder Beschreibung der Magischen Prophecey des Hochberümbten Doctor Philippi Theophrasti Paracelsi Magici Anno 1549," in *Prognosticon. Theophrasti Paracelsi. Daß ist Gewisse Propheceiung so der Weitberümbte vnd vornehme Herr Philipp Theophrastus Paracelsus sonst Bombast genandt Anno 1549. geschrieben* (no place, 1631); VD17 39:165542U; Sudhoff, *Bibliographia Paracelsica*, 548–549 n° 338) [**version 3a**]
- **1631** ("Extract Vnnd gründlicher Bericht oder Beschreibung der Prophecey deß Hochberühmten D. Philippi Theophrasti Paracelsi/ Anno 1546," in *Propheceyung/ Doctoris Philippi Theophrasti Paracelsi Anno 1546. Vom Löwen auß Mitternacht* (no place, 1631); VD17 23:254344T; Sudhoff, *Bibliographia Paracelsica*, 550–551 n° 340) [**version 3b**]
- **1673** ("Mein Phil. Paracelsi geheimes Inventarium, dem Erfinder treulichen geoffenbaret," in Johannes Praetorius, *Des Rübezahls Dritter und gantz Nagel-neuer Historischer-Theil* (Leipzig: Johann Bartholomäus Oehler, 1673), 93–106; VD17 23:325034S; not in Sudhoff, *Bibliographia Paracelsica*)[381] [***Geheimes Inventarium***]

380 See *Bibliotheca Esoterica. Catalogue annoté et illustré* (Paris, 1938), 370 n° 3530.

381 Praetorius's source is a manuscript from the library of Georg Tissenius. On Tissenius, see Ulrich Bubenheimer, "Schwarzer Buchmarkt in Tübingen und Frankfurt. Zur Rezeption nonkonformer Literatur in der Vorgeschichte des Pietismus," *Rottenburger Jahrbuch für Kirchengeschichte*, 13 (1994), 149–163, on 156, 175.

Modern editions:

- *Magische Unterweisungen des edlen und hochgelehrten Philosophi und Medici Philippi Theophrasti Bombasti von Hohenheim Paracelsus genannt*, ed. Franz Spunda (Leipzig, 1924), 14–21 [from Wien, ÖNB, Cod. 11266].[382]

Historical Manuscript Catalogues: Widemann, *Verzeichnisse* (Kassel), n° v, 93, 275; *Catalogus* Pfeffer (1683), 136; von Linden (1788), part 2, 38 n° 32

Essential bibliography: Sudhoff, *Bibliographia Paracelsica*, 697 n° 47; Sudhoff, *Paracelsus-Handschriften*, 798 n° v/3; Sudhoff, "Vorwort," in Paracelsus, *Sämtliche Werke*, I/14: XXXII; CP 2: 318; CP 3: 930.

Further bibliographical references:

Peuckert, *Pansophie* (1956), 493.

Peuckert, *Rosenkreutz* (1973), 39, 93, 116, 377.

Gilly, *Adam Haslmayr* (1994), 207, 214.

Carlos Gilly, "The 'Midnight Lion', the 'Eagle' and the 'Antichrist': Political, religious and chiliastic propaganda in the pamphlets, illustrated broadsheets and ballads of the Thirty Years War," *Nederlands Archief voor Kerkgeschiedenis*, 80 (2000), 46–77.

Carlos Gilly, "Der 'Löwe von Mitternacht', der 'Adler' und der 'Endchrist': Die politische, religiöse und chiliastische Publizistik in den Flugschriften, illustrierten Flugblättern und Volksliedern des Dreissigjährigen Krieges", in *Rosenkreuz als europäisches Phänomen im 17. Jahrhundert* (Amsterdam, 2002), 249–268 (with plates, 234–248).

Kathrin Pfister and Barbara Schmidt-Thieme, "'Der Löwe aus Mitternacht' – eine pseudoparacelsische Prophezeiung und ihr Fortleben," in *Neue Facetten zum Bild des Paracelsus* (Vienna, 2008) (*Salzburger Beiträge zur Paracelsusforschung*, 41), 38–68.

6.4 *Verzeichnis etlicher Schätze, auch Silber und Goldes, auch Silber- und Goldbergwerke* also: *Bergwerks- und Schatzbücher*

1. *Printing History, Manuscripts.* One manuscript.
2. *Editions.* Not edited by Huser or Sudhoff.
3. *Relationship between different versions.* Only one known version.
4. *Structure, genre/form, perspective, style.* A list of 32 places (in Austria) with ore occurrences and hidden treasures. The places are marked with stones, crosses and other signs.

382 Also printed separately: "Mein Philippi Paracelsi Geheimes Inventarium. Dem Erfinder treulich geoffenbahrt," in *Das Buch der Wolkenwanderer 1924* (Leipzig, 1924), 26–30.

5. *Relationship to other texts.* The pseudo-paracelsian *Occulta Philosophia* (§ 5.7) has a chapter "Von den Schätzen vnd verborgnem Gutt/ inn vnd vnter der Erden" (Huser, 9: 350–354), which may be a reason for the attribution of the *Verzeichnis* to Paracelsus. The whole text reminds of the "Walenbücher."[383]
6. *Authenticity, authorship.* The *Verzeichnis* cites Basilius Valentinus, therefore it was written in the 17th century. It may be based, however, on older lists.
7. *Time of writing.* Further study required.

Word count:

Incipit: "Im Enßthal ab dem stein khelber am Gryming ..."

Manuscripts:
- Wien, *Österreichische Nationalbibliothek*: Cod. 11295, f. 15–20

First printed: not printed

Essential bibliography: Sudhoff, *Paracelsus-Handschriften*, 696–698 n° 144, 798 n° v/4.

Further bibliographical references:
Richard Smekal, "Phantastische Prophezeiungen. Schatzgräberwinke in Paracelsus-Schriften," *Neues Wiener Journal*, 33, n° 11.281 (18 April 1925), 7; reprinted in *Preßburger Presse* (Bratislava), 30, n° 1483 (31 January 1927).

7 Alchemical Processes

7.1 *Animation des Kupfers, wenn sie zur Tinctur soll gebraucht werden*

Instructions for a process on copper. Related to a process in *De natura rerum* (§ 2.2), book 5 (Huser, 6 (1590): 287–288); book 5 is one of the books that may be authentic.

Title: *Die animation des kupfers wenn sie zur tinctur soll gebraucht werden. Ist des Theophrasti Paracelsi arbeit*

383 See Will-Erich Peuckert, "Walen und Venediger," *Mitteilungen der Schlesischen Gesellschaft für Volkskunde*, 30 (1929), 205–247; Will-Erich Peuckert, "Der Walenwegweiser zur Abendburg," in E. Boehlich, W. Jungandreas, W-E. Peuckert, eds., *Das älteste schlesische Walenbuch* (Breslau, 1938), 21–37.

Incipit: "Recipe dünn lamellirtes kupfer zehn, zwanzig oder dreißig mark …"

Word count: ca. 300 words

Manuscripts:
- Hamburg, *Staats- und Universitätsbibliothek*: Cod. alchim. 664, 79–80

First printed: not printed

7.2 *Arcanum lunam in solem mutare*

Instructions for the preparation of a transmutative substance.

Title: *Arcanum Theophrasti Parac[elsi] lunam uno die et tres aut quatuor horis in solem mutare*

Incipit: "Recipe pars cum parte solis et lunae [,] amalgama[?] nach gemeinem brauch mit mercurio …"

Word count: ca. 300 words

Manuscripts:
- Hamburg, *Staats- und Universitätsbibliothek*: Cod. alchim. 668, f. 198v–199r

First printed: not printed

7.3 *Dealbatio lunifica* also: *Tinctura veneris ad Lunam*

Instructions for a transmutative substance. Not authentic according to Sudhoff.[384]

Title: *Tinctura Veneris ad Lunam Theoph[rasti] Para[celsi] etc.*

Incipit: "Recipe 1 M. gebrant luna zerlaß in einem tigel …"

Word count:

Manuscripts:
- Kassel, *Landesbibliothek*: 4° Ms. chem. 18, f. 1

384 Sudhoff, *Paracelsus-Handschriften*, 735: "Als Paracelsisch kann ich diese Vorschrift […] nicht ansehen."

First printed: not printed

Essential bibliography: Sudhoff, *Paracelsus-Handschriften*, 734–735, 780 n° 11/9.

Further bibliographical references:
Broszinski, *Manuscripta chemica in quarto* (2011), 56.

7.4 *Ex secretis Theophrasti Paracelsi*

Instructions for some alchemical processes, partly written in the first person.

Title: *Ex secretis Theophrasti Paracelsi*

Incipit: "Nim in gottes namen eine drachme vnd ein quint feines gekorntes gold …"

Word count:

Manuscripts:
– Hamburg, *Staats- und Universitätsbibliothek*: Cod. alchim. 772, 202–?

First printed: not printed

Ex Theophrasto Bartholomaeus Korndörffer: see *Secretum magnum* (§ 7.19)

7.5 *Extract aus einem uralten alchymischen Buch* [and other processes]

Of no value according to Sudhoff.[385]

Title: *Extract aus einem vhralten Theophrastisten geschribenen alchymischen buech, welches vast ganz mit disem Proceß einstimmet*

Manuscripts:
– Wien, *Österreichische Nationalbibliothek*: Cod. 11266, f. 164r

Essential bibliography: Sudhoff, *Paracelsus-Handschriften*, 678, 780 n° 11/27.

385 Sudhoff, *Paracelsus-Handschriften*, 191: "ohne dass ich dem Werth beizulegen vermöchte."

7.6 *Gemeinsame Arbeit von Theophrastus mit seinem Vater*

For a list of more texts connected with Wilhelm von Hohenheim, see *Brief von Wilhelm von Hohenheim an seinen Sohn* (§ 4.20).

Title: *Eine Arbeit Auff die luna do D Theophrastus mit Seiem Vater Zu Vllach hatt gearbeit*

Incipit: "Nimm ein lot gefilt luna, ein lot roten schwefel, ein lot zinnober, ein lot calciirt vitriolum auff die rotenn …"

Word count: ca. 100 words

Manuscripts:
- Wolfenbüttel, *Herzog August Bibliothek*: Cod. Guelf. 151 Noviss. 8°, f. 36v

First printed: not printed

7.7 *Labor et opus lapidis philosophorum*

Instructions for the preparation of the *lapis philosophorum*, allegedly completed by Paracelsus on 28 January 1532. According to a note the process was acquired in Salzburg in 1618.

Title: *Celeberrimi D. Philippi Theophrasti Paracelsi labor et opus quo ipse usus sua aetate et anno* [*15*]*32 28. Januarii perfecte complevit*

Incipit: "Recipe jm nahmen Gottes creatoris deß weissen vnverbrennlichen sulphuris philosophorum einen theil …"

Word count: ca. 900 words

Manuscripts:
- Kassel, *Landesbibliothek*: 4° Ms. chem. 58, f. 116r–118r

First printed: not printed

Essential bibliography: Sudhoff, *Paracelsus-Handschriften*, 727, 780 n° 11/16.

Further bibliographical references:
Broszinski, *Manuscripta chemica in quarto* (2011), 193.

7.8 *Metalle in die primam materiam zu bringen*
also: *Process des Theophrastus für seinen Vater*

For a list of more texts connected with Wilhelm von Hohenheim see *Brief von Wilhelm von Hohenheim an seinen Sohn* (§ 4.20).

TITLE: *Die Mettallen in die Primam materiam id est terram philosophien zu bringen welches Theophrastus seinem Vatter zugeschrieben* [Kassel]

INCIPIT: "Theofrastus hatt seinem vatter zugeschriben vnd gelert den aller edelsten spiritum zu machen ..." [Kassel]; "Theophrastus hat seinem Vatter zugeschrieben die alleredeleste Tinctur [...] zu machen. Ex autographo ipsius Theophrasti sigillato ..." [Halle]

WORD COUNT: ca. 350 words

MANUSCRIPTS:
- KASSEL, *Landesbibliothek*: 4° Ms. chem. 72, f. 339v–340r[386]
- HALLE, *Universitäts- und Landesbibliothek*: Ms. 25 G 22, f. 164v–165v

FIRST PRINTED: not printed

7.9 *Oleum vitrioli*

Latin text. Instructions for preparing an "oleum vitrioli" that transmutes mercury and silver into gold.

TITLE: *Oleum vitrioli Theophr[asti]*

INCIPIT: "Recipe vitriolem ungaricum aut romanum ..."

WORD COUNT: ca. 150 words

MANUSCRIPTS:
- HAMBURG, *Staats- und Universitätsbibliothek*: Cod. alchim. 668, f. 146

FIRST PRINTED: not printed

386 Not mentioned in Broszinski, *Manuscripta chemica in quarto* (2011).

7.10 *Particularia des Wilhelm von Hohenheim für seinen Sohn*

Two processes with mercury. Three manuscripts, but only the London manuscript has the second process. The London manuscript is a mix of Latin and German. – For a list of more texts connected with Wilhelm von Hohenheim see *Brief von Wilhelm von Hohenheim an seinen Sohn* (§ 4.20).

INCIPIT: "Nimm solis ein lot, mercurii zwei lott / Recipe solis unum lot, mercurii duo lot, fac amalgama ..."

INCIPIT: "Recipe aquam fortem optime zwei theil, in qua solve vitrioli optimi ..."

WORD COUNT: ca. 350 words [first process], ca. 200 words [second process]

MANUSCRIPTS:
- KASSEL, *Landesbibliothek*: 4° Ms. chem. 46, f. 80r–80v
- LONDON, *Wellcome Library*: MS 518, f. 18v–20r[387]
- WERL, *Stadtarchiv*: Familienarchiv v. Papen-Lohe, Akte A III g 1, Kasten 215f. 104r–104v

FIRST PRINTED: not printed

FURTHER BIBLIOGRAPHICAL REFERENCES:
Peuckert, *Pansophie* (1956), 316.
Telle, "Paracelsus in pseudoparacelsischen Briefen" (2006–2007), 155 n. 27.
Broszinski, *Manuscripta chemica in quarto* (2011), 170.

7.11 *Practica martem in venerem transmutare*

Instructions for the preparation of a transmutative substance.

TITLE: Practica Theophrasti martem in venerem transmutare ope mercurii

INCIPIT: "Recipe reines eisen dabey kein ander metall sey ein pfund ..."

---

387 This is the manuscript mentioned by Sudhoff, "Einleitendes," in Paracelsus, *Sämtliche Werke* I/1: XXVII: "In einem Alchemistischen Sammelkodex, der von Wien aus in den Handel kam, traf ich unter Nr. 158 'Zwei Partikularia Wilhelmi Theophrasti, so er seinen Sun gegen Salzburg geschickt hat, aus seiner eigen Handschrift'." The manuscript was sold by a Vienna bookseller and announced in several catalogues, e.g. Gilhofer & Ranschburg (Vienna), *Catalogue No. 209 comprising an Important Private Library* (Vienna, no date [ca. 1928/1930]), 55–56 n° 96: "18b Zwey Particularia Wilhelmj Theophrastj so er seinen Sun gegen Saltzburg geschickt hat auß seiner aignen Handschrift."

Word count: ca. 350 words

Manuscripts:
- Hamburg, *Staats- und Universitätsbibliothek*: Cod. alchim. 668, f. 198r–198v

First printed: not printed

7.12 *Practica von des roten Löwens Blut und weißen Adlers Glut*

Rather vague instructions for an alchemical process.

Title: *Practica Th[eophrast]i. Von des rohten lewens blut vnd weißen adlers glutt*

Incipit: "Prius extrahatur sein liquorischer saurer sulphur …"

Word count: ca. 100 words

Manuscripts:
- Hamburg, *Staats- und Universitätsbibliothek*: Cod. alchim. 668, f. 202v

First printed: not printed

7.13 *Praeparatio salis circulati minoris*

Instructions for the preparation of the "sal circulatum minor", a substance mentioned by Paracelsus in the *Archidoxis*.

Title: *Praeparatio salis circulati minoris Paracelsi*

Incipit: "Man nimbt geläutert salem marinem als die gründliche vnd ursprüngliche matrix …"

Word count: ca. 300 words

Manuscripts:
- Hamburg, *Staats- und Universitätsbibliothek*: Cod. alchim. 671, 208–209

First printed: not printed

7.14 *Process aus seiner eigenen Handschrift* also: *Theophrastischer Process*

Instructions for a process on antimony.

TITLE: *Theophrasti proces auß seiner eigen handschrift* [4° Ms. chem. 28]

INCIPIT: "Nimb antimonium crudum ein pfund vnd eisen feilung ein pfund …" [4° Ms. chem. 28]

WORD COUNT: ca. 300 words

MANUSCRIPTS:
- KASSEL, *Landesbibliothek*: 4° Ms. chem. 8, f. 326r–327r
- KASSEL, *Landesbibliothek*: 4° Ms. chem. 28, f. 148v–149r

FIRST PRINTED: not printed

ESSENTIAL BIBLIOGRAPHY: Sudhoff, *Paracelsus-Handschriften*, 746, 780 n° 11/10.

FURTHER BIBLIOGRAPHICAL REFERENCES:
Broszinski, *Manuscripta chemica in quarto* (2011), 27, 92.

Process des Theophrastus für seinen Vater: see *Metalle in die primam materiam zu bringen* (§ 7.8)

7.15 *Process im Schlaftrunk einem verehrt* also: *Tinctur im Kloster Murrhardt hinterlassen*

Instructions for a process for the preparation of *mercurius philosophorum*. The text continues in a narrated dialogue between the first-person narrator and Paracelsus. Only in 4° Ms. chem. 58 is inserted the following story: The narrator was allegedly given the process while having a nightcap with Paracelsus and taking him off the pants (!).

Allegedly written down in the Kloster Murrhardt on 24 August ("vff Bartholomaei") 1537.

The manuscript 4° Ms. chem. 58 is in the hand of Benedictus Figulus, 4° Ms. chem. 39 in the hand of Raphael Egli.

A manuscript in the Uffenbach Library (1720) had the following title: "Philippi Theophrasti Paracelsi Tinctur, so alle menschliche und metallische Cörper durchdringet, aus seinem eigenen Mund Anno 1537. im Closter Murhard im Würtemberger Land excipirt, gerecht und bewerth."[388]

388 Bibliotheca Uffenbachiana MSSta seu catalogus et recensio MSStorum codicum qui in bibliotheca Zachariae Conradi ab Uffenbach […] adservantur (Halle, 1720), col. 668 n° xxv; see also Catalogus manuscriptorum codicum bibliothecae Uffenbachianae (Frankfurt a.M.: Balthasar Diehl, 1747), 76–77.

TITLE: *Folget Paracelsi proceß jm schlafftrunck einem verehret* [4° Ms. chem. 58]

INCIPIT: "Recipe gold künstlich purgiert …"

WORD COUNT: ca. 450 words

MANUSCRIPTS:
- KASSEL, *Landesbibliothek*: 4° Ms. chem. 39, f. 42v–43v
- KASSEL, *Landesbibliothek*: 4° Ms. chem. 58, f. 43r–43v

FIRST PRINTED: not printed

ESSENTIAL BIBLIOGRAPHY: Sudhoff, *Paracelsus-Handschriften*, 724, 780 n° 11/11.

FURTHER BIBLIOGRAPHICAL REFERENCES:
Hartmut Broszinski, "Spuren klösterlicher Alchemie in Kasseler Handschriften des 15. und 16. Jahrhunderts," in Andreas Bihrer and Elisabeth Stein, eds., *Nova de veteribus. Mittel- und neulateinische Studien für Paul Gerhard Schmidt* (Leipzig, 2004), 823–847, on 830.
Broszinski, *Manuscripta chemica in quarto* (2011), 139, 192.

7.16 *Processus auf Luna den halben Teil Gold zu machen*
Instructions for an alchemical process.

TITLE: *Processus Philippi Paracelsi et ignotii auf luna den halben theil gold zu machen*

INCIPIT: "Nimb antimonium quod vocatur tincturali sieben loht …"

WORD COUNT: ca. 250 words

MANUSCRIPTS:
- HAMBURG, *Staats- und Universitätsbibliothek*: Cod. alchim. 671, 339–340

FIRST PRINTED: not printed

7.17 *Projectio der roten Tinctur*
Instructions for the use of a transmutative substance.

MANUSCRIPTS:
- WOLFENBÜTTEL, *Herzog August Bibliothek*: Cod. Guelf. 264.31 Extrav., on 745

Essential bibliography: Sudhoff, *Paracelsus-Handschriften*, 221, 779 n° 11/8.

7.18 *Schöne und herrliche Kunst*

In Low German (?). Instructions for the preparation of Paracelsus's "Tinctur." Allegedly found in the papers of Paracelsus and tried by Dr. Johannes Hesse, physician in Tübingen. – Of no importance, according to Sudhoff.[389]

Title: *Ein schon och herllich Konst welchen ist nach dem hochberünten Herren Theofrasto gefunden, welchen ist sein tinctur gewesen*

Incipit: "Man muß nemen 4 Pundt engelsch tinnen ..."

Manuscripts:
- Kopenhagen, *Det Kongelige Bibliotek*: GKS 259 folio, 159–162

Essential bibliography: Sudhoff, *Paracelsus-Handschriften*, 666, 780 n° 11/24.

7.19 *Secretum magnum*
also: *Ex Theophrasto Bartholomaeus Korndörffer*

Instructions for the preparation of an "aqua ignea". At the end: "Ex Theophrasto Bartholomaeus Korndörffer." – For a list of texts associated with Korndörffer see *Wie man die sieben Metalle clarificieren soll* (§ 4.9).

Title: *Secretum magnum*

Incipit: "Cum ista aqua ignea ingredimur materiam lapidis ... / Nun nimb einen praecipitierten antimonium der nicht mehr flüchtig ist ..."

Expl.: "Diese lunam fixam lege in das vorbemelte oleum et habebis aurum bonum. Laus Deo. Ex Theophrasto Bartholom[aeus] Korndörffer."

Word count: ca. 200 words

Manuscripts:
- Hamburg, *Staats- und Universitätsbibliothek*: Cod. alchim. 668, f. 203v

First printed: not printed

389 Sudhoff, *Paracelsus-Handschriften*, 666: "Bedeutung ist diesem Bruchstück nicht beizumessen."

7.20 *Theophrasti Tinctur*
also: *Tinctur für König Ferdinand*

Instructions for the preparation of Paracelsus's "Tinctur." Associated with King Ferdinand of Bohemia. According to Sudhoff, unreadable since it uses unusual symbols for the substances. Certainly not written by Paracelsus according to Sudhoff.[390]

MANUSCRIPTS:
- WIEN, *Österreichische Nationalbibliothek*: Cod. 11428, f. 174v

ESSENTIAL BIBLIOGRAPHY: Sudhoff, *Paracelsus-Handschriften*, 201, 779 n° II/6.

Theophrastischer Process: see *Process aus seiner eigenen Handschrift* (§ 7.14)

Tinctur für König Ferdinand: see *Theophrasti Tinctur* (§ 7.20)

7.21 *Tinctura, so Theophrastus seinem Vetter gegeben*

Instructions for the preparation of a universal solvent. Whether the Wolfenbüttel manuscript is the same text as the Hamburg manuscript remains to be ascertained. Both are associated with Paracelsus's "vetter", which may refer either to a male relative (uncle, cousin) or a friend.

TITLE: *Tinctura so Theophrastus seinem vetter gegeben silber in gold tingiren* [Hamburg]; *Tinctura Magna, so Theoph*[*rastus*] *seinem Vettern zugeschickt hatt vnd tingirt 1 Theil 4000. teil lunae in solem* [Wolfenbüttel]

INCIPIT: "Recipe ein halbes pfund salarmoniac ein pfund gefloßen saltz reibs zart untereinander …"

WORD COUNT: ca. 200 words (Hamburg manuscript)

MANUSCRIPTS:
- HAMBURG, *Staats- und Universitätsbibliothek*: Cod. alchim. 664, 162–163
- WOLFENBÜTTEL, *Herzog August Bibliothek*: Cod. Guelf. 264.31 Extrav., 701–702

FIRST PRINTED: not printed

ESSENTIAL BIBLIOGRAPHY: Sudhoff, *Paracelsus-Handschriften*, 221, 779 n° II/7.

---

390 Sudhoff, *Paracelsus-Handschriften*, 201: "[stammt] sicher nicht von Hohenheim her […]."

7.22 *Tinctura physicorum*

Instructions for the preparation of a transmutative oil. Tried and tested by Simon Gischgar of Olmütz/Moravia on 20 October 1595. Not related to Paracelsus according to Sudhoff.[391]

Title: *Tinctura physicorum Theophrasti Paracelsi Bombasts etc.*

Incipit: "Recipe mercurium antimonii. Theile den in zwei theill ..."

Word count: ca. 200 words

Manuscripts:
- Kassel, *Landesbibliothek*: 4° Ms. chem. 58, f. 113r–113v

First printed: not printed

Essential bibliography: Sudhoff, *Paracelsus-Handschriften*, 727, 780 n° 11/15.

Further bibliographical references:
Broszinski, *Manuscripta chemica in quarto* (2011), 193.

7.23 *Tinctura physicorum et lapis philosophorum*

Instructions for the preparation of the "lapis philosophorum." Attributed to Paracelsus by a later hand. Not authentic according to Sudhoff.[392]

Title: Tinctura physicorum et lapis philosophorum

Incipit: "Man soll nemben von dem gerainigten flüchtigem stain ein vntz ..."

Word count: ca. 1.000 words

Manuscripts:
- Wien, *Österreichische Nationalbibliothek*: Cod. 11259, f. 70r–74r

First printed: not printed

391 "Hat mit Hohenheim nichts zu thun:" Sudhoff, *Paracelsus-Handschriften*, 727.

392 Sudhoff, *Paracelsus-Handschriften*, 191: "Ich kann dem späteren Bearbeiter nicht beistimmen, dass diese Anweisung zur Bereitung des Steins der Weisen von Paracelsus herstamme."

Essential bibliography: Sudhoff, *Paracelsus-Handschriften*, 191, 780 n° 11/13.

Tinctura veneris ad Lunam: see *Dealbatio lunifica* (§ 7.3)

7.24 *Verissimum aurum potabile Theophrasti Paracelsi* [and other processes]

Instructions for alchemical processes, possibly taken from or inspired by texts like *Thesaurus thesaurorum* (§ 4.52) and *Manuale* (§ 4.40) that are quoted in the text. Associated with the name "Johannes Waltz."

Title: *Verissimum aurum potabile Theop[hrasti] Paracelsi* [f. 115r]; *Volgt die warhaftige tinctura welche D. Theophrastus mit verwickleten vnndt vmbschweiffigen wortten im thesauro thesaurorum obscure begriffen* [f. 116v]; *Lapis philosophicus Theophrasti Paracelsi genant manuale* [f. 122r]

Incipit: "Nimb das allerfeinste goldt das drei mal durch den antimonium purgirt sey …"

Word count: ca. 2.000 words

Manuscripts:
- Hamburg, *Staats- und Universitätsbibliothek*: Cod. alchim. 767, f. 115r–129v

First printed: not printed

7.25 *Wahrhafte Kunst der chymischen Alchimei*

According to the preface, a transcript of an old manuscript given by a friend of Theophrastus [Paracelsus] to Christoph Grindtl who was Münzmeister (mint master) at the time of Paris von Lodron, archbishop of Salzburg (d. 1653).[393] Contains no texts or processes connected with Paracelsus, according to Sudhoff.[394]

Manuscripts:
- Salzburg, *Salzburg Museum*: Hs 2189 (*olim* Paracelsiana 1211); 154 pages

393 Grindtl is mentioned in Franz Valentin Zillner, *Geschichte der Stadt Salzburg* (Salzburg, 1885), 354.

394 Sudhoff, *Paracelsus-Handschriften*, 229: "Von Paracelsus findet sich kein Wort weiter in dem Büchlein; weder werden bestimmte Processe als von ihm herstammend gegeben, noch ist das Ganze als Paracelsisch aufzufassen."

Essential bibliography: Sudhoff, *Paracelsus-Handschriften*, 228–229, 780 n° II/22.

7.26 *Wie die anima sulfuris saturni clarificirt wird*

Allegedly written in 1537. Instruction for the preparation of "anima sulfuris saturni." Associated with the name of Isaac Hollandus. There is a parallel in *De natura rerum* (§ 2.2), book 5, according to Sudhoff.

Title: *Wie die Anima Sulfuris Saturni auß Ihrem Corpus Clarificiret und daß in ein laufend Wasser gemacht wirdt wie Anno 1537 der Theure Mann Theophrastus Paracelsus Saturnj in Solem zu Transmutiren.*

Incipit: "Kanstu nicht selber Bleiweiß oder Minien machen …"

Manuscripts:
- Wolfenbüttel, *Herzog August Bibliothek*: Cod. Guelf. 264.31 Extrav., 942–944

Essential bibliography: Sudhoff, *Paracelsus-Handschriften*, 221–222, 780 n° II/12.

8 Collections of Recipes, Processes and Experimenta

8.1 *Alchimica Paracelsi*

1. *Printing History, Manuscripts.* Not printed. One manuscript.
2. *Editions.* Not edited by Huser or Sudhoff.
3. *Relationship between different versions.* Only one known version.
4. *Structure, genre/form, perspective, style.* A collection of medical recipes, astrological notes and alchemical excerpts.
5. *Relationship to other texts.*
6. *Authenticity, authorship.* Not authentic according to Sudhoff.[395]
7. *Time of writing.* Further study required.

Word count:

Incipit: "Der natürliche wahrhafftige borras wirdtt gemacht aus borrasstein …"

---

395 Sudhoff, *Paracelsus-Handschriften*, 219: "Ich kann hierin und im darauf Folgenden nichts anderwärts als Paracelsisch Ueberliefertes oder wahrscheinlich von ihm Herstammendes finden."

MANUSCRIPTS:

- LEIDEN, *Universiteitsbibliotheek*: Voss. Chym. F.26, f. 6–101

FIRST PRINTED: not printed

ESSENTIAL BIBLIOGRAPHY: Sudhoff, *Paracelsus-Handschriften*, 219, 779 n° 1/28.

FURTHER BIBLIOGRAPHICAL REFERENCES:
Boeren, *Codices Vossiani Chymici* (1975), 76.

8.2 *Arcana*

1. *Printing History, Manuscripts.* Not printed. One manuscript from the library of Bonaventura Hahn, bishop of Breslau.[396]
2. *Editions.* Not edited by Huser or Sudhoff.
3. *Relationship between different versions.* Only one known version.
4. *Structure, genre/form, perspective, style.* A collection of medical recipes and experimenta, some of them explicitly naming Paracelsus as author, some anonymous, some naming other persons: Paul Hess, Andreas Rindfleisch, Pancratius Freudenberg, Elias Rhediger ("Rydyger"), Andreas Berthold.
5. *Relationship to other texts.*
6. *Authenticity, authorship.* Sudhoff considered the *Arcana* of no importance.[397]
7. *Time of writing.* Further study required.

WORD COUNT:

INCIPIT: "Nim hirschen bluett, so warmb du es immer gehaben magst ..."

MANUSCRIPTS:

- LEIDEN, *Universiteitsbibliotheek*: Voss. Chym. Q.14, f. 1–20 [followed by the collection *Secreta secretorum* (§ 8.13)]

FIRST PRINTED: not printed

ESSENTIAL BIBLIOGRAPHY: Sudhoff, *Paracelsus-Handschriften*, 208–209, 778 n° 1/11.

---

396 Bonaventura Hahn was a "patron" of Johann Huser (CP 2: 413). See also Telle, "Johann Huser in seinen Briefen" (1992), 170–172.

397 Sudhoff, *Paracelsus-Handschriften*, 217: "Die Arcana im Leidener Codex 14 scheinen mir keinerlei Beachtung zu verdienen; sie sind ihm entweder irrthümlich zugeschrieben oder absichtliches Blendwerk."

Further bibliographical references:

Boeren, *Codices Vossiani Chymici* (1975), 135.

Joachim Telle, "Johann Huser in seinen Briefen. Zum schlesischen Paracelsismus im 16. Jahrhundert," in Joachim Telle, ed., *Parerga Paracelsica* (Suttgart, 1992), 159–248, on 172 n. 54.

Laura Balbiani, "Zum Nachleben des Paracelsus am Sulzbacher Hof (1656–1708)," in *Neue Facetten zum Bild des Paracelsus* (Salzburg, 2006) (*Salzburger Beiträge zur Paracelsusforschung*, 41), 69–83, on 75 n. 14.

8.3 *Arcana quaedam chymica* also: *Verwendung des Oleum salis*

1. *Printing History, Manuscripts.* Not printed. Three manuscripts.
2. *Editions.* Not edited by Huser or Sudhoff.
3. *Relationship between different versions.* The title *Arcana quaedam chymica* is found only in the Hamburg manuscript, while the text seems to have no title in the other manuscripts. – The Kassel manuscript roughly comprises the second half of the Hamburg manuscript; some portions have a different order. We have not been able to study the Wien manuscript.
4. *Structure, genre/form, perspective, style.* A collection of medical and alchemical recipes.
5. *Relationship to other texts.* Some of the manuscripts contain also the *Secreta secretorum* (§ 8.13) that also has some textual parallels with this collection. There *may* be parallels to the *Antwortbrief an Wilhelm von Hohenheim* (§ 4.18). – One of the processes of the *Arcana quaedam chymica* was printed as "Secretum & Arcanum Magnum, wie man den Mercurium Philosophorum und das Oleum Salis, auch endlich die rechte wahre Tinctur, auf alle Metalle, solche in recht gut beständig Gold zu tingiren, machen und praepariren soll, an einen vornehmen Doctor geschrieben, von Bartel Veltenberger; den 22. Sept. 1553" in *Fontani Hortulani, Fata chymica, In Beschreibung Der Wahren und falschen Chymie, Mit Beygefügten acht raren Tractaetgen und zum Theil alten kostbahren Manuscriptis* (Frankfurt a.M.: David Jacob Cronau, 1737), 60–63. Further study required.
6. *Authenticity, authorship.* Allegedly from a manuscript of Paracelsus ("ex m[anu] s[cripto] Theophrast[i] Paracelsi"). One recipe ("Virtus olei salis") mentions a transmutation in the presence of Emperor Maximilian. A recipe for an "oleum solis" has a note written in the first person about the healing of a man suffering from dropsy (Wassersucht), that "I Theophrastus have effected in the Steiermark as well as in the Upper Alsace with the help of God. [...] I have also successfully cured a man of epilepsy in the hospital of Niederpoda."[398] Another recipe for an

398 "Niden pota" in Wien Cod. 11259, "Nideren" In Wien Cod. 11428 according to Sudhoff, *Paracelsus-Handschriften*, 190. Also in the *Secreta secretorum*; see Büttner, *Die pseudo-paracelsischen* Secreta Secretorum (2001), 47, 92.

"aurum potabile" is attributed to the Graf von Hardeck. An entry in Pierre Borel's *Bibliotheca chimica*, already cited by Sudhoff in a derogatory notice ("Pierre Borel is never to be trusted"), states: "Tinctura Solis, quam Paracelsus propria manu scriptam Comiti de Hardek dedit."[399]

7. *Time of writing.* Further study required.

Word count: ca. 3.700 words (Hamburg manuscript); ca. 1.500 words (Kassel manuscript)

Incipit: "Recipe reinen lehm zwei pfund der geschlemmet ist ..."

Manuscripts:
- Hamburg, *Staats- und Universitätsbibliothek*: Cod. alchim. 651, 502–515
- Kassel, *Landesbibliothek*: 4° Ms. chem. 33[10, f. 3r–7v [extracts]
- Wien, *Österreichische Nationalbibliothek*: Cod. 11259, f. 51–64

First printed: not printed

Essential bibliography: Sudhoff, *Paracelsus-Handschriften*, 190, 778 n° 1/8; CP 2: 461.

Further bibliographical references:
Büttner, *Die pseudoparacelsischen "Secreta Secretorum"* (2001), 91–95.
Broszinski, *Manuscripta chemica in quarto* (2011), 112.

8.4 *Arzneistücke*

1. *Printing History, Manuscripts.* Not printed. One manuscript.
2. *Editions.* Not edited by Huser or Sudhoff.
3. *Relationship between different versions.* Only one known version.
4. *Structure, genre/form, perspective, style.* A collection of medical and alchemical recipes and "historiae" about successful healings. Some parts written in the first person ("Ich Theophrastus"), but mostly in the third person ("Theophrastus spricht").
5. *Relationship to other texts.*

399 Pierre Borel, *Bibliotheca chimica. Seu catalogus librorum philosophicorum hermeticorum* (Paris: Charles du Mesnil and Thomas Jolly, 1654), 181; also: (Heidelberg: Samuel Broun, 1656), 172; quoted (with misspellings) by Sudhoff, *Paracelsus-Handschriften*, 23.

6. *Authenticity, authorship.* Allegedly from the autograph of Paracelsus ("Auß Theophrasti Paracelsi Aigner Handt geschriben").
7. *Time of writing.* Further study required.

WORD COUNT:

INCIPIT: "Recipe vitriol wol vertrocknet aufs aller weisest …"

MANUSCRIPTS:
- WIEN, *Österreichische Nationalbibliothek*: Cod. 11259, f. 85r–109

FIRST PRINTED: not printed

ESSENTIAL BIBLIOGRAPHY: Sudhoff, *Paracelsus-Handschriften*, 192–196, 778 n° 1/9.

FURTHER BIBLIOGRAPHICAL REFERENCES:

Auserlesenes Korn: see *Secreta secretorum* (§ 8.13)

8.5 *Centum quindecim experimenta curationesque* also: *Hundert und vierzehn Experimenta*

1. *Printing History, Manuscripts.* First printed in Latin, edited by Bernard Gilles Penot. According to the title page the Latin text was translated from German. First printed in German in 1606, edited by Johann Walch[400] and translated (or re-translated?) from Latin. In 1680 the German text was reprinted in a *Hauß-Kunst- und Wunder-Buch.*[401] First edited in English by John Hester ca. 1584 (no

400 On Johann Walch see Joachim Telle, "Zum 'Filius Sendivogii' Johann Hartprecht," in Christoph Meinel, ed., *Die Alchemie in der europäischen Kultur- und Wissenschaftsgeschichte* (Wiesbaden, 1986), 119–136, on 128. Walch was the author of *Decas fabularum* (1609) and of a commentary on Johann Grasse's *Kleiner Bauer.* More details on Walch can be found in the unpublished PhD thesis by Thomas Lederer, *Der Kölner Kurfürst Herzog Ernst von Bayern (1554–1612) und sein Rat Johann Grasse (um 1560–1618) als Alchemiker der frühen Neuzeit. Ein Beitrag zur Geschichte des Paracelsismus*, Diss. phil. (Heidelberg, 1992), 135–139.

401 Des vortrefflichen Herrn Johann Baptista Portae von Neapolis. Hauß- Kunst- und Wunder-Buchs/ Ander Theil, Abraham Lichtenthaler (Sulzbach) für Johann Zieger (Nürnberg), 1680. Not in Sudhoff, *Bibliographia Paracelsica.* The Wunder-Buch was reprinted in 1713 and 1715. – See Balbiani, "Zum Nachleben des Paracelsus am Sulzbacher Hof" (2008), 79–80.

date given in the printed book). One Latin manuscript, copied indirectly from the 1582 edition.[402]

2. *Editions.* Not edited by Huser or Sudhoff.
3. *Relationship between different versions.*
4. *Structure, genre/form, perspective, style.* A collection of "historiae" of successful healings. In contrast to the title, there are only 114 sections, which is the reason why the first German edition is called "Hundert vnd vierzehen Experimenta."
5. *Relationship to other texts.* Jost Weyer assumed that the *Experimenta* of Samuel Schlegel (§ 8.8) were printed in 1606 as the *Centum quindecim experimenta curationesque.*[403] According to Kühlmann-Telle this is not the case.[404]
6. *Authenticity, authorship.* Allegedly written by Paracelsus himself and found in Paracelsus's papers by his "famulus," Conrad Steinberg (otherwise unknown[405]). Penot states that he received the text from a German prince and translated it into Latin. – According to Penot, it is obvious enough that this collection of short relations of medical cures was written by Paracelsus.[406] According to Sudhoff the *Centum quindecim experimenta curationesque* are not authentic and rather written by a Paracelsist.[407]
7. *Time of writing.* Further study required.

Word count: ca. 3.500 words (Latin text), ca. 5.000 words (German text)

Incipit: "Quidam baro perpetuo male affectus aliqua passione ... / Ein freyherr hat ein immerwährend grimmen im leyb ..."

402 Sudhoff, *Paracelsus-Handschriften*, 692: "Das sinnlose 'ijs' ist in der Ueberschrift bei Penot nur Druckfehler für 115, welcher am Ende in den Erratis verbessert wird; in unserm Mscr. ist er gedankenlos nach geschrieben und beweist, dass die Handschrift auf diesem Drucke beruht, jedoch nicht unmittelbar davon abgeschrieben ist, weil einige Lesefehler vorkommen, die handschriftliche Zwischenglieder wahrscheinlich machen. Auch andere Varianten kommen vor ohne grosse Bedeutung, ausserdem einige erklärende Einschiebungen." – CP 3: 580: "weitgehend Abschrift des Penotschen Drucks."

403 Weyer, *Graf Wolfgang II. von Hohenlohe* (1992), 478 n° 3.10: "Druck erst 1606; siehe Sudh. Nr. 275."

404 CP 1: 378–379: "ein Textvergleich [ergab], daß die Angabe von Weyer [...] unzutreffend ist."

405 See CP 1: 96.

406 Penot, To the Reader, 24: "Tractatus hic experimentorum Paracelsicorum (amice lector) ex dono cuiusdam Illustrißimi Principis Germaniae ad me peruenit. Hanc [*sic*] quoniam Paracelsi esse satis luculenter constat, duxi mihi soli non reseruandum, sed in publicum edendum, quoniam absconditi thesauri nulla est vtilitas."

407 Sudhoff, *Bibliographia Paracelsica*, 330: "Ich kann dieselben nicht für paracelsisch halten. [...] Vermuthlich ist ein Paracelsist [...] der Verfasser."

MANUSCRIPTS:
- WOLFENBÜTTEL, *Herzog August Bibliothek*: Cod. Guelf. 47.21 Aug. 4°, f. 1r–14v [Latin]
- CARPENTRAS, *Bibliothèque Inguimbertine*: Ms. 310, f. 94–? [French]
- GLASGOW, *University Library*: Sp Coll Ferguson Am-b.9, 1–21 [English]

FIRST PRINTED:
- **1582** (*Philippi Aureoli Theophrasti Paracelsi* [...] *centum quindecim curationes experimentaque* (Genève: Jean Lertout, 1582); Sudhoff, *Bibliographia Paracelsica*, 328–331 n° 190)
- **1584 ca.** (*A hundred and fourtene experiments and cures of the famous Phisition Philippus Aureolus Theophrastus Paracelsus*, ed. by I.H. [John Hester] (no place or date [ca. 1584]); Sudhoff, *Bibliographia Paracelsica*, 684–685 n° 505)
- **1606** (*Hundert vnd vierzehen Experimenta vnd allerhand treffliche vnd bewehrte stuck der Artzeney/ Theophrasti Paracelsi*, ed. Johann Walch (Strasburg: Anton Bertram, 1606); VD17 39:142903C; Sudhoff, *Bibliographia Paracelsica*, 471–472 n° 275)

HISTORICAL MANUSCRIPT CATALOGUES: *Catalogus* Pfeffer (1683), 136

ESSENTIAL BIBLIOGRAPHY: Sudhoff, *Bibliographia Paracelsica*, 695 n° 9; Sudhoff, *Paracelsus-Handschriften*, 691–692, 779 n° I/29; CP 1: 96, 378–379.

FURTHER BIBLIOGRAPHICAL REFERENCES:

Jost Weyer, *Graf Wolfgang II. von Hohenlohe und die Alchemie. Alchemistische Studien in Schloß Weikersheim 1587–1610* (Sigmaringen, 1992), 478 n° 3.10.

Laura Balbiani, "Zum Nachleben des Paracelsus am Sulzbacher Hof (1656–1708)," in *Neue Facetten zum Bild des Paracelsus* (Salzburg, 2006) (*Salzburger Beiträge zur Paracelsusforschung*, 41), 69–83, on 79–80.

8.6 *Collectanea ex Theophrasto et aliis*

1. *Printing History, Manuscripts.* Not printed. One manuscript.
2. *Editions.* Not edited by Huser or Sudhoff.
3. *Relationship between different versions.* Only one known version.
4. *Structure, genre/form, perspective, style.* A collection of medical recipes, some attributed to Paracelsus, some to Paracelsists identified by their names. Some recipes are also present in the *Secreta secretorum* (§ 8.13).
5. *Relationship to other texts.*
6. *Authenticity, authorship.* According to Broszinski, compiled by Jacob Mosanus.
7. *Time of writing.* Further study required.

Word count:

Incipit: "Aurum potabile, cuius dosis scrupulum unum, dicitur cum aurum cum alijs spiritibus …"

Manuscripts:
- Kassel, *Landesbibliothek*: 4° Ms. chem. 54; 181 fols.

First printed: not printed

Essential bibliography:

Further bibliographical references: Sudhoff, *Paracelsus-Handschriften*, 739–743.
Büttner, *Die pseudoparacelsischen "Secreta Secretorum"* (2001), 15, 67.
Broszinski, *Manuscripta chemica in quarto* (2011), 184–185.

8.7 *Collectanea Oporini*

1. *Printing History, Manuscripts.* Not printed. One manuscript. The full title is: *Collectanea Oporini quae ipse vivente Theophrasto (cum ipsius discipulus esset) ex ore ipsius conscripsit.*
2. *Editions.* Not edited by Huser or Sudhoff.
3. *Relationship between different versions.* Only one known version.
4. *Structure, genre/form, perspective, style.* A collection of ca. 160 recipes, allegedly compiled by Johannes Oporinus while he was in the service of Paracelsus.
5. *Relationship to other texts.*
6. *Authenticity, authorship.* Further study required.
7. *Time of writing.* The whole manuscript contains copies of various texts by the Paracelsian Georg Phaedro (d. after 1577), dated 1590 and 1591. At what time the manuscript was written (either at these dates or later) is yet to be determined.

Word count:

Incipit: "So dz roß im aller besten fuetter essen ist …"

Manuscripts:
- Erlangen, *Universitätsbibliothek*: MS 999 (*olim* Irm. 1172), 473–504

First printed: not printed

Essential bibliography: CP 2: 94.

Further bibliographical references:
Carlos Gilly, *Die Manuskripte in der Bibliothek des Johannes Oporinus* (Basel, 2001), 34.

8.8 *Experimenta*

1. *Printing History, Manuscripts.* Not printed. One manuscript.
2. *Editions.* Not edited by Huser or Sudhoff.
3. *Relationship between different versions.* Only one known version.
4. *Structure, genre/form, perspective, style.* A collection of recipes, at some time compiled or owned by Samuel Schlegel.[408]
5. *Relationship to other texts.*
6. *Authenticity, authorship.* Schlegel was a court physician and was in 1564 in close relationship with Adam von Bodenstein.[409] Kurfürst August von Sachsen commissioned Schlegel to procure Paracelsian books,[410] and it is known that August received at an unknown date a "Büchlin ettlicher Experimenten" (a small book with many *experimenta*) from Schlegel. This may have been the same collection as the *Experimenta* in the Hamburg manuscript.[411]
7. *Time of writing.* Compiled before 1586 when the only existing manuscript copy was entered into a library catalogue. Further study required.

Word count:

Incipit: "Recipe zwei lot mercurij mit essig vnnd saltz wol gereinigt …"

Manuscripts:
– Hamburg, *Staats- und Universitätsbibliothek*: Cod. alchim. 192, f. 281r–308r

First printed: not printed

Historical Manuscript Catalogues: Arpe (1726), 123;

Essential bibliography: CP 1: 378–379.

408 Paulus, "Eine unbeachtete Paracelsus-Handschrift aus Franken" (1991), 146–147.
409 On Schlegel see CP 1: 373, 378, 630; CP 2: 200, 201, 965; CP 3: 45, 364.
410 Ariane Bartkowski, Fürstliche Laborportner in der alchemistischen Praxis. Das Netzwerk des Kurfürstenpaares August und Anna von Sachsen ([Görlitz] 2017), 177.
411 CP 1: 378.

FURTHER BIBLIOGRAPHICAL REFERENCES:

Julian Paulus, "Eine unbeachtete Paracelsus-Handschrift aus Franken," in Joachim Telle, ed., *Parerga Paracelsica* (Stuttgart, 1991), 141–147, on 146–147.

Jost Weyer, *Graf Wolfgang II. von Hohenlohe und die Alchemie. Alchemistische Studien in Schloß Weikersheim 1587–1610* (Sigmaringen, 1992), 215, 478 n° 3.10.

Hundert und vierzehn Experimenta: see *Centum quindecim experimenta curationesque* (§ 8.5)

8.9 *Manualia duo*
also: *Wanderbücher/Reiseaufzeichnungen/Primum manuale, Secundum manuale*

1. *Printing History, Manuscripts.* Latin text with some smaller parts in German. First printed in 1582 by Pietro Perna in Basel.
2. *Editions.* Edited by Huser, *Chir.* 1605, Appendix, 1–58. Edited by Sudhoff in Paracelsus, *Sämtliche Werke*, I/13: 425–539.
3. *Relationship between different versions.*
4. *Structure, genre/form, perspective, style.* Two collections of recipes (one chemical, one medicinal).
5. *Relationship to other texts.*
6. *Authenticity, authorship.* Edited from the autograph of Paracelsus according to the printer Perna. In the preface to the 1605 Appendix, Huser explained that they are certainly not by Paracelsus: he may have written them down, thus only copying recipes collected by him from other people.[412] Sudhoff casted serious doubts on this hypothesis: according to him, although Huser had obtained several texts from Johannes Homelius in Pettau, this is the only time that Huser does not refer to an autograph in the possession of Homelius but rather writes that Homelius has the autograph manuscript "bey sich gehabt," which would mean, still according to Sudhoff, that the autograph manuscript of these texts was already gone or

412 Huser to the Reader, in Huser, *Chir.* 1605, Appendix, sig. Qqq 1v: "Letzlich/ belangend die zwey Manualia/ (welche hier in der Ordnung zum ersten gesetzt) soll der günstige Leser wissen/ das sie zwar beyde in Theophrasti eigner handschrifft also gefunden worden (deren Originalia der hochgelehrte Herr D. Joh. Homelius Secundus zu Pelaw [*printer's error for 'Petaw'*] in der Steyrmarck bey sich gehabt:) Seind aber doch nicht vnbillich von andern/ des Theophrasti genuinis Operibus abgesondert worden/ weil es nicht sein werck oder Composition ist/ Sondern nur zusammen geschriebene proceß vnd Recepten/ welche er ohne zweiffel in seiner Jugent (als er in solchen dingen noch nicht so hoch geübt gewesen) von andern künstlern hin vnd wider gehört oder abgeschrieben/ vnd also in seiner wanderschafft als kleine Manualia oder vade mecum mit sich geführet." – On Homelius see CP 3: 834–837.

lost at the time Huser prepared this edition.[413] Although Sudhoff did not deem the two *Manualia* authentic, he preferred to include them in his edition, just in case.[414]

7. *Time of writing.* Further study required.

WORD COUNT: ca. 30.000 words (*Primum Manuale*), ca. 6.000 words (*Secundum Manuale*).

INCIPIT: "Recipe tartari calcinati libras duo ..."

MANUSCRIPTS: no manuscripts known

FIRST PRINTED:
- 1582 (in: *Aureoli Theophrasti Paracelsi Archidoxorum, Seu de Secretis Naturae mysteriis, libri decem* (Basel: Pietro Perna, 1582); VD16 P 399; Sudhoff, *Bibliographia Paracelsica*, 332–333 n° 191)

ESSENTIAL BIBLIOGRAPHY: Sudhoff, *Bibliographia Paracelsica*, 462; Sudhoff, *Paracelsus-Handschriften*, 777; Sudhoff, "Vorwort," in Paracelsus, *Sämtliche Werke*, I/13: XIII–XVI; CP 3: 814, 835–836.

FURTHER BIBLIOGRAPHICAL REFERENCES:

Karl Sudhoff, "Die Wanderbücher Hohenheims," *Münchener Medizinische Wochenschrift*, 54 (1907), vol. 2, n° 42, on 2110; reprinted in *Verhandlungen der Gesellschaft deutscher Naturforscher und Ärzte, 79. Versammlung zu Dresden 1907*, part 2/2 (Leipzig, 1908), 97.

Edwin Rosner, "Hohenheims große Wanderung von Ferrara nach Schwaz (*peregrinatio magna* 1516–1521)," in *Nachlese zum Jubliäumskongreß 500 Jahre Paracelsus* (Vienna, 1995) (*Salzburger Beiträge zur Paracelsusforschung*, 28), 45–62, on 55.

8.10 *Nota ex Theophrasti manu*

1. *Printing History, Manuscripts.* Not printed. One manuscript.
2. *Editions.* Not edited by Huser or Sudhoff.
3. *Relationship between different versions.* Recipes for diverse diseases, including a "historia" of a healing performed by Paracelsus.
4. *Structure, genre/form, perspective, style.*
5. *Relationship to other texts.*

---

413 Sudhoff, "Die Wanderbücher Hohenheims" (1907), 2110.

414 See Sudhoff's lengthy report in Paracelsus, *Sämtliche Werke*, I/13: XIII–XVI.

6. *Authenticity, authorship.* Authenticity unclear. The text is preceded by *Von der Scotomia oder Hauptsucht* (§ 3.13) and followed by two alchemical texts in the manuscript (an excerpt of the pseudo-Paracelsian *Thesaurus alchimistarum* (§ 4.52) and the *Tractatus de oleo stibii* (§ 4.27) sometimes attributed to Paracelsus, but ascribed to Roger Bacon in this manuscript). Mentions Johann Hiller[415] and Carolus Pirlinger.
7. *Time of writing.* Further study required.

WORD COUNT:

INCIPIT: "So ain mensch vnrichtigk wirt so nimb oleum lunae ..."

MANUSCRIPTS:
- KASSEL, *Landesbibliothek*: 2° Ms. chem. 3, f. 167r–168r

FIRST PRINTED: not printed

ESSENTIAL BIBLIOGRAPHY: Sudhoff, *Paracelsus-Handschriften*, 705–706.

8.11 *Praeparationes*

1. *Printing History, Manuscripts.* First edited in Latin by Gerhard Dorn in 1570, appended to the chapter on Persicaria of the *Herbarius*, but not as a separate work: the beginning seems as if it were the rest of the same chapter ("There are also other magnetic herbs with the same properties [...]").[416] Then follow other herbs, then recipes including some on antimony and vitriol.
2. *Editions.* Not edited by Huser or Sudhoff.
3. *Relationship between different versions.* Only one known version.
4. *Structure, genre/form, perspective, style.* A collection of recipes.
5. *Relationship to other texts.*
6. *Authenticity, authorship.* These supplementary *Praeparationes* were designed by Dorn (or Perna?) to complete the incomplete text of Paracelsus's *Praeparationes* (or rather the chapter on Persicaria from Paracelsus's *Herbarius*). Dorn explained p. 327 that several *Praeparationes* were lost, but he was able to recover some of them. He printed there his Latin translation of two fragments from *Von den natürlichen Dingen*, then the chapter on Persicaria from the *Herbarius* – followed by the supplementary *Praeparationes*. As stated by Sudhoff, these

415 On Hiller see CP 2: 201; CP 3: 43–47.

416 *Archidoxorum Aureoli Ph. Theophrasti Paracelsi* (1570), 383: "Sunt et aliae similium virium herbae magneticae, prout Solidago minor, Dracunculus alter, etc. [...]."

supplementary pages were not printed by Huser and every doubt can be raised on their authenticity.[417]

7. *Time of writing.* Possibly written or compiled by Dorn himself in the course of publication. Further study required.

WORD COUNT: ca. 2.000 words

INCIPIT: "Sunt et aeliae similium virium herbae magneticae ..."

MANUSCRIPTS: no manuscripts known

FIRST PRINTED:

- 1570 (in: *Archidoxorum Aureoli Ph. Theophrasti Paracelsi de secretis naturae mysteriis libri decem*, ed. and translated by Gerhard Dorn (Basel: Pietro Perna, 1570), 383–398; VD16 P 398; Sudhoff, *Bibliographia Paracelsica*, 206 n° 123)

ESSENTIAL BIBLIOGRAPHY: Sudhoff, *Bibliographia Paracelsica*, 206, 695 n° 7.

8.12 *Secreta Paracelsica*

1. *Printing History, Manuscripts.* Not printed. Three manuscripts.
2. *Editions.* Not edited by Huser or Sudhoff.
3. *Relationship between different versions.*
4. *Structure, genre/form, perspective, style.* A collection of medical and alchemical recipes.
5. *Relationship to other texts.*
6. *Authenticity, authorship.*
7. *Time of writing.* According to Broszinski, two of the Kassel manuscripts contain a chronogram which resolves to 1567.

WORD COUNT: ca. 1000 words

INCIPIT: "Recipe des feinstes golds, gebornet, gefeilet oder durch bleirauch calcinirt ..."

417 Sudhoff, *Bibliographia Paracelsica*, 206: "verschiedenerlei Verwendungsvorschriften anderer Kräuter und Darstellungsvorschriften von Arzneimitteln, welche sich keineswegs alle bei Huser nachweisen lassen und über deren Echtheit z[um] Th[eil] recht erhebliche Zweifel bestehen trotz Dorn's ausdrücklicher Versicherung." – *ibid.*, 695: "Praeparationes, welche von Dorn dem Paracelsus zugeschrieben werden."

MANUSCRIPTS:
- KASSEL, *Landesbibliothek*: 4° Ms. chem. 28, 74r–93r
- KASSEL, *Landesbibliothek*: 4° Ms. chem. 59, f. 104r–133r
- KASSEL, *Landesbibliothek*: 4° Ms. chem. 85, f. 44r–60r

FIRST PRINTED: not printed

ESSENTIAL BIBLIOGRAPHY: Sudhoff, *Paracelsus-Handschriften*, 745, 746, 779 n° 1/21.

FURTHER BIBLIOGRAPHICAL REFERENCES:
Broszinski, *Manuscripta chemica in quarto* (2011), 91–92, 199–200, 358.

8.13 *Secreta secretorum* also: *Auserlesenes Korn*

1. *Printing History, Manuscripts.* Nine manuscripts. A critical edition was published by Stephanie Büttner (now Stephanie Neufeldt) in 2001.[418]
2. *Editions.* Not edited by Huser or Sudhoff.
3. *Relationship between different versions.* Five manuscripts have the complete text (of which there are enlarged versions). One manuscript has a shortened version. Another manuscript has only part of the text. Three manuscripts have only single recipes.
4. *Structure, genre/form, perspective, style.* A collection of medical recipes, divided into 18 chapters. A nineteenth chapter found in a table of contents ("De peste") is missing in all manuscripts.[419] The alternative title, "Auserlesenes Korn," refers to the separation of grain from chaff.
5. *Relationship to other texts.*
6. *Authenticity, authorship.*
7. *Time of writing.* Probably compiled in the second half of the 16th century, possibly in Silesia.

WORD COUNT: ca. 15.000 words

INCIPIT: "Recipe olej vitrioli per se rectificatj zwei theil ..."

MANUSCRIPTS:

Complete versions
- WIEN, *Österreichische Nationalbibliothek*: Cod. 11428, f. 193r–209v

418 Büttner, *Die pseudoparacelsischen* Secreta Secretorum (2001), 75–200.
419 *Ibid.*, 68.

- KASSEL, *Landesbibliothek*: 4° Ms. chem. 33[7, f. 7r–26v

Shortened version

- WROCŁAW/BRESLAU, *Biblioteka Uniwersytecka*: Ms. 6610, f. 1r–6r (*olim* Görlitz, Oberlausitzische Bibliothek der Wissenschaften)

Enlarged versions

- GLASGOW, *University Library*: MS Ferguson 42, f. 1r 104v
- LEIDEN, *Universiteitsbibliotheek*: Voss. Chym. Q.14, f. 41v–135r [preceded by the collection *Arcana* (§ 8.2)]

Single recipes

- KASSEL, *Landesbibliothek*: 2° Ms. chem. 3, f. 167v, 168r [part of the collection *Nota ex Theophrasti manu* (§ 8.10)]
- KASSEL, *Landesbibliothek*: 4° Ms. chem. 8, f. 139r–140v, 185r, 199r
- KASSEL, *Landesbibliothek*: 4° Ms. chem. 54, f. 9v, 73r, 132r [part of a collection *Collectanea ex Theophrasto et alijs* (§ 8.6)]
- WIEN, *Österreichische Nationalbibliothek*: Cod. 11259, f. 56v–59r, 85r–108r[420]

FIRST PRINTED: not printed before Büttner (2001)

ESSENTIAL BIBLIOGRAPHY: Sudhoff, *Paracelsus-Handschriften*, 201–215, 705, 720–721, 737–739, 743, 778 n° 1/10 and n° 1/12; CP 2: 461, 481; CP 3: 445.

FURTHER BIBLIOGRAPHICAL REFERENCES:

Boeren, *Codices Vossiani Chymici* (1975), 135–136.
Büttner, *Die pseudoparacelsischen "Secreta Secretorum"* (2001).
Broszinski, *Manuscripta chemica in quarto* (2011), 24, 25,110–111, 185.

8.14 *Tincturae Paracelsicae*

1. *Printing History, Manuscripts.* First printed in 1598. One manuscript.
2. *Editions.* Not edited by Huser or Sudhoff.
3. *Relationship between different versions.* The Kassel manuscript has only two of the processes.
4. *Structure, genre/form, perspective, style.* A small collection of alchemical processes.
5. *Relationship to other texts.*

420 *Ibid.*, 61–68.

6. *Authenticity, authorship.* Allegedly written by Johannes Trithemius, then given to Paracelsus and copied from his autograph. According to Sudhoff there is probably nothing authentic in these alchemical processes.[421]
7. *Time of writing.* Further study required.

Word count: ca. 3.000 words

Incipit: "Die reinigung beschehe deß leibs, deß geists ein jegklichs für sich selbst ..."

Manuscripts:
- Kassel, *Landesbibliothek*: 4° Ms. chem. 2, f. 25v–27v [fragments]

First printed:
- 1598 (in: *Aurei Velleris Oder Der Guldin Schatz vnd Kunstkammer Tractatus II* (Rorschach, 1598); VD16 T 1961; Sudhoff, *Bibliographia Paracelsica*, 424–427 n° 244)

Essential bibliography: Sudhoff, *Bibliographia Paracelsica*, 696 n° 13; Sudhoff, *Paracelsus-Handschriften*, 196, 779 n° II/5.
Broszinski, *Manuscripta chemica in quarto* (2011), 7.

8.15 [*Untitled collection*]

1. *Printing History, Manuscripts.* Not printed. One manuscript.
2. *Editions.* Not edited by Huser or Sudhoff.
3. *Relationship between different versions.* Only one known version.
4. *Structure, genre/form, perspective, style.* A collection of medical recipes. At least nine of them are attributed to "Theophrastus."
5. *Relationship to other texts.*
6. *Authenticity, authorship.*
7. *Time of writing.* Further study required.

Word count:

Incipit: "Oleum salarmoniaci dieses olichs etzlich tropffen vnd im weissenn wein gemischt ..."

Manuscripts:
- Kassel, *Landesbibliothek*: 4° Ms. chem. 23, f. 1r–10v

421 Sudhoff, *Bibliographia Paracelsica*, 426: "Etwas Echtes ist wohl nicht unter diesen alchemistischen Processen."

FIRST PRINTED: not printed

ESSENTIAL BIBLIOGRAPHY: Broszinski, *Manuscripta chemica in quarto* (2011), 75–76.

Verwendung des Oleum salis: see *Arcana quaedam chymica* (§ 8.3)

9 Single Medical Recipes and Experimenta

Aqua zu alten Schäden: see *Praecipitatum, Mercurius Saturni* ... (§ 9.20)

9.1 *Aquila seu theriaca*

German text with parts in Latin. Recipe for a remedy against morbus gallicus, dropsy and podagra. Some reader wrote in the margin: "Paracels. tom. 5. p. 45. Mercurius optime praeparatus et Naturalis Theriaca."

TITLE: *Aquila seu theriaca Paracelsi secundum descriptionem Belbrum*

INCIPIT: "Recipe mercurii facti ex regulo antimonii cum mercurio sublimato ..."

WORD COUNT: ca. 300 words

MANUSCRIPTS:
- HAMBURG, *Staats- und Universitätsbibliothek*: Cod. alchim. 661, 314–316

FIRST PRINTED: not printed

9.2 *Arcanum vitrioli*

Instructions for a remedy against epilepsia and paralysis. Part of a collection of recipes by Marx (Marcus) Bernhard[422] from Glogau/Silesia. The collection is found in two Kassel manuscripts.

422 On Marx Bernhard see Ernst-Heinz Lemper, "Görlitz und der Paracelsismus," *Deutsche Zeitschrift für Philosophie*, 18 (1970), 347–360, on 354. Lemper points to a Wolfenbüttel manuscript (Cod. Guelf. 160 Extrav.) written in 1567 in Görlitz and bearing the initials "M B" (Sudhoff, *Paracelsus-Handschriften*, 271). This identification of "M.B." is repeated by Siegfried Wollgast, *Philosophie in Deutschland zwischen Reformation und Aufklärung (1550–1650)*, 2nd ed. (Berlin, 1993), 664 n. 211 and Ernst-Heinz Lemper, "Anfänge akademischer Sozietäten in Görlitz und Bartholomäus Scultetus (1540–1614)," in Klaus Garber, ed., *Europäische Sozietätsbewegung und demokratische Tradition* (Tübingen, 1996), 1152–1178, on 1163–1164.

TITLE: *Arcanum vitrioli Theophrasti*

INCIPIT: "Recipe vitriolum so viel man wil mit gemeinem bron wasser in gleichem gewicht …"

WORD COUNT:

MANUSCRIPTS:
- KASSEL, *Landesbibliothek*: 4° Ms. chem. 60[4,1, f. 9v
- KASSEL, *Landesbibliothek*: 4° Ms. chem. 85, f. 16r–16v

FIRST PRINTED: not printed

ESSENTIAL BIBLIOGRAPHY: Sudhoff, *Paracelsus-Handschriften*, 743, 779 n° 1/20.

FURTHER BIBLIOGRAPHICAL REFERENCES:
Broszinski, *Manuscripta chemica in quarto* (2011), 239, 358.

9.3 *Archicatolic purge*

In English. Instructions for the preparation of a purgative. Connected with the name of Georg Phaedro. Allegedly taken from the "Paradoxes" (i.e. the *Archidoxis*).

TITLE: *Archicatholik purge described and suked out of Paracelsus Paradoxes, composed of vitriol Crocus and cremor tartari sana.*

INCIPIT: "The vitriol is to be purified best from his waterish flegma …"

WORD COUNT:

MANUSCRIPTS:
- LONDON, *British Library*: Sloane MS 665, f. 7r–8r

FIRST PRINTED: not printed

ESSENTIAL BIBLIOGRAPHY: Sudhoff, *Paracelsus-Handschriften*, 218–219, 779 n° 1/26.

9.4 *Aurum diaphoreticum*

First printed in 1608. No manuscripts known. The text printed by Figulus follows two other (though unrelated) texts, *Figierung auf den Mercurium Saturni* and *Borras*

Theophrasti (see § 4.31). – "Aurum diaphoreticum" was a remedy devised by Paracelsus in the *Große Wundarznei* and in other places.

TITLE: *Folget aurum diaphoreticum Theophrasti von Hohenheimb, Germani Monarchae & Principis Medicorum*

INCIPIT: "Nimb sal gemmae, das auffs reineste zu pulver gestossen ..."

WORD COUNT:

MANUSCRIPTS:

FIRST PRINTED:
- 1608 (in: *Rosarium novum olympicum et benedictum. Das ist: Ein newer Gebenedeyter Philosophischer Rosengart*, ed. Benedictus Figulus (Basel, 1608), 27–28; VD17 12:133068G; Sudhoff, *Bibliographia Paracelsica*, 482–485 n° 283)

ESSENTIAL BIBLIOGRAPHY: Sudhoff, *Bibliographia Paracelsica*, 483.

9.5 *Aurum potabile*

A recipe also found in the *Secreta secretorum* (§ 8.13).[423] – The initials "G.S." in the title may refer to Georg Schwalenberg,[424] a canonicus in Fritzlar who was in contact with Benedictus Figulus and many alchemists.[425]

TITLE: *Aurum potabile Theophrasti G.S.*

INCIPIT: "Recipe oleum vitrioli per se rectificirs zwei mal ..."

WORD COUNT: ca. 250 words

MANUSCRIPTS:
- KASSEL, *Landesbibliothek*: 4° Ms. chem. 8, f. 139r

FIRST PRINTED: not printed

423 See Büttner, *Die pseudoparacelsischen* Secreta Secretorum (2001), 15, 66, 75–76 § 1.1.

424 Broszinski, *Manuscripta chemica in quarto* (2011), 116.

425 See Joachim Telle, "Benedictus Figulus. Zu Leben und Werk eines deutschen Paracelsisten," *Medizinhistorisches Journal*, 22 (1987), 303–326, on 310.

Essential bibliography: Sudhoff, *Paracelsus-Handschriften*, 720.

Further bibliographical references:
Büttner, *Die pseudoparacelsischen "Secreta Secretorum"* (2001), 15, 66, 75–76 § 1.1.
Broszinski, *Manuscripta chemica in quarto* (2011), 24.

9.6 *Balneum*

Instructions for a healing bath. Authenticity not denied by Sudhoff, who merely writes that he has not seen this "Balneum" anywhere else.[426]

Title: *Teophrasti Paracelsi Balneo*

Incipit: "Ob diese nachvolgende speties schon gering anzusehen vndt nicht auß Arabia vndt India kommen …"

Word count: ca. 200 words

Manuscripts:
- Kassel, *Landesbibliothek*: 4° Ms. chem. 19, f. 48r–48v

First printed: not printed

Essential bibliography: Sudhoff, *Paracelsus-Handschriften*, 731–732, 778 n° 1/17.

Further bibliographical references:
Broszinski, *Manuscripta chemica in quarto* (2011), 60.

9.7 *Bei krummen und lahmen Gliedern*

A recipe also found in the *Secreta secretorum* (§ 8.13). Associated in the manuscript with "J. Langehan," not identified by Büttner, but identical with Jacob Langenhain according to Broszinski.

Title: *Theophrasti secretum krumen vndt lamen glidern wider zu recht helffen*

Incipit: "Das oleum vitrioli per se in aqua rorida destilata …"

Word count: ca. 110 words

---

426 Sudhoff, *Paracelsus-Handschriften*, 732: "Schriftstück […], das mir sonst nicht vorgekommen ist."

MANUSCRIPTS:
- KASSEL, *Landesbibliothek*: 4° Ms. chem. 8, f. 139r–139v

FIRST PRINTED: not printed

ESSENTIAL BIBLIOGRAPHY: Sudhoff, *Paracelsus-Handschriften*, 720.

FURTHER BIBLIOGRAPHICAL REFERENCES:
Broszinski, *Manuscripta chemica in quarto* (2011), 24.
Büttner, *Die pseudoparacelsischen "Secreta Secretorum"* (2001), 15, 66, 84 § 1.5.

9.8 *Catholicon contra colicam*

Latin with a few words in German and Italian. Instructions for the preparation of a purgative. According to Sudhoff this recipe is not the same as some recipes for the same indication in the authentic *Archidoxis* and therefore is rather not authentic.[427]

TITLE: *Catholicon Theophrastj contra colicam et in omnibus purgationibus secretis*

INCIPIT: "Recipe aloes succotrini unciam unam ..."

WORD COUNT:

MANUSCRIPTS:
- WIEN, *Österreichische Nationalbibliothek*: Cod. 11259, f. 91r(?)

FIRST PRINTED: not printed

ESSENTIAL BIBLIOGRAPHY: Sudhoff, *Paracelsus-Handschriften*, 200, 779 n° 11/25.

9.9 *Conservativum coitius*

A recipe also found in the *Secreta secretorum* (§ 8.13).

TITLE: *Conservativum coitius Theoph[rasti]*

---

427 Sudhoff, *Paracelsus-Handschriften*, 200: "Dies 'Catholicon' stimmt weder mit dem 'Elixir proprietatis' noch mit dem 'Specificum Purgativum' der Archidoxen; ein ähnliches Recept gibt Croll in der Basilica Chymica S. 139. Echt Paracelsische Ueberlieferung haben wir kaum hier vor uns."

Incipit: "Recipe galgantt, bibergeil, muscatt blütt, neglein, anis, langen pfeffer, muscatt, borras ana uncias quinque dz ist fünf quint …"

Word count: ca. 80 words

Manuscripts:
- Kassel, *Landesbibliothek*: 4° Ms. chem. 8, f. 140r–140v

First printed: not printed

Essential bibliography: Sudhoff, *Paracelsus-Handschriften*, 721.

Further bibliographical references:
Büttner, *Die pseudoparacelsischen "Secreta Secretorum"* (2001), 15, 66, 170 § 18.6.
Broszinski, *Manuscripta chemica in quarto* (2011), 24.

9.10 *De auro potabili*

First edited in 1608 by Benedictus Figulus. No manuscripts. Detailed recipe for the preparation of an "aurum potabile." Usage not given. – Figulus received the text as a work of Paracelsus from his "dear and intimate friend" Johann Georg Cressius in Heidelberg. Figulus himself doubted its authenticity, "since so many processes and texts are circulating under his [Paracelsus's] name among the lovers of alchemy that he [Paracelsus] never would have written."[428] According to Sudhoff without any traces of authenticity.[429]

Title: *De auro potabili Theophrasti Paracelsi*

Incipit: "Recipe einen alten vngerischen wein, der nit abgestochen ist …"

Word count: ca. 650 words

---

428 Figulus, To the Reader, in *Pandora* (1608), 273: "Vielgeliebter Freund Bruder vnd Leser/ da diß Büchlein zum End lauffen thet sub praelo Argentinensi, habe ich eben diß nachfolgende Discription Auri Potabilis von meinem Lieben vnnd Hochvertrawten Freund Johanne Georgio Cressio Cive Heydelbergensi Philochemico, etc. bekommen/ welches Theophrasti sein soll/ zum beschluß anhängen wollen. Ob dem aber also sey/ daß es vnsers Thewren Praeceptoris Paracelsi sein solle oder nicht/ kan ich nit für gewiß sagen sintemal viel processus vnd scripta also vnder seinem Nahmen bey den Liebhabern Alchimiae herumb fahren/ die er doch nie zubeschreiben in Sinn genommen/ auff solche weiß/ Es sey nun dem wie es wolle/ so verhoffe ich daß dir solches nicht vnannemblich sein werde."

429 Sudhoff, *Bibliographia Paracelsica*, 482: "ohne Spuren der Echtheit."

MANUSCRIPTS: no manuscripts known

FIRST PRINTED:
- 1608 (in: *Pandora Magnalium Naturalium Aurea Benedicta, De Benedicto Lapidis Philosoph. Mysterio*, ed. Benedictus Figulus (Strasburg: Lazarus Zetzner, 1608), 275–278; VD17 23:277709W; Sudhoff, *Bibliographia Paracelsica*, 480–482 n° 282)

ESSENTIAL BIBLIOGRAPHY: Sudhoff, *Bibliographia Paracelsica*, 696 n° 19; CP 2: 462.

9.11 *In caduco arcanum ex auro*

A recipe for a remedy against epilepsia prepared from gold.

Not directly attributed to Paracelsus but inserted between excerpts from *De praeparationibus* and *Herbarius.*

TITLE: *In Caduco Arcanum ex Auro*

INCIPIT: "Recipe aurum durch antimonium am höchstenn gefinirtt …"

WORD COUNT: ca. 60 words

MANUSCRIPTS:
- KASSEL, *Landesbibliothek*: 4° Ms. chem. 17, f. 138v–139r

FIRST PRINTED: not printed before Sudhoff (1899)

ESSENTIAL BIBLIOGRAPHY: Sudhoff, *Paracelsus-Handschriften*, 712, 778 n° I/13.

FURTHER BIBLIOGRAPHICAL REFERENCES:
Broszinski, *Manuscripta chemica in quarto* (2011), 54.

In pleuresi: see *Praecipitatum, Mercurius Saturni …* (§ 9.20)

9.12 *Laudanum*

A recipe also found in the *Secreta secretorum* (§ 8.13). It contains a "historia" of the healing of a man suffering from constipation.[430]

430 For a modern adoption of this "historia" see Screamin' Jay Hawkins, *Constipation Blues* (1970). Cf. Frank Geerk, "Paracelsus' Bedeutung in der Musikgeschichte," in Frank Geerk, ed., *Paracelsus – Arzt unserer Zeit. Leben, Werk und Wirkungsgeschichte des Teophrastus* [sic] *von Hohenheim* (Zurich, 1992), 397–402, on 402: "Ob sich für die Avantgarde der

TITLE: *Dz laudanum des hocherleüchten mans Theophrasty*

INCIPIT: "Auf eine zeit hat ein herr jn langer zeit keyn stulgang gehapt ..."

WORD COUNT: ca. 400 words

MANUSCRIPTS:
- KASSEL, *Landesbibliothek*: 4° Ms. chem. 8, f. 199r–200v

FIRST PRINTED: not printed

ESSENTIAL BIBLIOGRAPHY: Sudhoff, *Paracelsus-Handschriften*, 721.

FURTHER BIBLIOGRAPHICAL REFERENCES:
Büttner, *Die pseudoparacelsischen "Secreta Secretorum"* (2001), 15, 66, 133–134 § 14.7.
Broszinski, *Manuscripta chemica in quarto* (2011), 24–25.

9.13 *Laudanum Theophrasti*

Instructions for the preparation of laudanum, one of the most famous of Paracelsus's remedies.

TITLE: *Laudanum Theophrasti*

INCIPIT: "Recipe salpetrum in einem schmeltz tiegel zerlaßen ..."

WORD COUNT: ca. 250 words

MANUSCRIPTS:
- KASSEL, *Landesbibliothek*: 4° Ms. chem. 45, f. 52r–52v
- KASSEL, *Landesbibliothek*: 4° Ms. chem. 60[4,1, f. 10r–10v
- KASSEL, *Landesbibliothek*: 4° Ms. chem. 85, f. 18r–18v

FIRST PRINTED: not printed

ESSENTIAL BIBLIOGRAPHY: Sudhoff, *Paracelsus-Handschriften*, 743–744, 778 n° 1/16; CP 2: 478–483.

modernen Musik ähnlich starke Bezüge zu paracelsischen Überlieferungen herstellen ließen wie bei der bildenden Kunst, wäre noch zu untersuchen."

Further bibliographical references:
Broszinski, *Manuscripta chemica in quarto* (2011), 164, 239, 358.

9.14 *Laudanum opiatum Theophrasti*

Instructions for the preparation of laudanum from opium.

Title: *Laudanum opiatum Theophrasti Paracelsi*

Incipit: "Recipe opii theobacii optimi [...] schneide im auf das dünnest ..."

Word count:

Manuscripts:
- Kassel, *Landesbibliothek*: 4° Ms. chem. 50, f. 80r–80v, 82v–83r [the same recipe, written in two different hands]
- New Haven, *Beinecke Library*: Osborn fb255, f. 56 [English, written by Robert Paston, Earl of Yarmouth (1631–1683)]

First printed:
- 1615 (Conrad Khunrath, *Medulla destillatoria et medica sextum aucta & renovata* (Hamburg: Froben, 1638), 2: 463–464; VD17 7:711346T)

Essential bibliography: Sudhoff, *Paracelsus-Handschriften*, 692, 742; CP 2: 480, 507, 1000.

Further bibliographical references:
Broszinski, *Manuscripta chemica in quarto* (2011), 78.

9.15 *Magenpulver*

A recipe for treatment of the stomach.

Edited by Volker Wendland in 1972.

The recipe was given to the compiler of the manuscript by a man called Goller. It is part of a collection of recipes for Herzogin Dorothea Susanne von Sachsen-Weimar written in Amberg in 1573.

Title: *Ain guet pulver hat mir der herr Goller vom Doctor Theofrastis fur den magenn gebenn*

Incipit: "Nembt seniff pletter vier lot, jmber vier lot ..."

Word count: ca. 180 words

Manuscripts:
– Heidelberg, *Universitätsbibliothek*: Cod. Pal. germ. 182, f. 164r

First printed: not printed before Wendland (1972)

Further bibliographical references:
Volker Wendland, "Paracelsus in Heidelberger Handschriften," *Medizinische Monatsschrift*, 26 (1972), n° 10, 471–476.
Ulrike Schofer, *Katalog der deutschen medizinischen Handschriften der Universitätsbibliothek Heidelberg aus dem Besitz von Kurfürst Ludwig VI. von der Pfalz (1539–1583)* (Heidelberg, 1999), 157–162.
Matthias Miller and Karin Zimmermann, *Die Codices Palatini germanici in der Universitätsbibliothek Heidelberg (Cod. Pal. germ. 182–303)* (Wiesbaden, 2005), 4.

Mercurius praecipitatus diaphoreticus: see *Praecipitatum diaphoreticum* (§ 9.21)

Mercurius Saturni: see *Praecipitatum, Mercurius Saturni* ... (§ 9.20)

Oleum sive extractio plumbi: see *Praecipitatum, Mercurius Saturni* ... (§ 9.20)

9.16 *Oleum antimonii*

Some owner later added: "Von G: S:" to the title, which may refer to Georg Schwalenberg, see *Aurum potabile* (§ 9.5).[431]

Title: *In Chyrurgia Magna Theophrasti / Oleum Antimonij Jacob Langenhain*

Incipit: "Recipe sal geme, antimonii ana ein pfund, nota eß stehet sonst alcool antimonii ..."

Word count:

Manuscripts:
– Kassel, *Landesbibliothek*: 4° Ms. chem. 8, f. 142r–142v

First printed: not printed

431 Broszinski, *Manuscripta chemica in quarto* (2011), 116.

Essential bibliography: Sudhoff, *Paracelsus-Handschriften*, 721.

Further bibliographical references:
Büttner, *Die pseudoparacelsischen "Secreta Secretorum"* (2001), 15, 66, 118–119 § 12.3.
Broszinski, *Manuscripta chemica in quarto* (2011), 24.

9.17 *Oleum tartari*

A recipe also found in the *Secreta secretorum* (§ 8.13).

Title: *Oleum tartari Theophrasti*

Incipit: "Recipe tartarum calcinatum ad album, solvir vnd coagulir ihn vier mal …"

Word count: ca. 50 words

Manuscripts:
- Kassel, *Landesbibliothek*: 4° Ms. chem. 8, f. 184v

First printed: not printed

Essential bibliography: Sudhoff, *Paracelsus-Handschriften*, 721.

Further bibliographical references:
Büttner, *Die pseudoparacelsischen "Secreta Secretorum"* (2001), 15, 66, 118–119 § 12.3.
Broszinski, *Manuscripta chemica in quarto* (2011), 24.

9.18 *Oleum vitrioli*

With an illustration showing a glass receptacle.

According to Sudhoff and Broszinski, related to a text edited by Huser (Huser, 6: 253). We have not been able to see any similarities, however.

Title: *Ein oleum vitrioli Theophrasti Paracelsi*

Incipit: "Recipe vitrioli sovill du wilt, reib in khlein auf einem porphyrio …"

Word count: ca. 300 words

Manuscripts:
- Kassel, *Landesbibliothek*: 4° Ms. chem. 22, f. 136r–137r
- Kassel, *Landesbibliothek*: 4° Ms. chem. 23, f. 9v

First printed: not printed

Essential bibliography: Sudhoff, *Paracelsus-Handschriften*, 733, 778 n° I/18.

Further bibliographical references:
Broszinski, *Manuscripta chemica in quarto* (2011), 71, 76.

9.19 *Paradieswasser*

Part of an alchemical tract without title (the first chapter is "Von dem weisen rosengarten der Kunst Alchimia belangent"). Contains a recipe for a "paradise water" allegedly found by Paracelsus "in der Pfaltz in dem Kloster Lorch." Sudhoff doubts its authenticity.[432]

Allegedly written in 1532.

Title: *Von dem paradeis wasser*

Incipit: "Nimm zwei theyl unseres fixen corporis also bereiteter und vollendeter ..."

Word count:

Manuscripts:
- Wien, *Österreichische Nationalbibliothek*: Cod. 11266, f. 143v

First printed: not printed before Spunda (1924)

Historical Manuscript Catalogues: von Linden (1786), 32 n° 90; von Linden (1788), part 2, 38 n° 32.

Modern editions:
- *Magische Unterweisungen des edlen und hochgelehrten Philosophi und Medici Philippi Theophrasti Bombasti von Hohenheim Paracelsus genannt*, ed. Franz Spunda (Leipzig, 1924), 34–62, on 58–62 [from Wien, ÖNB, Cod. 11266]

Essential bibliography: Sudhoff, *Paracelsus-Handschriften*, 677, 780 n° II/26.
Richard Smekal, "Phantastische Prophezeiungen. Schatzgräberwinke in Paracelsus-Schriften," *Neues Wiener Journal*, vol. 33, n° 11.281 (18 April 1925), 7.

432 Sudhoff, *Paracelsus-Handschriften*, 677: "Etwas Paracelsisches ist nicht an diesem Paradieswasser zu finden."

9.20 *Praecipitatum, Mercurius Saturni, Oleum sive extractio plumbi, In pleuresi, Aqua zu alten Schäden*

Diverse recipes found in a manuscript volume once owned by Nicolaus Albinus.[433] According to Sudhoff none of these recipes is to be found in the known works of Paracelsus.[434]

Title: *Theophrastus*

Incipit: "Mache ein aquam regis darinn solvire gemeinen mercurium crudum …"

Title: *Mercurium saturni zemachen in die edlnen vnguenten zuprauchen Theophrastj*

Incipit: "Recipe plumbum vnd solvir das jn ein aqua fort …"

Title: *Oleum sive extractio plumbi Theophrastj in seine emplastra et vnguenta*

Incipit: "Recipe silber gleth, reibs khlain thues jn ein glas …"

Title: *In pleuresj a Theophrasto probatum*

Incipit: "Recipe sal absinthij drachmam unam cum aceto accipiat …"

Title: *Aqua Theophrasti zu althen schäden*

Incipit: "Recipe ein schwefel kyß, den distillir per alembicum …"

Word count:

Manuscripts:
- Wien, *Österreichische Nationalbibliothek*: Cod. 11548, f. 498r–506v

First printed: not printed

Essential bibliography: Sudhoff, *Paracelsus-Handschriften*, 178–179, 779 n° 1/27.

433 On Nicolaus Albinus see Limbeck, "Paracelsus in einer frühneuzeitlichen Historiensammlung" (1994), 18–19.

434 Sudhoff, *Paracelsus-Handschriften*, 179: "In dieser Form finden sich diese Recepte und Darstellungsvorschriften nicht bei Hohenheim, doch mag ja mancherlei derart überliefert sein, was aus den Kreisen der Schüler stammte."

Further bibliographical references:
Sven Limbeck, "Paracelsus in einer frühneuzeitlichen Historiensammlung. Die *Rhapsodia vitae Theophrasti Paracelsi* von Peter Payngk," in Joachim Telle, ed., *Analecta Paracelsica* (Stuttgart, 1994), 1–58, on 18–19.

9.21 *Praecipitatum diaphoreticum* also: *Mercurius praecipitatus diaphoreticus*

Latin text. According to the heading, taken from Paracelsus. The second part of this process is indeed related to a paragraph in *De natura rerum* (§ 2.2), book 5 ("De morte rerum naturalium"; Huser, 6: 288).

Title: *Praecipitatum diaphoreticum Paracelsi in libro de morte rerum et chirurgia magna*

Word count: ca. 120 words

Manuscripts:
- Hamburg, *Staats- und Universitätsbibliothek*: Cod. alchim. 665, on 91

First printed: not printed

9.22 *Tugenden des olei sulphuris*

According to Sudhoff, a list of fifty-six usages of the *oleum sulphuris* "no doubt influenced by Paracelsian medicine, but not containing anything actually from Paracelsus."[435]

Another list of usages of the "oleum sulphuris" is found in Kassel, *Landesbibliothek*: 4° Ms. chem. 60[8,2, f. 120r–121r.

Title: *Die tugenten deß olei sulphurris, welches grosses lobs würdig in alchimei vnd artzney, wie ich dan solches im Königreich Behe mit grossem lob, ruhme vnd ehre erfahren.*

Incipit: "Das öhl genommen mit malvasier vnnd bibenellen wasser …"

Word count:

Manuscripts:
- Kassel, *Landesbibliothek*: 4° Ms. chem. 33[7, f. 2r–4r

First printed: not printed

435 Sudhoff, *Paracelsus-Handschriften*, 737: "Zweifellos von der Paracelsischen Medicin beeinflusst, aber nichts wirklich Paracelsisches enthaltend."

Essential bibliography: Sudhoff, *Paracelsus-Handschriften*, 736–737, 778 n° 1/19

Further bibliographical references:
Broszinski, *Manuscripta chemica in quarto* (2011), 110, 286.

9.23 *Zu Erhaltung des menschlichen Lebens*

Instructions for a remedy against various diseases.

The heading of the recipe is *Secretum secretorum Theophrasti* and it is closely related to a recipe found both in the *Manuale medicum* (§ 8.9) and in the *Secreta secretorum* (§ 8.13).[436]

Title: *Secretum secretorum Theophrasti. Zuerhalltung das manschlich leben jnn guter lanngwiriger gesundthait, bis auf die zukhommendte bestimbte zeit, die ainem von Gott auferlegt jst, jst dies potabile medicamentum zugeprauchen.*

Incipit: "Recipe aquae melissae, arthemisiae, valerianae, pimpinellae ana ..."

Word count: ca. 250 words

Manuscripts:
- Wolfenbüttel, *Herzog August Bibliothek*: Cod. Guelf. 20 Aug. 2°, f. 186r

First printed: not printed

Further bibliographical references:
Büttner, *Die pseudoparacelsischen "Secreta Secretorum"* (2001), 15, 66, 118–119 § 12.3.

10 Miscellaneous Texts

Poems

10.1 *Aus mir zeuchstu ein flüssigs Feuer* also: *Antimon-Rollengedicht*

A poem originally not attributed to Paracelsus and first edited in 1608 by Benedictus Figulus. The first letters of each line form the word "ANTIMONIVM".

First attributed to Paracelsus by Johann Joachim Becher in 1682 who quoted only verses 1 and 2.

436 See Büttner, *Die pseudoparacelsischen* Secreta Secretorum (2001), 21, 145–146 ("Ad Juuentutem Conseruandum"). This manuscript not noted by Büttner.

Incipit: "Auß mir zeuchstu ein flüssigs Fewr / Nichts solls ohne den Mercurium thewer …"

Word count: 10 verses, ca. 70 words

Manuscripts:
- Hamburg, Staats- und Universitätsbibliothek: Cod. alchim. 668, f. 203v–204r
- Hannover, *Gottfried Wilhelm Leibniz Bibliothek/Niedersächsische Landesbibliothek*: Ms. IV 341, on 617
- Kassel, *Landesbibliothek*: 2° Ms. chem. 19[3, f. 60v
- Kassel, *Landesbibliothek*: 4° Ms. chem. 22, f. 262r
- Kassel, *Landesbibliothek*: 4° Ms. chem. 60[5,2, f. 165r
- Kassel, *Landesbibliothek*: 4° Ms. chem. 60[6,5, f. 13v
- Kassel, *Landesbibliothek*: 4° Ms. chem. 60[8,1, f. 161ra
- Kassel, *Landesbibliothek*: 4° Ms. chem. 86, f. 35v, 40v
- Kassel, *Landesbibliothek*: 4° Ms. chem. 96, 269
- Kassel, *Landesbibliothek*: 8° Ms. chem. 8, f. 167r
- Leiden, *Universiteitsbibliotheek*: Voss. Chym. Q.24, f. 1*
- Wien, *Österreichische Nationalbibliothek*: Cod. 11314, f. 74r [attributed to a Georg Fischer]

First printed:
- **1606** (in: *Cabala Chymica. Concordantia Chymica. Azot Philosoph. Solificatum. Drey vnnterschiedliche/ Nützliche/ vnd zuvor nie außgegangene Tractätlein*, ed. Franz Kieser (Mühlhausen: Martin Spieß for Johann Spieß, 1606), 116; VD17 12:650901T; Sudhoff, *Bibliographia Paracelsica*, 473–474 n° 277)
- **1608** (in: *Rosarium novum olympicum et benedictum. Das ist: Ein newer Gebenedeyter Philosophischer Rosengart*, ed. Benedictus Figulus (Basel, 1608), 65, 72; VD17 12:133068G; Sudhoff, *Bibliographia Paracelsica*, 482–485 n° 283)
- **1625** (in: Herbrandt Jamsthaler, *Viatorium spagyricum* (Frankfurt a.M.: Lucas Jennis, 1625), 104–105; VD17 1:634657R)
- **1682** (in: Johann Joachim Becher, *Chymischer Glücks-Hafen* (Frankfurt a.M.: Schiel, 1682), 443; VD17 23:238915Z)

Modern edition:
- Telle, "'Wo sind meine Vers'?'" (2008), 335.

Further bibliographical references:

Rudolf Schmitz and Adolf Winkelmann, "Über die alchemistischen Geheimschriften im Briefwechsel des Landgrafen Moritz von Hessen-Kassel (1592–1627)," *Pharmazeutische Zeitung*, 106 (1961), n° 13, 374–378, on 376.

Boeren, *Codices Vossiani Chymici* (1975), 156.
Telle, "'Wo sind meine Vers'?'" (2008).
Joachim Telle, "Alchemiker im frühneuzeitlichen Kassel – Partisanen der Wissenschaftlichen Revolution oder Repräsentanten menschlicher Narrheit?," *Zeitschrift des Vereins für hessische Geschichte*, 116 (2011), 71–91, on 82.
Broszinski, *Manuscripta chemica in quarto* (2011), 74, 250, 262, 281, 361, 384.

10.2 *Carmen Theophrasti ab Hohenheim Trismegisti*

Latin; 12 verses

INCIPIT: "Tramite artis scibiles Musarum …"

MANUSCRIPTS:
- MÜNCHEN, *Bayerische Staatsbibliothek*: clm 25108, f. 75

FURTHER BIBLIOGRAPHICAL REFERENCES:
Telle, "'Wo sind meine Vers'?'" (2008), 335.

10.3 *Das Bergwerk gibt einen roten Ton*

16 verses, ca. 100 words

INCIPIT: "Das Bergwerg giebt ein rothen Dohn, / Wer jhn recht kennt, der hat groß Lohn …"

First printed in 1639, not attributed to Paracelsus. Later (in 1749) said to have been inscribed by Paracelsus himself on the wall of an old mine near Gastein to mark the place where a red substance is to be found.

Modern edition by Joachim Telle in *Paracelsus im Gedicht*.

FIRST PRINTED:
- 1639 (Johannes Agricola, *Ander Theil Commentariorum, Notarum, Observationum & Animadversionum in Johannis Poppii Chymische Medicin* (Leipzig: Thomas Schürers Erben und Matthias Götze, 1639), 288; VD17 23:290569A)
- 1749 (Carl Herrmann Gravel, *Fontina Bernhardi Revelata: Oder: Das, den Filiis Artis und dem Publico zum besten getreulich eröffnete Königliche Wunder-Baad* (Erlangen: Johann Carl Tetschner, 1749; 2nd printing 1750), 126)

MODERN EDITION:
- Telle, "'Wo sind meine Vers'?'" (2008), 334–335.

FURTHER BIBLIOGRAPHICAL REFERENCES:
Telle, "'Wo sind meine Vers'?'" (2008).

10.4 *Das ganze Werk in Reimen*

10 verses, ca. 80 words

INCIPIT: "Eß ist ein Stein vnnd doch kein Stein, / In dem da stehet die Kunst allein …"

First printed in 1718. Several manuscripts. Critical edition by Joachim Telle in 1994.

Originally part of a larger alchemical poem, the 10 verses of *Das ganze Werk in Reimen* form part of David Beuther's alchemical treatise *Opus philosophicum* (anonymous in many manuscripts). However, the copy in Karlsruhe, written by Paul Nagel, has another context.

Allegedly taken from the autograph of Paracelsus ("des Paracelsi eigene Handschrift"). The text quotes the "red lion" found in various pseudo-Paracelsian texts, expressly names "mercurius" and "sulphur," while the first verses may refer to "salt": "It is a stone and not a stone […] found in valleys and mountains." Thus there is a loose connection to the Paracelsian *tria prima* doctrine.

Written in the second half of the sixteenth century.

MANUSCRIPTS:

independent copies

- KARLSRUHE, *Badische Landesbibliothek*: Hs. Allerheiligen 3, f. 36r
- KASSEL, *Landesbibliothek*: 4° Ms. chem. 60[6,5, f. 29v

Part of David Beuther, "Opus philosophicum"

- DRESDEN, *Sächsische Landesbibliothek*: Mscr.Dresd.N.117.c
- HAMBURG, *Staats- und Universitätsbibliothek*: Cod. alchim. 678, f. 14v
- HAMBURG, *Staats- und Universitätsbibliothek*: Cod. alchim. 704, p. 408
- HAMBURG, *Staats- und Universitätsbibliothek*: Cod. alchim. 766, f. 82r
- HEIDELBERG, *Universitätsbibliothek*: Cod. Pal. germ. 434, f. 40v
- NEW HAVEN, *Beinecke Library*: Mellon MS 97
- TOURNAI, *Bibliothèque du Séminaire*: Cod. 138, f. 22r–22v

FIRST PRINTED:

- 1718 (David Beuther, *Universal und Particularia*, ed. Johann Christoph Sprögel (Hamburg: Samuel Heyl, 1718), 26–27, 51; VD18, 14402769; not in Sudhoff, *Bibliographia Paracelsica*)

MODERN EDITION:

- Joachim Telle, "Paracelsus im Gedicht" (1982), 558–559
- Joachim Telle, "*Vom Stein der Weisen*" (1994), 193–194

ESSENTIAL BIBLIOGRAPHY: CP 2: 202–203 (on Beuther).

FURTHER BIBLIOGRAPHICAL REFERENCES:

Joachim Telle, "Paracelsus im Gedicht. Materialien zur Wirkungsgeschichte Theophrasts von Hohenheim im 16. und 17. Jahrhundert," in Gundolf Keil, ed., *Fachprosa-Studien. Beiträge zur mittelalterlichen Wissenschafts- und Geistesgeschichte* (Berlin, 1982), 552–573.

Joachim Telle, "*Vom Stein der Weisen.* Eine alchemoparacelsistische Lehrdichtung des 16. Jahrhunderts," in Joachim Telle, ed., *Analecta Paracelsica* (Stuttgart, 1994), 167–212. – Reprinted in Telle, *Alchemie und Poesie* (2013), 407–459.

Ein tinctur man machen soll: see *Von der Tinctur* (§ 10.10)

10.5 *In utrumque Antichristum* also: *Theophrastus adversus utrumque Antichristum*

36 verses, ca. 250 words

INCIPIT: "Sey wie jhm sey, So will der eigen Gewin, / Nicht weichen von dieser Brüder sinn ..."

First printed in 1619. One manuscript, written by Karl Widemann and dated 1597. The author is a protestant spiritualist Paracelsian.[437]

MANUSCRIPTS:

- WOLFENBÜTTEL, *Herzog August Bibliothek*: Cod. Guelf. 28.4 Aug. 4°, f. 67r–67v

FIRST PRINTED:

- 1619 (in: *Liber Sermonum in Antichristos et Pseudoprophetis Veteris & Noui Testamenti. Das ist: Ein Büchlein wider die Antichristen* [...] *Philippi Theophrasti Paracelsi, ab Hohenheim* (Frankfurt a.M.: Lucas Jennis, 1619), 51–52; VD17 1:057553M; Sudhoff, *Bibliographia Paracelsica*, 520–521 n° 311)

ESSENTIAL BIBLIOGRAPHY: Sudhoff, *Bibliographia Paracelsica*, 697, n° 42; Sudhoff, *Paracelsus-Handschriften*, 552.

FURTHER BIBLIOGRAPHICAL REFERENCES:

Telle, "'Wo sind meine Vers'?'" (2008), 333.

10.6 *Instruktion der Universaltinktur*

6 verses, ca. 50 words

437 Telle, "'Wo sind meine Vers'?'" (2008), 333: "verfaßt von einem Angehörigen der paracelsistischen Spiritualistenfraktion des protestantischen Dissidentenlagers."

Incipit: "Nimm mich schön klar, im Feuer mich figir, / In Lufft stell mich laß Sonn Reiff, Regen, nit zu mir …"

First printed in 1720 by J.J. Chymiphilus under the title "Kurtze warhaffte und gantz deutlich- klar- verständige Instruction der Universal-Tinctur" and labelled as "Theophrastische Zeilen". The initial letters of the six lines form the word "NITRVM".

Edited by Joachim Telle in 1982.

First printed:

- 1720 (in: J.J. Chymiphilus, *Der wahren Chymischen Weisheit Offenbahrung*, no place 1720, p. 280)

Further bibliographical references:

Joachim Telle, "Paracelsus im Gedicht. Materialien zur Wirkungsgeschichte Theophrasts von Hohenheim im 16. und 17. Jahrhundert," in Gundolf Keil, ed., *Fachprosa-Studien. Beiträge zur mittelalterlichen Wissenschafts- und Geistesgeschichte* (Berlin, 1982), 552–573, on 559.

10.7 *Meines Sohns Name ist doch wahr* and *O wehe bring mir ein nackends Weib*

Von Bereitung des gebenedeiten philosophischen Steins (§ 4.53), also attributed to Caspar Hartung vom Hoff and first printed in 1603 in Paracelsus's *Opera* edited by Ps.-Huser, contains two poems that are therefore indirectly attributed to Paracelsus.

(1) 8 verses, ca. 50 words

Incipit: "Meines sohns nam ist doch wahr / Vnd kam mit jhm selben alter dahin …"

(2) 5 verses, ca. 40 words

Incipit: "O wehe bring mir ein nackents weib / Wann vnselig war mein erster leib …"

The second one is a fragment of a longer and well known alchemical poem edited by Joachim Telle in 1980.

Further bibliographical references:

Joachim Telle, *Sol und Luna. Literar- und alchemiegeschichtliche Studien zu einem altdeutschen Bildgedicht* (Hürtgenwald, 1980), 126–127.

Telle, "'Wo sind meine Vers'?'" (2008), 335.

10.8 *Nimm ein Ding welches sei gar fix*

12 verses, ca. 80 words

Incipit: "Nimb ein Ding welches sey gar fix / Sonst ist dein Arbeit gar für nix …"

First edited in 1617 by Radtichs Brotofferr (Christoph Barbarossa).[438] Modern edition by Joachim Telle in *Paracelsus im Gedicht.*

First printed:
- 1617 (in: Radtichs Brotofferr [Christoph Barbarossa], *Elucidarius major. Oder Erleuchterunge vber die Reformation der gantzen weiten Welt/ F.C.R.* (Lüneburg: Stern, 1617), 162; VD17 39:116498E)
- 1682 (in: Johann Joachim Becher, *Chymischer Glücks-Hafen* (Frankfurt a.M.: Schiel, 1682), 179; VD17 23:238915Z)

Further bibliographical references:

Jiří Pražák, *Katalog rukopisů křižovnické knihovny, nyní deponovaných ve Státní knihovně České socialistické republiky v Praze* (Prague, 1980), 166, 282.

Gilly, *Paracelsus in der BPH* (1993), 24, 61–62.

Carlos Gilly, *Cimelia Rhodostaurotica. Die Rosenkreuzer im Spiegel der zwischen 1610 und 1660 entstandenen Handschriften und Drucke* (Amsterdam, 1995), 126–127.

Telle, "'Wo sind meine Vers'?'" (2008), 334.

O wehe bring mir ein nackends Weib: see *Meines Sohns Name ist doch wahr* (§ 10.7)

Theophrastus adversus utrumque Antichristum: see *In utrumque Antichristum* (§ 10.5)

10.9 *Vom roten Löwen*

14 verses, ca. 90 words

Incipit: "Vom rotten Löwen verstehe mich recht, / Wiß er sey ein Meister vnd kein Knecht …"

One manuscript, dated 1633. The headline says: "Dieses soll des Theophrasti sein" ("this is said to be by Theophrastus").

Manuscripts:
- Wien, *Österreichische Nationalbibliothek*: Cod. 11323c, f. 12r

438 On Barbarossa (1560–1623), a protestant minister in Lüneburg, Oldenburg and Altenbruch, and superintendent of Land Hadeln, see Gilly, *Paracelsus in der BPH* (1993), 24, 61–62; Gilly, *Cimelia Rhodostaurotica* (1995), 101, 126–127. – "Radtichs Brotoffer" is an anagram of "Christoff Rodebart."

FURTHER BIBLIOGRAPHICAL REFERENCES:
Telle, "'Wo sind meine Vers'?'" (2008), 334.

10.10 *Von der Tinctur* also: *Ein tinctur man machen soll*

30 verses, ca. 190 words

INCIPIT: "Ein tincktur mann machen soll, / Wie man macht öl von vitriol ..."
First printed in 1582 as part of the *Primum manuale* (§ 8.9). One manuscript. Edited by Sudhoff in Paracelsus, *Sämtliche Werke*, 1/13: 470–471.

MANUSCRIPTS:
- WIEN, *Österreichische Nationalbibliothek*: Cod. 11466[439]

FIRST PRINTED:
- 1582 (in: *Aureoli Theophrasti Paracelsi Archidoxorum, Seu de Secretis Naturae mysteriis, libri decem* (Basel: Pietro Perna, 1582), 236–237; VD16 P 399; Sudhoff, *Bibliographia Paracelsica*, 331–334 n° 191)

ESSENTIAL BIBLIOGRAPHY: Sudhoff, *Paracelsus-Handschriften*, 156.

FURTHER BIBLIOGRAPHICAL REFERENCES:
Hermann Menhardt, *Verzeichnis der altdeutschen literarischen Handschriften der Österreichischen Nationalbibliothek*, vol. 3 (Berlin, 1961), 1226.
Telle, "'Wo sind meine Vers'?'" (2008), 334.

Dialogues

Chrysorrhemon / Chrysorrhoas: see *Güldin Flüss* (§ 10.14)

10.11 *Colloquium Dr. Wolfgangi Kappleri et Christ. Schaffneri cum Theophrasto Paracelso*

1. *Printing History, Manuscripts.* One manuscript. Edited in Sudhoff, *Paracelsus-Handschriften*, 183–185. Critical edition by Joachim Telle in *Paracelsus im Gedicht*.
2. *Editions.* Not edited by Huser.

439 According to Sudhoff, transcribed and translated from Huser, *Chir.* 1618.

3. *Relationship between different versions.* Only one known version.
4. *Structure, genre/form, perspective, style.* A dialogue in rhymed verses between Dr. Wolfgang Kapler, Dr. Christoph Schaffner and Dr. Theophrastus.
5. *Relationship to other texts.*
6. *Authenticity, authorship.* Wolfgang Kappler (1493–1567) is a well-known personality.[440] The question whether the dialogue contains any reliable facts is controversial, especially whether there has ever been a personal encounter between Kapler and Paracelsus. – According to Telle, the author is Matthias Koch (d. 1579), Kappler's son-in-law.
7. *Time of writing.* The manuscript was copied by Karl Widemann in Augsburg in 1595. He received the text from Johann Vogler in Strasbourg. Probably written in the 1570s.

Word count: ca. 600 words

Incipit: "Herr Christoff seht dort disen Mann / Der vmbvagiert vnd lacht vns ahn …"

Manuscripts:
- Leiden, *Universiteitsbibliotheek*: Voss. Chym. Q.17, f. 137v–139v

First printed: not printed before Sudhoff (1899) and Telle (2008)

Modern Editions
- Sudhoff, *Paracelsus-Handschriften* (1899), 183–185.
- Telle, *Paracelsus im Gedicht* (2008), 65–67.

Essential bibliography: Sudhoff, *Paracelsus-Handschriften*, 183–185, 780 n° 11/18.

Further bibliographical references:

Boeren, *Codices Vossiani Chymici* (1975), 145.

Rudolf Werner Soukup, *Chemie in Österreich von den Anfängen bis zum Ende des 18. Jahrhunderts*, vol. 1: *Bergbau, Alchemie und frühe Chemie* (Vienna, 2007), 222–223.

Joachim Telle, *Paracelsus im Gedicht. Theophrastus von Hohenheim in der Poesie des 16. bis 21. Jahrhunderts* (Hürtgenwald, 2008), 65–67, 260–262.

440 Soukup, *Chemie in Österreich* (2007), 1: 222–223.

10.12 *Colloquium Ferdinandi Regis cum Theophrasto Paracelsi* also: *Ein Gespräch des Königs Ferdinand mit Paracelsus*

1. *Printing History, Manuscripts.* One manuscript, written in Czech. A German translation was published by Georg Sticker in 1941. It is, however, deficient, according to Tschinkel.[441]
2. *Editions.* Not edited by Huser or Sudhoff.
3. *Relationship between different versions.* Only one known version.
4. *Structure, genre/form, perspective, style.*
5. *Relationship to other texts.*
6. *Authenticity, authorship.* The question whether the dialogue is fictitious or based on historical facts is controversial. Tschinkel supports the latter and dates the encounter between Paracelsus and King Ferdinand ca. 1540. Fellmeth deems the encounter rather unlikely.
7. *Time of writing.* According to Tschinkel the author uses information gathered in 1580 and later. The *Colloquium* was probably written in the late 1580s or in the 1590s.

WORD COUNT:

INCIPIT: Wssiezknj lidé Pod ☉ 3 země drozenj rozličním

MANUSCRIPTS:
- WIEN, *Österreichische Nationalbibliothek*: Cod. 11133, f. 11v–16v

FIRST PRINTED: not printed

ESSENTIAL BIBLIOGRAPHY: Sudhoff, *Paracelsus-Handschriften*, 187, 780 n° II/19.

FURTHER BIBLIOGRAPHICAL REFERENCES:
Georg Sticker, *Ein Gespräch des Königs Ferdinand mit Paracelsus* (Halle/Saale, 1941).
Ildefons Betschart, "Ein neugefundenes Paracelsus-Porträt," *Nova Acta Paracelsica*, 8 (1957), 126–127.

441 Tschinkel, "Paracelsus bei König Ferdinand I." (1975), 335–336: "Die von Sticker veröffentlichte Übersetzung ist mangelhaft. Der Philologe, den man in Wien darum bemüht hat, dürfte sich in der tschechischen Sprache nicht ganz ausgekannt haben. Die Schrift hat er stellenweise falsch entziffert, was sich bei den Eigennamen als schwerwiegend erweist, denn sie sind es vornehmlich, die für die Glaubwürdigkeit des Ms. einstehen sollen."

Otto Nowotny, "*De auro potabili*," in Lavoslav Glesinger, *Der Humanist Andreas Dudith im Kreise der Antiparacelsisten* (Vienna, 1967) (*Salzburger Beiträge zur Paracelsusforschung*, 7), 12–23, on 19–20.

Augustin Tschinkel, "Paracelsus bei König Ferdinand I.," in Sepp Domandl, ed., *Paracelsus. Werk und Wirkung. Festgabe für Kurt Goldammer zum 60. Geburtstag* (Vienna, 1975) (*Salzburger Beiträge zur Paracelsusforschung*, 13), 335–344.

Ulrich Fellmeth, "Im Reich und außerhalb – Wirkungsstätten und Reiserouten des Paracelsus," in *Paracelsus und das Reich* (Salzburg, 2007) (*Salzburger Beiträge zur Paracelsusforschung*, 40), 28–44, on 40.

Johannes Lang, "Der sagenhafte Paracelsus," in *Paracelsus gestern und heute* (Salzburg, 2012) (*Salzburger Beiträge zur Paracelsusforschung*, 44), 68–91, on 83.

Jaroslava Hausenblasová and Ivo Purš, "Simon Thadeas Budek and his Contacts at the Court of Rudolf II", in Ivo Purš and Vladimír Karpenko, eds., *Alchemy and Rudolf II: Exploring the Secrets of Nature in Central Europe in the 16th and 17th centuries* (Prague, 2016), 607–624, on 616.

Ivo Purš, "The Habsburgs on the Bohemian Throne and Their Interest in Alchemy and the Occult Sciences", in *Alchemy and Rudolf II*, op. cit., 93–127, on 104–105.

10.13 *Dialogus occulti Theophrasti, alchimistae et naturae*

1. *Printing History, Manuscripts.* First printed in 1616. Two manuscripts, probably copied from the printed edition. – The *Dialogus* was printed along with or as a companion to Marcus Eugenius Bonacina, *Compendiolum de praeparatione auri potabilis veri* (Olomouc: Paulus Schrammius, 1616) of which there are at least six manuscripts, all of them, however, without the *Dialogus*.[442]
2. *Editions.* Not edited by Huser or Sudhoff.
3. *Relationship between different versions.* Only one version known. The Hamburg manuscript has the text of the printed edition; it is, however, obscured by heavy use of abbreviations and alchemical signs.
4. *Structure, genre/form, perspective, style.* Written in Latin in the form of a dialogue between an old man ("Senex"), two alchemists ("Alchymist", "Consocius"), the "Occultum Theophrasti", and Nature. The participants discuss "what this *Secret*

442 Manuscripts of the *Compendiolum de praeparatione auri potabilis veri* without the *Dialogus*: HAMBURG, *Staats- und Universitätsbibliothek*: Cod. alchim. 661; NEW HAVEN, *Beinecke Library*: Mellon MS 131, f. 42r–105v; PRAHA, *Národní muzeum*: Ms. XI C 42; PRAHA, *Národní muzeum*: Ms. XI H 61; Überlingen, Leopold Sophien Bibliothek, Ms. 171; WROCŁAW/BRESLAU, *Biblioteka Uniwersytecka*: R 2318. – Another manuscript (current location unknown) was in the library of Ádám Mányoki (1673–1757); see Enikő Buzási, "Mányoki Ádám könyvtára és képzőművészeti hagyatéka" [Ádám Mányoki's Library and Art Collection], in *Művészettörténeti értesítő* 46 (1997), 181–206, on 199 n° 251: „Ein Manuscript, sub rubrica Compendiolum De praeparatione Auri Potabilis."

thing of Theophrastus may be, with which one could make and prepare the stone of the sages".[443] The "Occultum Theophrasti" takes part in the discussion.

5. *Relationship to other texts.* The *Dialogus* is a *plagium literarium* of Michael Sendivogius's *Dialogus Mercurii, Alchymistae et Naturae* (Cologne, 1607).
6. *Authenticity, authorship.* Written by the Milanese physician Marcus Eugenius Bonacina (b. ca. 1570, d. 1621) who was active in Moravia.
7. *Time of writing.* Probably written in or around the year of publication (1616).

Word count: ca. 3.000 words

Incipit: "In exitu compendioli de praeparatione auri potabilis chimistae convenerunt …"

Manuscripts:
- Hamburg, Staats- und Universitätsbibliothek: Cod. alchim. 575, f. 35r–43v
- Wrocław/Breslau, *Biblioteka Uniwersytecka*: R 378

First printed:
- **1616** (*Dialogus Marci E. Bonacinae D. Medici, Nobilis Mediolauensis* (Olomouc: Paulus Schrammius, 1616); not in VD17; not in Sudhoff, *Bibliographia Paracelsica*)

Essential bibliography: Sudhoff, "Vorwort," in Paracelsus, *Sämtliche Werke*, 1/1: XXVII.

Further bibliographical references:

Jiří Šmeral and Martin Holý, "Bonacina Marcus Eugenius", in *Biografický slovník českých zemí* 6 (Praha, 2007), 44, online (http://biography.hiu.cas.cz/Personal/index.php/BONACINA_Marcus_Eugenius_1570-20.3.1621, accessed 15 June 2021).

Ivo Purš, "Das alchemistische Traktat des Marcus Eugenius Bonacina von 1616 über Trinkgold", in *Studia Rudolphina*, 8 (2008), 77–90, on 78.

Ivo Purš and Vladimír Karpenko, "Alchemy at the Aristocratic Courts of the Lands of the Bohemian Crown", in Ivo Purš and Vladimír Karpenko, eds., *Alchemy and Rudolf II: Exploring the Secrets of Nature in Central Europe in the 16th and 17th centuries* (Prague, 2016), 47–92, on 91.

Miloš Jesenský, *Dějiny alchymie v českých zemích, v Polsku a na Slovensku* (Bratislava, 2016), 244.

443 "quale istud occultum Theophrasti esset, cum quo lapidem sapientum facere & praeparare possent" (f. A3r).

10.14 *Güldin Flüss* also: *Chrysorrhoas sive de arte chymica dialogus/Chrysorrhemon*

1. *Printing History, Manuscripts.* First printed in Latin in 1559. First printed in German in 1598. One German manuscript. An early French translation, now lost (*Dialogue en défense de la chrysopoie ou maniere de faire de l'or*), was prepared before 1584 by the Huguenot Jacques Trouillard, sieur de La Boulaye, a physician to Henri de Navarre and teacher in philosophy at the Collège de Lescar until 1584.[444] An eighteenth-century French translation (*Dialogue de l'art chymique*) is preserved in Glasgow (*Glasgow University Library*, MS Ferguson 328).
2. *Editions.* Not edited by Huser or Sudhoff.
3. *Relationship between different versions.* Sudhoff suggests that the Latin version might have been translated from German. One indication may be the title of the German manuscript version: *Das Gulden Vliess* ("The golden fleece"): there might have been a confusion between "Vließ" (fleece) and "Flüß" (river), leading to the Greek title *Chrysorrhoas*, i.e. "the river of gold" (the Pactolus: see pseudo-Plutarchus, *De fluviis*, VII). On the other hand, the title of the printed version (*Das Güldin Flüss*) – which is not consistent with the collection *Aureum Vellus* where it is published – seems to indicate that this German version was a translation from the Latin. In any case, Guglielmo Gratarolo, the editor of the second edition of 1561, was neither the author, nor the translator – if the work was indeed translated from the German, which is still to be decided.[445]
4. *Structure, genre/form, perspective, style.* A dialogue between Theophrastus [i.e. Paracelsus] and Chrysophilus [i.e. an alchemist].
5. *Relationship to other texts.*
6. *Authenticity, authorship.* Not ascribed to Paracelsus in the original Latin version where the title is only: *Chrysorrhoas. Sive. De arte Chymica, Dialogus.* The first page of the text (fol. A2r) has simply the heading: "Chrysophilus. Theophrastus." However, the character "Theophrastus" is clearly identified: "Sed video obambulantem ilhic Theophrastum ab Hohenheim, hominem, si quem Germania vnquam tulit, huius artis absolutum artificem" ("But I see here walking Theophrast von Hohenheim, a consummate artist in this art if ever Germany bore one," fol. A2v). – The German version was published as the first item in the "Tractatus II" (second part) of the *Aureum Vellus* collection. This second part was, according to the title page, a collection of texts by (1) Paracelsus and

444 CP 1: 611, citing a private communication by Didier Kahn. The translation was mentioned in 1584 by François Grudé, sieur de La Croix Du Maine, *Bibliotheque Françoise*. See the ed. by Rigoley de Juvigny (Paris, 1772), 1: 354; and see Alan Boase, *Vie de Jean de Sponde* (Geneva and Lille, 1977), 19.

445 Kahn, *Alchimie et paracelsisme en France* (2007), 25–26, 105, 142.

(2) Bartholomäus Korndörffer. The title of the text was now: "Das güldin Flüß Theophrasti Paracelsi/ ist ein Gespräch von der Alchimey/ da der Theophrastus den Chrisophilum [...] vnderrichtet". Is Paracelsus's authorship implied there, or is not Paracelsus merely acknowledged as an acting part of the dialog?

7. *Time of writing.* Written before 1559, most probably in the 1550s.

Word count: ca. 4.500 words [Latin], ca. 7.000 words [German]

Incipit: "Χρυσόφιλος. Perpetuam illam ac communem hominum miseriam deplorare non satis possum ..." / "Chrisophilus redet erstlich mit sich selbs, sprechende. Ich kan nicht genugsam verklagen ..."

Manuscripts:
- Kopenhagen, *Det Kongelige Bibliotek*: GKS 244 folio, f. 155r–171v [German]
- Glasgow, *Glasgow University Library* : MS Ferguson 328 [French]

First printed:
- 1559 (in: *Chrysorrhoas. Sive. De arte Chymica, Dialogus. Qui preclarissimas hujus scientiæ actiones, rationi ac naturalibus principiis consentaneas esse demonstrat. Impostorum vero & sycophantarum somnia ac nugas retegit* (Köln: Maternus Cholin, 1559); in Latin; not in VD16 nor in Sudhoff, *Bibliographia Paracelsica*); reprinted by G. Gratarolus in *Verae Alchemiae Artisque Metallicae* (Basel, 1561; VD16 G 2915) and in *Alchemiae, quam vocant, artisque metallicae, doctrina* (Basel, 1572; VD16 G 2883); reprinted under the title "Χρυσορρημων" in Morienus, *De transfiguratione metallorum* (Hanau, 1593; VD16 M 6354); reprinted again as "Chrysorrhoas" in *Theatrum Chemicum*, vol. 2 (Strasburg, 1602; VD17 23:292821P) and in the subsequent editions of the *Theatrum.*
- 1598 (in: *Aurei Velleris Oder Der Guldin Schatz vnd Kunstkammer Tractatus II* (Rorschach, 1598); VD16 T 1961; Sudhoff, *Bibliographia Paracelsica*, 424–427 n° 244); reprinted in the subsequent editions of 1599, 1708 and 1718.

Essential bibliography: Sudhoff, *Bibliographia Paracelsica*, 425–426, 695 n° 11; Sudhoff, *Paracelsus-Handschriften*, 223, 779 n° II/2; CP 1: 611; CP 2: 667, 700, 701–702; CP 3: 451.

Further bibliographical references:

Didier Kahn, *Alchimie et paracelsisme en France à la fin de la Renaissance* (Geneva, 2007), 25–26, 105 and 142.

Prayers

10.15 *Theophrasti Paracelsi tägliches Gebet*

Paracelsus's Daily Prayer was first printed as part of *Von Bereitung des gebenedeiten philosophischen Steins* (§ 4.53; attributed to Caspar Hartung) in 1603 in Paracelsus's *Opera* edited by Ps.-Huser. Afterwards it was printed many times. – The Hannover manuscript, written by Karl Widemann, has an enlarged version.

Edited by Kühlmann/Telle in CP 1: 450–451.

WORD COUNT: ca. 65 words [ed. 1603], ca. 480 words [Hannover manuscript]

INCIPIT: "O Heyliger Geist / weise mich, das ich nit weiß ..."

MANUSCRIPTS: no manuscripts known
- HANNOVER, *Gottfried Wilhelm Leibniz Bibliothek/Niedersächsische Landesbibliothek*: Ms. IV 341, 2

FIRST PRINTED: not printed before Ps.-Huser (1603)
- 1603 (in: Ps.-Huser, *Opera Bücher vnd Schrifften* (1603), 687; VD17 12:168390P)

ESSENTIAL BIBLIOGRAPHY: Sudhoff, *Bibliographia Paracelsica*, 105, 625, 650, 697 n° 53; CP 1: 450–451.

Others

10.16 *Klage Theophrasti über seine eigenen Discipel*

1. *Printing History, Manuscripts.* First printed in 1594.
2. *Editions.* Not edited by Huser or Sudhoff.
3. *Relationship between different versions.* Only one known version.
4. *Structure, genre/form, perspective, style.*
5. *Relationship to other texts.*
6. *Authenticity, authorship.* An anthology of warnings and critical statements taken from authentic Paracelsian works. Obviously collected in opposition to the widespread alchemical and medical forgeries ascribed to Paracelsus.[446]
7. *Time of writing.* Probably compiled in the early 1590s.

---

446 Gilly, *Paracelsus in der BPH* (1993), 61.

Word count:

Incipit: "Theophrastus Paracelsus sagt im vierdten theil deß zehenden buchs der grosen wundartzney …"

Manuscripts: no manuscripts known

First printed:
- **1594** (*Klage Theophrasti/ vber seine eigene Discipel/ vnnd leichtfertige Ertzte* (no place, 1594); VD16 P 383; Sudhoff, *Bibliographia Paracelsica*, 414–415 n° 233; another edition: VD16 P 384, Sudhoff, *Bibliographia Paracelsica*, 415–416 n° 234)

Essential bibliography: Sudhoff, *Bibliographia Paracelsica*, 414–416.

Further bibliographical references:
Gilly, *Paracelsus in der BPH* (1993), 61.

11 Appendix: *Kurzfassungen* of Authentic Theological Works

1. *Printing History, Manuscripts.* Only the short version of *De confessione, poenitentia et remissione* was printed (in 1618). Manuscripts of the short versions are found in Leiden, Görlitz, Kopenhagen, Fulda, Wrocław/Breslau, Greifswald, Halle, Wolfenbüttel.
2. *Editions.* 16 shortened versions edited by Goldammer in Paracelsus, *Sämtliche Werke*, II/Supplement (1973). 3 more shortened versions noticed by Gantenbein.[447]
3. *Relationship between different versions.* Some theological works by Paracelsus exist in two versions: a longer version ("Langfassung") that is considered authentic and a shortened version ("Kurzfassung") that is the result of excerptation, shortening and paraphrasing the original text by one or more Paracelsists (see Goldammer, *passim*). Czifra points out that already Johann Huser was confronted with the problem of shortened versions of some of Paracelsus's writings on natural philosophy.[448] Actually the volume entitled *Philosophia magna* edited by Balthasar Flöter in 1567 already contained excerpts from the tracts of

447 Gantenbein, "Einleitung," in *Neue Paracelsus-Edition* (2008), 86.

448 Czifra, "Paracelsus' Abendmahlschriften" (2016), 71. – Huser, "Lectori S[alutem]" (Preface to the Reader), in Huser, 9 (1590): sig. ***3v: "es seind an vielen Enden auß diesen Büchern nur kurtze Summarische Extract/ auch etliche sonst Stückweiß und Imperfect abgeschriben/ vnd also in Truck verfertiget worden."

Paracelsus later published by Huser in vol. 9 of his edition. Huser was able to recover the full version of only some of these treatises.

4. *Structure, genre/form, perspective, style.*
5. *Relationship to other texts.* "Apart from shortenings and paraphrasings the excerpted versions keep the sequence of thought of the longer versions for about two thirds of the text. In some places the shortened versions follow the longer versions almost word by word. Where the sequence of thought is altered in the shortened versions this is done in such a way that the shortened version is often clearer in its thoughts than the longer version" (Kircher, "Das Verhältnis," LXVII). All shortened versions are preserved in at least two or three manuscripts (Weimann, "Einleitendes," XIX) dating from ca. 1570–1595 (*ibid.*, XX).
6. *Authenticity, authorship.*
7. *Time of writing.* The production of these shortened versions may have taken place in the 1570s and was possibly connected with the Silesian paracelsist Johannes Montanus.

FURTHER BIBLIOGRAPHICAL REFERENCES:

Karl-Heinz Weimann, "Einleitendes zur Text- und Überlieferungsgeschichte," in *Religiöse und sozialphilosophische Schriften in Kurfassungen*, Supplement to Kurt Goldammer, ed., Paracelsus, *Sämtliche Werke*, 2. Abteilung: *Theologische und religionsphilosophische Schriften* (Wiesbaden, 1973), XIX–XXVII.

Kurt Goldammer, "Aus der Werkstatt der Paracelsisten des 16. und 17. Jahrhunderts. Bemerkungen zu den Kurzfassung der theologischen Schriften des Paracelsus," *ibid.*, XXIX–LXIV.

Norbert Kircher, "Das Verhältnis der Kurzfassungen L_1 F_2 B_1 zur Langfassung H_1 und das Verhältnis der Kurzfassungen untereinander," *ibid.*, LXV–XCII.

Urs Leo Gantenbein, "Einleitung," in Gantenbein, *Neue Paracelsus-Edition* (2008), 3–121, on 86.

Nikolaus Czifra, "Paracelsus' Abendmahlschriften im Spiegel ihrer Überlierferung. Fragen der Paracelsus-Philologie," *Nova Acta Paracelsica*, N.F. 27 (2016), 67–82, on 71, 75.

11.1 *De baptismate* (Kurzfassung) also: *Vom tauf der Christen*

Edited by Goldammer in 1973 in Paracelsus, *Sämtliche Werke*, II/Suppl.: 113–123 n° 13.

WORD COUNT: ca. 3.200 words

INCIPIT: "Sacrament: das wort hat vil dolmetschung ..."

MANUSCRIPTS:
- GÖRLITZ, *Oberlausitzische Bibliothek der Wissenschaften*: Th VI 146, 286–295
- KOPENHAGEN, *Det Kongelige Bibliotek*: Thott 119 kvart, f. 63v–68r
- LEIDEN, *Universiteitsbibliotheek*: Voss. Chym. F.24, f. 335v–340r
- WOLFENBÜTTEL, *Herzog August Bibliothek*: Cod. Guelf. 28.4 Aug. 4°, f. 7r–15v
- WROCŁAW/BRESLAU, *Biblioteka Uniwersytecka*: R 333, f. 123r–129r

FIRST PRINTED: not printed before Goldammer (1973).

ESSENTIAL BIBLIOGRAPHY: Sudhoff, *Paracelsus-Handschriften*, 785 n° III/32.

11.2 *De confessione, poenitentia et remissione* (Kurzfassung)

Edited by Goldammer in 1973 in Paracelsus, *Sämtliche Werke*, II/Suppl.: 125–137 n° 14.

WORD COUNT: ca. 3.400 words

INCIPIT: "Beuchten ist, so ein mensch übels tut und bekents …"

MANUSCRIPTS:
- GOTHA, *Forschungsbibliothek*: Chart. B 31, f. 158r–170r
- LEIDEN, *Universiteitsbibliotheek*: Voss. Chym. F.24, f. 258r–264r
- WROCŁAW/BRESLAU, *Biblioteka Uniwersytecka*: R 334 [probably lost], f. 165v–167r
- HALLE, *Franckesche Stiftungen*: Cod. B 63, 4°, 15–37

FIRST PRINTED:
- 1618 (in: *Philosophia Mystica, Darinn begriffen Eilff vnterschidene Theologico-Philosophische/ doch teutsche Tractätlein/ zum theil auß Theophrasti Paracelsi, zum theil auch M. Valentini Weigelii* (Newstadt [Frankfurt a.M.]: Lucas Jennis, 1618), 22–32; VD17 3:604297L; Sudhoff, *Bibliographia Paracelsica*, 513–516 n° 306)

ESSENTIAL BIBLIOGRAPHY: Sudhoff, *Bibliographia Paracelsica*, 696 n° 35; Sudhoff, *Paracelsus-Handschriften*, 784 n° III/28.

FURTHER BIBLIOGRAPHICAL REFERENCES:

Peuckert, *Pansophie* (1956), 455.

Astrid Carigiet, "*De confessione et penitentia et remissione*. Paracelsus' Beichtkonzept zwischen Katholizismus und Lutherischer Reform," *Nova Acta Paracelsica*, N.F. 13 (1999), 101–128.

11.3 *De ecclesiis veteris et novi testamenti* (Kurzfassung)

Edited by Goldammer in 1973 in Paracelsus, *Sämtliche Werke*, II/Suppl.: 19–32 n° 3.

Word count: ca. 4.400 words

Incipit: "Nachdem Adam und Eva aus dem paradeis getriben …"

Manuscripts:
- Fulda, *Hochschul- und Landesbibliothek*: Cod. 22a (2), 2°, f. 213r–220r
- Leiden, *Universiteitsbibliotheek*: Voss. Chym. F.24, f. 381r–387r
- Wrocław/Breslau, *Biblioteka Uniwersytecka*: R 334 [probably lost], f. 121r–124r

First printed: not printed before Goldammer (1973).

Essential bibliography: Sudhoff, *Paracelsus-Handschriften*, 782 n° III/8.

Further bibliographical references:
Urs Leo Gantenbein, "Einleitung," in Gantenbein, *Neue Paracelsus-Edition* (2008), 3–121, on 96 n° A13.

11.4 *De felici liberalitate* (Kurzfassung)

Edited by Goldammer in 1973 in Paracelsus, *Sämtliche Werke*, II/Suppl.: 83–92 n° 10.

Word count: ca. 3.200 words

Incipit: "Niemandt kan ausgeben, allein der reiche …"

Manuscripts:
- Fulda, *Hochschul- und Landesbibliothek*: Cod. 22a (2), 2°, f. 170r–175r
- Leiden, *Universiteitsbibliotheek*: Voss. Chym. F.24, f. 405v–410r
- Wrocław/Breslau, *Biblioteka Uniwersytecka*: R 334 [probably lost], f. 167r–168v

First printed: not printed before Goldammer (1973).

Essential bibliography: Sudhoff, *Paracelsus-Handschriften*, 781 n° III/3.

Further bibliographical references:
Urs Leo Gantenbein, "Einleitung," in Gantenbein, *Neue Paracelsus-Edition* (2008), 3–121, on 89 n° A4.

11.5 *De honestis utrisque divitiis* (Kurzfassung)

Edited by Goldammer in 1973 in Paracelsus, *Sämtliche Werke*, II/Suppl.: 93–101 n° 11.

Word count: ca. 2.300 words

INCIPIT: "Unser vernunft und weisheit mag nit erdenken ..."

MANUSCRIPTS:
- FULDA, *Hochschul- und Landesbibliothek*: Cod. 22a (2), 2°, f. 175r–179r
- LEIDEN, *Universiteitsbibliotheek*: Voss. Chym. F.24, f. 175r–179r
- WROCŁAW/BRESLAU, *Biblioteka Uniwersytecka*: R 334 [probably lost], f. 273r–273v

FIRST PRINTED: not printed before Goldammer (1973).

ESSENTIAL BIBLIOGRAPHY: Sudhoff, *Paracelsus-Handschriften*, 781 n° III/4.

FURTHER BIBLIOGRAPHICAL REFERENCES:
Urs Leo Gantenbein, "Einleitung," in Gantenbein, *Neue Paracelsus-Edition* (2008), 3–121, on 89–90 n° A5.

11.6 *De martyrio Christi et nostris deliciis* (Kurzfassung)

Edited by Goldammer in 1973 in Paracelsus, *Sämtliche Werke*, II/Suppl.: 49–55 n° 6.

WORD COUNT: ca.2.100 words

INCIPIT: "Der irdische geist der erden soll nichts ..."

MANUSCRIPTS:
- GÖRLITZ, *Oberlausitzische Bibliothek der Wissenschaften*: Th VI 146, 311–317
- GREIFSWALD, *Universitätsbibliothek*: Ms. 1002 (*olim* theol. quart. 8), f. 65v–70v
- LEIDEN, *Universiteitsbibliotheek*: Voss. Chym. F.24, f. 307v–310v
- MÜNCHEN, *Bayerische Staatsbibliothek*: cgm 4416(25), f. 1r–6r
- WROCŁAW/BRESLAU, *Biblioteka Uniwersytecka*: R 334 [probably lost], f. 119r–120v

FIRST PRINTED: not printed before Goldammer (1973).

ESSENTIAL BIBLIOGRAPHY: Sudhoff, *Paracelsus-Handschriften*, 783 n° III/15.

FURTHER BIBLIOGRAPHICAL REFERENCES:
Urs Leo Gantenbein, "Einleitung," in Gantenbein, *Neue Paracelsus-Edition* (2008), 3–121, on 98–99 n° A16.

11.7 *De nupta, de alienis rebus non concupiscendis* (Kurzfassung)

Edited by Goldammer in 1973 in Paracelsus, *Sämtliche Werke*, II/Suppl.: 103–111 n° 12.

Word count: ca. 3.000 words

Incipit: "Es ist nit volkomben aus dem liecht der natur zu leben ..."

Manuscripts:
- Görlitz, *Oberlausitzische Bibliothek der Wissenschaften*: Th VI 146, 351–359
- Greifswald, *Universitätsbibliothek*: Ms. 1002 (*olim* theol. quart. 8), f. 94v–101v
- Leiden, *Universiteitsbibliotheek*: Voss. Chym. F.24, f. 326v–330v

First printed: not printed before Goldammer (1973).

Essential bibliography: Sudhoff, *Paracelsus-Handschriften*, 782 n° III/13.

Further bibliographical references:
Urs Leo Gantenbein, "Einleitung," in Gantenbein, *Neue Paracelsus-Edition* (2008), 3–121, on 93 n° A10.

11.8 *De officiis, beneficiis et stipendiis* (Kurzfassung)

Edited by Goldammer in 1973 in Paracelsus, *Sämtliche Werke*, II/Suppl.: 41–48 n° 5.

Word count: ca.2.400 words

Incipit: "Gott hat uns ein zal ämpter gesätzt ..."

Manuscripts:
- Görlitz, *Oberlausitzische Bibliothek der Wissenschaften*: Th VI 146, 332–339
- Greifswald, *Universitätsbibliothek*: Ms. 1002 (*olim* theol. quart. 8), f. 83v–89r
- Leiden, *Universiteitsbibliotheek*: Voss. Chym. F.24, f. 317r–320v

First printed: not printed before Goldammer (1973).

Essential bibliography: Sudhoff, *Paracelsus-Handschriften*, 782 n° III/5.

Further bibliographical references:
Urs Leo Gantenbein, "Einleitung," in Gantenbein, *Neue Paracelsus-Edition* (2008), 3–121, on 91 n° A7.

11.9 *De potentia et potentiae gratia Dei* (Kurzfassung)

Not edited by Goldammer.

Word count:

Incipit: "Alle ding kummen auß der wurtz vnd gehen wider inn sie …"

Manuscripts:
- Görlitz, *Oberlausitzische Bibliothek der Wissenschaften*: Th VI 146, 305–310
- Greifswald, *Universitätsbibliothek*: Ms. 1002 (*olim* theol. quart. 8), f. 61r–65r
- Wolfenbüttel, *Herzog August Bibliothek*: Cod. Guelf. 28.4 Aug. 4°, f. 81v–86r

First printed: not printed

Essential bibliography: Sudhoff, *Paracelsus-Handschriften*, 255, 555, 587, 782 n° III/10.

Further bibliographical references:
Urs Leo Gantenbein, "Einleitung," in Gantenbein, *Neue Paracelsus-Edition* (2008), 3–121, on 86, 97–98 n° A15.

11.10 *De re templi ecclesiastica* (Kurzfassung)

Edited by Goldammer in 1973 in Paracelsus, *Sämtliche Werke*, II/Suppl.: 33–40 n° 4.

Word count: ca. 2.400 words

Incipit: "Der falsch, so in einem gerechten ding eingefuert wird …"

Manuscripts:
- Görlitz, *Oberlausitzische Bibliothek der Wissenschaften*: Th VI 146, 324–331
- Greifswald, *Universitätsbibliothek*: Ms. 1002 (*olim* theol. quart. 8), f. 77r–83r
- Leiden, *Universiteitsbibliotheek*: Voss. Chym. F.24, f. 313v–317r

First printed: not printed before Goldammer (1973).

Essential bibliography: Sudhoff, *Paracelsus-Handschriften*, 782 n° III/9.

Further bibliographical references:
Urs Leo Gantenbein, "Einleitung," in Gantenbein, *Neue Paracelsus-Edition* (2008), 3–121, on 97 n° A14.

11.11 *De religione perpetua* (Kurzfassung)

Not edited by Goldammer.

Word count:

Incipit: "Alle gaben so vns von Gott geben sein, sollen wir seeliglich brauchen …"

Manuscripts:
- Görlitz, *Oberlausitzische Bibliothek der Wissenschaften*: Th VI 146, 279–280
- Kopenhagen, *Det Kongelige Bibliotek*: Thott 119 kvart, f. 59v–60r

First printed: not printed

Essential bibliography: Sudhoff, *Paracelsus-Handschriften*, 248, 591, 782 n° III/7.

Further bibliographical references:
Urs Leo Gantenbein, "Einleitung," in Gantenbein, *Neue Paracelsus-Edition* (2008), 3–121, on 86, 94 n° A12.

11.12 *De remissione peccatorum* (Kurzfassung)

Edited by Goldammer in 1973 in Paracelsus, *Sämtliche Werke*, II/Suppl.: 57–63 n° 7.

Word count: ca.2.300 words

Incipit: "So ein schwach geschöpf sein wir …"

Manuscripts:
- Görlitz, *Oberlausitzische Bibliothek der Wissenschaften*: Th VI 146, 317–324
- Greifswald, *Universitätsbibliothek*: Ms. 1002 (*olim* theol. quart. 8), f. 70v–76v
- Halle, *Franckesche Stiftungen*: Cod. B 63, 4°, 1–15
- Leiden, *Universiteitsbibliotheek*: Voss. Chym. F.24, f. 310r–313v

First printed: not printed before Goldammer (1973).

Essential bibliography: Sudhoff, *Paracelsus-Handschriften*, 783 n° III/16.

Further bibliographical references:
Urs Leo Gantenbein, "Einleitung," in Gantenbein, *Neue Paracelsus-Edition* (2008), 3–121, on 99–100 n° A17.

11.13 *De resurrectione et glorificatione corporum et de resurrectione mortuorum* (Kurzfassung)

Edited by Goldammer in 1973 in Paracelsus, *Sämtliche Werke*, II/Suppl.: 73–81 n° 9.

WORD COUNT: ca.2.600 words

INCIPIT: "Zwei herkomben haben wir menschen …"

MANUSCRIPTS:
- GÖRLITZ, *Oberlausitzische Bibliothek der Wissenschaften*: Th VI 146, 339–341
- GREIFSWALD, *Universitätsbibliothek*: Ms. 1002 (*olim* theol. quart. 8), f. 59v–60v
- LEIDEN, *Universiteitsbibliotheek*: Voss. Chym. F.24, f. 320v–321v / 299v–302r
- WROCŁAW/BRESLAU, *Biblioteka Uniwersytecka*: R 334 [probably lost], f. 126r–127r

FIRST PRINTED: not printed before Goldammer (1973).

ESSENTIAL BIBLIOGRAPHY: Sudhoff, *Paracelsus-Handschriften*, 783–784 n° III/18–19.

FURTHER BIBLIOGRAPHICAL REFERENCES:
Urs Leo Gantenbein, "Einleitung," in Gantenbein, *Neue Paracelsus-Edition* (2008), 3–121, on 101–102 n° A19.

11.14 *De summo et aeterno bono* (Kurzfassung)

Edited by Goldammer in 1973 in Paracelsus, *Sämtliche Werke*, II/Suppl.: 9–18 n° 2.

WORD COUNT: ca. 3.100 words

INCIPIT: "Das höchste gut kan nit sein hier auf erden …"

MANUSCRIPTS:
- FULDA, *Hochschul- und Landesbibliothek*: Cod. 22a (2), 2°, f. 165r–170r
- LEIDEN, *Universiteitsbibliotheek*: Voss. Chym. F.24, f. 402r–405v
- WROCŁAW/BRESLAU, *Biblioteka Uniwersytecka*: R 334 [probably lost], f. 124r–126r

FIRST PRINTED: not printed before Goldammer (1973).

ESSENTIAL BIBLIOGRAPHY: Sudhoff, *Paracelsus-Handschriften*, 781 n° III/2.

FURTHER BIBLIOGRAPHICAL REFERENCES:
Urs Leo Gantenbein, "Einleitung," in Gantenbein, *Neue Paracelsus-Edition* (2008), 3–121, on 88–89 n° A3.

11.15 *De sursum corda* (Kurzfassung)

Not edited by Goldammer.

Word count:

Incipit: "Wir sein auß der Erden, aber Gott der Schöpffer …"

Manuscripts:
- Görlitz, *Oberlausitzische Bibliothek der Wissenschaften*: Th VI 146, 394–404
- Wrocław/Breslau, *Biblioteka Uniwersytecka*: R 334 [probably lost], f. 144v–145v

First printed: not printed

Essential bibliography: Sudhoff, *Paracelsus-Handschriften*, 268, 530, 784 n° III/21.

Further bibliographical references:
Urs Leo Gantenbein, "Einleitung," in Gantenbein, *Neue Paracelsus-Edition* (2008), 3–121, on 86, 102–103 n° A20.

11.16 *De venerandis sanctis* (Kurzfassung)

Edited by Goldammer in 1973 in Paracelsus, *Sämtliche Werke*, II/Suppl.: 65–72 n° 8.

Word count: ca. 2.700 words

Incipit: "Wir wissen unser herkomben durch gott den vatter …"

Manuscripts:
- Fulda, *Hochschul- und Landesbibliothek*: Cod. 22a (2), 2°, f. 193r–197v
- Leiden, *Universiteitsbibliotheek*: Voss. Chym. F.24, f. 450r–453v
- Wrocław/Breslau, *Biblioteka Uniwersytecka*: R 334 [probably lost], f. 135v–137r

First printed: not printed before Goldammer (1973).

Essential bibliography: Sudhoff, *Paracelsus-Handschriften*, 783 n° III/17.

Further bibliographical references:
Urs Leo Gantenbein, "Einleitung," in Gantenbein, *Neue Paracelsus-Edition* (2008), 3–121, on 100 n° A18.

11.17 *Prologus in vitam beatam* (Kurzfassung)

Edited by Goldammer in 1973 in Paracelsus, *Sämtliche Werke*, II/Suppl.: 1–7 n° 1.

WORD COUNT: ca. 2.000 words

INCIPIT: "Ein arzet braucht vil an einem kranken …"

MANUSCRIPTS:
- GÖRLITZ, *Oberlausitzische Bibliothek der Wissenschaften*: Th VI 146, 281–286
- KOPENHAGEN, *Det Kongelige Bibliotek*: Thott 119 kvart, f. 60v–63r
- LEIDEN, *Universiteitsbibliotheek*: Voss. Chym. F.24, f. 342r–345r

FIRST PRINTED: not printed before Goldammer (1973).

ESSENTIAL BIBLIOGRAPHY: Sudhoff, *Paracelsus-Handschriften*, 781 n° III/1.

FURTHER BIBLIOGRAPHICAL REFERENCES:

Urs Leo Gantenbein, "Einleitung," in Gantenbein, *Neue Paracelsus-Edition* (2008), 3–121, on 87–88 n° A2.

11.18 *Sermo "Date Caesari quae sunt Caesaris"* (Kurzfassung)

Edited by Goldammer in 1973 in Paracelsus, *Sämtliche Werke*, II/Suppl.: 153–165 n° 16.

WORD COUNT: ca. 3.900 words

INCIPIT: "Haec dei sunt: von unser substanz sollen wir gott geben […]" / "Niemandt gehört kein zins, allein dem keiser …"

MANUSCRIPTS:
- FULDA, *Hochschul- und Landesbibliothek*: Cod. 22a (2), 2°, f. 226v–232v
- LEIDEN, *Universiteitsbibliotheek*: Voss. Chym. F.24, f. 393r–399r
- WROCŁAW/BRESLAU, *Biblioteka Uniwersytecka*: R 334 [probably lost], f. 10r–11v

FIRST PRINTED: not printed before Goldammer (1973).

ESSENTIAL BIBLIOGRAPHY: Sudhoff, *Paracelsus-Handschriften*, 787 n° III/54.

11.19 *Sermones der erkanntnus ad Clementem VII* (Kurzfassung)

Edited by Goldammer in 1973 in Paracelsus, *Sämtliche Werke*, II/Suppl.: 139–152 n° 15.

Word count: ca. 3.900 words

Incipit: "Besser ist ruhe dann unrue …"

Manuscripts:
- Fulda, *Hochschul- und Landesbibliothek*: Cod. 22a (2), 2°, f. 220v–226v
- Halle, *Franckesche Stiftungen*: Cod. B 63, 4°, 38–62
- Leiden, *Universiteitsbibliotheek*: Voss. Chym. F.24, f. 387v–393v
- Wolfenbüttel, *Herzog August Bibliothek*: Cod. Guelf. 28.4 Aug. 4°, f. 135r–141r
- Wrocław/Breslau, *Biblioteka Uniwersytecka*: R 334 [probably lost], f. 162r–164v

First printed: not printed before Goldammer (1973).

Essential bibliography: Sudhoff, *Bibliographia Paracelsica*, 696 n° 33; Sudhoff, *Paracelsus-Handschriften*, 786 n° 111/45.

Index of Titles

Index of Incipits

Index of Manuscripts

Seitenstetten, Stiftsbibliothek
Cod. 69 [lost] § 5.1
Solothurn, Zentralbibliothek
Cod. S I 185 § 4.1
Strasbourg, Bibliothèque nationale et universitaire
ms. 199 § 4.25, 5.23
ms. 2.034 § 4.48
Stuttgart, Hauptstaatsarchiv
A47, Bü 12 § 6.3

Tartu, Ülikooli Raamatukogu [University Library]
Mscr. 655 § 2.7
Tournai, Bibliothèque du Séminaire
Cod. 138 § 10.4
Tovačov, Zámecká knihovna
R 328 § 2.2
R 331 § 4.42
Třeboň/Wittingau, Státní oblastní archiv
z Rožmberka 25, no shelfmark § 4.9
Trogen, Kantonsbibliothek
CM Ms. 20 § 4.3
Tübingen, Universitätsbibliothek
Jf IX 133 H § 5.29

Überlingen, Leopold Sophien Bibliothek
Ms. 177 § 4.25
Ms. 269 § 4.2
Uppsala, Universitetsbiblioteket
Cod. Ups. Palmsk. 172 § 6.3

Washington, Smithsonian Libraries
MSS 000939 B § 4.33
Weimar, Herzogin Anna Amalia Bibliothek
Q 286/20 § 1.22, 1.25, 1.25
Q 456 § 4.14
Q 462 § 4.25, 4.40
Werl, Stadtarchiv
Familienarchiv v. Papen-Lohe, Akte A III g 1, Kasten 215 § 4.21, 7.10
Wien, Österreichische Nationalbibliothek
Cod. 5230 § 4.35
Cod. 11114 § 5.31
Cod. 11133 § 10.12
Cod. 11200 § 3.3
Cod. 11259 § 2.7, 4.9, 5.5, 5.8, 5.11, 7.23, 8.3, 8.4, 8.13, 9.8
Cod. 11266 § 5.10, 5.19, 5.25, 5.29, 7.5, 9.19
Cod. 11266, f. 83r–91v § 6.3
Cod. 11295 § 6.4
Cod. 11314 § 10.1
Cod. 11323b § 4.8
Cod. 11323c § 10.9
Cod. 11330 § 5.23, 5.29
Cod. 11343 § 4.30
Cod. 11347 § 4.9
Cod. 11355 § 4.35
Cod. 11428 § 4.53, 7.20, 8.13
Cod. 11466 § 4.35, 10.10
Cod. 11548 § 9.20
Cod. Ser. n. 4135 § 4.3
Wien, Schottenstift
Cod. 722 (*olim* Hübl 539) § 4.49
Wolfenbüttel, Herzog August Bibliothek
Cod. Guelf. 11 Aug. 2° § 4.52
Cod. Guelf. 20 Aug. 2° § 9.23
Cod. Guelf. 37.2 Aug. 2° § 1.3
Cod. Guelf. 18.13 Aug. 4° § 4.35
Cod. Guelf. 28.4 Aug. 4° § 1.6, 10.5, 11.1, 11.9, 11.19
Cod. Guelf. 30.4 Aug. 4° § 4.29
Cod. Guelf. 33.6 Aug. 4° § 5.8
Cod. Guelf. 47.21 Aug. 4° § 8.5
Cod. Guelf. 48.1 Aug. 4° § 4.8
Cod. Guelf. 50 Aug. 4° § 1.9
Cod. Guelf. 51.3 Aug. 4° § 1.16, 5.23
Cod. Guelf. 60.6 Aug. 8° § 2.2, 4.39
Cod. Guelf. 74 Blank. § 4.1
Cod. Guelf. 94 Blank. § 4.25
Cod. Guelf. 115 Blank. § 5.1
Cod. Guelf. 194.3 Extrav. § 1.4
Cod. Guelf. 264.31 Extrav. § 4.1, 4.9, 4.45, 4.1, 7.17, 7.21, 7.26
Cod. Guelf. 264.32 Extrav. § 6.3
Cod. Guelf. 9 Noviss. 2° § 4.12, 4.13, 6.3
Cod. Guelf. 49 Noviss. 8° § 4.18, 4.20
Cod. Guelf. 151 Noviss. 8° § 7.6
Wrocław/Breslau, Biblioteka Kapitulna/Dombibliothek
Ms. 87 [lost] § 4.5
Wrocław/Breslau, Biblioteka Uniwersytecka
Ms. 6610 (*olim* Görlitz) § 5.7, 8.13
R 333 § 1.19, 11.1
R 334 [probably lost] § 1.7, 11.2, 11.3, 11.4, 11.5, 11.6, 11.13, 11.14, 11.15, 11.16, 11.18, 11.19
R 378 § 10.13
R 2316 § 1.13, 5.22, 5.26
R 2317 § 4.29, 4.30, 5.1

Index of Names

Index of Alleged Dates of Writing

Index of Probable Dates of Writing

1530s
- *Brief an einen Gelehrten* § 1.6
- *Consilium für Bernhard Rechlinger* § 3.3
- *De mumia* (?) § 2.1
- *Die diebus Aegyptiacis* § 1.7
- *Für Pestilenz* (?) § 3.9
- Liber imaginibus (?) § 5.17

1540s
- *Augmentum Solis* (?) § 4.4
- *Beschreibung der deutschen Alchimia* (?) § 4.7

1550s
- *Ars Gaballiae* (?) § 5.2
- *Güldin Flüss* § 10.14
- *Title* § 3.11
- *Von Bereitung des gebenedeiten philosophischen Steins* § 4.53
- *Von der Heilung zauberischer Schäden* § 5.29

1560s
- *Archidoxis magica* § 5.1
- *Aurora philosophorum* § 4.5
- *Brief an Augustin Sattler* § 4.8
- *Brief an die Wittenberger Theologen* § 1.4
- *De antimonio* (?) § 4.27
- *De iustitia* § 1.9
- *De natura rerum* § 2.2
- *De occulta philosophia* § 5.7
- *De philomela* § 5.8
- *De secretis creationis* § 1.12
- *De spiritibus metallorum* § 4.29
- *De tinctura physicorum* § 4.30
- *Holzbüchlein* § 3.10
- *Kurze Auslegung der zehn Gebote Gottes* § 1.18
- *Liber principiorum* § 5.18
- *Liber vexationum* § 4.39
- *Manuale de lapide philosophico medicinali* § 4.40
- *Philosophia ad Athenienses* (?) § 2.6
- *Praeparationes* § 8.11
- *Ratio extrahendi ex omnibus metallis mercurium* § 4.45
- *Secreta Paracelsica* § 8.12
- *Thesaurus thesaurorum* § 4.52
- *Von dem Vitriol* § 3.12
- *Von der ewigen Heimlichkeit* (?) § 5.27

1570s
- *Anatomia corporum adhuc viventium* § 3.1
- *Brief an eine geistliche Person / Practica des verkehrten Mondes* § 4.13
- *Buch mit der Sackpfeife* § 4.21
- *Chymischer Psalter* § 4.23
- *Colloquium Dr. Wolfgangi Kappleri ...* § 10.11
- *De magia veterum* § 5.6
- *De pestilitate* § 3.5
- *De quinta essentia* § 3.6
- *De septem stellis* § 5.9
- *Dritter Teil der großen Wundartzney* § 3.7
- *Kurzfassungen* § 11
- *Lapis philosophorum* § 4.37
- Liber imaginibus (?) § 5.17
- *Liber mercuriorum* § 4.38
- *Zenechdon* (?) § 5.31

1580s
- *Colloquium Ferdinandi Regis cum Theophrasto Paracelsi* § 10.12
- *De lapide philosophorum, qui lapis amoris appellatur* (?) § 5.5
- *De lumine naturae* § 4.28
- *Kleine Handbibel und Einführung* § 1.17
- *Liber Azoth* § 2.5
- *Mysterium lapidis philosophorum* § 4.43
- *Psalterium magicum* § 5.21
- *Speculum sive lumen naturale* § 5.24

1590s
- *Eine andere Prognostication* § 6.1
- *Aus einem Brief an Bartholomäus Korndörffer* § 4.10
- *Auslegung des 91. Psalms* (?) § 1.3
- *Brief an Bartholomäus Korndörffer* § 4.9
- *Brief an einen Gelehrten* § 1.6
- *Brief an Theodor* § 4.17
- *Brief von Salomon Trismosin an Paracelsus* § 4.19
- *De tribus facultatibus* § 2.4
- *Erklärung über Lukas 1* § 1.13
- *Klage Theophrasti über seine eigenen Discipel* § 10.16
- *Secretum magicum* § 2.7

17th century
- *Beschreibung des wunderbaren Bechers* § 5.3
- *Glücksrute zu Paracelsi chymischem Schatz* § 4.33
- *Hundertachtunddreißig ganz neu entdeckte Geheimnisse* § 5.14
- *Oraculum Oneiromanticum* § 5.19

Index of First Printings

1608 (cont.)
Figierung auf den Mercurium Saturni § 4.31
Theoria aus den cabalischen Gezeugnissen § 4.51
Von der Heilung zauberischer Schäden § 5.29
Von der Materia lapidis philosophorum § 4.55
1612
Brief von Salomon Trismosin an Paracelsus § 4.19
1613
Eine Andere Prognostication (*German*) § 6.1
Clavis vel praxis manualis § 4.26
1615
Laudanum opiatum Theophrasti § 9.14
1616
Dialogus occulti Theophrasti, alchimistae et naturae § 10.13
1617
Nimm ein Ding welches sei gar fix § 10.8
1618
Astronomia Olympi Novi § 1.2
Brief an die Wittenberger Theologen § 1.4
De confessione (Kurzfassung) § 11.2
De iustitia § 1.9
Erklärung über Lukas 1 § 1.13
Das ganze Werk in Reimen § 10.4, 10.4
Theologia Cabalistica § 1.23
Zehntes Buch der Archidoxen § 4.56
1619
In utrumque Antichristum § 10.5
1620
Krankes Römisches Reich § 6.2
1622
Prophezeiung von den drei Schätzen (*version 1*) § 6.3
1623
Clavis chymiae § 3.2
1624
Clavis oder Das zehnte Buch der Archidoxen § 4.25
1625
Mysterium lapidis philosophorum § 4.43
Prophezeiung von den drei Schätzen (*version 2*) § 6.3
1629
Schatzkammer der Natur § 5.23
1631
Prophezeiung von den drei Schätzen (*version 3*) § 6.3
1639
Das Bergwerk gibt einen roten Ton § 10.3
1673
Geheimes Inventarium / Prophezeiung von den drei Schätzen § 6.3
1676
Wahrhaftes Laudanum metallicum § 3.14
1679
Glücksrute zu Paracelsi chymischem Schatz § 4.33
1686
Vom Modus des Gebrauchs mit der Tinctur § 5.25
1700
De magia veterum § 5.6
1705
Speculum sapientiae § 4.49
1717
Hundertachtunddreißig ganz neu entdeckte Geheimnisse § 5.14
1720
Instruktion der Universaltinktur § 10.6
1738
Geheimes und vollständiges Wunsch-Hütlein § 4.32
1749
Kurzer Weg den Lapidem Philosophorum zu bereiten § 4.36
1756
Heimlichkeit des Herrn Wildwold § 4.35
1771
Chymischer Psalter § 4.23
1780
Arcana arcanorum arcanissima § 4.2
1894
Short Catechism of Alchemy § 4.47
1899
Colloquium Dr. Wolfgangi Kappleri ... § 10.11
Consilium für Bernhard Rechlinger § 60
In caduco arcanum ex auro § 9.11
ca. 1900
Heiliges Sales-Büchlein § 5.13

Index of First Huser Editions

Index of Sudhoff's "Uebersichten" (in *Bibliographia Paracelsica*, 695-697 and *Paracelsus-Handschriften*, 771-798)

n° 16 *Laudanum Theophrasti* § 9.13
n° 17 *Balneum* § 9.6
n° 18 *Oleum vitrioli* § 9.18
n° 19 *Tugenden des olei sulphuris* § 9.22
n° 20 *Arcanum vitrioli* § 9.2
n° 21 *Secreta Paracelsica* § 8.12
n° 22 *Figierung auf den Mercurium Saturni* § 4.31
n° 23 *Clavis chymiae* § 3.2
n° 24 *Secretum Theophrasti* § 4.46
n° 25 *Catholicon contra colicam* § 9.8
n° 26 *Archicatolic purge* § 9.3
n° 27 *Praecipitatum* § 9.20
n° 29 *Centum quindecim experimenta curationesque* § 8.5

Alchemistisches
n° 1/I *Zehntes Buch der Archidoxen* § 4.56
n° 1/II *Clavis oder Das zehnte Buch der Archidoxen* § 4.25
n° 2 *Güldin Flüss* § 10.14
n° 4 *Brief an Bartholomäus Korndörffer* § 4.9
n° 5 *Tincturae Paracelsicae* § 8.14
n° 6 *Theophrasti Tinctur* § 7.20
n° 7 *Tinctura so Theophrastus seinem Vetter gegeben* § 7.21
n° 8 *Proiectio der roten Tinctur* § 7.17
n° 9 *Dealbatio lunifica* § 7.3
n° 10 *Process aus seiner eigenen Handschrift* § 7.14
n° 11 *Process im Schlaftrunk einem verehrt* § 7.15
n° 12 *Wie die anima sulfuris saturni clarificirt wird* § 7.26
n° 13 *Tinctura physicorum et lapis philosophorum* § 7.23
n° 14 *Lapis philosophorum* § 4.37
n° 15 *Tinctura physicorum* [process] § 7.22
n° 16 *Labor et opus lapidis philosophorum* § 7.7
n° 17 *Methodus chymica* § 4.41
n° 18 *Colloquium Dr. Wolfgangi Kappleri ...* § 10.11
n° 19 *Colloquium Ferdinandi Regis cum Theophrasto Paracelsi* § 10.12
n° 20 *Buch mit der Sackpfeife* § 4.21
n° 21 *Brief an einen Gönner und Freund* § 4.14
n° 22 *Wahrhafte Kunst der chymischen Alchimei* § 7.25
n° 23 *Beschreibung der deutschen Alchimia* § 4.7
n° 24 *Schöne und herrliche Kunst* § 7.18
n° 26 *Paradieswasser* § 9.19
n° 27 *Extract aus einem uralten alchymischen Buch* § 7.5

Theologisches
n° 1 *Prologus in vitam beatam* (Kurzfassung) § 11.17
n° 2 *De summo et aeterno bono* (Kurzfassung) § 11.14
n° 3 *De felici liberalitate* (Kurzfassung) § 11.4
n° 4 *De honestis utrisque divitiis* (Kurzfassung) § 11.5
n° 5 *De officiis* (Kurzfassung) § 11.8
n° 7 *De religione perpetua* (Kurzfassung) § 11.11
n° 8 *De ecclesiis veteris et novi testamenti* (Kurzfassung) § 11.3
n° 9 *De re templi ecclesiastica* (Kurzfassung) § 11.10
n° 10 *De potentia et potentiae gratia Dei* (Kurzfassung) § 11.9
n° 13 *De nupta* (Kurzfassung) § 11.7
n° 15 *De martyrio Christi* (Kurzfassung) § 11.6
n° 16 *De remissione peccatorum* (Kurzfassung) § 11.12
n° 17 *De venerandis sanctis* (Kurzfassung) § 11.16
n° 18 *De resurrectione et glorificatione corporum* (Kurzfassung) § 11.13
n° 19 *De resurrectione mortuorum* (Kurzfassung) § 11.13
n° 21 *De sursum corda* (Kurzfassung) § 11.15
n° 25 *De praedestinatione et libera voluntate* § 1.10
n° 28 *De confessione, poenitentia et remissione* (Kurzfassung) § 11.2
n° 31 *De iustitia* § 1.9

Article: *Dane T. Daniel*

Abstract

The paper is an evaluation of differing opinions regarding the authenticity of two of Paracelsus' works, the *Astronomia Magna* (1537/1538) and *Brief an die Wittenberger Theologen* (1525). Karl Sudhoff – the exceptionally erudite and prolific Paracelsus scholar and editor – considered Paracelsus' letter to Luther and the Wittenberg theologians to be spurious. Others have questioned the extent to which the *Astronomia Magna* (often considered to be Paracelsus' *Meisterstück*) should be placed among the genuine texts of the Paracelsian corpus. Based on historical and conceptual data, and focusing on Paracelsus' theology, the author argues that both works under consideration are authentic. The study thus highlights the need to develop a diagnostic rubric for evaluating the authenticity of works in the Paracelsian *oeuvre* and suggests ideas toward this end on the basis of theological motifs prevalent within Paracelsus' clearly authentic works.

Keywords

Paracelsus – Paracelsians – Martin Luther – Karl Sudhoff – history of science – alchemy – chymistry – history of medicine – Radical Reformation

Article: *Didier Kahn*

Abstract

The pseudo-Paracelsian *Philosophia ad Athenienses* (1564) draws upon many of the ideas of Paracelsus but combines them with many other elements not found in the genuine works of the Swiss physician. After discussing the details of its edition, we summarize its content, then examine its possible sources, including the authentic texts of Paracelsus by which the unknown author will have taken inspiration. We discuss specifically the question of uncreated prime matter and provide a number of criteria allowing us to consider the treatise as spurious. Finally, we argue that the unknown author wrote this text and ascribed it to Paracelsus in order to produce a more Platonist Paracelsian cosmology than that of the genuine treatises of Paracelsus, which apparently disappointed the unknown author in this regard.

Keywords

Paracelsus – Birckmann printing press – pseudepigraphy – cosmology – Creation – uncreated prime matter – separation – Neoplatonism – elemental beings – prophecy

Article: *Tobias Bulang*

Abstract

This article introduces the pseudo-Paracelsian treatise *Aurora Philosophorum* (last third of the sixteenth century) and focusses on the contexts of a specific genealogy of knowledge presented in this text. It reaches from a divine origin of human knowledge to the universities of the author's times. Here, the transfer of science throughout history is described as a process of continuous decline. The concept of knowledge negotiated by this text is conceptualized not as theory alone: knowledge has a genuine instrumental and practical aspect. The article also presents the principles of the forthcoming edition and commentary of the handwritten and printed copies of the *Aurora Philosophorum*.

Keywords

Paracelsus – *Aurora Philosophorum* – pseudepigraphy – Gerhard Dorn – genealogy of knowledge – Adamitic science – edition – commentary

Article: *Hiro Hirai*

Abstract

One of the most popular writings ascribed to Paracelsus, *De natura rerum* appeared in 1572. That was when the movement of forgery production reached its climax, in parallel with the multiple editions of his genuine work *Archidoxis*. This article aims to place the genesis of *De natura rerum* in the context of publication history. It will first reconstruct a "library" by surveying the works ascribed to Paracelsus which could serve as instruments for the "author/reworker/editor" of *De natura rerum*. Then it will examine the evolution of this forgery production by focusing on the divergent editions of *Archidoxis* from 1569 to 1572.

Keywords

Paracelsus – forgery – alchemy – publication – edition – library – *De natura rerum* – *Archidoxis*

Article: *Charles D. Gunnoe, Jr.*

Abstract

While *De Pestilitate* is generally regarded as pseudepigraphic by modern scholarship, the treatise occupied a prominent place in Johann Huser's definitive edition of Paracelsus' *Bücher und Schrifften* (1589–1591). The text offers a compelling and generally reliable guide to Paracelsian plague theory with clear resemblances to the authentic *Zwey Bücher von der Pestilentz und ihren zufällen* and *De Peste Libri tres.* The text emphasizes the astrological transmission of the disease, describes a large role for divine retribution and demonological agency, and promotes the utility of folkloric cures. While the plague theory reflects Paracelsus' own thought and is well synthesized with his *tria prima* chemical philosophy, there are sufficient dissonances between the text and the corpus of his accepted writings to maintain its pseudepigraphic status, even if it displays some indications of a reworking of genuine Paracelsica.

Keywords

Paracelsus – plague – spurious works – cosmology – magical imagination – folkloric cures – witchcraft – astrology

Article: *Martin Žemla*

Abstract

The *Astronomia Olympi novi* and the *Theologia Cabalistica* were published as part of the *Philosophia Mystica* (1618). This influential collection of Paracelsian and Weigelian texts was among the first to include a publication of the *theologica* of Paracelsus. Both of these short pseudo-Paracelsian works were written by Adam Haslmayr (1560–1630), the propagator of the "Theophrastia Sancta," a philosophical theology of Paracelsus mixed with Weigelian and alchemical influences. These works, taken in the *Philosophia*

Mystica, are among the very few of his texts that appeared in print in his lifetime. In this paper, the content of both pseudepigrapha is analyzed and related to ideas found in both the authentic and the pseudepigraphic works of Paracelsus, as well as in relevant works of some Paracelsians (Khunrath, Figulus, Weigel, Croll, Siebmacher and Mögling). Specifically, the notions of the "*Olympus novus*" and the "cabalistic principles" are discussed in a wider context.

Keywords

Paracelsus – Valentin Weigel – Adam Haslmayr – Johann Siebmacher – Benedictus Figulus – *Theophrastia Sancta* – cabala – alchemy – universal reform – Rosicrucianism

Article: *Lawrence M. Principe*

Abstract

Early modern alchemical literature is full of pseudonymous corpora. One of the most famous of these is connected with the name Basil Valentine, a supposed Benedictine monk and master of both medicinal and transmutational chymistry. Accreted over a period of nearly a century, the Valentine corpus is complex and heterogeneous. This paper endeavors to organize and recount the construction of the corpus by an array of authors, editors, publishers, and bibliographers, to sort out some of its strata, and to trace the origins and modifications of some of its texts. This exercise will be useful not only for further investigations of Basil Valentine and other pseudonymous chymical corpora, but also for broader studies of forgery and book history.

Keywords

Basil Valentine – Johann Thoelde – pseudepigraphy – alchemy – decipherment – Paracelsus – antimony

Index Nominum

Compiled by Claire Carpentier

www.ingramcontent.com/pod-product-compliance
Lightning Source LLC
Chambersburg PA
CBHW071232230926
51849CB00001B/76
9789004503373